The Ecological Genomics of Fungi

The Ecological Genomics of Fungi

Editor
FRANCIS MARTIN

WILEY Blackwell

Editorial Offices
1606 Golden Aspen Drive, Suites 103 and 104, Ames, Iowa 50010, USA
The Atrium, Southern Gate, Chichester, West Sussex, PO19 8SQ, UK
9600 Garsington Road, Oxford, OX4 2DQ, UK

For details of our global editorial offices, for customer services and for information about how to apply for permission to reuse the copyright material in this book please see our website at www.wiley.com/wiley-blackwell.

Library of Congress Cataloging-in-Publication Data

The ecological genomics of fungi / editor, Francis Martin.
pages cm
Includes bibliographical references and index.
ISBN 978-1-119-94610-6 (cloth : alk. paper) – ISBN 978-1-118-72970-0 (epub) – ISBN 978-1-118-72971-7 (epdf) – ISBN (invalid) 978-1-118-72972-4 (emobi) – ISBN 978-1-118-73589-3 (ebook) 1. Fungi–Genetics. 2. Genomics. 3. Ecology. I. Martin, Francis, 1954– editor of compilation.
QK602.E26 2014
571.5′92–dc23

2013029869

A catalogue record for this book is available from the British Library.

Wiley also publishes its books in a variety of electronic formats. Some content that appears in print may not be available in electronic books.

Cover images: © Francis Martin
Cover design by Nicole Teut

Set in 10.5/12pt Times by SPi Publisher Services, Pondicherry, India
Printed and bound in Malaysia by Vivar Printing Sdn Bhd

1 2014

Contents

Contributors

Ólafur S. Andrésson
Institute of Life and Environmental Sciences
University of Iceland
Reykjavik, Iceland

Scott E. Baker
Pacific Northwest National Laboratory
Richland, Washington

Isabelle Benoit
CBS-KNAW Fungal Biodiversity Centre
Utrecht, The Netherlands

Gabriele Berg
Institute for Environmental Biotechnology
Graz University of Technology
Graz, Austria

Teun Boekhout
CBS-KNAW Fungal Biodiversity Centre
Utrecht, The Netherlands

Gaëtan Burgaud
Laboratoire Universitaire de Biodiversité et Ecologie Microbienne
Université Européenne de Bretagne
Université de Brest
ESIAB
Technopôle Brest-Iroise
Plouzané, France

Nicolas Corradi
Canadian Institute for Advanced Research
Department of Biology
University of Ottawa
Ontario, Canada

Dan Cullen
Forest Products Laboratory
Madison, Wisconsin

Tom Dawson
Procter & Gamble Co.
Cincinnati, Ohio

Ronald P. de Vries
CBS-KNAW Fungal Biodiversity Centre
Utrecht, The Netherlands

Irina S. Druzhinina
Research Area Biotechnology and Microbiology
Institute of Chemical Engineering
Vienna University of Technology
Vienna, Austria
and
Austrian Center of Industrial Biotechnology
Institute of Chemical Engineering
Vienna University of Technology
Vienna, Austria

Sébastien Duplessis
Laboratory of Excellence ARBRE
UMR 1136 INRA-Université de Lorraine
Interactions Arbres-Microorganismes
INRA-Nancy
Champenoux, France

Paul S. Dyer
School of Biology
University of Nottingham
Nottingham, United Kingdom

Laurence Fraissinet-Tachet
Ecologie Microbienne, UMR CNRS 5557 – USC INRA 1364
Université de Lyon
Université Lyon 1, Villeurbanne, France

Stephen B. Goodwin
USDA-ARS
Crop Production and Pest Control Research Unit
Department of Botany and Plant Pathology
Purdue University
West Lafayette, Indiana

Igor V. Grigoriev
US Department of Energy
Joint Genome Institute
Walnut Creek, California

Martin Grube
Institut für Pflanzenwissenschaften
Karl-Franzens-Universität Graz
Graz, Austria

Ferry Hagen
Department of Medical Microbiology
and Infectious Diseases
Canisius Wilhelmina Hospital
Nijmegen, The Netherlands

Joseph Heitman
Department of Molecular Genetics
and Microbiology
Duke University Medical Center
Durham, North Carolina

Sue A. Karagiosis
Pacific Northwest National Laboratory
Richland, Washington

Patrick J. Keeling
Canadian Institute for Advanced Research
Department of Botany
University of British Columbia
Vancouver, Canada

Annegret Kohler
Laboratory of Excellence ARBRE
UMR 1136 INRA-Université de Lorraine
Interactions Arbres-Microorganismes
INRA-Nancy
Champenoux, France

James Kronstad
Michael Smith Laboratories
Department of Microbiology
and Immunology
University of British Columbia
Vancouver, Canada

Christian P. Kubicek
Research Area Biotechnology and
Microbiology
Institute of Chemical Engineering
Vienna University of Technology
Vienna, Austria
and
Austrian Center of Industrial
Biotechnology
Institute of Chemical Engineering
Vienna University of Technology
Vienna, Austria

Cheryl R. Kuske
Environmental Microbiology Team
Bioscience Division
Los Alamos National Laboratory
Los Alamos, New Mexico

Jessy Labbé
BioSciences Division
Oak Ridge National Laboratory
Oak Ridge, Tennessee

Marc-André Lachance
Department of Biology
University of Western Ontario
London, Ontario, Canada

Thomas Le Calvez
Université de Rennes 1, CNRS,
UMR6553 EcoBio
Observatoire Des Sciences de l'Univers de Rennes (OSUR)
Campus de Beaulieu
Rennes, France

Björn D. Lindahl
Swedish University of Agricultural Sciences
Department of Forest Mycology and Plant Pathology
Uppsala, Sweden

Patricia Luis
Ecologie Microbienne, UMR CNRS 5557 – USC INRA 1364
Université de Lyon
Université Lyon 1, Villeurbanne, France

Stéphane Mahé
Université de Rennes 1, CNRS
UMR6553 EcoBio
Observatoire Des Sciences de l'Univers de Rennes (OSUR)
Campus de Beaulieu
Rennes, France

Roland Marmeisse
Ecologie Microbienne, UMR CNRS 5557 – USC INRA 1364
Université de Lyon
Université Lyon 1, Villeurbanne, France

Francis Martin
Laboratory of Excellence ARBRE
UMR 1136 INRA-Université de Lorraine
Interactions Arbres-Microorganismes
INRA-Nancy
Champenoux, France

Vivian P.W. Miao
Department of Microbiology
and Immunology
University of British Columbia
Vancouver, Canada

Claude Murat
Laboratory of Excellence ARBRE
UMR 1136 INRA-Université de Lorraine
Interactions Arbres-Microorganismes
INRA-Nancy
Champenoux, France

Thibaut Payen
Laboratory of Excellence ARBRE
UMR 1136 INRA-Université de Lorraine
Interactions Arbres-Microorganismes
INRA-Nancy
Champenoux, France

Denis Petitpierre
Laboratory of Excellence ARBRE
UMR 1136 INRA-Université de Lorraine
Interactions Arbres-Microorganismes
INRA-Nancy
Champenoux, France

Vanessa Rédou
Laboratoire Universitaire de Biodiversité et
Ecologie Microbienne
Université Européenne de Bretagne
Université de Brest
ESIAB
Technopôle Brest-Iroise
Plouzané, France

Charles Saunders
Procter & Gamble Co.
Cincinnati, Ohio

Jan Schirawski
Microbial Genetics
Aachen Biology and Biotechnology
RWTH Aachen University
Aachen, Germany

Pietro D. Spanu Department of Life Sciences
Imperial College London
London, United Kingdom

Raymond J. St. Leger Department of Entomology
University of Maryland
College Park, Maryland

Sheng Sun Department of Molecular Genetics
and Microbiology
Duke University Medical Center
Durham, North Carolina

Philippe Vandenkoornhuyse Université de Rennes 1, CNRS
UMR6553 EcoBio
Observatoire Des Sciences de l'Univers de
Rennes (OSUR)
Campus de Beaulieu
Rennes, France

Oddur Vilhelmsson Department of Natural Resource Sciences
University of Akureyri
Borgir vid Nordurslod
Akureyri, Iceland

Chengshu Wang Key Laboratory of Insect Developmental
and Evolutionary Biology
Institute of Plant Physiology and Ecology
Shanghai Institutes for Biological
Sciences
Chinese Academy of Sciences
Shanghai, China

Jun Xu Procter & Gamble Co.
Cincinnati, Ohio

Lucie Zinger Laboratoire d'Ecologie Alpine
UMR CNRS 5553
Université Joseph Fourrier
Grenoble, France

Preface

Fungi have been divided into discrete ecological guilds, such as leaf-litter decomposers, humus saprobes, white- and brown-rot wood decayers, plant or animal parasites, endophytes, and mutualistic symbionts. However, the actual functional properties of individual species and the synergistic effects among them are often obscure. Tremendous progress has been made in recent years on genomics of these fungi as approximately 250 genome sequences have been released, and these genetic blueprints are providing new highlights on the gene networks evolved by fungi to interact with their biotic and abiotic environments. We have entered a new era of molecular ecology research in which high-throughput molecular tools for documentation of fungal diversity and genetic variation are increasingly combined with population genetics, phylogenomics, population genomics, and community ecology to provide deeper insights into the role and function of fungi in situ. The present book aims to act as a catalyst for future research, bridging fungal genomics, metagenomics, and metatranscriptomics by bringing together a collection of contributions on genomes across a range of lifestyles and ecological traits (saprotrophism, pathogenesis [biotrophs, hemibiotrophs, necrotrophs], and symbiosis). Authors have been encouraged to explore how the massive streams of fungal sequences could be exploited to get a better understanding of the evolution of fungi and their ecological roles through ecological genomics.

The book combine a series of chapters written by leading scientists who have established cutting-edge research programs in genomics and metagenomics involving a diversity of fungal systems. Such a broad-ranging approach should provide a unique insight and a better understanding of the functions of fungi in various ecosystems, from soil to plant to human. The research that the specialists included in the volume discuss are far-reaching extensions of their current or past work and propose cross-cutting research questions whenever possible. By exploring this new field of research—ecological genomics—there are tremendous opportunities for novel discovery of key molecular mechanisms controlling plant-microbe interactions, evolution of fungi, and developmentally and ecologically relevant traits. This new research field should provide important new insights into host and habitat factors driving host specificity and community dynamics.

I am especially indebted and grateful to these authors for the high quality of their contributions. Thanks to their effort, we have produced the most complete and up-to-date treatment of fungal genomics. This book should provide guidance for future research.

I hope that this book will serve as a primary research reference for researchers and research managers working in the expanding field of mycology, fungal genomics, and ecological genomics, as well as plant-microbe interactions. It should provide a useful resource for experienced as well as new researchers and students that are moving into the field each year.

Francis Martin

Section 1
Sequencing Fungal Genomes

1 A Changing Landscape of Fungal Genomics

Igor V. Grigoriev

US Department of Energy Joint Genome Institute, Walnut Creek, California

Introduction

Fungi play an important role in nature and the economy. Being an important source of food, medicine, and enzymes, fungi can also cause human disease, threaten agriculture, and damage buildings. In nature, fungi can efficiently decompose dead organic matter and recycle nutrients, enhance plant growth as mutualists, or attack other organisms as pathogens. The kingdom Fungi represents one of the largest branches of the Fungal Tree of Life with more than 1 billion years of evolutionary history and more than 1.5 million species of which about 100,000 are known. Despite this tremendous range of lifestyles, little is known about genomic diversity and evolution of fungi, and more practically, about the rich catalog of enzymes encoded in fungal genomes and the metabolites produced by these enzymes.

Sequencing of the first fungal genome, *Saccharomyces cerevisiae*, led to an unprecedented development of the baker's yeast as a model organism and to the building of an entire collection of tools to explore fungal and eukaryotic biology. Large-scale genomic initiatives led to generating a critical mass of data for comparative genomics. Initially expensive and time consuming, sequencing technology has made giant leaps in the last few years to not only become affordable and available to many labs, but to also enable scientists to ask new types of questions, to look at many genomes at once, and to explore metagenomes of complex communities. New types and large amounts of data posed new challenges for bioinformatics but also opened doors to new applications. This chapter explores the changing landscape of fungal genomics, the challenges in experimental and computational technologies, and the transformations they lead to in biological science.

Genome Sequencing Evolution

In 1977 Frederic Sanger proposed a technique later named after him: Sanger sequencing (Sanger, Nicklen, et al., 1977). Based on separation of fluorescently

The Ecological Genomics of Fungi, First Edition. Edited by Francis Martin.

labeled DNA fragments by length on a polyacrylamide gel, bases at the end of each fragment were determined by the dye with which they interacted. Next, he proposed the method of "shotgun" sequencing using random DNA fragments from a genome as primers for its polymerase chain reaction (PCR) amplification. The amplified overlapping regions of DNA were assembled into fully resolved nucleotide sequences called "contigs" and then linked into scaffolds using "mate-paired" reads. The first shotgun-sequenced genome was of a 48.5 Kbp bacteriophage, which heralded large-scale whole-genome sequencing projects.

The 12.5 Mb *S. cerevisiae* genome was the first fungal genome sequenced and published in 1996 by the European consortium (Goffeau, Barrell, et al., 1996). Sequencing a model organism was a step toward the Human Genome Project (HGP), which is a joint effort of the US Department of Energy (DOE) and the National Institutes of Health (NIH), and provided important information not only to boost *S. cerevisiae* exploration but also launch fungal genomics. Shortly after, two more ascomycetes—a fission yeast *Schizosaccharomyces pombe* (Wood, Gwilliam, et al., 2002) and a model filamentous fungus *Neurospora crassa* (Galagan, Calvo, et al., 2003)—were sequenced. The first basidiomycete genome was of a white rot fungus *Phanerochaete chrysosporium* (Martinez, Larrondo, et al., 2004). More than 200 sequenced fungal genomes have been deposited to GenBank in the past 16 years (Fig. 1.1), each project involving series of experimental and computational tasks dependent on quickly changing technologies (Fig. 1.2).

Sanger sequencing dominated genomics for more than two decades until the "Big Bang" of the next-generation sequencing (NGS) tools, which offered diverse technologies of pyrosequencing, sequencing by synthesis,

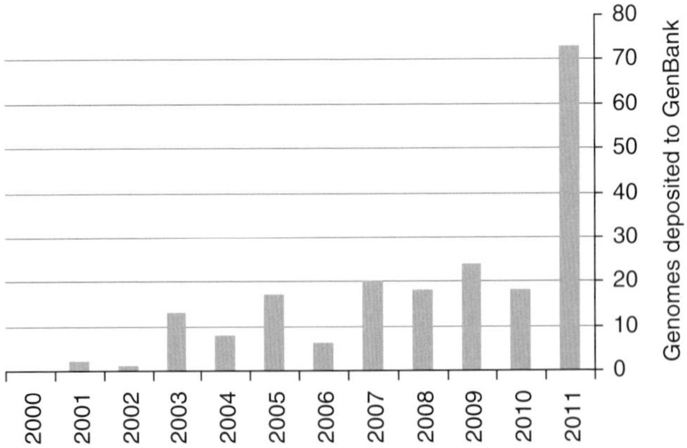

Figure 1.1 More than 200 completed fungal genomes with sequence data submitted to GenBank (http://www.ncbi.nlm.nih.gov/genbank).

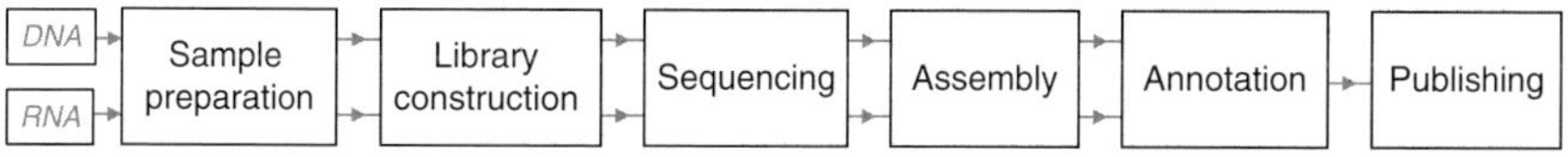

Figure 1.2 Work flow of a genome project consists of genome and transcriptome sequencing in parallel for better genome annotation. Specific steps, tools, and algorithms depend on the platforms used for sequencing.

single-molecule sequencing, real-time sequencing, and others, which were all cheaper and faster than Sanger (Metzker, 2010; Fig. 1.3). Most of these methods use DNA amplification and do not rely on bacterial clone libraries. These NGS techniques are different in the chemistry used and hence in read length, GC bias, and the amount and accuracy of produced data. For example, the Illumina HiSeq 2000 can produce nearly half a Terabase of sequence in a single run with reads up to 150-bp long. Its younger sister, MiSeq, can produce longer reads up to 2×250 bp faster, at lower throughput, and approach about 450 bp of effective read length using overlapping reads. Roche/454 pyrosequencing produces longer reads (up to 1 Kbp with its latest XLR version) than Illumina but at a higher cost and with challenging homopolymers-related errors. A relatively recent addition to NGS family, Pacific Biosciences (PacBio) machines produce substantially longer reads (up to several Kbp), with more frequent (up to 15 percent) but randomly distributed errors along the read length, and can lead to new applications using single DNA molecule sequencing. Finally, Oxford Nanopores recently announced a new machine that offers USB chip-like interface for single molecule sequencing (Pennisi, 2012)

These approaches, each with its own deficiencies and strengths, can also be combined to produce better assemblies. A number of hybrid assemblies were produced by combining data from 454, Illumina, and occasionally Sanger fosmid paired-end reads until the use of new assemblers such as AllPaths-LG (Gnerre, Maccallum, et al., 2011) led to Illumina-only assemblies, which had better scaffolding than even early Sanger draft assemblies. The combination of Illumina and PacBio data can be used for genome improvement. With Sanger, the traditional recipe for genome sequencing included a combination of paired-end libraries with different insert sizes: 3–5 Kbp, 6–10 Kbp, and fosmids (~40 Kbp) to achieve a cumulative read coverage of 6×–10×. Shorter or more error-prone 454 or Illumina require coverage of about 30× and 100×, respectively. Obtaining the mate pairs required for scaffolding is often challenging, especially for longer insert size libraries and depends on DNA quality.

Data from new NGS platforms can be used to gradually improve existing genome assemblies. These improvements, called *genome finishing*, can be accomplished either by adding read coverage over the entire genome or in a targeted fashion to expand contigs, connecting them into scaffolds,

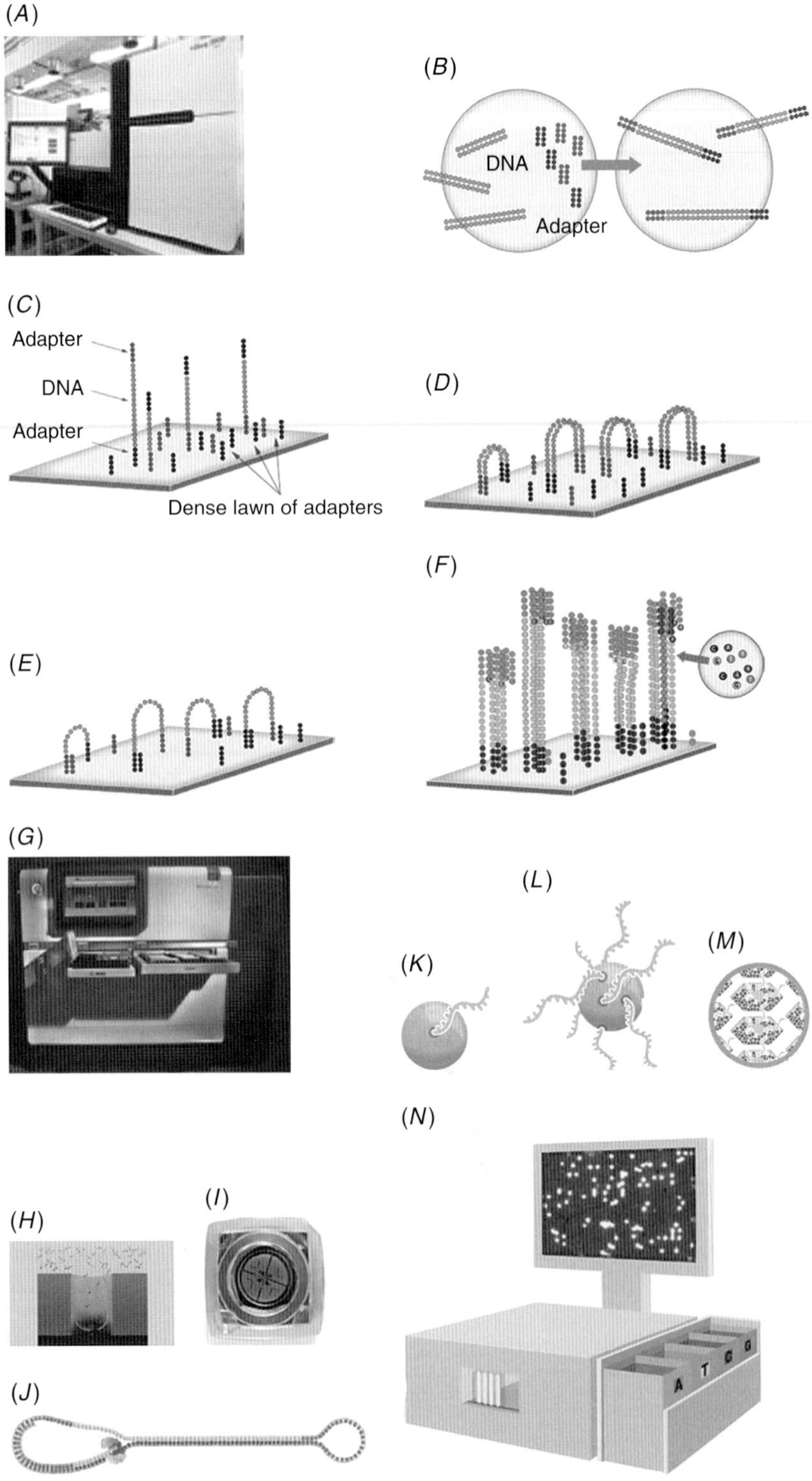
(A)
(B)
DNA
Adapter
(C)
Adapter
DNA
Adapter
Dense lawn of adapters
(D)
(E)
(F)
(G)
(L)
(K)
(M)
(N)
(H)
(I)
(J)
A
T
C
G

filling the gaps, and improving confidence of individual base calls. The genome of *S. cerevisiae* was not only the first fungal genome sequenced, but also the first fully finished genome. It was followed by a number of finished small size yeast genomes. But it took 13 years before the first finished genomes of filamentous fungi were reported in 2011 (Berka, Grigoriev, et al., 2011) despite a large number of nearly complete genomes, such as *N. crassa* (Galagan, Calvo, et al., 2003). The challenges include repeats, polymorphism, and recent segmental duplications, which are difficult to assemble; G + C and other sequencing biases often leave certain portions of genomic sequences lacking in sequence read coverage and make it difficult to finish genomes according to the latest standards (Chain, Grafham, et al., 2009).

A complete genome assembly offers considerably more information than a draft assembly: correct gene order, confidences in gene presence or absence, and distinctions between genes and pseudogenes. Repeated genes from subtelomeric regions can be incorrectly assembled or omitted from draft assemblies but carry important biological information, often related to ecological specialization like virulence genes in pathogens. Although most of the gene space of a 40-Mb genome can be revealed in a fraction of an Illumina lane, its completeness is uncertain until the genome is finished and thoroughly annotated. As traditional finishing using Sanger becomes cost prohibitive, NGS offers new possibilities for genome improvement, such as using a combination of Illumina sequencing with PacBio. Even though long PacBio reads have an error rate of 15 percent, these unbiased errors can be corrected using Illumina data and can serve as a framework for scaffolding short Illumina contigs and closing gaps between them. Between 30× and 50× coverage in PacBio reads may be sufficient for finishing microbial or fungal assemblies (Copeland, personal communication). In addition, genome maps—physical, genetic or optical—offer additional resources to build better assemblies by connecting smaller fragments into bigger chromosome-sized scaffolds and

←

Figure 1.3 Examples of the next-generation sequencing includes (*A–F*) sequencing by synthesis using Illumina; (*G–J*) single-molecule sequencing using Pacific Biosciences; and (*K–N*) Roche's 454 pyrosequencing. (*A*) Illumina Hi-Seq process includes (*B*) adapter ligation to both ends of the random DNA fragments, (*C*) binding single-stranded DNA fragments to the inside surface of flow cell channels, (*D–E*) bridge formation and amplification, and (*F*) sequencing by detecting signals from four-labeled reversible terminators incorporated by DNA polymerase enzyme. (*G*) Pacific Biosciences uses (*H*) SMRT cells, each containing (*I*) a single DNA polymerase attached to the bottom, filled with diffused nucleotides with fluorescent markers on the terminal phosphate, which are illuminated during the reaction time; (*J*) double-stranded DNA linear structures (200–10,000 base pairs long) attached to the SMRT adapters produce a topologically closed circle and enable consensus sequencing of the same template. (*K*) Roche's 454 sequencing uses single-stranded DNA attached to beads, (*L*) amplified using emulsion polymerase chain reaction, (*M*) loaded into hexagonal wells of a fiber-optic slide, and (*N*) pyrosequenced by capturing light flashes from incorporation of each base (A, T, C, or G). For color detail, please see color plate section.

validating their order. For instance, optical maps based on restriction fragment measurements were successfully used for several fungal genomes (Samad, Huff, et al., 1995; Coleman, Rounsley, et al., 2009).

Bioinformatics Challenges

Along with the significant advances in sequencing throughput, meaningful analysis of massive amounts of data became the biggest bottleneck. Besides faster and cheaper sequencing of reference genomes, NGS has opened doors to many new applications, each requiring specialized pipelines such as of single nucleotide polymorphism (SNP) analysis, quantification and comparison of expression profiles, and assemblies of metagenomes and metatranscriptomes. The speed with which huge data volumes are processed becomes important, which requires smarter algorithms, parallel processing, and large amounts of memory and storage. Thus, as data production becomes simpler and cheaper, analysis requires more powerful and expensive computational infrastructure, shifting the budget of genomics operations.

Efficient mapping the millions of short reads produced by even a single Illumina lane to a reference genome has required the development of new algorithms, distinct for genomics reads (e.g., BWA, Li & Durbin, 2009) versus EST reads (e.g., BowTie, Langmead & Salzberg, 2012). The latter requires gapped alignment to map introns and exons and for higher efficiency can combine both approaches: initial non-gapped alignment of all mappable reads followed by gapped alignment of reads not mapped at the first stage.

Assembling millions of short reads de novo is an even bigger challenge. NGS platforms quickly replace each other, producing new types of data; new versions of the same platform offer datasets with dramatically different characteristics. The assembly of a genome requires taking into account unique combinations of read length, quality, and coverage. Waterman's theoretical model (Lander & Waterman, 1988) suggests 15× coverage of 100-bp reads, while in practice it takes more than 100× coverage to account for sequencing errors and biases and a combination of different size mate-pair reads to resolve repeats (Gnerre, Maccallum, et al., 2011). A general approach to assemble NGS short reads is based on *de Bruijn* graph, in which reads are converted into k-mers and then assembled first into contigs and scaffolds (Pevzner, Tang, et al., 2001). Error correction and filtering are critical steps of data preprocessing. Several assembly packages were broadly used with NGS data and recently benchmarked against each other during the Assemblathon and Genome Assembly Gold-Standard Evaluation (GAGE) to demonstrate significant differences in performance (Earl, Bradnam, et al., 2011; Salzberg, Phillippy, et al., 2012). Most of them are publicly available tools that can be run in every lab but may need special requirements such as high-memory computers.

Finally, genome annotation challenges, always present in eukaryotic genomes, have been multiplied with fragmented assemblies that result in partial genes or gene fragments; genome-centric studies have been replaced by comparative and functional genomics to bring with them new complexities.

A Single Genome Story

The first genome sequencing projects were undertakings of a grand scale. A consortium of 74 different laboratories was formed in 1989 to sequence and analyze the first eukaryotic genome of the baker's yeast *S. cerevisiae*, which was published in 1996 (Goffeau, Barrell, et al., 1996). The effect of obtaining the first genetic blueprint of a eukaryote on biological research was astonishing, going far beyond yeast labs, to allow the use of yeast genetics to study protein functions and interactions (e.g., Foury 1997; Winzeler, Shoemaker, et al., 1999; Primig, Williams, et al., 2000; Bennett, Lewis, et al., 2001). Even today, experts argue about the accurate gene count in this relatively small and compact genome (e.g., Brachat, Dietrich, et al., 2003) and continue to update them in GenBank/EMBL/DDBJ on a regular basis. Therefore, manual curation of genes, functions, and available literature remains to be critical in genome analysis. The results of such curation are evident in the Saccharomyces Genome Database (SGD; Cherry, Hong, et al., 2012) and MIPS Comprehensive Yeast Genome Database (CYGD; Güldener, Münsterkötter, et al., 2005), both being examples of rich resources of genomic data devoted to this single genome with a huge user base and advanced manual curation tools.

Prediction of genes from genome sequence, particularly in eukaryotes in which complex intron-exon structure is typical, poses unique challenges. Predicting genes was relatively straightforward in *S. cerevisiae* because most of them had only a single exon, allowing detection simply as an open-reading frame (ORF) above a certain size. This advantage disappeared in the second yeast genome *S. pombe* with its multi-exon genes (Wood, Gwilliam, et al., 2002) and since then has remained a challenge for every genome of a filamentous fungus. Manual curation of gene structure became important and required analysis of several alternative gene models from different predictors in comparison with experimental data, such as transcriptomics or computed features such as genome conservation, to select the most accurate of these predicted models. The three major approaches to predict the complex intron-exon structure of eukaryotic genes includes: (a) EST-based, (b) protein homology-based, and (c) ab initio. EST-based methods have benefitted the most from NGS, which allowed the majority of genes to be predicted using de novo or genome-based assembled transcriptomes (e.g., PASA, Haas, Delcher, et al., 2003; Trinity, Grabherr, Haas, et al., 2011;

Cufflinks, Roberts, Pimentel, et al., 2011) from, for example, Illumina RNA-Seq data. Homology-based methods (e.g., GeneWise, Birney, Clamp, et al., 2004) rely on close protein homologs from other organisms. Ab initio methods use nucleotide signals derived from a set of known genes and predict them for the entire genome. They require training for each new genome using a set of reliable gene models. GeneMark-ES (Ter-Hovhanissyan, Lomsadze, et al., 2008) is a self-training algorithm, which uses specific features of fungal introns. The other algorithms can be universally used for different eukaryotes when trained for each of them. Because these approaches complement each other, they perform best when combined using various filtering procedures to pick the most feasible model for a locus (e.g., Combiner, Allen, Pertea, et al., 2004).

The complex problems of visually curating gene models in the context of a genome sequence, full-length mRNAs, and homology to sequence databases has led to the development of genome browsers. Initially developed for HGP (Guigo, Flicek, et al., 2006), genome browsers enabled visualization and comparison of multiple predicted models and have since emerged as a centerpiece of manual curation tools. Several genome browsers are available now as open source projects (e.g., GBrowse, Donlin 2009; JBrowse, Skinner, Uzilov, et al., 2009). The DOE's Joint Genome Institute (JGI) was one of the HGP partners charged with sequencing chromosomes 5, 16, and 19. The JGI Genome Portal (Grigoriev, Nordberg, et al., 2012), equipped with web-based manual curation tools, was based on the previous version of the UCSC Genome Browser (Fujita, Rhead, et al., 2011) with a configurable selection of tracks to display predicted gene models and annotations along with different lines of evidence in support of these predictions (e.g., gene and protein expression profiles). More than 4,000 human genes on the three chromosomes were predicted and manually curated. In-house curation also appeared to be useful for a number of model organisms but is not scalable to larger genome projects and is therefore difficult to fund (Howe, Costanzo, et al., 2008). However, some tools developed for manual curation were redirected to user-community curation. Using these tools, JGI developed a community annotation model, which was unique across sequencing centers and which engaged users in collective analysis and improvement of genome annotations. The first genomes thus taught us that accurate gene prediction required a combination of multiple approaches, the integration of predicted models and experimental lines of evidence, web-based visual interfaces and manual curation tools, and an active participation of research community.

Comparative Genomics

Sequencing fungal genomes one at a time has created a critical mass of data to explore in a comparative fashion. These initial analyses in turn suggested new phylogeny-driven approaches for designing new comparative studies The first yeast genome was followed by comparative studies of Kellis, Patterson, et al. (2003) and Dujon, Sherman, et al. (2004). In 2000, the Broad Institute initiated discussions of the Fungal Genomics Initiative (FGI; Cuomo & Birren, 2010). Nominations from the fungal research community led to a series of four white papers to the National Human Genome Research Institute (NHGRI) to cover the scope of FGI, which at this point has delivered around 50 genomes (http://www.broadinstitute.org/scientific-community/science/projects/fungal-genome-initiative). The resulting comparative analyses (e.g., Galagan, Henn, et al., 2005) have shown a bias toward a single phyla Ascomycota and fungi related to human health (Fig. 1.4).

Responding to this imbalance, JGI sequenced its Basidiomycete first fungus in 2004 (Martinez, Larrondo, et al., 2004). In 2009, JGI started the Fungal Genomics Program (http://jgi.doe.gov/fungi) to focus on fungi that were important to energy and the environment. Its first large-scale comparative project, called the Genomic Encyclopedia of Fungi (Grigoriev, Cullen, et al., 2011), has started with several "chapters" aligned with DOE missions in bioenergy production, biogeochemistry, and carbon cycling. Combining new sequencing technologies and comparative genomics analyses, JGI aimed to survey the broad phylogenetic and ecological diversity of fungi and capture genomic variation in natural populations and engineered strains. Comparison of the first genomes of white rot (*P. chrysosporium*) and brown rot (*Postia placenta*) fungi (Martinez, Larrondo, et al., 2004; Martinez, Challacombe, et al., 2009) has revealed dramatic differences in mechanisms of lignocellulose degradation between these two closely related fungi. This led to the sequencing of 30 wood decay fungi concentrated in the Agaricomycetes class and resulted in the first published chapter of the encyclopedia by Floudas, Binder, et al. in 2012. It presents the most comprehensive catalog of lignocellulolytic enzymes and reconstructions of white and brown rot evolution. In parallel, a study of 25 mycorrhizal fungi (more than half of them sequenced to date-*see* Chapter 8) started from observing dramatic differences between the first two sequenced symbionts *Laccaria bicolor* (Martin, Aerts, et al., 2008) and black truffle (*Tuber melanosporum*; Martin, Kohler, et al., 2010). Following the saprotroph and symbiont genomes, more than a dozen of *Dothideomycete* plant pathogens were sequenced for the largest comparative study of its type accompanied by several focused in-depth analyses and functional genomics of their subsets (Ohm, Feau, et al., 2012; *see* Chaper 6). In addition, several groups of industrially related fungi (Saccharomycetes, *Aspergilli*, *Trichoderma*) are being explored in depth (for example, *see* Chapter 4). Finally, the sequencing of a number of divergent fungi was an initial attempt to complete the picture and gave the start to a much larger scale exploration of fungal genomes across the Fungal Tree of Life (Martin, Cullen, et al., 2011). Through all these efforts

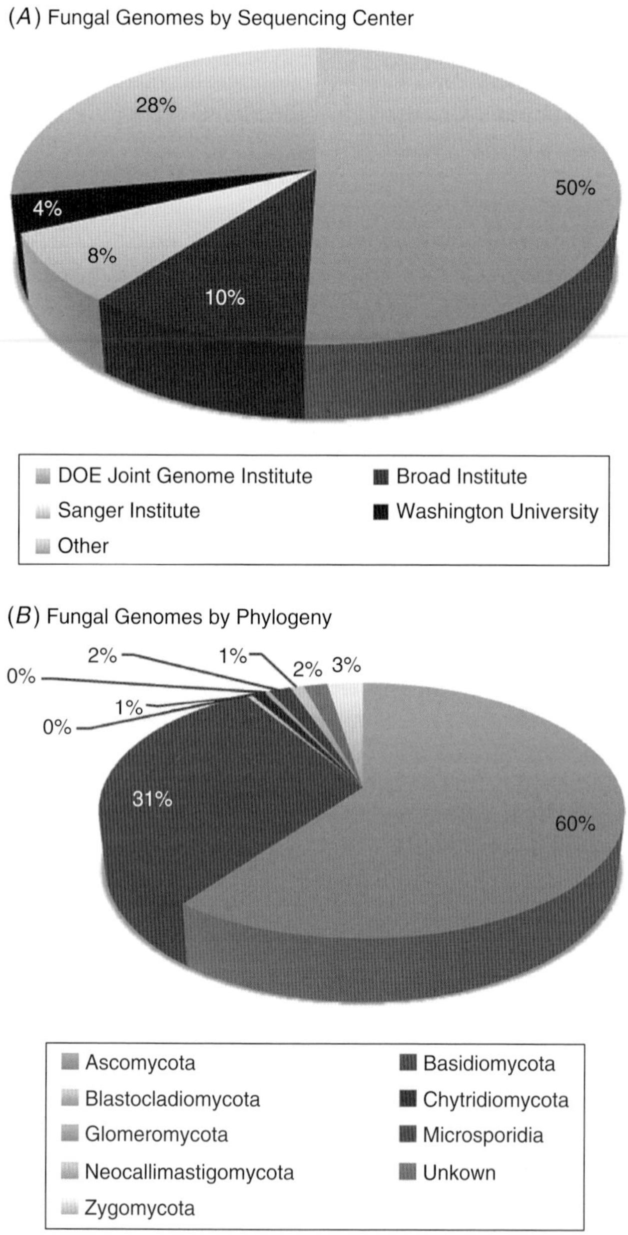

Figure 1.4 More than 1000 ongoing fungal genomic projects have been registered in GOLD database. They are grouped by (*A*) sequencing center and (*B*) represented phyla. For color detail, please see color plate section.

the total number of ongoing genome projects has increased significantly (*see* Fig. 1.4).

Comparative genomics requires data integration and can be a challenge when data have been produced by different centers and labs. In February 2010, a group of fungal biologists and bioinformaticians met in Alexandria, Virginia, United States, to call for the integration of all fungal genomes and analytical tools in one place to enable efficient comparative analyses. Released in March 2012, JGI MycoCosm (http://jgi.doe.gov/fungi) was one of the first responses to this call. It brought together fungal genomics data and interactive analytical tools for diverse fungi from JGI and its users to promote user-community participation in data submission, annotation, and analysis (Grigoriev, Nordberg, et al., 2012). More than 250 newly sequenced and annotated fungal genomes from JGI and elsewhere are available to the public through MycoCosm, and new annotated genomes are being added to this resource on completion of annotation. Nodes of the MycoCosm tree represent different groups of sequenced fungi and moving from one node to another redefines the search and analysis space, from a single organism to a group or the entire list of fungal genomes. These groups of genomes are linked to comparative tools. Gene family expansions or contractions can be identified using side-by-side comparison of each genome's functional profiles (according to the Gene Ontology (Ashburner, Ball, et al., 2000), KEGG pathways (Ogata, Goto, et al., 1999), and KOG clusters of orthologs (Koonin, Fedorova, et al., 2004) classifications) or with analysis of gene clusters produced by the MCL protein sequence-clustering algorithm (Enright, Van Dongen, et al., 2002). Analysis of structural genome organization using VISTA Point tools for pairwise DNA alignments recently led to the understanding of an interesting phenomenon of mesosynteny in Dothideomycetes (Ohm, Feau, et al., 2012).

Data integration also raises a question of data consistency and comparability of genome annotations. Eukaryotic genome annotation is challenging, requires a combination of different approaches, and lacks the standards developed, for example, for bacterial genomes. Use of the same tools for different genomes makes gene annotations comparable despite possible inaccuracies. Until the recent introduction of MAKER (Cantarel, Korf, et al., 2008), no fully automated pipeline for eukaryotic annotation was available, in contrast to several solutions for prokaryotes. Each genome center has developed its own "production" annotation pipeline using similar combination of gene prediction and annotation tools (Grigoriev, Martines, et al., 2006; Haas, Zeng, et al., 2011). The JGI annotation pipeline (Grigoriev, Martines, et al., 2006), for example, was used to annotate more than 100 fungal genomes so far and to achieve data consistency at least within this data set (http://jgi.doe.gov/fungi). Even though similar approaches have been used in different annotation pipelines, differences in parameters, pre- and post-processing lead to differences in gene count numbers by ,for instance, including or excluding transposons and pseudogenes.

Using a comparative genomics approaches can further improve quality of genome annotation and consistency at least across related genomes (e.g., Arnaud, Cerqueira, et al., 2012). For example, using comparative approach even in the first comparative study, 40 new *S. cerevisiae* genes with less than 100 amino acids in length were predicted based on their conservation among the sequenced species, and 500 predicted genes were suggested as dubious protein-coding genes because of lack of such conservation or lack of supporting experimental data available at the time (Kellis, Patterson, et al., 2003).

Thus, progress in genome sequencing has introduced comparative genomics as a powerful analytical tool. What was critical for annotation of the first genomes—manual curation—remains important but not achievable given the rapidly growing number of sequences. Scalable genome annotation demands robust automated pipelines for annotating genomes in a consistent manner. Comparative pipelines offer better accuracy than genome-centric ones because evolutionary information serves as an additional line of evidence in predicting and validating predicted genes. Research communities can help validate automatically predicted genes and functions, whereas distributed curation offers the only scalable option. Large-scale genomics thus depends on consistency of data and tools, their integration, research community coordination, and new comparative genomics tools.

New Genomics

Democratization of sequencing has moved sequencing of a single microbial genome mostly outside of the domain of large genomics centers, which historically were focused on these tasks. Instead, genome centers embraced projects of large scale and complexity and involve broad research communities for grand scale initiatives such as the 1000 genomes, 1001 Arabidopsis genomes (http://1001genomes.org), 1000 fungal genomes (http://1000.fungalgenomes.org), Genomic Encyclopedia of Bacteria and Archae (GEBA), drawing a grand picture of genomics for the near future (Weigel & Mott, 2009; 1000 Genomes Project Consortium, 2010; Wu, Hugenholtz, et al., 2009). These large- scale grand initiatives will require both data-integration and research-community coordination at unprecedented levels. Accelerated generation of sequence data dramatically increases the gap between sequence and functional information. New high-throughput experimental methods are needed for assigning functions to each of the predicted genes. System-wide studies of individual organisms are critical and should go beyond studies of a few models, such as *S. cerevisiae* or *N. crassa*. Relatively simple two- or three-component systems of interacting organisms, such as lichens (symbiosis of algae and fungus) or mycorrhizal tissues of plants (symbiosis of plant roots and mutualistic fungi), should lead to better understanding of gene networks within and subsequently between organisms with the ultimate goal of exploring more complex interactions in metagenomes.

The 1000 fungal genomes project is one of the latest large-scale genomic initiatives. More than a million species in the kingdom Fungi have evolved over millions of years to occupy diverse ecological niches and have accumulated an enormous but yet undiscovered natural arsenal of potentially useful innovations. Although the number of fungal genome sequencing projects continues to increase, the phylogenetic breadth of current sequencing targets is extremely limited. Exploration of phylogenetic and ecological diversity of fungi by genome sequencing is therefore a potentially rich source of valuable metabolic pathways and enzyme activities that will remain undiscovered and unexploited until a systematic survey of phylogenetically diverse genome sequences is undertaken.

At the same time, the ability to sample environments for complex fungal metagenomes is rapidly becoming a reality, while capabilities to accurately analyze these data relies on well-characterized, foundational reference data of fungal genomes. To bridge this gap in the understanding of fungal diversity, an international research team in collaboration with JGI has embarked on a 5-year project to sequence 1000 fungal genomes from across the Fungal Tree of Life. The overall plan is to fill in gaps in the Fungal Tree of Life by sequencing at least two reference genomes from the more than 500 recognized families of Fungi. With 14 principal investigators from different labs around the world, several culture collections participating, and a growing interest from entire mycological community, this project aims to provide genomic references to inform research on plant-microbe interactions and environmental metagenomics. Instead of coordinating individual groups of researchers focused on the analysis of a single genome or a relatively small group of genomes, larger-scale comparisons bring community coordination to the top as critical for target selection, coordinated analysis, and publishing.

Metagenomics offers another way to explore diversity of fungi (*see* Chapters 13, 14, and 15). Because many fungi cannot be isolated in pure culture, the analysis of microbial communities can shed light on fungal diversity in natural habitats. For example, soil is important for understanding biogeochemical cycles, while the human gut microbiome is crucial for human health. System-wide analysis of these complex communities can also determine types of interactions between different organisms or subsystems and their responses to environmental changes. Currently metagenomics of prokaryotic components of microbial communities, including soil and HGP, is relatively straightforward. In contrast, the analysis of eukaryotic metagenomes is challenging because of much larger genome sizes, complex gene and genome structures, often insufficient amount and quality of DNA material, and need for increasing the number of fungal reference genomes (*see* 1000 Fungal Genomes Project). Therefore, instead of deciphering metagenomes (i.e., assembling genomic sequences), analysis of fungal metatranscriptomes (i.e., assembling and counting spliced transcripts) appears to be an easier target. Although effectively eliminating the need to predict exon-intron gene structure in incomplete genomic data sets, this approach explores spliced transcripts, or ORFs, adds functional information, and determines quantitative expression values of genes. However, in complex microbial communities, the fraction of eukaryotic transcriptome can be small, or communities can be so complex that neither metagenomics nor metatranscriptomics approaches will be feasible. Therefore, often the first step in analysis of fungal communities is analysis of species diversity using fungal markers. The internal transcribed spacers (ITS) 1 and 2 of the ribosomal RNA genes provide powerful tools for identifying species because DNA sequences of the ITS region diverge rapidly between species, whereas concerted evolution maintains a high level of sequence uniformity within species (*see* Chapter 13). This approach, however, is also dependent on availability of reference sequences.

Functional genomics fills the growing gap between the large number of predicted genes from sequenced genomes and their functions. *S. cerevisiae* is the best example of a genome sequence leading to a comprehensive metabolic reconstruction and ultimately to modeling different processes in the organism. This made it not only an excellent experimental framework for plugging in new processes, but also an industrial workhorse. Recently, it was engineered to ferment xylose, one of the key components of plant biomass and currently underused in biofuel production. Initially, genes discovered in *Pichia stipitis* (Jeffries, Grigoriev, et al., 2007), the most potent xylose fermenter, were engineered into strains of *S. cerevisiae* to use xylose. However, their xylose fermentative capacity pales in comparison with glucose fermentation, limiting the economic feasibility of industrial fermentations. Comparative genomics and transcriptomics of 14 xylose-utilizing and xylose-fermenting fungi suggested additional genes and processes involved in xylose assimilation. Several of these genes significantly improved xylose use when engineered into *S. cerevisiae*, demonstrating the power of comparative methods to rapidly identify genes for biomass conversion (Wohlbach, Kuo, et al., 2011).

Xylose and glucose are some of the building blocks of the heterogeneous plant cell wall collectively called *lignocellulose*. Lignocellulose is composed mainly of cellulose, hemicellulose, pectin, and lignin and is a promising resource for producing biofuels. During their long evolutionary history, plants developed lignocellulose to resist microbial attacks. At the same time, wood decay fungi developed enzymes for lignocellulose degradation. Despite significant efforts analyzing enzymes involved in wood decay and more than 40 wood decay fungi genomes sequenced and annotated (*see* Chapter 3), relatively little is known about how the process works at the molecular level. Wood decay fungi are mostly not experimentally tractable organisms, and a new model basidiomycete is needed to study wood decay mechanisms. *Schizophyllum commune* is one of few transformable wood decay fungi to be further explored at JGI (Ohm, Jong, et al., 2010).

Similar to what has been seen in sequencing technologies, a revolution in functional genomics technologies is needed to understand the functions of a majority of genes encoded in sequenced genomes (as opposed to the relatively small fraction we currently understand). Initiatives such as human ENCODE Project Consortium (2004) give promise of developing high-throughput techniques that could be applicable to fungi and other organisms. These may include identification and quantification of RNA species, mapping of protein-coding regions, delineation of chromatin and DNA accessibility and structure with nucleases and chemical probes, mapping of histone modifications and transcription factor (TF) binding sites by chromatin immunoprecipitation (ChIP), measurement of DNA methylation, examining long-range chromatin interactions, localizing binding proteins on RNA, identifying transcriptional silencer elements, and understanding detailed promoter sequence architecture.

Conclusion

In contrast to the early days of sequencing when the first genomes were done by consortia of sequencing centers, new sequencing technologies are available to many small laboratories, each producing tons of data. Efficient data processing and integration, as well as new comparative tools, have become critical for large-scale genomics to answer a broad range of biological questions, analyzing genome organization across fungi, and scrutinizing features of individual genes along their evolution history. With this avalanche of data produced by different groups around the world independently of each other, communication and coordination becomes important to minimize duplication of effort and to integrate all data into one big picture. In the world of thousands of sequenced genomes, a single gene function still matters and can be critical for our understanding of complex biological processes. However, the method of determining a gene's function is changing and may well start with sequencing an entire genome or several of them.

Genomics became a new tool in the toolkit of modern mycology. Today's biology depends on genomics of large scale and high complexity involving large scientific communities and applied functional genomics. This in turn requires new type of bioinformatics, capable of addressing the scale of the data produced and the complexity of the questions raised.

References

1000 Genomes Project Consortium. 2010. A map of human genome variation from population-scale sequencing. *Nature*. 467:1061–1073.

Allen JE, Pertea M, et al. 2004. Computational gene prediction using multiple sources of evidence. *Genome Res*. 14(1):142–148.

Arnaud MB, Cerqueira GC, et al. 2012. The Aspergillus Genome Database (AspGD): Recent developments in comprehensive multispecies curation, comparative genomics and community resources. *Nucl Acids Res*. 40(Database issue): D653–D659.

Ashburner M, Ball CA, et al. 2000. Gene ontology: Tool for the unification of biology. *Nat Genet*. 25:25–29.

Bennett CB, Lewis LK, et al. 2001. Genes required for ionizing radiation resistance in yeast. *Nat Genet*. 29(4):426–434.

Berka RM, Grigoriev IV, et al. 2011. Comparative genomic analysis of the thermophilic biomass-degrading fungi *Myceliophthora thermophila* and *Thielavia terrestris*. *Nat Biotechnol*. 29(10):922–927.

Birney E, Clamp M, et al. 2004. GeneWise and Genomewise. *Genome Res*. 14(5): 988–995.

Brachat S, Dietrich FS, et al. 2010. Reinvestigation of the *Saccharomyces cerevisiae* genome annotation by comparison to the genome of a related fungus: *Ashbya gossypii*. *Nat Rev Genet* 11: 31–46

Cantarel BL, Korf I, et al. 2008. MAKER: An easy-to-use annotation pipeline designed for emerging model organism genomes. *Genome Res*. 18(1):188–196.

Chain PS, Grafham DV, et al. 2009. Genome project standards in a new era of sequencing. *Science*. 326(5950): 236–237.

Cherry JM, Hong EL, et al. 2012. Saccharomyces Genome Database: The genomics resource of budding yeast. *Nucl Acids Res*. 40(Database issue):D7 00–D705.

Coleman JJ, Rounsley SD, et al. 2009. The genome of *Nectria haematococca*: Contribution of supernumerary chromosomes to gene expansion. *PLoS Genet*. 5(8):e1000618.

Cuomo CA & Birren BW. 2010. The fungal genome initiative and lessons learned from genome sequencing. *Methods Enzymol*. 470:833–855.

Donlin MJ. 2009. Using the Generic Genome Browser (GBrowse). *Curr Protoc Bioinformatics*. Chapter 9: Unit 9.9

Dujon B, Sherman D, et al. 2004. Genome evolution in yeasts. *Nature*. 430(6995): 5–44.

Earl D, Bradnam K, et al. 2011. Assemblathon 1: Acompetitive assessment of de novo short read assembly methods.*Genome Res*. 21(12):2224–2241.

ENCODE Project Consortium. 2004. The ENCODE (ENCyclopedia Of DNA Elements) Project. *Science* 306: 636–640.

Enright AJ, Van Dongen S, et al. 2002. An efficient algorithm for large-scale detection of protein families. *Nucl Acids Res*. 30(7):1575–1784.

Floudas D, Binder M, et al. 2012. The Paleozoic origin of enzymatic lignin decomposition reconstructed from 31 fungal genomes. *Science*. 336(6089): 1715–1719. doi: 10.1126/science.1221748.

Foury F. 1997. Human genetic diseases: A cross-talk between man and yeast. *Gene*. 195(1): 1–10.

Fujita PA, Rhead B, et al. 2011. The UCSC Genome Browser database: Update 2011. *Nucl Acids Res*. 39: D876–D882.

Galagan JE, Calvo SE, et al. 2003. The genome sequence of the filamentous fungus *Neurospora crassa*. *Nature*. 422(6934): 859–868.

Galagan JE, Henn MR, et al. 2005. Genomics of the fungal kingdom: Insights into eukaryotic biology. *Genome Res*. 15(12):1620–1631.

Gnerre S, Maccallum I, et al. 2011. High-quality draft assemblies of mammalian genomes from massively parallel sequence data. *Proc Natl Acad Sci USA*. 108(4): 1513–15188.

Goffeau A, Barrell BG, et al. 1996. Life with 6000 genes. *Science*. 274 (5287): 546, 563–567,

Grabherr MG, Haas BJ, et al. 2011. Full-length transcriptome assembly from RNA-Seq data without a reference genome. *Nat Biotechnol*. 29(7): 644–652

Grigoriev IV, Cullen D, et al. 2011. Fueling the future with fungal genomics. *Mycology*. 2(3): 192–209.

Grigoriev IV, Martines, DA, et al. 2006. Fungal genomic annotation. In Applied Mycology and Biotechnology (eds. DK Aurora, RM Berka, et al.), Vol. 6, *Bioinformatics*, 123–142. Philadelphia: Elsevier.

Grigoriev IV, Nordberg H, et al. 2012. The Genome Portal of the Department of Energy Joint Genome Institute. *Nucl Acids Res*. 40(1): D26–D32.

Guigó R, Flicek P, et al. 2006. EGASP: The human ENCODE Genome Annotation Assessment Project. *Genome Biol*. 7 Suppl 1:S2.1–31.

Güldener U, Münsterkötter M, et al. 2005. CYGD: The Comprehensive Yeast Genome Database. *Nucl Acids Res*. 33(Database issue): D364–D368.

Haas BJ, Delcher AL, et al. 2003. Improving the Arabidopsis genome annotation using maximal transcript alignment assemblies. *Nucl Acids Res*. 31(19): 5654–5666.

Haas BJ, Zeng Q, et al. 2011. Approaches to fungal genome annotation. *Mycology*. 2(3): 118–141.

Howe D, Costanzo M, et al. 2008. Big data: The future of biocuration. *Nature*. 455(7209): 47–50.

Jeffries TW, Grigoriev IV, et al. 2007. Genome sequence of the lignocellulose-bioconverting and xylose-fermenting yeast *Pichia stipitis*. *Nat Biotechnol*. 25:319–326.

Kellis M, Patterson N, et al. 2003. Sequencing and comparison of yeast species to identify genes and regulatory elements. *Nature*. 423(6937): 241–254.

Koonin EV, Fedorova ND, et al. 2004. A comprehensive evolutionary classification of proteins encoded in complete eukaryotic genomes. *Genome Biol*. 5: R7.

Lander ES, Waterman MS.1988. Genomic mapping by fingerprinting random clones: A mathematical analysis. *Genomics*. 2(3): 231–239.

Langmead B, Salzberg SL. 2012. Fast gapped-read alignment with Bowtie 2. *Nat Methods*. 9(4): 357–359/

Li H & Durbin R. 2009. Fast and accurate short read alignment with Burrows-Wheeler transform. *Bioinformatics*. 25(14): 1754–1760.

Martin F, Aerts A, et al. 2008. The genome of Laccaria bicolor provides insights into mycorrhizal symbiosis. *Nature*. 452, 88–92.

Martin F, Cullen D, et al. 2011. Sequencing the fungal tree of life. *New Phytol*. 190(4):818–821.

Martin F, Kohler A, et al. 2010. Périgord black truffle genome uncovers evolutionary origins and mechanisms of symbiosis. *Nature*. 464(7291): 1033–1038.

Martinez D, Challacombe J, et al. 2009. Genome, transcriptome, and secretome analysis of wood decay fungus *Postia placenta* supports unique mechanisms of lignocellulose conversion. *Proc Natl Acad Sci USA*. 106:1954–1959.

Martinez D, Larrondo LF, et al. 2004. Genome sequence of the lignocellulose degrading fungus *Phanerochaete chrysosporium* strain RP78. *Nat Biotechnol*. 22(6): 695–700.

Metzker ML. 2010. Sequencing technologies—the next generation. *Nat Rev Genet*. 11(1): 31–46

Ogata H, Goto S, et al. 1999. KEGG: Kyoto Encyclopedia of Genes and Genomes. *Nucl Acids Res*. 27: 29–34.

Ohm RA, Feau N, et al. 2012 Diverse lifestyles and strategies of plant pathogenesis encoded in the genomes of eighteen *Dothideomycetes* fungi. *PLoS Pathog*. 8(12): e1003037. doi: 10.1371/journal.ppat.1003037.

Ohm RA, Jong JF, et al. 2010. Formation of mushrooms and lignocellulose degradation encoded in the genome sequence of *Schizophyllum commune*. *Nat Biotechnol*. 28(9): 957–963.

Pennisi E. 2012. At long last, nanopore sequencing seems poised to leave the lab, promising a new and better way to decode DNA. *Science*. 336: 534–537.

Pevzner PA, Tang H, et al. 2001. An eulerian path approach to DNA fragment assembly. *Proc Natl Acad Sci USA* 98(17): 9748–9753.

Primig M, Williams RM, et al. 2000. The core meiotic transcriptome in budding yeasts. *Nat Genet*. 26(4): 415–423.

Roberts A, Pimentel H, et al. 2011. Identification of novel transcripts in annotated genomes using RNA-Seq. *Bioinformatics*. 27(17): 2325–2329.

Salzberg SL, Phillippy AM, et al. 2012. GAGE: A critical evaluation of genome assemblies and assembly algorithms. *Genome Res*. 22(3): 557–567.

Samad A, Huff EF, et al. 1995. Optical mapping: A novel, single-molecule approach to genomic analysis. *Genome Res*. 5(1): 1–4.

Sanger F, Nicklen S, et al. 1977. DNA sequencing with chain-terminating inhibitors. *Proc Natl Acad Sci USA*. 74(12): 5463–5467.

Skinner ME, Uzilov AV, et al. 2009. JBrowse: A next-generation genome browser. *Genome Res*. 19(9):1630–1638.

Ter-Hovhannisyan V, Lomsadze A, et al. 2008. Gene prediction in novel fungal genomes using an ab initio algorithm with unsupervised training. *Genome Res*. 18(12): 1979–1990.

Weigel D & Mott R. 2009. The 1001 genomes project for Arabidopsis thaliana. *Genome Biol*. 10(5):107.

Winzeler EA, Shoemaker DD, et al. 1999. Functional characterization of the *S. cerevisiae* genome by gene deletion and parallel analysis. *Science*. 285(5429): 901–906.

Wohlbach DJ, Kuo A, et al. 2011. Comparative genomics of xylose-fermenting fungi for enhanced biofuel production. *Proc Natl Acad Sci USA*. 108(32): 13212–13217.

Wood V, Gwilliam R, et al. 2002. The genome sequence of *Schizosaccharomyces pombe*. *Nature*. 415(6874): 871–880.

Wu D, Hugenholtz P, et al. 2009. A phylogeny-driven genomic encyclopaedia of bacteria and archaea. *Nature*. 462(7276): 1056–1060.

2 Repeated Elements in Filamentous Fungi with a Focus on Wood-Decay Fungi

Claude Murat[1], Thibaut Payen[1], Denis Petitpierre[1], and Jessy Labbé[2]

[1]*Laboratory of Excellence ARBRE, UMR 1136 INRA-Université de Lorraine, Interactions Arbres-Microorganismes, INRA-Nancy, Champenoux, France*
[2]*BioSciences Division, Oak Ridge National Laboratory, Oak Ridge, Tennessee*

Introduction

In the last decade, the genome of several dozen filamentous fungi have been sequenced. Interestingly, vast diversity in genome size was observed (Fig. 2.1) with 14-fold differences between the 9 Mb of the human pathogenic dandruff fungus (*Malassezia globosa*; Xu, Saunders, et al., 2007) and the 125 Mb of the ectomycorrhizal black truffle of Périgord (*Tuber melanosporum*; Martin, Kohler, et al., 2010). Recently, Raffaele and Kamoun (2012) highlighted that the genomes of several lineages of filamentous plant pathogens have been shaped by repeat-driven expansion. Indeed, repeated elements are ubiquitous in all prokaryote and eukaryote genomes; however, their frequencies can vary from just a minor percentage of the genome to more that 60 percent of the genome. Repeated elements can be classified in two major types: satellites DNA and transposable elements. In this chapter, the different types of repeated elements and how these elements can impact genome and gene repertoire will be described. Also, an intriguing link between the transposable elements richness and diversity and the ecological niche will be highlighted.

Satellites DNA

Satellites are tandem repetitions of motifs ranging from one to thousands of nucleotides in length. The length of the motif is used to classify satellites in three groups: microsatellites (1–6 nucleotides), minisatellites (7–100 nucleotides), and satellites (more than 100 nucleotides). However this classification

The Ecological Genomics of Fungi, First Edition. Edited by Francis Martin.

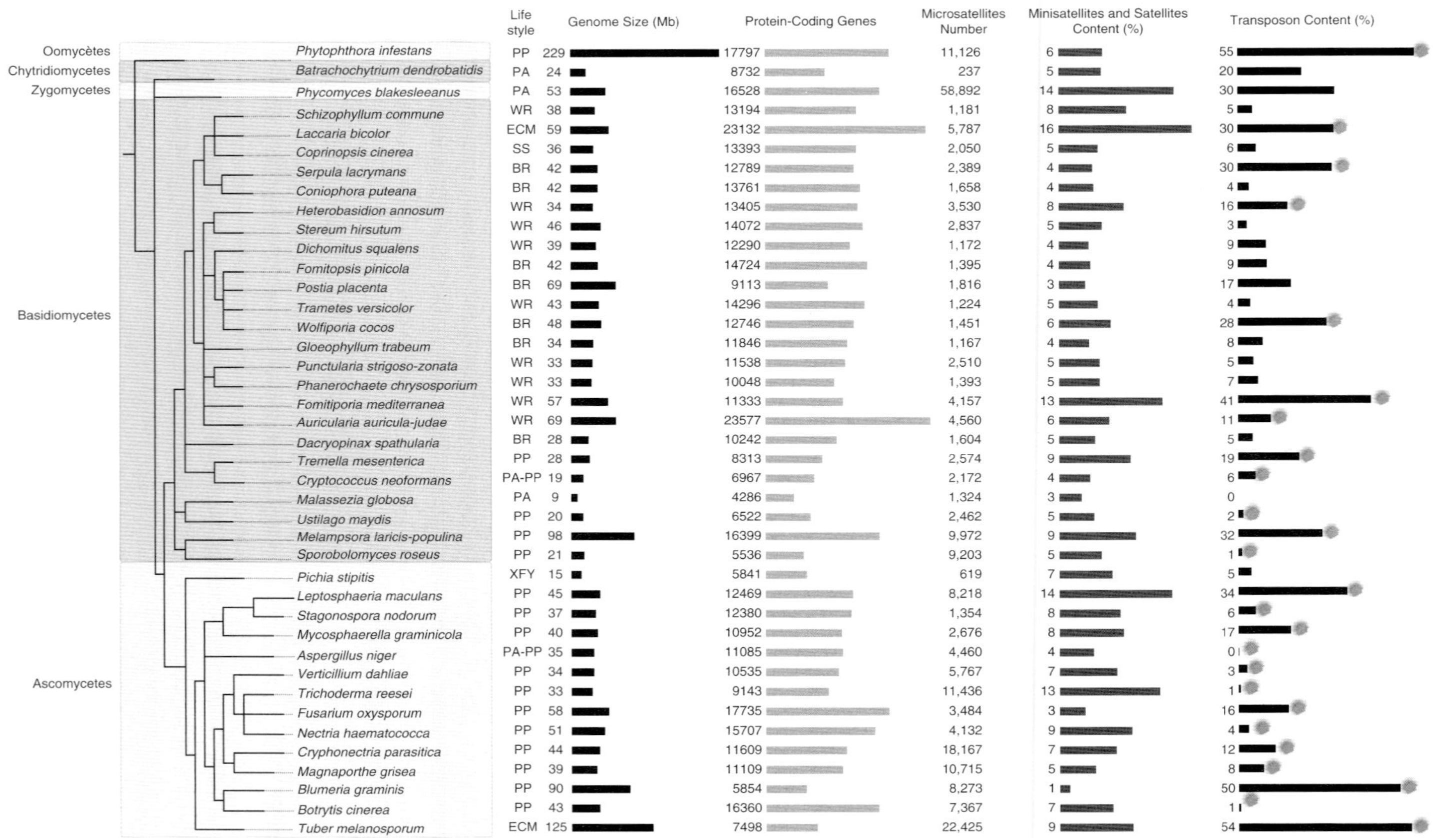

Figure 2.1 Main feature of some filamentous fungi sequenced belonging to plant pathogen, saprotroph, and mycorrhizal species. The phylogeny was generated using the Interactive Tree of Life (iTOL) with National Center for Biotechnology Information (NCBI) taxonomy identifiers (branch lengths are arbitrary). The life style (BR, brown rot; ECM, ectomycorhizal; PA, animal pathogen; PP, plant pathogen; SS, soil saprotroph; WR, white rot; XFY, xylose fermentating yeast), genome size, protein content, number of microsatellites, percentage of genome coverage for minisatellites and satellites and transposable elements are indicated. The species interacting with living plants are indicated with a green star near the transposon content. Transposable element identification procedure is described in the supplementary data of Foudas, Binder, et al. (2012).

can vary according to authors (e.g., minisatellites is considered to be between 7 and 100 bp [Vergnaud & Denoeud 2000] or 10 and 50 bp [Jeffreys, Wilson, et al., 1985]).

Microsatellites

Microsatellites are the most studied tandem repeats, and several studies comparing microsatellite richness in filamentous fungi are available (Pannebakker, Niehuis, et al., 2010; Labbé, Murat, et al., 2011; Murat, Riccioni, et al., 2011). Murat, Riccioni, et al. (2011) compared the microsatellites of 48 fungal species highlighting large differences in the number of microsatellites among fungal species that vary from 224 elements for *Batrachochytrium dendrobatidis* to 56,846 elements for *Phycomyces blakesleeanus*. For all Ascomycetes and Basidiomycetes, microsatellites cover less than 1 percent of their genome at the exception of Mucoromycotina species: *Rhizopus oryzae* (1.5 percent) and *P. blakesleeanus* (2.5 percent), and Saccharomycotina species: *Lodderomyces elongisporus* (3.6 percent), *Candida tropicalis* (2.4 percent) and *Candida albicans* (1.9 percent). Interestingly the microsatellites do not seem to contribute significantly to genome size because no correlation between number and relative abundance of microsatellites with genome size was found (Murat, Riccioni, et al., 2011). Microsatellites are not distributed equally in the genome. In general, microsatellites seem more frequent in introns and intergenic regions, but for *Laccaria bicolor*, they are also frequent in transposable elements (Labbé, Murat, et al., 2011). Although microsatellites may not have a clear effect on genome size, it has been shown that they modify protein function when localized in exons and can also modify gene expression when localized in promoter regulatory regions (Verstrepen, Jansen, et al., 2005; Riley & Krieger, 2009; Vinces, Legendre, et al., 2009; Rudd, Antoniw, et al., 2010). Microsatellites in coding regions have been extensively studied within the context of human diseases, revealing abundant evidence on their contribution to neuronal diseases and cancers (Ashley & Warren, 1995). Microsatellite instability (MSI) is a hypermutable phenotype caused by the loss of DNA mismatch repair activity. MSI is detected in about 15 percent of all colorectal cancers (Boland & Goel, 2010). Microsatellite expansions or contractions in protein-coding regions can lead to a gain or loss of gene function via frameshift mutation or expanded toxic mRNA (Li, Korol, et al., 2004).

Verstrepen, Jansen, et al. (2005) found that in the genome of *Saccharomyces cerevisiae*, 75 percent of the gene models containing microsatellites in their coding region coded for cell surface proteins. Moreover, several gene models with microsatellites in their coding region have been implicated in plant infection of pathogenic fungi because they could be involved during the formation of intercellular hyphae or could conceivably function as effectors during symptomless host plant colonization (Rudd, Antoniw, et al., 2010). Riley and Krieger (2009) highlighted that dinucleotide repeats in the untranslated region (UTR) of human genes are mainly involved in regulation of gene expression. This was shown for the microsatellites occurring in promoter regions of *S. cerevisiae* genes, where variations in repeat length can promote changes in expression (Vinces, Legendre, et al., 2009). In *T. melanosporum*, polymorphic microsatellites were found in the UTR of fruiting bodies and ectomycorrhiza-regulated genes, suggesting that they can have an effect on gene expression in fungi as in human, but no evidence of this effect is actually available (Murat, Riccioni, et al., 2011). The increasing interest of microsatellites studies and publications of the last years may come from their role in human disease and the multiplication of genomic resources.

Beside the potential effect of microsatellites on phenotype, microsatellites are also among the most popular molecular markers for population genetics in all organisms because they are assumed to be neutral and have a high level of polymorphism (Jarne & Lagoda, 1996). In human microsatellites, mutation rates of microsatellites are as frequent as 10^{-3} to 10^{-4} per locus per generation (Weber & Wong, 1993) compared with a rate of 10^{-8} per generation for single-nucleotide substitutions (Drake, Charlesworth, et al., 1998). For *T. melanosporum*, only the number of repetition for dinucleotide was correlated with the number of alleles. Dutech, Enjalbert, et al. (2007) found a similar result with a correlation of the mean repeat number for the dinucleotide and the number of alleles in several fungal species, in birds, insects, and fish but not in mammals and angiosperms. The mutation leading to the formation of a new allele can occur through different mechanisms, including errors during recombination, unequal crossing over and polymerase slippage during DNA replication, or from repair (Oliveira, Pádua, et al., 2006). If the accumulation of errors in a microsatellite continues indefinitely, then microsatellite will be large. This is not what was observed in the fungal genomes because longer microsatellites covered few hundred bp (e.g., 180, 204, 210, and 312 bp in *L. bicolor*, *P. blakesleeanus*, *Melampsora laricii-populina*, and *T. melanosporum*, respectively). There is therefore probably some selection acting against long microsatellites or a specific mechanism that eliminates long microsatellites. One hypothesis is that these sequences are particularly prone to big deletion or to single nucleotide substitution (Chambers & MacAvoy, 2000). This would generate small alleles that represent the "death" of the microsatellite under its life cycle (Hancock, Goldstein, et al., 1999; Chambers & MacAvoy, 2000). The smaller size of microsatellites in fungi suggest that in this group of organisms the phenomenon of single-sequence repeat (SSR) "death" did not allow the SSR to growth indefinitely, although in other eukaryotes some microsatellites spanning until 2 kbp have been observed (Sharma, Grover, et al., 2007).

The use of microsatellites to analyze fungal population genetics is not new but as highlighted by Dutech, Enjalbert, et al. (2007) microsatellites makers were not available for many species of fungi compared to other organisms. However, the availability of genome sequences coupled with new sequencing technology makes it easier now to characterize polymorphic microsatellites in non-model species. In fully sequenced genomes, microsatellites were characterized, using bioinformatic tools such as MISA or MAGELLAN, in several species such as *T. melanosporum* (Murat, Riccioni, et al., 2011), *L. bicolor* (Labbé, Murat, et al., 2011), and *M. laricii-populina* (Xhaard, Andrieux, et al., 2009). The new sequencing technologies become important to characterize microsatellites with two approaches using 454 pyrosequencing: enriched library sequencing or shotgun sequencing. The enriched library pyrosequencing was used by Malausa, Gilles, et al. (2011) to characterize microsatellites in insects, fungi, Oomycetes, and plants. However this requires information from enriched libraries, and this is why shotgun sequencing with 454 pyrosequencing seems more promising. Microsatellites can be identified directly in pyrosequencing reads without a preliminary assembly. This approach was used to characterize microsatellites in animals (e.g., termites, Singham, Vargo, et al., 2012), plants (e.g., *Acacia harpophylla*, Lepais & Bacles, 2011) and fungi (e.g., *Peltigera dolichorhiza* complex, Magain, Forest, et al., 2010; Burgundy Truffle [*T. aestivum*], Molinier, 2013). Beside pyrosequencing, Illumina sequencing was recently used to identify microsatellites in bird and snakes (Castoe, Poole, et al., 2012). Perl scripts named PAL_FINDER_v0.02.03 were used to extract reads with perfect microsatellites. The same pipeline PAL_FINDER_v0.02.03 allowed the design of primers to characterize new polymorphic markers with these microsatellites. These authors highlighted the usefulness of Illumina reads and concluded that Illumina "Seq-to-SSR" is effective, inexpensive, and reliable even for species that have few microsatellites loci.

Minisatellites and Satellites

Often, because of the lack of a precise definition of minisatellites, these structures are usually not annotated in genomic sequence data. There are few minisatellites studies relative to microsatellites focused studies. However, as for microsatellites, length variations of minisatellites have been found to be involved in several diseases such as diabetes, epilepsy, or cancer, and there is evidence for other contributions to genome function (see Buard & Jeffreys 1997; Vergnaud & Denoeud, 2000). The study of minisatellites in human genomics is not new because in 1985, Jeffreys, Wilson, et al. demonstrated that minisatellites with a repeat of 10 to 15 nucleotides could provide an individual-specific DNA "fingerprint" of general use in human genetic analysis. In fungi few studies focused on minisatellites markers. Recently, Bally, Grandaubert, et al. (2010) defined a pipeline, FONZIE, aimed to provide a set of specific primer sequences for polymerase chain reaction (PCR) amplification of single-locus micro- and minisatellites markers. This pipeline was successfully used to characterize minisatellite markers for the pathogenic fungi *Leptosphaeria maculans* (Dilmaghani, Gladieux, et al., 2012). The advantage to studying minisatellites is that their size (motifs more than 7 nucleotides) allows the analysis of amplicons to be conducted directly on agarose gels, thereby reducing costs.

Unlike microsatellites, there is no study available, to our knowledge, comparing minisatellite and satellite patterns in fungal genomes. Using the software *tandem repeat finder* (Benson, 1999), minisatellites and satellites were identified in 40 fungal genomes belonging to Ascomycetes, Basidiomycetes, Zygomycetes, and Chytridiomycetes, as well as in one Oomycete (*Phytophthora infestans*) (*see* Fig. 2.1). In fungi, the number of minisatellites ranged from 1,631 for *B. dendrobatidis* to 110,404 for *T. melanosporum* (data not shown). The genome coverage of minisatellites ranged from 1.27 percent in *Blumeria graminis* to 13.4 percent in *L. maculans*. The number of satellites ranged from 28 for *M. globosa* to 3,772 for *L. bicolor* and the genome coverage was for all species less than 2 percent except for *L. bicolor* (4.17 percent). The majority (90 percent) of the motifs were smaller than 40 nucleotides and the most represented length was 21 nucleotides with a mean of 4,342 minisatellites of this size per genome (data not shown). However, the sequence of the 21 bp minisatellites is different from a species to another.

Unlike microsatellites, a correlation between the genome size and the number of minisatellites and satellites was found ($r^2=0,55$ and $r^2=0,53$, respectively). This suggests that minisatellites and satellites contribute to genome expansion; this seems particularly true for *L. bicolor*, *L. maculans*, *Fomitiporia mediterranea*, *Trichoderma reesei*, and *P. blakesleeanus* for which minisatellites and satellites represented more than 10 percent of their genome. The role of minisatellites and satellites in fungal genome is actually unknown. But that they could be important for gene regulation, such as microsatellites, and genomic rearrangement, such as transposable elements, can not be excluded.

In conclusion, satellites DNA are frequent in the fungal genomes and for some species this particular repeated sequences could represent more than 10 percent of the genome. Until now mainly microsatellites have been analyzed because of their usefulness as molecular markers. But the availability of many fungal species, as well as the easy tools existing to identify these sequences, make possible to define their pattern in fungal genomes now. Minisatellites and satellites could contribute to the expansion of genome size; another type of repeated sequences, transposable elements, will now be the focus.

Transposable Elements

The Different Families of Transposable Elements

Transposable elements (TEs) are short, mobile, conserved segments of DNA that can replicate and randomly insert copies within genomes of all species of the three domains of life: eubacteria, archaeabacteria, and eukaryotes. Although TEs were first identified in fungi in the yeast *S. cerevisiae* (Boeke, 1989) and known to exist in bacteria, plant, and animals since the 1970s, conventional genetic studies with *Ascobolus immersus* mutants established their existence in filamentous fungi beginning of the 1980s (Decaris, Francou, et al., 1978; Berg & Howe, 1989; Craig 2002). TEs are viewed as having an important influence on the evolution of eukaryote genomes and as central agents in the evolutionary restructuration of fungal genomes (Kidwell & Lisch, 2000). Given their abundance, these elements often constitute a large proportion of eukaryotic genomes (e.g., ~45% of the human genome, Lander, Linton, et al., 2001); 50 to 80 percent of some grass genomes (Meyers, Tingey, et al., 2001); and more than 50 percent in some fungal species (Martin, Kohler, et al., 2010; Spanu, Abbott, et al., 2010). Their dynamics include different mechanisms, such as transposition (normal or aberrant), ectopic recombination, horizontal transmission, amplification bursts, degradation, and epigenetic inactivation. Moreover, the examination of TE distribution in natural populations provided valuable information concerning ecological and epidemiological considerations (Daboussi & Capy, 2003). An enormous increase in the understanding of the biology of fungal TEs has occurred in the past decade because of the diversity of fungal research in organisms playing an important role in agriculture, medicine, and biotechnology and has been supported by the sequencing of more than 50 genomes. Here, the genomic features of TEs in filamentous fungi are reviewed, with a particular focus on their abundance, distribution, and importance in genome structure.

Eukaryotic TEs are divided into two classes, depending on their mode of transposition (for a review see Wicker, Sabot, et al., 2007; Nakayashidi, 2011): Class I elements or retroelements or also retrotransposons, which mobilize via a "copy-and-paste" mechanism that uses a RNA intermediate and class II elements or DNA transposons, which mobilize via a cut-and-paste mechanism that use a DNA intermediate. These two classes are composed of five major types: long terminal repeat (LTR) retrotransposons, non-LTR retrotransposons, cut-and-paste DNA transposons, rolling-circle DNA, and self-synthesizing DNA transposons. Each type of TE is composed of a number of superfamilies or clades based on length and target site features, with each superfamily consisting of numerous families. The retrotransposons (class I elements) are the most common TE in fungi (Boeke, Stoye, et al., 1997). As noted, retrotransposons can be classified into two types: LTR retrotransposons and non-LTR retrotransposons (encompassing LINE elements), depending whether they possess or lack LTRs at both ends, tyrosine recombinase retroelements (YR; subdivided in three families, *DIRS*, *Ngaro*, and *VIPER*), Penelope-like retrotransposable elements, and short interspersed nuclear elements (SINEs). The LTR retrotransposons, which have a LTR at their extremities, have been divided into superfamilies: vertebrate retroviruses (Retroviridae), hepadnaviruses, caulimoviruses, *Ty1-Copia-like* (Pseudoviridae), *Ty3-Gypsy-like* (Metaviridae), and *Pao-BEL-like*, depending on their sequence similarity and the type of gene products they encode. The two main superfamilies of LTR retrotransposons found in fungi are *Gypsy* and *Copia*, which differ in the order of reverse transcriptase (RT), ribonuclease H (RH), and integrase (IN) domains in the virus-like polyprotein (POL; *Gypsy*: PR-RT-RH-INT, *Copia*: PR-INT-RT-RH). The DNA transposons (class II elements) have terminal inverted repeats (TIRs) or a rolling-circle replicon mechanism (e.g., *Helitrons* elements), similar to some known prokaryotic transposition mechanism or self-synthesizing DNA transposons (*Polintons*). Members of both classes are found in the genomes of filamentous fungi (Wicker, Sabot, et al., 2007).

Are Transposable Elements Impacting the Genome?

TEs have a remarkable potential to cause a variety of changes in the genome of their hosts. By transposing into or near genes, class I and class II TEs contribute to partial or total gene inactivation. Insertion may also place a gene under the control of TE regulatory sequences. Resulting from their ability to excise from a given site, class II transposons can generate a wide degree of variation in DNA sequence and phenotype. In addition, TEs have the ability to rearrange genomic information. DNA rearrangements may be local or associated with large-scale chromosomal modification. The range of transposon-associated genetic changes is well documented in many organisms (Kidwell & Lisch, 2002). Davière, Langin, et al. (2001) and Daboussi and Capy (2001) analyzed the impact of TEs in the rapid reorganization of the *Fusarium oxysporum* genome. Karyotypic variation is a common feature in natural fungal isolates, especially in those lacking the sexual cycle (Fierro & Martin, 1999; Kistler & Miao, 1992). Extensive analysis of chromosome length polymorphism has provided evidence that they include translocations, deletions of large chromosomal fragments, and much duplication. These studies showed that the high level of chromosome-length polymorphism of some chromosomes correlates with the high density of TEs (Davière, Langin, et al., 2001) and that the occurrence of chromosomal rearrangements is frequently associated with clustering of TEs on chromosomes (Hua-Van, Davière, et al., 2000). These findings suggest that they probably result from ectopic recombination between TEs scattered throughout the genome. The range of karyotypic changes observed in some species during mitosis without phenotypic changes indicates that many of them are probably genetically neutral (Davière, Langini, et al., 2001), at least under laboratory conditions (Kistler & Miao, 1992). However, some rearrangements can be beneficial and may play an important role in the evolution of the host, as reported for wine yeast strains (Perez-Ortin, Querol, et al., 2002). Such events could lead to new gene linkages that may be advantageous for adaptation to new environments (e.g., the translocation-associated *Tox1* locus of *Cochliobolus heterostrophus* [Kodama, Rose, et al., 1999]) and the appearance of new virulent alleles in *Magnaporthe grisea* as a result of the rearrangements in unstable subtelomeric regions with nested repeated sequences (Orbach, Farrall, et al., 2000).

Filamentous fungi show a large variability in genome sizes (Raffaele & Kamoun, 2012). Filamentous fungi typically have small genomes in the 10- to 40-Mb range, usually with limited amounts of repetitive DNA (Baker, Thykaer, et al., 2008) and thus the Ascomycota and Basidiomycota appear to have a tendency toward streamlined genomes. The majority of these taxa contain no more than 10 to 15 percent repetitive DNA (Wöstemeyer & Kreibich, 2002). However some filamentous fungi are rich in non-coding DNA and display an irregular architecture, with an uneven distribution of genes and repetitive elements across and between chromosomes (Novikova, Fet, et al., 2009; Ma, van der Does, et al., 2010; Labbé, Murat, et al., 2012). Some species have genomes with an extremely high proportion of repetitive DNA, reaching 64 percent in *B. graminis* (Spanu, Abbott, et al., 2010). Typically, the expansion of filamentous fungal genomes can be largely accounted for by a proliferation of repetitive DNA. As seen previously, the satellites' DNA can contribute to expansion of genome size; however, this contribution is not sufficient to explain the entire genome size, and often TEs are the main responsible for genome expansion (*see* Fig. 2.1).

How Can Genomes Control Transposable Elements' Diversity and Expansion?

Diversity of TEs and their copy number depends on the evolutionary history of a particular species or a cluster of closely related species, their population structure, and ecological features. There are several main processes that could affect the copy number and diversity of TEs in fungal genomes: (a) stochastic loss of elements as described for mariner-like elements (Lohe, Mriyama, et al., 1995); (b) burst of transposition (e.g., in *T. melanosporum* two independent burst of Gypsy retrotransposons were highlighted [Martin, Kohler, et al., 2010]); (c) the limitation of copy number increase by natural selection, which removes deleterious insertions—the effect of deleterious insertion is difficult to evaluate but a study was realized with *Drosophila melanogaster* to assess the effect of P element insertion on fitness (Mackay, Lyman, et al., 1992); (d) passive and active inactivation of repetitive sequences—these mechanisms will be described more in detail; and (e) self-regulation of transposition (decrease of the transposition rate when the copy number increases, Johnson, 2007)—this mechanism was proposed for MAGGY in *M. grisea* (Murata, Kadotani, et al., 2007). The population structure and dynamics, as well as mating mode and environmental conditions, also play an important role in the TEs' evolution. The transposition of the LTR retrotransposon MAGGY of *M. grisea* was shown to be particularly enhanced during mating and under abiotic stress (Eto, Ikeda, et al., 2001; Ikeda, Nakayashiki, et al., 2001). Similarly, numerous RTs and transposases were over-expressed in *T. melanosporum*–fruiting bodies, suggesting their activation during the sexual reproduction phase (Martin, Kohler, et al., 2010). Interestingly, Duplessis, Spanu, et al. (*see* Chapter 7) suggest that the frequency of sex during host infection for pathogenic fungi can impact the TE invasion.

The inactivation of repeated sequences is an important factor, which leads to the shifts in diversity and copy number of TEs, especially in fungi. The known mechanisms of repeat sequences inactivation include the repeat induced point (RIP) mutation, methylation induced premeiotically (MIP), and quelling. RIP was the first genome defense mechanism identified in eukaryotes, discovered in *Neurospora crassa* (Selker, Cambareri, et al., 1987). RIP occurs only during sexual cycle by introducing C:G to A:T mutations into two copies of duplications greater than about 400 bp with more than approximately 80 percent of nucleotide identity. In *Neurospora*, the RIP mutations are preferentially occurring in CpA dinucleotide (Cambareri, Jensern, et al., 1989). Actually RIP-like mechanisms were detected in *Podospora anserina* (Graia, Lespinet, et al., 2000), *M. grisea* (Ikeda, Nakayashiki et al., 2002), *L. maculans* (Idnurm & Howlett, 2003), and *Nectria haematococca* (Coleman, Rounsley, et al., 2009). Recently, Clutterbuck (2011) investigated the genome of 49 filamentous Ascomycetes to examine the evidence of multiple C/T transitions typical of RIP. The results highlighted that RIP-like activity varied greatly in extent of mutation as well as in dinucleotide context of C/T transition. Interestingly only *Chaetomium globosum* showed no evidence of directional mutation. In Basidiomycetes, RIP-like accumulation was described for *Puccinia graminis*, *M. larici-populina*, *Microbotrytium lychnidis-dioicae*, and *Rhodotorula graminis* (Hood, Katawczik, et al., 2005; Horns, Petit, et al., 2012) in which the target site seems to be the trinucleotide TpCpG. Horns, Petit, et al. (2012) did not find RIP-like hyperaccumulation mutation in four species of Agaromycotina and Ustilagomycotina, suggesting that RIP-like process is conserved within Puccinomycotina subphylum. Galagan and Selker (2004) highlighted that RIP not only impacted the genome via repeated sequences inactivation, but also that this mechanism was available to control gene duplication, which is considered as crucial for genome evolution. RIP could also be a mechanism promoting gene divergence as for *L. maculans* effectors (Rouxelle, Grandaubert, et al., 2011).

The MIP was first described in *A. immersus* (Goyon & Faugeron, 1989). MIP was also detected in the Basidiomycete *Coprinopsis cinereus* (Freedman & Pukkila, 1993). MIP follows the same rules as RIP; that is, duplications are inactivated prior to meiosis, but results in cytosine methylation without mutation, and consequently this mechanism is reversible. These similarities suggested that RIP evolved from MIP (Selker, 2002). The methylation caused by MIP can block transcription elongation, resulting in gene silencing (Barry, Faugeron, et al., 1993). In *T. melanosporum*, the genes involved in RIP were not identified (Martin, Kohler, et al., 2010), although a strong preference for transition in CpG dinucleotide was observed (Clutterbuck, 2011). A possible explanation could be the presence of MIP that can increase the mutation of the methylated cytosines as documented for mammalian DNA (Kricker, Drake, et al., 1992).

The third mechanism is the quelling described in *N. crassa* that resembles posttranscriptional gene silencing in plant (Irelan & Selker, 1996). Quelling recognizes mRNA from repeated sequences in the vegetative tissues and targets them for degradation.

The Impact of Transposable Elements on Genes and Phenotypes

The investigation of the localization of TE in genomes can provide interesting information. The distribution of TE in the genome is different from a species to another; some species have TE located in clusters or nests of several hundred Kbp such as in the genome of *Trametes versicolor* (Floudas, Binder, et al., 2012). Often the regions rich in TEs are telomeric and centromeric. For example in *Agaricus bisporus*, 66 percent of the TEs are located in telomeric and centromeric regions (Foulongne-Oriol et al., unpublished). Other genomes have TEs all along their genome; this is the case for *F. mediterranea* (Floudas, Binder, et al., 2012), *T. melanosporum* (Martin, Kohler, et al., 2010) and *B. graminis* (Spanu, Abbott, et al., 2010). These last species have a high percentage of their genome corresponding to TEs (41.28 percent, 57.73 percent, and 64 percent for the three species, respectively). In general the TE rich regions are poor in genes but as will be seen, genes can be present in these regions impacting greatly their evolution and expression. In *Verticillium* spp., the observed biased TE insertion in gene-rich regions within an individual genome and the "patchy" distribution among different strains suggested that TE could be a major generator of *Verticillium* intra- and interspecific genomic variation (Amyotte, Tan, et al., 2012). One other example is the finding of particular TE associated with mating type locus in *Neurospora* spp. (Gioti, Mushegian, et al., 2012). This finding suggested that these elements could have contributed to the shift from heterothallic ancestors to homothallic species by direct transposition of neighboring genes and facilitating unequal crossovers between unrelated intergenic regions of opposite mating types.

Repeat-rich genomic regions frequently coincide with synteny breakpoints, having evolved at accelerated rates compared with the rest of the genome. This is the case for the macrosynteny between *A. bisporus* and *C. cinereus* (Fig. 2 in Morin, Kohler, et al., 2012). Similarly, synteny breakpoints between *Sclerotinia scerotiorum* and *Botrytis cinerea* are marked by an increased density of repetitive elements in *S. scerotiorum* (Amselem, Cuomo, et al., 2011). For example, several synteny loss in species-specific secondary metabolism cluster loci appeared to be associated with the presence of TEs in one genome. Interestingly, between *Paxillus involutus* and *Pisolithus tinctorius*, some break of synteny are the result of a block of repeated sequences present at the same position in both species, suggesting that it appears in the common ancestor of these species (data not shown).

Such repeat rich regions tend to harbor genes that are implicated notably in virulence and host adaptation or as effector genes. For example in *L. maculans*, a plant pathogen, AT-rich blocks originated from RIP on repeated sequences, is enriched in effector-like (Rouxelle, Grandaubert, et al., 2011). The presence of effector-like genes in TE-rich regions for mildew species is discussed in Chapter 7 by Duplessis and colleagues (pp. 149–168). This particular environment induces rapid-sequence divergence and promotes the potentiality to adapt rapidly to new host constraint. TEs are also able to modify phenotype as shown for *Phytophthora ramorum* (Kasuga, Kozanitas, et al., 2012). These authors observed a burst of TE expression in oak isolates of *P. ramorum* in concomitance with phenotypic alterations, suggesting that TE derepression correlated with diversity in expression profiles leading to the phenotypic alteration.

TEs are generally distributed throughout the fungal genomes and could be major contributors to the genesis of new genes or to the adaptation of existing genes, notably via mechanisms such as molecular domestication, ectopic recombination, and gene retrotransposition. Molecular domestication, also known as the process of TE recruitment by the host genome, is the co-opted use by the organism of a function carried by a TE. Because TEs encode proteins that can, for example, bind, copy, break, join, or degrade nucleic acids, they have been repeatedly domesticated during eukaryotic evolution (Miller, McDonald, et al., 1999). As another mechanism, retrotransposon-mediated ectopic recombination results from the physical occurrence of retrotransposon insertions at particular sites in the genome and can imply various genomic rearrangements, such as duplications, deletions, and translocations. Gene retrotransposition is also another mechanism that can rearrange genes. Gene retrotransposition operates during the retrotransposition process itself and only duplicates gene sequences but no retrotansposons sequence. The genome sequences have revealed a lot of new information about the evolution of filamentous fungi and the genomic features that underlie their success. Most strikingly, several lineages of filamentous fungi are remarkable in displaying an evolutionary trend toward bigger, TE-rich genomes.

Are Fungal Species Rich in Transposable Elements Interacting with Living Plants?

Recently, Raffaelle and Kamoun (2012) discussed why some filamentous plant pathogens have convergent evolution toward large genomes infested with repetitive elements. For pathogenic fungi, the plasticity conferred by the TEs is thought to be adaptive because they increased recombination rate driven by TE activity. Consequently they adapt faster during coevolution with their host. These authors proposed that clade selection opposes the advantages conferred by smaller compact genomes because lineage with less adaptable genomes have an increased probability of extinction. Raffaelle and Kamoun (2012) focused on pathogenic filamentous fungi and Oomycetes, but is it possible to draw similar conclusion with fungi having different life strategy? Interestingly, the two ectomycorrhizal fungal genomes, *L. bicolor* and *T. melanosporum*, have large genomes rich in repeated sequences (*see* Fig. 2.1). Recently, Floudas, Binder, et al. (2012) performed a comparative analysis of 31 fungal genomes suggesting that lignin-degrading peroxidases expanded in the lineage leading to the ancestor of the Agaricomycetes. To gain information about the effect of TEs on wood decay fungi genome size and to investigate a putative link between life style and TE richness, the repeated elements were characterized in these genomes (*see* supplementary data of Floudas, Binder, et al., 2012).

The TE genome coverage varied from 0 percent for *M. globosa* to 41.42 percent for the white rot *F. mediterranea* (*see* Fig. 2.1 and Table S5 in Floudas, Binder, et al., 2012). Repeated sequences have not fully disappeared from *M. globosa* genome because 3 percent of its genome corresponds to satellite DNA (*see* Fig. 2.1). For all the 31 genomes, a correlation between genome size and TE richness was found (*see* Fig. S4 in Floudas, Binder, et al., 2012). In most genomes, the Gypsy retrotransposons and not categorized elements are the most frequent (*see* Fig. S1 in Floudas, Binder, et al., 2012). In *F. mediterranea*, the Gypsy retrotransposons covered more than 20 percent of its genome and almost 30 percent of the *T, melanosproum* genome (Martin, Kohler, et al., 2010). To gain more information on Gyspy-like retrotransposons diversity in wood decay fungi, a specific identification of reverse transcriptase (RT) was realized (Payen, Murat, et al., unpublished data). This analysis did not identify RT in the genome of only five (*Aspergillus niger*, *Pichia stipidis*, *Stagonospora nodorum*, *T. reesei*, and *Ustilago maydis*) out of the 31 species. Almost 30 percent of the RT identified have no homology with known Gypsy retrotransposon families and therefore can correspond to new families. For the other RTs most of them belong to the Chromovirus clade with the exception of

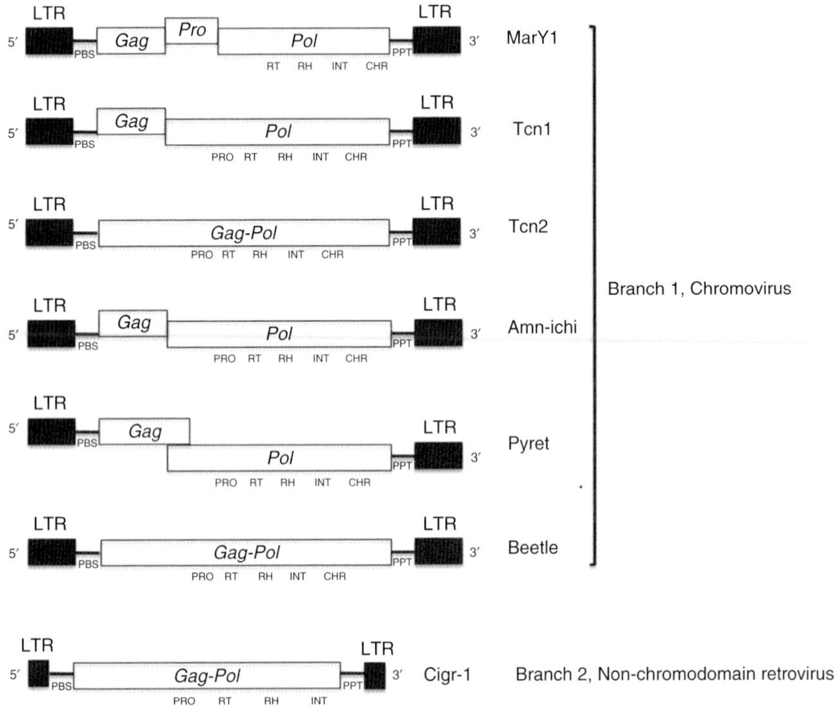

Figure 2.2 Structural organization of full-length LTR Gypsy retrotransposons found in filamentous fungi. The families presented are the most frequent found by reverse transcriptase (RT) screening in the genome species included in Floudas, Binder, et al. (2012). The gypsy retrotransposon diversity was assessed by RT identification using a RPS-Blast search (Altschul, 1990) with the reverse transcriptase 1 motif (pfam00078). For each species, the putative RT sequences were isolated and clustered together using Usearch (Edgar, 2010) and 90 percent of similarity on at least 90 percent of the sequence (Gorinsek, Gubensek, et al., 2004). For each of these clusters a sequence was taken randomly. All clusters were aligned with the reference sequences coming from gypsyDB and others families (Gorinsek, Gubensek, et al., 2004). The alignment was done using Clustal Omega (Sievers, Wilm, et al., 2011). A phylogeny by neighbor joining was done using QuickTree (Howe, Bateman, et al., 2002) with a bootstrap value of 10,000. We have considered as belonging to a family of known Gypsy family all the clusters supported by at least a bootstrap of 40 percent. CHR, chromodomain; INT, integrase; LTR, long terminal repeat; PBS, putative primer-binding site; PPT-polypurine tract; PRO, proteinase; RH, ribonuclease H; RT, reverse transcriptase.

Cigr-1 family, which presents in *M. laricii-populina* and *Cryptococcus neoformans* (Fig. 2.2). Interestingly, Cigr-1 was considered as a plant-animal lineage Gypsy retrotransposon family and elements of this family was not yet found in fungi (Sormacheva & Blinov, 2011). Additional investigations are needed to know if these Cigr-1 RTs can get results from horizontal transfer because both species interact with plants. Indeed horizontal transfer was already suggested with LTR retrotransposons. Novikova, Smyshlyaev, et al.

(2010) proposed a horizontal transfer of Tcn1 gypsy retrotransposons between fungi and non-seed plant. Chromovirus were identified in all 26 genomes, and among them MarY1 family was the most frequent with 52.8 percent of the RT sequences (*see* Fig. 2.2). MarY1 is a Chromovirus initially characterized in the genome of *Tricholoma matsutake* (Murata & Yamada, 2000) known to be widespread in fungus. The second frequent Gypsy retrotransposon family is Tcn2 known to be specific of Basidiomycetes (*see* Fig. 2.2; Novikova, Smyshlyaev, et al., 2010). Elements of this family were not identified in Ascomycetes. With the exception of *Postia placenta*, all species with high number of RTs (more than 200) interact with living plants. According to the present results and previous data on plant symbiotic and pathogenic fungi and Oomycetes, it seems that fungi rich in TEs interact with living plants (*see* Fig. 2.1; Martin, Aerts, et al., 2008; Haas, Kamoun, et al., 2009; Martin, Kohler, et al., 2010; Spanu, Abbott, et al., 2010; Raffaelle & Kamoun 2012). How can this observation be explained?

One hypothesis is that species interacting with living plants need to evolve rapidly and TE can give plasticity to the genome. Recently, Zeh, Zeh, et al. (2009) proposed the "epi-transposon equilibrium" hypothesis in which TEs play the role of "punctuated equilibria." The punctuated equilibrium means that the evolution proceeds through rapid morphological change and speciation followed by long-term stasis. These authors proposed that the punctuated equilibria result from an evolutionary tug-of-war between host genomes and TEs. According to the epi-transposon equilibrium hypothesis, stresses associated with climatic changes or colonization of new habitat or ecological niches result in TE reactivation, via disruption of epigenetic controls (e.g., MIP). TE can then rapidly modify the genome and gene expression, allowing the adaptation to the new conditions. It cannot be excluded that by changing their ecological niche (e.g., interaction with dead wood to living plant, transition between saprotrophic to symbiotic status) some fungal species activated the TE present in their genome. Interestingly for the two ectomycorrhizal fungi with genome sequenced to date, *L. bicolor* and *T. melanosporum*, RIP was not found. This absence of irreversible genome defense could explain the high proportion of TE in these genomes and the possibility to reactive their TE, and the question is if the RIP was not lost during evolution by ectomycorrhizal fungi to increase genome plasticity? A second hypothesis is that symbiotic fungi have a particular ecological niche linked with their host plants and therefore their populations could be more limited. These small populations could promote the TE expansion. Among the different mechanisms of TE invasion, the horizontal transfer is a possibility of new element to colonize a genome. Richards (2011) highlighted that the horizontal transfer could be linked with ecological niche; consequently it cannot be excluded that fungi interacting with living plants are subject to transfer of TE from the plant. This hypothesis needs to be taken in consideration for the future analyses. There is no doubt that the genome projects aimed to sequence more mycorrhizal genomes as well as plant genomes will have more information on the possible link between TE abundance and the interaction with living plants.

Acknowledgments

We are grateful to Francis Martin, François Le Tacon, Emmanuelle Morin, Marc-Henri Lebrun, and Joëlle Amselem for the numerous critical discussions. We would like to thanks also Francis Martin for the critical comments of the manuscript. This work was supported by grants from the National Institute of Agricultural Research, the Région Lorraine Council, Lab of Excellence ARBRE, ANR SYSTERRA SYSTRUF (ANR-09-STRA-10), TUBEREVOL project of the Genoscope and the Plant-Microbe Interfaces Scientific Focus Area project at Oak Ridge National Laboratory (ORNL) sponsored by Office of Biological and Environmental Research at the United States Department of Energy Office of Science. ORNL is managed by UT-Battelle, LLC, under contract DE-AC05-00OR22725 for the United States Department of Energy. The wood decay work conducted by the US Department of Energy Joint Genome Institute was supported by the Office of Science of the US Department of Energy under contract DE-AC02-05CH11231.

References

Altschul S. 1990. Basic local alignment search tool. *J Mol Biol.* 215: 403–410.

Amyotte SG, Tan X, et al. 2012. Transposable elements in phytopathogenic *Verticillium* spp.: Insights into genome evolution and inter- and intra-specific diversification. *BMC Genomics.* 13: 314.

Amselem J, Cuomo CA, et al. 2011. Genomic analysis of the necrotrophic fungal pathogens Sclerotinia scerotiorum and Botrytis cinerea. *PLoS Genet.* 7: e1002230.

Ashley CT & Warren ST. 1995. Trinucleotide repeat expansion and human disease. *Annu Rev Genet.* 29: 703–728.

Baker SE, Thykaer J, et al. 2008. Fungal genome sequencing and bioenergy. *Fungal Biol Rev.* 22: 1–5.

Bally P, Grandaubert J, et al. 2010. FONZIE: An optimized pipeline for minisatellite marker discovery and primer design from large sequence data sets. *BMC Res Notes.* 3: 322.

Barry C, Faugeron G, et al. 1993. Methylation induced premeiotically in *Ascobolus*: coextension with DNA repeat lengths and effect on transcript elongation. *Proc Natl Acad Sci USA* 90: 4557–4561.

Benson G. 1999. Tandem repeats finder: A program to analyze DNA sequences. *Nucl Acids Res.* 27: 573–580.

Berg DE & Howe MM. 1989. *Mobile DNA*. Washington DC: American Society for Microbiology Press.

Boeke JD. 1989. Transposable elements in *Saccharomyces cerevisiae*. In *Mobile DNA* (eds. DE Berg & MM Howe), 335–374. Washington DC: American Society for Microbiology Press.

Boeke JD, Stoye JP. 1997. Retrotransposons, endogenous retroviruses, and the evolution of retroelements, retroviruses. In *Retroviruses* (eds. JM Coffin, SH Hughes, et al.), 343–435. Cold Spring Harbor, NY: Cold Spring Harbor Laboratory Press.

Boland CR & Goel A. 2010. Microsatellite instability in colorectal cancer. *Gastroenterology*. 138: 2073–2087.
Buard J & Jeffreys A. 1997. Big, bad minisatellites. *Nat Genet.* 15:327–328.
Cambareri EB, Jensern BC, et al. 1989. Repeat-induced G-C to A-T mutations in *Neurospora*. *Science.* 244: 1571–1575.
Castoe TA, Poole AW, et al. 2012. Rapid Microsatellite Identification from Illumina Paired-End Genomic Sequencing in Two Birds and a Snake. *PLoS One.* 7(2): e30953.
Chambers GK & MacAvoy ES. 2000. Microsatellites: Consensus and controversy. *Comp Biochem Physiol B Biochem Mol Biol.* 126: 455–476.
Clutterbuck AJ. 2011. Genomic evidence of repeat-induced point mutation (RIP) in filamentous ascomycetes. *Fungal Genet Biol.* 48: 306–326.
Coleman JJ, Rounsley SD, et al. 2009. The genome of *Nectria haematococca*: contribution of supernumerary chromosomes to gene expansion. *PLoS Genet.* 5: e1000618.
Craig NL. 2002. *Mobile DNA II*. Washington, DC: American Society for Microbiology Press.
Daboussi MJ & Capy P. 2003. Transposable elements in filamentous fungi. *Annu Rev Microbiol.* 57: 275–299.
Davière JM, Langin T, et al. 2001. Potential role of transposable elements in the rapid reorganization of the *Fusarium oxysporum* genome. *Fungal Genet Biol.* 34(3): 177–192.
Decaris B, Francou F, et al. 1978. Unstable ascospore color mutants of *Ascobolus immerses*. *Mol Gen Genet.* 81: 69–81.
Dilmaghani A, Gladieux P, et al. 2012. Migration patterns and changes in population biology associated with the worldwide spread of the oilseed rape pathogen *Leptosphaeria maculans*. *Mol Ecol.* 21: 2519–2533.
Drake JW, Charlesworth B, et al. 1998. Rates of spontaneous mutation. *Genetics.* 148: 1667–1686.
Dutech C, Enjalbert J, et al. 2007. Challenges of microsatellite isolation in fungi. *Fungal Genet Biol.* 44: 933–949.
Edgar RC. 2010. Search and clustering orders of magnitude faster than BLAST. *Bioinformatics.* 26: 2460–2461.
Eto Y, Ikeda K, et al. 2001. Comparative analyses of the distribution of various transposable elements in *Pyricularia* and their activity during and after the sexual cycle. *Mol Gen Genet.* 264: 565–577.
Fierro F & Martin JF. 1999. Molecular mechanisms of chromosomal rearrangement in fungi. *Crit Rev Microbiol.* 25: 1–17.
Floudas D, Binder M, et al. 2012. The oaleozoic origin of enzymatic lignin decomposition reconstructed from 31 fungal genomes. *Science.* 336: 1715–1719.
Freedman T & Pukkila PJ. 1993. De novo methylation of repeat sequences in *Coprinus cinerus*. *Genetics.* 135: 357–366.
Galagan JE & Selker EU. 2004. RIP: The evolutionary cost of genome defense. *Trends Genet.* 20: 417–423.
Gioti A, Mushegian AA, et al. 2012. Unidirectional evolutionary transitions in fungal mating systems and the role of transposable elements. *Mol Biol Evol.* 29: 3215–3226.
Gorinsek B, Gubensek F, et al. 2004. Evolutionary genomics of chromoviruses in eukaryotes. *Mol Biol Evol.* 21: 781–798.
Goyon C & Faugeron G. 1989. Targeted transformation of *Ascobolus immersus* and de novo methylation of the resulting duplicated DNA sequences. *Mol Cell Biol.* 9: 2818–2827.
Graia F, Lespinet O, et al. 2001. Genome quality control: RIP (repeat-induced point mutation) comes to *Podospora*. *Mol Microbiol.* 40: 586–595.
Haas BJ, Kamoun S, et al. 2009. Genome sequence and analysis of the Irish potato famine pathogen *Phytophthora infestans*. *Nature.* 461: 393–398.
Hancock JM, Goldstein DB, et al. 1999. Microsatellites and other simple sequences: genomic context and mutational mechanisms. In *Microsatellites: Evolution and Applications* (eds. DB Goldstein & C Schlotterer), 1–9. Oxford: Oxford University Press.
Hood ME, Katawczik M, et al. 2005. Repeat-induced point mutation and the population structure of transposable elements in *Microbotryum violaceum*. *Genetics.* 170: 1081–1089.

Horns F, Petit E, et al. 2012. Patterns of repeat-induced point mutation in transposable elements of basidiomycete fungi. *Genome Biol Evol.* 4: 240–247.

Howe K, Bateman A, et al. 2002. QuickTree: Building huge Neighbour-Joining trees of protein sequences. *Bioinformatics.* 18: 1546–1547.

Hua-Van A, Davière JM, et al. 2000. Genome organization in *Fusarium oxysporum*: Clusters of class II transposons. *Curr Genet.* 37:339–347.

Jarne P & Lagoda PJL. 1996. Microsatellites, from molecules to populations and back. *Trends Ecol Evol.* 11: 424–429.

Idnurm A & Howlett BJ. 2003. Analysis of loss pathogenicity mutants reveals that Repeat-induced point mutation can occur in the Dothideomycete *Leptoshaeria maculans. Fungal Genet Biol.* 39: 31–37.

Ikeda K, Nakayashiki H, et al. 2001. Heat shock, copper sulfate and oxidative stress activate the retrotransposon MAGGY resident in the plant pathogenic fungus *Magnaporthe grisea. Mol Genet Genomics.* 266: 318–325.

Ikeda K, Nakayashiki H, et al. 2002. Repeat-induced point mutation (RIP) in *Magnaporthe grisea*: implication for its sexual cycle in the natural field context. *Mol Microbiol.* 45: 1355–1364.

Irelan JT & Selker EU. 1996. Gene silencing in filamentous fungi: RIP, MIP and quelling. *J Genetics.* 3: 313–324.

Jeffreys AJ, Wilson V, et al. 1985. Hypervariable "minisatellite" regions in human DNA. *Nature.* 314: 67–73.

Johnson LJ. 2007. The genome strikes back: The evolutionary importance of defence against mobile elements. *Evol Biol.* 34: 121–129.

Kasuga T, Kozanitas M, et al. 2012. Phenotypic diversification is associated with host-induced transposon derepression in the Sudden Oak death pathogen *Phytophthora ramorum. PLoS One.* 7(4): e34728.

Kidwell MG & Lisch DR. 2000. Transposable elements and host genome evolution. *Trends Ecol Evol.* 15: 95–99.

Kidwell MG & Lisch DR. 2002. Transposable elements as sources of genomic variation. In: *Mobile DNA II* (ed. NL Craig), 59–90. Washington, DC: American Society for Microbiology Press.

Kistler HC & Miao VP. 1992. New modes of genetic change in filamentous fungi. *Annu Rev Phytopathol.* 30: 131–152.

Kricker MC, Drake JW, et al. 1992. Duplication-targeted DNA methylation and mutagenesis in the evolution of eukaryotic chromosomes. *Proc Natl Acad Sci USA.* 89: 1075–1079.

Kodama M, Rose MS, et al. 1999. The translocation-associated Tox1 locus of *Cochliobolus heterostrophus* is two genetic elements on two different chromosomes. *Genetics.* 151: 585–596.

Labbé J, Murat C, et al. 2011. Survey and analysis of simple sequence repeats in the *Laccaria bicolor* genome, with development of microsatellite markers. *Curr Genet.* 57: 75–88.

Labbé J, Murat C, et al. 2012. Characterization of transposable elements in the ectomycorrhizal fungus *Laccaria bicolor. PLoS One.* 7: e40197.

Lander ES, Linton LM, et al. 2001. Initial sequencing and analysis of the human genome. *Nature.* 409: 860–921.

Lepais O & Bacles CFE. 2011. Comparison of random and SSR-enriched shotgun pyrosequencing for microsatellite discovery and single multiplex optimization in *Acacia harpophylla* F, Muell, *Ex* Bebth. *Mol Ecol Res.* 11: 711–724.

Li Y-C, Korol AB, et al. 2004. Microsatellites within genes: Structure, function, and evolution. *Mol Biol Evol.* 21: 991–1007.

Lohe AR, Mriyama EN, et al. 1995. Horizontal transmission, vertical inactivation and stochastic loss of mariner-like transposable elements. *Mol Biol Evol.* 12: 62–72.

Ma L-J, van der Does HC, et al. 2010. Comparative genomics reveals mobile pathogenicity chromosomes in *Fusarium. Nature.* 464: 367–373.

Mackay TFC, Lyman RF, et al. 1992. Effects of P-element insertions on quantitative traits in *Drosophila melanogaster. Genetics.* 130: 315–332.

Magain N, Forrest LL, et al. 2010. Microsatellite primers in the *Peltigera dolichorhiza* complex (lichenized ascomycete, Peltigerales). *Am J Bot.* 97. e102–e104.
Malausa T, Gilles A, et al. 2011. High-throughput microsatellite isolation through 454 GS-FLX Titanium pyrosequencing of enriched DNA libraries. *Mol Ecol Res.* 11: 638–644.
Martin F, Aerts A, et al. 2008. The genome of *Laccaria bicolor* provides insights into mycorrhizal symbiosis. *Nature.* 452: 88–92.
Martin F, Kohler A, et al. 2010. Périgord black truffle genome uncovers evolutionary origins and mechanisms of symbiosis. *Nature.* 464: 1033–1038.
Meyers BC, Tingey SV, Morgante M. 2001. Abundance, distribution, and transcriptional activity of repetitive elements in the maize genome. *Genome Res.* 11: 1660–1676.
Miller WJ, McDonald J F, et al. 1999. Molecular domestication-more than a sporadic episode in evolution. *Genetica.* 107: 197–207.
Molinier V, Murat C, et al. 2012. First identification of polymorphic microsatellite markers in the Burgundy truffle, *Tuber aestivum* (Tuberaceae). *Application in Plant Sciences.* 1(2): 1200220. doi: http://dx.doi.org/10.3732/apps.1200220.
Morin E, Kohler A, et al. 2012. Genome sequence of the button mushroom Agaricus bisporus reveals mechanisms governing adaptation to a humic-rich ecological niche. *Proc Natl Acad Sci USA.* 109: 17501–17506.
Murat C, Riccioni C, et al. 2011. Distribution and localization of microsatellites in the Perigord black truffle genome and identification of new molecular markers. *Fungal Genet Biol.* 48: 592–601.
Murata T, Kadotani N, et al. 2007. siRNA-dependent and -independent post-transcriptional cosuppression of the LTR-retrotransposon MAGGY in the phytopathogenic fungus *Magnaporthe grisea. Nucl Acid Res.* 35: 5987–5994.
Murata H & Yamada A. 2000. marY1, a Member of the gypsy group of long terminal repeat retroelements from the ectomycorrhizal basidiomycete *Tricholoma matsutake. Appl Environ Microbiol.* 66: 3642–3645.
Nakayashiki H. 2011. The trickster in the genome: Contribution and control of transposable elements. *Genes Cells.* 16: 827–841.
Novikova O, Fet V, et al. 2009. Non-LTR retrotransposons in fungi. *Funct Integr Genomics.* 9: 27–42.
Novikova O, Smyshlyaev G, et al. 2010. Evolutionary genomics revealed interkingdom distribution of Tcn1-like chromodomain-containing Gypsy LTR retrotransposons among fungi and plants. *BMC Genomics.* 11:231.
Oliveira EJ, Pádua JG, et al. 2006. Origin, evolution and genome distribution of microsatellites. *Genet Mol Biol.* 29: 294–307.
Orbach MJ, Farrall L, et al. 2000. A telomeric avirulence gene determines efficacy for the rice blast resistance gene Pi-ta. *Plant Cell.* 12: 2019–2032.
Pannebakker BA, Niehuis O, et al. 2010. The distribution of microsatellites in the *Nasonia parasitoid* wasp genome. *Insect Mol Biol.* 19: 91–98.
Perez-Ortin JE, Querol A, et al. 2002. Molecular characterization of a chromosomal rearrangement involved in the adaptive evolution of yeast strains. *Genome Res.* 12: 1533–1539.
Raffaele S & Kamoun S. 2012. Genome evolution in filamentous plant pathogens: Why bigger can be better. *Nature Rev Microbiol.* 10: 417–430.
Richards TA. 2011. Genome evolution: Horizontal movements in the fungi. *Curr Biol.* 21: 166–167.
Riley DE & Krieger JN. 2009. UTR dinucleotide simple sequence repeat evolution exhibits recurring patterns including regulatory sequence motif replacements. *Gene.* 429: 80–86.
Rouxelle T, Grandaubert J, et al. 2011. Effector diversification within compartment of the *Leptosphaeria maculans* genome affected by repeat-induced point mutation. *Nature Commun.* 2: doi:10,1038/ncomms1189.
Rudd JJ, Antoniw J, et al. 2010. Identification and characterisation of *Mycosphaerella graminicola* secreted or surface-associated proteins with variable intragenic coding repeats. *Fung Genet Biol.* 47: 19–32.

Selker EU, Cambareri EB, et al. 1987. Rearrangement of duplicated DNA in specialized cells of *Neurospora*. *Cell.* 51: 741–752.

Selker EU. 2002. Repeat induced gene silencing in fungi. *Adv Genet.* 46: 439–450.

Sharma PC, Grover A, et al. 2007. Mining microsatellites in eukaryotic genomes. *Trends Biotechnol.* 25: 490–498.

Sievers F, Wilm A, et al. 2011. Fast, scalable generation of high-quality protein multiple sequence alignments using Clustal Omega. *Mol Syst Biol.* 7: 539.

Singham GV, Vargo EL, et al. 2012. Polymorphic microsatellite moci from an indigenous Asian fungus-growing termite, *Macrotermes gilvus* {(Blattodea:} Termitidae) and cross amplification in related taxa. *Environ Entomol.* 41: 426–431.

Sormacheva ID & Blinov AG. 2011. LTR retrotransposons in plants. *Russ J Genet Appl Res.* 1: 540–564.

Spanu PD, Abbott JC, et al. 2010. Genome expansion and gene loss in powdery mildew fungi reveal tradeoffs in extreme parasitism. *Science.* 330: 1543–1546.

Vergnaud G & Denoeud F. 2000. Minisatellites: Mutability and genome architecture. *Genome Res.* 10: 899–907.

Verstrepen KJ, Jansen A, et al. 2005. Intragenic tandem repeats generate functional variability. *Nature Genet.* 37: 986–990.

Vinces MD, Legendre M, et al. 2009. Unstable tandem repeats in promoters confer transcriptional evolvability. *Science.* 324: 1213–1216.

Weber JL & Wong C. 1993. Mutation of human short tandem repeats. *Human Mol Genet.* 2: 1123–1128.

Wicker T, Sabot F, et al. 2007. A unified classification system for eukaryotic transposable elements. *Nature Rev Genet.* 8: 973–982,

Wöstemeyer J & Kreibich A. 2002. Repetitive DNA elements in fungi (Mycota): Impact on genomic architecture and evolution. *Curr Genet.* 41: 189–198.

Xhaard C, Andrieux A, et al. 2009. Characterization of 41 microsatellite loci developed from the genome sequence of the poplar rust fungus, *Melampsora larici-populina. Conservation Genet Resour.* 1: 21–25.

Xu J, Saunders CW, et al. 2007. Dandruff-associated *Malassezia* genomes reveal convergent and divergent virulence traits shared with plant and human fungal pathogens. *Proc Natl Acad Sci USA*. 104: 18730–18735.

Zeh DW, Zeh JA, et al. 2009. Transposable elements and an epigenetic basis for punctuated equilibria. *Bioessays.* 31: 715–726.

Section 2
Saprotrophic Fungi

3 Wood Decay

Dan Cullen

Forest Products Laboratory, Madison, Wisconsin

Introduction

A significant portion of global carbon is sequestered in forest systems. Specialized fungi have evolved to efficiently deconstruct woody plant cell walls. These important decay processes generate litter, soil bound humic substances, or carbon dioxide and water. This chapter reviews the enzymology and molecular genetics of wood decay fungi, most of which are members of the Agaricomycotina subphylum. This chapter emphasizes recent advances derived from a growing number of genome resources but otherwise directs interested readers to previously published reviews for additional information. Along these lines, background on wood cell wall polymer chemistry and the oxidative systems involved in their depolymerization have been extensively reviewed (Eriksson, Blanchette, et al., 1990; Cullen & Kersten, 2004).

Wood Composition and the Challenges Posed as Substrate

Cellulose, a linear polymer of anhydrocellobiose units linked by β-1, 4-glycosidic bonds, constitutes approximately 40 percent of the weight of wood. Through Van der Waals forces and hydrogen bonding, individual cellulose molecules are arrayed into microfibrils, each of which contains approximately 40 cellulose chains. Regions along the cellulose microfibrils are highly ordered and crystalline in diffraction measurements. In the primary cell wall, fibrils appear randomly oriented within a matrix of xyloglucan and pectic substances at the cell surface. In the S2 layer of the secondary wall (the bulk of wood weight) cellulose microfibrils are organized approximately parallel to the cell long axis. The cellulose microfibrils appear embedded in a matrix of hemicelluloses and lignin.

Making up 25 to 30 percent of wood weight, hemicelluloses are linear β-1,4-linked monosaccharide polymers with limited branching consisting of mono-, di-, or trisaccharides. Branches can be sugars, sugar acids and acetylated sugars, and sugar acid esters. In the major hemicellulose of hardwoods,

The Ecological Genomics of Fungi, First Edition. Edited by Francis Martin.

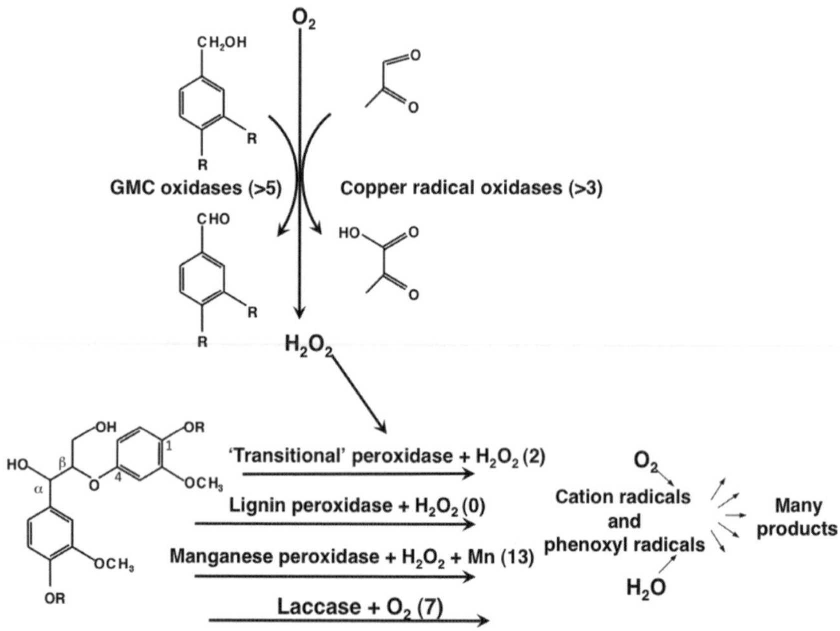

Figure 3.1 Schematic representation of extracellular processes involved in lignin degradation by the white rot fungus *Ceriporiopsis subvermispora*. The model emphasizes the central role of peroxide. For each enzyme class, the predicted gene number is shown in parentheses (Fernandez-Fueyo, Ruiz-Dueñas, et al., 2012).

O-acetyl-4-O-methylglucuronoxylan, approximately 7 of 10 xylosyl residues are acetylated, and about every tenth contains α-glucoronic acid. The same basic structure occurs in conifers, but without acetyl groups, with more glucuronic acid residues, and with α-arabinose residues on about every eighth xylosylresidue. A major hemicellulose of conifers, O-acetylgalactoglucomannan, contains galactose-to-glucose-to-mannose at a ratio of approximately 1:1:3. Lignin is covalently bonded via infrequent linkages to the hemicelluloses.

The third major component of wood, lignin, is comprised of carbon-carbon and ether bonds between phenylpropanoid residues. Formed through free radical-induced polymerization of monolignols *p*-coumaryl alcohol, coniferyl alcohol, and sinapyl alcohol, the structures are often referred to as *p*-hydroxyphenyl (H), guaiacyl (G), and syringyl (S) subunits, respectively (Ralph, Lundquist, et al., 2004). Proportions vary between species. Softwoods generally tend toward G-lignins with little or no H and S units, whereas hardwoods contain varying ratios of G/S lignins. Often referred to as β-O-4 linkage, the ether substructure shown (Fig. 3.1) is representative of the major linkage (about 90 percent) in lignins. A consequence of such ether bonds is that degradation involves oxidative mechanisms, as opposed to hydrolytic mechanisms.

Lignin polymers are stereoirregular and insoluble, properties that present significant experimental challenges. Model compounds, particularly those mimicking non-phenolic interunit linkages (*see* Fig. 3.1), have proven especially useful for characterizing high-oxidation potential enzymes. However, it should be noted that Cα-Cβ cleavage of such compounds, by itself, does not address the question of lignin depolymerization. Use of commercially available lignin as a substrate can be similarly misleading because the product typically contains various contaminants, and the lignin is partially depolymerized and modified (e.g., sulfonated). Compelling evidence for lignin degradation is gained from direct measurements of lignin mineralization in wood or metabolism of synthetic lignins. No microbe has been convincingly shown to use lignin as sole carbon source.

General Characteristics of Wood Decay Fungi

The principal wood-decay fungi lie within the Agaricomycetes (Basidiomycota), although representatives from other groups of Basidiomycota and Ascomycota have been documented (Eriksson, Blanchette, et al., 1990). Two broad categories of wood decay are recognized; white rot and brown rot. Phylogenetic and comparative genomic studies support the view that white rot is plesiomorphic in Agaricomycetes and that brown rot has evolved repeatedly (Hibbett & Donoghue, 2001; Floudas, Binder, et al., 2012). As of this date, genome analyses of 17 wood decay fungi have been published (Table 3.1).

Only white rot Basidiomycetes have been convincingly shown to efficiently mineralize lignin. This unique ability to completely degrade lignin is generally viewed as a strategy to gain access to carbohydrate polymers of plant cell walls for use as carbon and energy sources. Decay patterns may differ substantially among white rot species and strains (Eriksson, Blanchette, et al., 1990; Blanchette, 1991; Daniel, 1994), and two gross morphologies are recognized: simultaneous decay of cellulose, hemicelluloses, and lignin; and selective delignification, in which lignin and hemicelluloses are removed more rapidly than the cellulose. During simultaneous decay, erosion troughs appear beneath hyphae, the cell walls become gradually thinner, and holes appear between cells as decay advances. In contrast, cell walls retain their morphology during selective ligninolysis. Simultaneous and selective white rot fungi are exemplified by *Phanerochaete chrysosporium* and *Ceriporiopsis subvermispora*, respectively.

Brown rot fungi modify but do not remove bulk lignin. Instead, the lignin remains as a polymeric residue following removal of cellulose and hemicellulose (Blanchette, 1995; Worrall, Anagnost, et al., 1997; Niemenmaa, Uusi-Rauva, et al., 2007; Yelle, Ralph, et al., 2008). Brown rot residues resist further decay and contribute to the carbon pool in humic soils. Early in decay,

Table 3.1 Wood decay fungi with published genomes; taxonomic order, number of genes, and secreted proteins.[a]

White rot	Order	LiP	MnP	VP[b]	Lac	CROs	GLX	CDH	ALE	P450	GH61	GH6	GH7
Phach	Polyporales	10	5	0	0	6	1	1	2	149	15*	1*	6*
Cersu	Polyporales	0	13*	2	7*	3	0	1	0	222	9	1*	3*
Dicsq	Polyporales	0	9	3	11*	5*	5	1*	3*	187	15*	1*	4*
Trave	Polyporales	10	13*	2*	7*	5*	5*	1*	1*	190	18*	1*	4*
Stehi	Russulales	0	5	0	15	5*	3	1*	2*	215	16*	1*	3*
Hetan[c]	Russulales	0	8	0	13	5	0	1	1	144	10	1	1
Aurde	Auriculariales	0	5	0	7	7*	2	1	2	249	19*	2*	6*
Fomme	Hymenochaetales	0	16*	0	10*	4	0	1	1	130	13	2	2
Punst	Corticales	0	10*	0	12*	6*	3*	1*	2	144	14*	1*	5*
Schco[c]	Agaricales	0	0	0	2	2	0	1	2	115	22	1	2
Brown rot													
Pospl	Polyporales	0	0	0	3	3*	0	0	0	254	2	0	0
Fompi	Polyporales	0	0	0	5	*	0	0	1	190	4	0	0
Wolco	Polyporales	0	0	0	5*	4*	0	0	0	206	2	0	0
Conpu	Boletales	0	0	0	6	6	0	1*	1*	238	10	2	2
Serla[c]	Boletales	0	0	0	4	3	0	2	2	164	5	1	0
Glotr	Gloeophyllales	0	0	0	4	2*	0	1	2*	130	4*	0	0
Dacsp	Dacrymycetales	0	0	0	0	3	0	0	0	126	0	0	0

Aurde, *Auricularia delicata*; CDH, cellobiose dehydrogenase; Cersu, *Ceriporiopsis subvermispora*; Conpu, *Coniophora puteana*; CROs, copper radical oxidases; Dacsp, *Dacryopinax* sp; Dicsq, *Dichomitus squalens*; Fomme, *Fomotiporia mediterranea*; Fompi, *Fomitopsis pinicola*; Glotr, *Gloeophyllum trabeum*; GLX, Glyoxal oxidase; Hetan, *Heterobasidion annosum*; Lac, laccase; LiP, lignin peroxidase; MnP, manganese peroxidase; P450, cytochrome P450 and GH61, GH6, GH7, members of the Glycoside Hydrolase families 61, 6, and 7, respectively; Phach, *Phanerochaete chrysosporium*; Pospl, *Postia placenta*; Punst. *Punctularia strigo-zonata*; Schco, *Schizophyllum commune*; Serla, *Serpula lacrymans*; Stehi, *Stereum hirsutum*; Trave, *Trametes versicolor*; VP, versatile peroxidase; Wolco, *Wolfiporia cocos*.

[a]*NanoLC-MS/MS unambiguously identified at least one protein in media containing ground aspen as sole carbon source. See supplemental files published for *Postia placenta* (Martinez, Challacombe, et al., 2009), *Ceriporiopsis subvermispora* and *Phanerochaete chrysosporium* (Fernandez-Fueyo, Ruiz-Dueñas, et al., 2012), and refer to Floudas, Binder, et al., 2012 for others.

[b]VPs include two "transitional" peroxidases recently characterized in *C. subvermispora* (Fernandez-Fueyo, Ruiz-Dueñas, et al., 2012).

[c]Secretome data derived for media containing ball-milled aspen not available.

cellulose is depolymerized, a process that leads to rapid loss of wood strength (Kleman-Leyer & Kirk, 1992). This unusual depolymerization is markedly different from the more gradual cellulose degradation attributed to conventional hydrolytic enzymes.

Most species of wood-decaying Agaricomycetes exhibit characteristic substrate preference for either conifers (gymnosperms) or hardwoods (angiosperms). Some species are more or less restricted to one or a few wood species, and others appear to be true generalists. Typically, brown rot species are associated with conifer decay, but some have been isolated from hardwoods (Gilbertson, 1981; Hibbett & Donoghue, 2001). Certain wood decay Agaricomycetes can attack living trees (e.g., *Heterobasidion annosum*) or colonize freshly cut sapwood (e.g., *Phlebiopsis gigantea*), whereas many decay only dead trees (Blanchette, 1991).

Microscopic analysis of selective delignification (white rot) and cellulose depolymerization (brown rot) during incipient decay argue against direct interactions between enzymes and their polymeric substrates. Simply put, enzymes are too large to penetrate sound, intact wood. Blanchette, Krueger, et al. (1997) demonstrated this limited accessibility by showing that during *C. subvermispora* decay of pine, the walls only gradually became permeable to insulin (5.7 kDa), and then to myoglobin (17.6 kDa), but not to ovalbumin (44.3 kDa), even in relatively advanced stages of decay. Lignin-depolymerizing enzymes and many cellulases are in the same size range as ovalbumin, and it is generally thought therefore that small molecular weight, oxidizing species are generated and that these diffuse into the walls. The remainder of this chapter describes mechanisms of lignocellulose degradation with particular emphasis on insight gained from the genomes of wood decay fungi.

Mechanisms of Wood Decay

Peroxidases

Lignin peroxidase (LiP) catalyzes Cα-Cβ cleavage of propyl side chains of lignin and lignin models (*see* Fig. 3.1), hydroxylation of benzylic methylene groups, oxidation of benzyl alcohols to the corresponding aldehydes or ketones, phenol oxidation, and aromatic cleavage of nonphenolic lignin model compounds. A wide array of oxidations, all dependent on H_2O_2, has been demonstrated. In a mechanism described as "enzymatic combustion" (Kirk & Farrell, 1987), LiP oxidizes aromatic compounds by a single electron and the resulting aryl cation radicals undergo spontaneous reactions that yield many different products dependent on substrate structure (*see* Fig. 3.1). The biochemistry of peroxidases in ligninolysis has been reviewed (Hammel & Cullen, 2008).

Ten LiP-encoding genes have been identified in the white rot fungi *P. chrysosporium* and *Trametes versicolor* (*see* Table 3.1). Transcript levels of *P. chrysosporium* LiP are substantially altered by culture conditions (Stewart & Cullen, 1999). In soil cultures, transcript patterns are modulated in response to specific pollutants, for example, anthracene versus pentachlorophenol (Cullen, 2002).

More recent secretome and transcriptome studies revealed complex patterns of LiP gene expression in defined media (Vanden Wymelenberg, Sabat, et al., 2005; Vanden Wymelenberg, Sabat, et al., 2006; Vanden Wymelenberg, Gaskell, et al., 2009) and in more complex lignocellulose-containing media such as red oak (Sato, Feltus, et al., 2009) and ball milled aspen (BMA) suspended in basal medium (Vanden Wymelenberg, Gaskell, et al., 2010). Systematic studies of the *Phanerochaete carnosa* transcriptome show that wood species significantly impacts LiP transcript levels (Macdonald, Doering, et al., 2011; Macdonald & Master, 2012; MacDonald, Suzuki, et al., 2012).

Beyond the LiPs, evidence strongly supports a role for manganese peroxidase (MnP) involvement in lignin degradation by white rot fungi (Hammel & Cullen, 2008). Discovered in *P. chrysosporium* cultures, MnP oxidizes Mn^{2+} to Mn^{3+} using H_2O_2 as oxidant. Organic acids, such as oxalic acid, stimulate MnP through stabilization of Mn^{3+} and form diffusible oxidizing chelates. The interactions between oxalate and Mn as they relate to MnP activity and to peroxide generation in *C. subvermispora* cultures have been investigated (Urzua, Kersten, et al., 1998). MnPs lack sufficient oxidative potential to cleave the major non-phenolic units of lignin but can oxidize phenolic structures. Scission of non-phenolic structures within lignin may be mediated by lipid peroxidation mechanisms (Cullen & Kersten, 2004; Hammel & Cullen, 2008). This view is supported by recent analysis of the *C. subvermispora* genome (Fernandez-Fueyo, Ruiz-Dueñas, et al., 2012).

MnP genes are widely distributed among white rot fungi but are absent from brown rot genomes (see Table 3.1). Manganese concentration dramatically affects transcriptional regulation and may also influence MnP secretion, at least in *C. subvermispora* cultures (Mancilla, Canessa, et al., 2010). Putative metal response elements (MREs) have been implicated in the regulation of *P. chrysosporium mnps* but not that of *T. versicolor mnp2* (Cullen & Kersten, 2004). In nutrient limited medium, transcripts and peptides corresponding to *P. chrysosporium mnp1* accumulated, whereas *mnp2* transcripts were upregulated in nitrogen-starved cultures not in carbon-starved cultures (Ravalason, Jan, et al., 2008; Vanden Wymelenberg, Gaskell, et al., 2009). Differentially regulated transcription of *mnps* has also been observed in more complex substrates (Janse, Gaskell, et al., 1998; Stuardo, Vasquez, et al., 2004), and depletion of the polycyclic aromatic hydrocarbon (PAH) fluorine roughly correlates with transcript levels of *mnp1*, *mnp2*, and *mnp3* (Bogan, Schoenike, et al., 1996a). The latter observation supports lipid peroxidation mechanisms in *P. chrysosporium* (Watanabe, Tsuda, et al., 2010), as does the simultaneous upregulation of MnPs and putative lipid biosynthesis genes in *C. subvermispora* (Fernandez-Fueyo, Ruiz-Dueñas, et al., 2012).

Versatile peroxidases (VPs) oxidize Mn(II) as well as non-phenolic substrates (e.g., veratryl alcohol) in the absence of manganese (Mester & Field, 1998; Camarero, Sarkar, et al., 1999). These enzymes feature Mn-binding residues and a conserved Trp required for electron transfer. VP-encoding genes have not been observed in any brown rot fungi, but white rot species *Dichomitus squalens*, and *T. versicolor* feature three and two genes, respectively (*see* Table 3.1). Transcriptome studies have not yet been reported for these fungi, but VP-derived peptides have been identified in *T. versicolor* cultured on BMA medium (Floudas, Binder, et al., 2012).

Certain sequence deviations resist simple classification, such as LiP, MnP, or VP. Two *C. subvermispora* proteins were classified as LiP and VP genes based on homology modeling and conservation of catalytic residues. Predictably, the corresponding proteins were capable of oxidizing non-phenolic model compounds, but the putative VP was unable to oxidize Mn, and both enzymes exhibited catalytic properties intermediate between conventional LiPs and MnP (Fernandez-Fueyo, Ruiz-Dueñas, et al., 2012).

Possibly involved in ligninolysis, heme thiolate peroxidases (HTPs) and dye decolorization peroxidases (DyPs) (Hofrichter, Ullrich, et al., 2010) have received increasing attention. HTPs include chloroperoxidases and peroxygenases that can catalyze a variety of reactions such as oxidations of various aliphatic and aromatic compounds (Ullrich & Hofrichter, 2005; Gutierrez, Babot, et al., 2011). Recent studies have attributed high redox potentials for HTP from the white rot fungus *Auricularia auricula-judae* (Liers, Bobeth, et al., 2010). Multiple HTP-encoding genes occur in all wood decay genomes and peptides corresponding to *A. auricularia*, *Fomitopsis pini*, and *Dacryopinax* sp. have been identified in media containing BMA (*see* Table 3.1). Three predicted *P. chrysosporium* HTP genes exhibited differential regulation in response to substrate composition (Vanden Wymelenberg, Gaskell, et al., 2011).

DyP genes are irregularly distributed among genomes. Excluding *Gloeophyllum trabeum*, none were detected in brown rot genomes. Analysis of the white rot genomes showed HTP genes absent from *P. chrysosporium* and *C. subvermispora*, whereas *T. versicolor* and *D. squalens* featured two and one gene, respectively. Analysis of BMA culture filtrates suggest that a *D. squalens* DyP and a *T. versicolor* DyP protein are especially abundant, constituting 1.3 percent and 2.2 percent of the total spectra, respectively (Fernandez-Fueyo, Ruiz-Dueñas, et al., 2012).

Laccases

Phenolics, aromatic amines, and other electron-rich substrates are oxidized by laccases, members of the multicopper oxidase family (Hoegger, Kilaru, et al., 2006). The one-electron oxidation of the phenolic units in lignin generates phenoxy radicals that may lead to aryl-Cα cleavage (Kawai, Umezawa, et al., 1988), but the dominant non-phenolic substructures do not serve as substrates. These are only oxidized in the presence of auxiliary substrates, such as ABTS (2,2'-azino-bis-3-ethylthiazoline-6-sulfonate). In this context, the white rot fungi *Pycnoporus cinnabarinus* and *T. versicolor* produce small molecular weight compounds that may act as mediators for oxidation of non-phenolic lignin substructures (Eggert, Temp, et al., 1996; Johannes & Majcherczyk, 2000). Most white rot fungi secrete multiple laccase isozymes. In contrast, *P. chrysosporium* produces none indicating that the enzyme is not uniformly required for lignin degradation. Laccases have been reviewed (Giardina, Faraco, et al., 2010).

Excluding *P. chrysosporium*, families of structurally related genes encode laccases in wood decay fungi. This genetic multiplicity appears slightly reduced in brown rot fungi, and *Dacryopinax* sp. contains none (*see* Table 3.1). As a percentage of total mass spectra, a single laccase is the most abundant *T. versicolor* protein observed in BMA culture filtrates (2.8 percent). More modest estimates of laccase abundance were inferred from mass spectrometry analysis of *D. squalens*, *Fomotiporia mediterranea*, *Punctularia strigozonata*, *C. subvermispora*, and *Wolfiporia cocos*. Differential regulation of laccase genes is well established, and a potential ACE response may play a role in the copper induction of *C. subvermispora* laccases and MnP (Alvarez, Canessa, et al., 2009). Laccase regulation has been reviewed (Piscitelli, Giardina, et al., 2011).

Intracellular Enzymes Involved in Ligninolysis

Complete mineralization of many small molecular weight extractives and lignin-derived compounds requires intracellular metabolism. Intracellular systems also generate the secondary metabolites (e.g., veratryl alcohol and quinones) thought to support extracellular metabolism. Examples of important *P. chrysosporium* enzymes include methanol oxidase (Asada, Watanabe, et al., 1995); 1,4-benzoquinone reductase (Brock & Gold, 1996); methyltransferases (Jeffers, McRoberts, et al., 1997); L-phenylalanine ammonia-lyase (Hattori, Nishiyama, et al., 1999); 1,2,4-trihydroxybenzene 1,2-dioxygenase (Rieble, Joshi, et al., 1994); glutathione transferases (Dowd, Buckley, et al., 1997); superoxide dismutase (Ozturk, Bozhaya, et al., 1999); catalase (Kwon & Anderson, 2001); aryl alcohol dehydrogenase (Reiser, Muheim, et al., 1994);

and cytochrome P450s (Kullman & Matsumura, 1997; Yadav & Loper, 2000; Van Hamme, Wong, et al., 2003; Yadav, Soellner, et al., 2003).

Agaricomycotina genomes have revealed impressive genetic diversity and complex organization among cytochrome P450s (*see* Table 3.1). Most of these sequences have be assigned to families and clans but possible relationships to ecological roles have been not been ascertained. Functional characterization of these P450s has been limited, but many are assumed to be involved in the metabolism of aromatic compounds, including lignin breakdown products. Comparisons of aspen- versus pine-grown *P. placenta* showed differential regulation of 14 P450s (Vanden Wymelenberg, Gaskell, et al., 2011), and two *P. chrysosporium* P450s were shown upregulated in ligninolytic cultures (Shary, Kapich, et al., 2008).

Fenton Systems and Iron Homeostasis

Hydroxyl radicals have been repeatedly implicated as diffusible oxidants in brown rot (Xu & Goodell, 2001; Cohen, Jensen, et al., 2002), and to a lesser extent, in white rot. Fenton chemistry ($H_2O_2 + Fe^{2+} + H^+ \rightarrow H_2O + Fe^{3+} + OH$) is often invoked as the underlying system for generating the highly reactive radicals. Mechanisms controlling extracellular Fenton reactions are the subject of considerable debate (Goodell, 2003; Baldrian & Valaskova, 2008), and three overlapping models have been offered. In one case, the importance of cellobiose dehydrogenase (CDH) had been emphasized, but it is now clear that the efficient brown rot fungus *P. placenta* does not produce this enzyme (Martinez, Challacombe, et al., 2009). Another view stresses the role of low molecular weight glycopeptides that catalyze extracellular iron reduction (Tanaka, Yoshida, et al., 2007). The third mechanism involves extracellular quinone redox cycling (Varela & Tien, 2003; Shimokawa, Nakamura, et al., 2004; Suzuki, Hunt, et al., 2006) (Fig. 3.2). The cycle is complicated in oxalate-accumulating fungi such as *P. placenta* (Kaneko, Yoshitake, et al., 2005), because Fe^{3+}-oxalate chelates are poorly reduced by hydroquinones (Jensen, Houtman, et al., 2001). In such cases, laccases may be involved in hydroquinone oxidation (Wei, Houtman, et al., 2009). Irrespective of the precise mechanism(s), hydroxyl radical pretreatment of lignocellulose substrates clearly enhances enzymatic saccharification (Ratto, Ritschkoff, et al., 1997), and it is widely held that brown rot involves sequential oxidation and hydrolysis. Supporting hydroquinone importance in *P. placenta*, genes likely involved in their biosynthesis, transport and reduction are upregulated in a BMA medium relative to glucose-containing medium. Transcripts of *P. placenta* laccases, possibly supporting a redox system via oxidation of hydroquinones (Gomez-Toribio, Garcia-Martin, et al., 2009), also accumulated in BMA medium (Vanden Wymelenberg, Gaskell, et al., 2010).

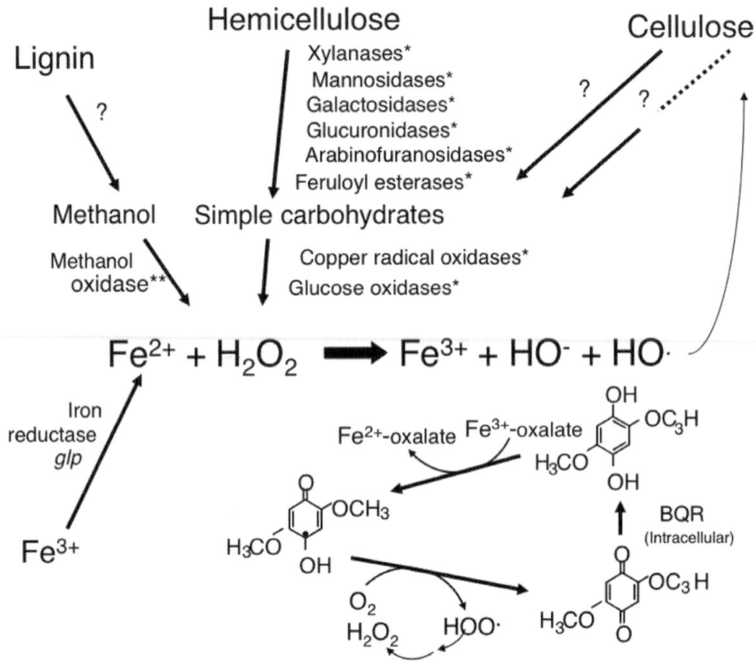

Figure 3.2 Schematic speculation related to systems for generating highly reactive extracellular hydroxyl radical. Enzymes followed by asterisks have been unambiguously identified in culture filtrates of the brown rot fungus, *Postia placenta* (Martinez, Challacombe, et al., 2009). BQR, benzoquinone reductase.

Extracellular Oxidoreductases

Other components commonly ascribed to wood decay systems include extracellular enzymes capable of generating peroxide. Copper radical oxidases and at least four flavin enzymes may be physiologically linked to peroxidases, and possibly to Fenton systems.

Consistent with a close physiological relation, glyoxal oxidase (GLX) is temporally correlated with peroxidases in ligninolytic cultures (Kersten & Kirk, 1987; Kirk & Farrell, 1987; Kersten, 1990), and activity is responsive to peroxidase, peroxidase substrates, and peroxidase products (Kersten, 1990; Kurek & Kersten, 1995). The enzyme will oxidize simple aldehyde, α-hydroxycarbonyl, and α-dicarbonyl compounds; some of which are likely lignin-derived metabolites. Interestingly, such copper radical oxidases (CROs) have two distinct one-electron acceptors, a Cu(II) metal center and an internal Cys-Tyr radical forming a metalloradical complex (Whittaker, 2002).

Initially discovered in *P. chrysosporium* cultures, GLX homologs have been identified in the genomes of most white rot fungi but none of the brown rot genomes (*see* Table 3.1). Coordinate increases in GLX and peroxidase expression were observed under nutrient starvation (Stewart, Kersten, et al., 1992; Vanden Wymelenberg, Minges, et al., 2006; Vanden Wymelenberg, Gaskell, et al., 2009), in colonized wood (Janse, Gaskell, et al., 1998; Sato, Feltus, et al., 2009), and in soil (Bogan, Schoenike, et al., 1996a, b). Seven structurally related copper radical oxidase genes (*glx*, *cro1-cro6*) have been identified in the *P. chrysosporium* genome, and *cro3*, *cro4*, and *cro5* lie within a LiP gene cluster (Cullen & Kersten, 2004). The clustering of *lip* and *cro* genes may also be related to a physiological connection between peroxidases and these peroxide-generating oxidases. *P. chrysosporium* CRO2 substrate preference differs from that of GLX (Vanden Wymelenberg, Sabat, et al., 2006), and this functional diversity may allow adaptation to shifting substrate accessibility and composition during cell wall degradation.

This view may also explain the absence of a GLX homolog in the selective lignin degrader, *C. subvermispora* (*see* Table 3.1). Perhaps functionally related CROs are better suited for a spectrum of small molecular weight substrates unique to the ligninolytic system of *C. subvermispora*. Supporting this, a *C. subvermispora cro2*-like gene and several MnP genes are upregulated in cultures containing BMA (Fernandez-Fueyo, Ruiz-Dueñas, et al., 2012). Little is known concerning the expression of GLX genes in other fungi, although *T. versicolor* and *P. strigo-zonata* GLX proteins were identified in BMA cultures (Floudas, Binder, et al., 2012).

Extracellular peroxide generation may also be supported by Glucose-Methanol-Choline (GMC) oxidases, a large family of flavin enzymes that includes aryl alcohol oxidase (AAO), methanol oxidase (MOX), and various sugar oxidases (Hernandez-Ortega, Ferreira, et al., 2012). AAOs oxidize benzyl alcohols to aldehydes, transferring the electrons to O_2, producing H_2O_2 (Muheim, Leisola, et al., 1990; Asada, Watanabe, et al., 1995). The AAO genes are widely distributed among wood decay fungi, but at least one white rot fungus, *Auricularia delicata*, and brown rot fungi *Coniophora puteana*, *W. coccos*, and *Dacryopinax* sp. have no detectable AAO gene (Floudas, Binder, et al., 2012). Transcript levels in nutrient-starved medium, Avicel medium, and BMA medium were modest for *C. subvermispora* and *P. chrysosporium* (Vanden Wymelenberg, Gaskell, et al., 2009). AAO-derived peptides have been identified in BMA cultures of *T. versicolor* and *D. squalens* (Floudas, Binder, et al., 2012). Hernandez-Ortega, Ferreira, et al. (2012) provide detailed analyses of 40 AAO genes.

MOX, highly expressed in the brown rot fungus *G. trabeum* (Daniel, Volc, et al., 2007), may be linked to the demethoxylation of lignin (Niemenmaa, Uusi-Rauva, et al., 2007) and thereby produce H_2O_2. High expression has also been observed in cultures of the white rot fungus *P. chrysosporium* (Vanden Wymelenberg, Gaskell, et al., 2010). In contrast to *P. chrysosporium*, no *D. squalens* and *T. versicolor* MOX proteins were detected by LC-MS/MS in

BMA medium (Floudas, Binder, et al., 2012). The inability to detect soluble MOX protein in filtrates should be cautiously interpreted because cell wall associations are likely (Daniel, Volc, et al., 2007).

Pyranose 2-oxidase genes have been isolated from *T. versicolor* (Nishimura, Okada, et al., 1996), *P. chrysosporium* (de Koker, Mozuch, et al., 2004), and *G. trabeum* (Dietrich & Crooks, 2009), but obvious homologs are lacking from most sequenced genomes. Transcripts of the *P. chrysosporium* gene are upregulated under ligninolytic conditions (de Koker, Mozuch, et al., 2004; Vanden Wymelenberg, Gaskell, et al., 2009), and the protein has been identified in carbon-starved cultures (Vanden Wymelenberg, Gaskell, et al., 2010) and in BMA medium (Vanden Wymelenberg, Gaskell, et al., 2011).

Another oxidoreductase—CDH—oxidizes cellodextrins, mannodextrins, and lactose. Electron acceptors include quinones, phenoxyradicals, and Fe^{3+}. The protein contains a dehydrogenase domain, a heme prosthetic group and a cellulose binding module (Hallberg, Bergfors, et al., 2000). CDH is widely distributed among fungi, including non-wood decay Ascomycotina. The precise role(s) remain uncertain (Zamocky, Ludwig, et al., 2006) but, as mentioned previously, involvement in hydroxyl radical generation has been proposed.

All white rot genomes have a single CDH gene, but the number varies in brown rot genomes, which have none (*P. placenta*, *Fomitopsis pinicola*, *W. cocos*), one (*C. puteana*, *G. trabeum*), or two (*Serpula lacrymans*) copies of the CDH gene. Sequences share a common architecture with separate flavin, heme, and cellulose-binding domains (CBD).

Upregulation of *P. chrysosporium cdh* has been demonstrated by Northern blots in cellulose-containing media, by competitive real time-polymerase chain reaction (RT-PCR) in colonized wood (Cullen & Kersten, 2004) and more recently by microarrays in media containing Avicel or BMA (Vanden Wymelenberg, Gaskell, et al., 2010) and by RNAseq in red oak medium (Sato, Feltus, et al., 2009). The corresponding peptides were identified in media that were nutrient starved (Vanden Wymelenberg, Gaskell, et al., 2009), containing Avicel, or containing complex lignocellulose substrates (Sato, Feltus, et al., 2009; Vanden Wymelenberg, Gaskell, et al., 2011). The wood species influence expression with higher transcript and protein levels in ball milled pine relative to BMA (Vanden Wymelenberg, Gaskell, et al., 2011).

CDH expression is typically coordinate with that of aldose 1-epimerase (ALE) (Vanden Wymelenberg, Sabat, et al., 2005; Sato, Feltus, et al., 2009; Vanden Wymelenberg, Gaskell, et al., 2011). This expression pattern may indicate a physiological relationship through generation of the cellobiose β-anomer, the preferred CDH substrate (Higham, Gordon-Smith, et al., 1994). In this connection, of five recently sequenced wood decay fungi (*T. versicolor*, *D. squalens*, *P. strigoso-zonata*, *Stereum hirsutum*, *C. puteana*), all but *P. strigoso-zonata* simultaneously secreted ALE and CDH in BMA medium. Co-expression of CDH and certain “hydrolases” was also observed. Initially classified as glycoside hydrolase family 61 (GH61) enzymes, many are now considered copper-dependent monooxygenases (Quinlan, Sweeney, et al., 2011; Westereng, Ishida, et al., 2011). Together, GH61s and CDH boost cellulose depolymerization (Harris, Welner, et al., 2010; Langston, Shaghasi, et al., 2011). In addition to cellulose, secretion of CDH and GH61 were also observed on xylan-containing medium (Hori, Igarashi, et al., 2011). The precise roles(s) and interaction(s) between these genes remain to be clarified.

Glycosyl Hydrolases and Related Carbohydrate Active Enzymes

Employing a battery of hydrolases, cellulose degradation by white rot fungi follows a strategy similar, but not identical, to a diverse array of microbes, particularly the heavily studied industrial ascomycete *Trichoderma reesei.* Key components include exocellobiohydrolase I (CBHI), exocellobiohydrolase II (CBHII), β-1,4-endoglucanase (EG), and β-glucosidase (β-Glu) (Kirk & Cullen, 1998; Baldrian & Valaskova, 2008). Crystalline cellulose is degraded through the synergistic action of the exo- and endo-hydrolases and the resulting oligo- and disaccharides are cleaved to monomers by β-glucosidase. In *T. reesei*, these hydrolases are encoded by relatively few genes principally assigned to glycoside hydrolase families GH6 (CBHII), GH7 (CBHI, EG), GH12 (EG), GH5 (EG), and β-Glu (GH1, GH3). In contrast, six *P. chrysosporium* genes are predicted to encode distinct CBH1 isozymes, and it has been suggested that such diversity reflects subtle functional differences (Munoz, Ubhayasekera, et al., 2001) that allow adaptation to changing environmental conditions during decay.

Multiple CBHI-encoding genes, and typically 1–2 CBHII genes, have been identified in other white rot fungi, an observation standing in contrast to brown rot fungi, which have few, if any, exocellobiohydrolases (*see* Table 3.1). CBH1 and CBHII proteins of white rot fungi are usually abundant in filtrates of BMA medium (*see* Table 3.1). Thus, the number and expression of white rot cellulases support a conventional hydrolytic attack on cellulose. But in brown rot fungi, the paucity of CBHI and CBHII genes (*see* Table 3.1) points toward the aforementioned oxidative depolymerization of cellulose. However, several putative GH5 β-1,4-endoglucanase genes have been identified in brown rot fungi (Floudas, Binder, et al., 2012). One such putative EG is highly expressed in *P. placenta* cultures containing BMA (Vanden Wymelenberg, Gaskell, et al., 2010), but it seems doubtful that the enzyme could efficiently depolymerize crystalline cellulose in the absence of CBHI or CBHII.

Complete breakdown of wood hemicelluloses requires the combined activities of an array of glycoside hydrolases, carbohydrate esterases (CEs), and polysaccharide lyases (PLs). These include endoxylanase, acetylxylan esterase, α-glucuronidase, β-xylosidase, α-arabinosidase, endomannanase, α-galactosidase, acetylglucomannan esterase, β-mannosidase, and β-glucosidase (Kirk & Cullen, 1998). As in the case of cellulases, these enzymes are often classified according to carbohydrate active enzyme (CAZy) family (Cantarel, Coutinho, et al., 2009; www.cazy.org/) but, with the possible exception of GH74, CE1, and CE12, the families are dispersed among taxa and show little relationship to ecology and decay patterns (Floudas, Binder, et al., 2012). Relatively little work has focused on the regulation of these genes, but transcript and secretome profiles in colonized aspen versus pine show significant differences for several *P. chrysosporium* genes, including a GH92 α-1,2-mannosidase, a GH27 α-galactosidase, a GH5 1,4 β-mannan endohydrolase, and two carbohydrate esterases (CE15) (Vanden Wymelenberg, Gaskell, et al., 2011). The latter esterases may attack hemicellulose-lignin linkages (Duranova, Spanikova, et al., 2009). The influence of wood species on regulation of these genes has also been shown for *P. carnosa* (Macdonald, Doering, et al., 2011).

Future Prospects

Enumeration and classification of wood decay genes provide considerable insight into mechanisms of carbon cycling by wood decay fungi. Transcriptome and secretome profiles offer additional clues and help focus research on potentially important components. Still, major obstacles remain and, among these, functional analysis of unknown or hypothetical proteins constitutes a major challenge. For perspective, nanoLC-MS/MS (Vanden Wymelenberg, Gaskell, et al., 2009) unambiguously assigned peptides to 55, 32, and 14 genes encoding unknown proteins in lignocellulose-containing cultures of *P. chrysosporium*, *P. placenta*, and *C. subvermispora*, respectively (Vanden Wymelenberg, Gaskell, et al., 2010, Fernandez-Fueyo, Ruiz-Dueñas, et al., 2012). Some of these unknown proteins are particularly intriguing, as is the case for several highly expressed *P. placenta* genes whose transcript levels are differentially regulated in response to wood species (Vanden Wymelenberg, Gaskell, et al., 2011). However, functional analysis has been hindered by the lack of genetic tools and difficulties performing detailed biochemical analysis on purified proteins. In this context, encouraging progress has been made in the development of techniques for targeted RNAi (Salame, Yarden, et al., 2010) and gene replacement (Salame, Knop, et al., 2012) for the white rot fungus *Pleurotus ostreatus*. Such genetic "toolboxes" might be applied to other wood decay fungi.

Another daunting challenge is to attain a deeper understanding of wood decay in more natural substrates including, ultimately, field conditions. Without doubt, white and brown rot fungi interact with bacteria and soft rot fungi during natural decay (Eriksson, Blanchette, et al., 1990), and interactions with humicolous fungi such as *Agaricus* and *Coprinopsis* at the soil interface likely influence and maintain soil properties. In turn, metabolic activities of these microbes are expected to substantially impact establishment of mycorrhizal systems.

In depth understanding of the structure, physiological activities, and interactions within these complex communities is needed, but here too, research has been stymied by the shortage of experimental tools. Recently however, high throughput metagenomic techniques have been brought to bear (Damon, Lehembre, et al., 2012; de Menezes, Clipson, et al., 2012), and in one case, litter and soil horizons fractions were examined for community composition and for transcript profiles by metagenome and metatranscriptome approaches, respectively (Baldrian, Kolarik, et al., 2012). Focusing on fungi, the genes and transcripts corresponding to CBHIs were also quantified (Baldrian, Kolarik, et al., 2012) and revealed higher numbers and diversity of cellulose decomposers in the litter. Extending such investigations to samples collected over time from multiple ecosystems will identify key species and processes involved in nutrient cycling and forest health.

References

Alvarez JM, Canessa P, et al. 2009. Expression of genes encoding laccase and manganese-dependent peroxidase in the fungus *Ceriporiopsis subvermispora* is mediated by an ACE1-like copper-fist transcription factor. *Fungal Genet Biol.* 46: 104–111.

Asada Y, Watanabe A, et al. 1995. Purification and characterization of an aryl-alcohol oxidase from the lignin-degrading basidiomycete *Phanerochaete chrysosporium. Biosci Biotech Biochem.* 59: 1339–1341.

Baldrian P & Valaskova V. 2008. Degradation of cellulose by basidiomycetous fungi. *FEMS Microbiol Rev.* 32: 501–521.

Baldrian P, Kolarik M, et al. (2012) Active and total microbial communities in forest soil are largely different and highly stratified during decomposition. *ISME J.* 6: 248–258.

Blanchette R. 1991. Delignification by wood-decay fungi. *Annu Rev Phytopath.* 29: 381–398.

Blanchette R. 1995. Degradation of the lignocellulose complex in wood. *Can J Bot.* 73: 999–1010.

Blanchette R, Krueger E, et al. 1997. Cell wall alterations in loblolly pine wood decayed by the white-rot fungus, *Ceriporiopsis subvermispora. J Biotechnol.* 53: 203–213.

Bogan B, Schoenike B, et al. 1996a. Manganese peroxidase mRNA and enzyme activity levels during bioremediation of polycyclic aromatic hydrocarbon-contaminated soil with *Phanerochaete chrysosporium. Appl Environ Microbiol.* 62: 2381–2386.

Bogan B, Schoenike B, et al. 1996b. Expression of *lip* genes during growth in soil and oxidation of anthracene by *Phanerochaete chrysosporium. Appl Environ Microbiol.* 62: 3697–3703.

Brock BJ & Gold MH. 1996. 1,4-Benzoquinone reductase from basidiomycete *Phanerochaete chrysosporium*: Spectral and kinetic analysis. *Arch Biochem Biophys.* 331: 31–40.

Camarero S, Sarkar S, et al. 1999. Description of a versatile peroxidase involved in the natural degradation of lignin that has both manganese peroxidase and lignin peroxidase substrate interaction sites. *J Biol Chem.* 274: 10324–10330.

Cantarel BL, Coutinho PM, et al. 2009. The Carbohydrate-Active EnZymes database (CAZy): An expert resource for Glycogenomics. *Nucl Acids Res.* 37: D233–D238.

Cohen R, Jensen KA, et al. 2002. Significant levels of extracellular reactive oxygen species produced by brown rot basidiomycetes on cellulose. *FEBS Lett.* 531: 483–488.

Cullen D. 2002. Molecular genetics of lignin-degrading fungi and their application in organopollutant degradation. In *The Mycota*, Vol. XI (ed. F Kempken), 71–90. Berlin: Springer-Verlag.

Cullen D & Kersten PJ. 2004. Enzymology and molecular biology of lignin degradation. In *The Mycota III Biochemistry and Molecular Biology* (eds. R Brambl & GA Marzulf), 249–273. Berlin: Springer-Verlag.

Damon C, Lehembre F, et al. 2012. Metatranscriptomics reveals the diversity of genes expressed by eukaryotes in forest soils. *PLoS One.* 7: e28967.

Daniel G. 1994. Use of electron microscopy for aiding our understanding of wood biodegradation. *FEMS Microbiol Rev.* 13: 199–233.

Daniel G, Volc J, et al. 2007. Characteristics of *Gloeophyllum trabeum* alcohol oxidase, an extracellular source of H_2O_2 in brown rot decay of wood. *Appl Environ Microbiol* 73: 6241–6253.

de Koker TH, Mozuch MD, et al. 2004. Pyranose 2-oxidase from *Phanerochaete chrysosporium*: isolation from solid substrate, protein purification, and characterization of gene structure and regulation. *Appl Environ Microbiol.* 70: 5794–5800.

de Menezes A, Clipson N, et al. 2012. Comparative metatranscriptomics reveals widespread community responses during phenanthrene degradation in soil. *Environ Microbiol.* 12: 2577–2588.

Dietrich D & Crooks C. 2009. Gene cloning and heterologous expression of pyranose 2-oxidase from the brown-rot fungus, *Gloeophyllum trabeum. Biotechnol Lett.* 31: 1223–1228.

Dowd CA, Buckley CM, et al. 1997. Glutathione S-transferases from the white-rot fungus, *Phanerochaete chrysosporium. Biochem J.* 324 (Pt 1): 243–248.

Duranova M, Spanikova S, et al. 2009. Two glucuronoyl esterases of *Phanerochaete chrysosporium. Arch Microbiol.* 191: 133–140.

Eggert C, Temp U, et al. 1996. A fungal metabolite mediates degradation of non-phenolic lignin structures and synthetic lignin by laccase. *FEBS Lett.* 391: 144–148.

Eriksson K-EL, Blanchette RA, et al. 1990. *Microbial and Enzymatic Degradation of Wood and Wood Components*. Berlin: Springer-Verlag.

Fernandez-Fueyo E, Ruiz-Dueñas FJ, et al. 2012. Comparative genomics of *Ceriporiopsis subvermispora* and *Phanerochaete chrysosporium* provide insight into selective ligninolysis. *Proc Natl Acad Sci USA.* 109: 5458–5463.

Floudas D, Binder M, et al. 2012. The Paleozoic origin of enzymatic lignin decomposition reconstructed from 31 fungal genomes. *Science.* 336: 1715–1719.

Giardina P, Faraco V, et al. 2010. Laccases: A never-ending story. *Cell Mol Life Sci.* 67: 369–385.

Gilbertson RL. 1981. North American wood-rotting fungi that cause brown rots. *Mycotoaxon* 12: 372–416.

Gomez-Toribio V, Garcia-Martin AB, et al. (2009) Induction of extracellular hydroxyl radical production by white-rot fungi through quinone redox cycling. *Appl Environ Microbiol.* 75: 3944–3953.

Goodell B. 2003. Brown rot fungal degradation of wood: our evolving view. *Wood deterioration and preservation* (eds. B Goodell, D Nicholas, et al.), 97–118. Washington, DC: American Chemical Society.

Gutierrez A, Babot ED, et al. 2011. Regioselective oxygenation of fatty acids, fatty alcohols and other aliphatic compounds by a basidiomycete heme-thiolate peroxidase. *Arch Biochem Biophys.* 514: 33–43.

Hallberg BM, Bergfors T, et al. 2000. A new scaffold for binding haem in the cytochrome domain of the extracellular flavocytochrome cellobiose dehydrogenase. *Structure Fold Des.* 8: 79–88.

Hammel KE & Cullen D. 2008. Role of fungal peroxidases in biological ligninolysis. *Curr Opin Plant Biol.* 11: 349–355.

Harris PV, Welner D, et al. 2010. Stimulation of lignocellulosic biomass hydrolysis by proteins of glycoside hydrolase family 61: Structure and function of a large, enigmatic family. *Biochemistry.* 49: 3305–3316.

Hattori T, Nishiyama A, et al. 1999. Induction of L-phenylalanine ammonia-lyase and suppression of veratryl alcohol biosynthesis by exogenously added L-phenylalanine in a white-rot fungus *Phanerochaete chrysosporium*. *FEMS Microbiol Lett.* 179: 305–309.

Hernandez-Ortega A, Ferreira P, et al. 2012. Fungal aryl-alcohol oxidase: a peroxide-producing flavoenzyme involved in lignin degradation. *Appl Microbiol Biotechnol.* 93: 1395–1410.

Hibbett DS & Donoghue MJ. 2001. Analysis of character correlations among wood decay mechanisms, mating systems, and substrate ranges in homobasidiomycetes. *Syst Biol.* 50: 215–242.

Higham CW, Gordon-Smith D, et al. 1994. Direct 1H NMR evidence for conversion of beta-D-cellobiose to cellobionolactone by cellobiose dehydrogenase from *Phanerochaete chrysosporium*. *FEBS Lett.* 351: 128–132.

Hoegger PJ, Kilaru S, et al. 2006. Phylogenetic comparison and classification of laccase and related multicopper oxidase protein sequences. *FEBS J.* 273: 2308–2326.

Hofrichter M, Ullrich R, et al. 2010. New and classic families of secreted fungal heme peroxidases. *Appl Microbiol Biotechnol.* 87: 871–897.

Hori C, Igarashi K, et al. 2011. Effects of xylan and starch on secretome of the basidiomycete *Phanerochaete chrysosporium* grown on cellulose. *FEMS Microbiol Lett.* 321: 14–23.

Janse BJH, Gaskell J, et al. 1998. Expression of *Phanerochaete chrysosporium* genes encoding lignin peroxidases, manganese peroxidases, and glyoxal oxidase in wood. *Appl Environ Microbiol.* 64: 3536–3538.

Jeffers MR, McRoberts WC, et al. 1997. Identification of a phenolic 3-O-methyltransferase in the lignin-degrading fungus *Phanerochaete chrysosporium*. *Microbiology.* 143 (Pt 6): 1975–1981.

Jensen KA Jr., Houtman CJ, et al. 2001. Pathways for extracellular Fenton chemistry in the brown rot basidiomycete *Gloeophyllum trabeum*. *Appl Environ Microbiol.* 67: 2705–2711.

Johannes C & Majcherczyk A. 2000. Natural mediators in the oxidation of polycyclic aromatic hydrocarbons by laccase mediator systems. *Appl Environ Microbiol.* 66: 524–528.

Kaneko S, Yoshitake K, et al. 2005. Relationship between production of hydroxyl radicals and degradation of wood, crystalline cellulose, and lignin-related compound or accumulation of oxalic acid in cultures of brown-rot fungi. *J Wood Sci.* 51: 262–269.

Kawai S, Umezawa T, et al. 1988. Degradation mechanisms of phenolic b-1 lignin substructure and model compounds by laccase of *Coriolus versicolor*. *Arch Biochem Biophys.* 262: 99–110.

Kersten PJ. 1990. Glyoxal oxidase of *Phanerochaete chrysosporium*: Its characterization and activation by lignin peroxidase. *Proc Natl Acad Sci USA.* 87: 2936—2940.

Kersten PJ & Kirk TK. 1987. Involvement of a new enzyme, glyoxal oxidase, in extracellular H_2O_2 production by *Phanerochaete chrysosporium*. *J Bacteriol.* 169: 2195–2201.

Kirk TK & Farrell RL. 1987. Enzymatic "combustion": The microbial degradation of lignin. *Annu Rev Microbiol.* 41: 465–505.

Kirk TK & Cullen D. 1998. Enzymology and molecular genetics of wood degradation by white-rot fungi. *Environmentally Friendly Technologies for the Pulp and Paper Industry* (eds. RA Young & M Akhtar), 273–308. New York: John Wiley and Sons.

Kleman-Leyer K & Kirk TK. 1992. Changes in the molecular size distribution of cellulose during attack by white-rot and brown-rot fungi. *Appl Environ Microbiol.* 58: 1266–1270.

Kullman SW & Matsumura F. 1997. Identification of a novel cytochrome P-450 gene from the white rot fungus *Phanerochaete chrysosporium*. *Appl Environ Microbiol.* 63: 2741–2746.

Kurek B & Kersten P. 1995. Physiological regulation of glyoxal oxidase from *Phanerochaete chrysosporium* by peroxidase systems. *Enzmol Microb Technol.* 17: 751–756.

Kwon SI & Anderson AJ. 2001. Catalase activities of *Phanerochaete chrysosporium* are not coordinately produced with ligninolytic metabolism: catalases from a white-rot fungus. *Curr Microbiol.* 42: 8–11.

Langston JA, Shaghasi T, et al. 2011. Oxidoreductive cellulose depolymerization by the enzymes cellobiose dehydrogenase and glycoside hydrolase 61. *Appl Environ Microbiol.* 77: 7007–7015.

Liers C, Bobeth C, et al. 2010. DyP-like peroxidases of the jelly fungus *Auricularia auricula-judae* oxidize nonphenolic lignin model compounds and high-redox potential dyes. *Appl Microbiol Biotechnol.* 85: 1869–1879.

Macdonald J & Master ER. 2012. Time-dependent profiles of transcripts encoding lignocellulose-modifying enzymes of the white rot fungus *Phanerochaete carnosa* grown on multiple wood substrates. *Appl Environ Microbiol.* 78: 1596–1600.

MacDonald J, Suzuki H, et al. 2012. Expression and regulation of genes encoding lignocellulose-degrading activity in the genus *Phanerochaete*. *Appl Microbiol Biotechnol.* 94: 339–351.

Macdonald J, Doering M, et al. 2011. Transcriptomic responses of the softwood-degrading white-rot fungus *Phanerochaete carnosa* during growth on coniferous and deciduous wood. *Appl Environ Microbiol.* 77: 3211–3218.

Mancilla RA, Canessa P, et al. 2010. Effect of manganese on the secretion of manganese-peroxidase by the basidiomycete *Ceriporiopsis subvermispora*. *Fungal Genet Biol.* 47: 656–661.

Martinez D, Challacombe J, et al. 2009. Genome, transcriptome, and secretome analysis of wood decay fungus *Postia placenta* supports unique mechanisms of lignocellulose conversion. *Proc Natl Acad Sci USA*. 106: 1954–1959.

Mester T & Field JA. 1998. Characterization of a novel manganese peroxidase-lignin peroxidase hybrid isozyme produced by *Bjerkandera* species strain BOS55 in the absence of manganese. *J Biol Chem.* 273: 15412–15417.

Muheim A, Leisola MSA, et al. 1990. Aryl-alcohol-oxidase and lignin-peroxidase from the white-rot fungus *Bjerkandera adusta* comparison with *Phanerochaete chrysosporium* lignin-peroxidase for reactivity with veratryl alcohol, homoveratric acid and alpha-benzyl veratryl alcohol. *J Biotechnol.* 13: 159–167.

Munoz IG, Ubhayasekera W, et al. 2001. Family 7 cellobiohydrolases from *Phanerochaete chrysosporium*: Crystal structure of the catalytic module of Cel7D (CBH58) at 1.32 A resolution and homology models of the isozymes. *J Mol Biol*. 314: 1097–1111.

Niemenmaa O, Uusi-Rauva A, et al. 2007. Demethoxylation of [O(14)CH (3)]-labeled lignin model compounds by the brown-rot fungi *Gloeophyllum trabeum* and *Poria (Postia) placenta*. *Biodegradation.* 19: 555–565.

Nishimura I, Okada K, et al. 1996. Cloning and expression of pyranose oxidase cDNA from *Coriolus versicolor* in *E. coli*. *J Biotechnol*. 52: 11–20.

Ozturk R, Bozhaya I, et al. 1999. Purification and characterization of superoxide dismutase from *Phanerochaete chrysosporium*. *Enzyme Mirobiol Technol.* 25: 392–399.

Piscitelli A, Giardina P, et al. 2011. Induction and transcriptional regulation of laccases in fungi. *Curr Genomics.* 12: 104–112.

Quinlan RJ, Sweeney MD, et al. 2011. Insights into the oxidative degradation of cellulose by a copper metalloenzyme that exploits biomass components. *Proc Natl Acad Sci USA*. 108: 15079–15084.

Ralph J, Lundquist K, et al. 2004. Lignins: Natural polymers from oxidative couplin of 4-hydroxyphenyl-propanoids. *Phytochem Rev.* 3: 29–60.

Ratto M, Ritschkoff A, et al. 1997. The effect of oxidative pretreatment on cellulose degradation by *Poria placenta* and *Trichoderma reesei*. *Appl Microbiol Biotechnol* 48: 53–57.

Ravalason H, Jan G, et al. 2008. Secretome analysis of *Phanerochaete chrysosporium* strain CIRM-BRFM41 grown on softwood. *Appl Microbiol Biotechnol.* 80: 719–733.

Reiser J, Muheim A, et al. 1994. Aryl-alcohol dehydrogenase from the white-rot fungus *Phanerochaete chrysosporium*: Gene cloning, sequence analysis, expression and purification of recombinant protein. *J Biol Chem.* 269: 28152–28159.

Rieble S, Joshi D, et al. 1994. Purification and characterization of a 1,2,4-trihydroxybenzene 1,2-dioxygenase from the basidiomycete *Phanerochaete chrysosporium*. *J Bacteriol.* 176: 4838–4844.

Salame TM, Yarden O, et al. 2010. *Pleurotus ostreatus* manganese-dependent peroxidase silencing impairs decolourization of Orange II. *Microb Biotechnol.* 3: 93–106.

Salame TM, Knop D, et al. 2012. A gene-targeting system for *Pleurotus ostreatus*: demonstrating the predominance of versatile-peroxidase (mnp4) by gene replacement. *Appl Environ Microbiol.* 78: 5341–5352.

Sato S, Feltus FA, et al. 2009. The first genome-level transcriptome of the wood-degrading fungus *Phanerochaete chrysosporium* grown on red oak. *Curr Genet.* 55:273–286.

Shary S, Kapich AN, et al. 2008. Differential expression in *Phanerochaete chrysosporium* of membrane-associated proteins relevant to lignin degradation. *Appl Environ Microbiol.* 74: 7252–7257.

Shimokawa T, Nakamura M, et al. 2004. Production of 2,5-dimethoxyhydroquinone by the brown-rot fungus *Serpula lacrymans* to drive extracellular Fenton reaction. *Holzforschung.* 58: 305–310.

Stewart P & Cullen D. 1999. Organization and differential regulation of a cluster of lignin peroxidase genes of *Phanerochaete chrysosporium. J Bacteriol.* 181: 3427–3432.

Stewart P, Kersten P, et al. 1992. The lignin peroxidase gene family of *Phanerochaete chrysosporium*: Complex regulation by carbon and nitrogen limitation, and the identification of a second dimorphic chromosome. *J Bacteriol.* 174: 5036–5042.

Stuardo M, Vasquez M, et al. 2004. Molecular approach for analysis of model fungal genes encoding ligninolytic peroxidases in wood-decaying soil systems. *Lett Appl Microbiol.* 38: 43–49.

Suzuki MR, Hunt CG, et al. 2006. Fungal hydroquinones contribute to brown rot of wood. *Environ Microbiol.* 8: 2214–2223.

Tanaka H, Yoshida G, et al. 2007. Characterization of a hydroxyl-radical-producing glycoprotein and its presumptive genes from the white-rot basidiomycete *Phanerochaete chrysosporium. J Biotechnol.* 128: 500–511.

Ullrich R & Hofrichter M. 2005. The haloperoxidase of the agaric fungus *Agrocybe aegerita* hydroxylates toluene and naphthalene. *FEBS Lett.* 579: 6247–6250.

Urzua U, Kersten PJ, et al. 1998. Kinetics of Mn3+–oxalate formation and decay in reactions catalyzed by manganese peroxidase of *Ceriporiopsis subvermispora. Arch Biochem Biophys.* 360: 215–222.

Van Hamme JD, Wong ET, et al. 2003. Dibenzyl sulfide metabolism by white rot fungi. *Appl Environ Microbiol.* 69: 1320–1324.

Vanden Wymelenberg A, Sabat G, et al. 2006. Structure, organization, and transcriptional regulation of a family of copper radical oxidase genes in the lignin-degrading basidiomycete *Phanerochaete chrysosporium. Appl Environ Microbiol.* 72:4871–4877.

Vanden Wymelenberg A, Gaskell J, et al. 2009. Transcriptome and secretome analysis of *Phanerochaete chrysosporium* reveal complex patterns of gene expression. *Appl Environ Microbiol.* 75: 4058–4068.

Vanden Wymelenberg A, Sabat G, et al. 2005. The *Phanerochaete chrysosporium* secretome: database predictions and initial mass spectrometry peptide identifications in cellulose-grown medium. *J Biotechnol.* 118: 17–34.

Vanden Wymelenberg A, Gaskell J, et al. 2010. Comparative transcriptome and secretome analysis of wood decay fungi *Postia placenta* and *Phanerochaete chrysosporium. Appl Environ Microbiol.* 76: 3599–3610.

Vanden Wymelenberg A, Gaskell J, et al. 2011. Significant alteration of gene expression in wood decay fungi *Postia placenta* and *Phanerochaete chrysosporium* by plant species. *Appl Environ Microbiol.* 77: 4499–4507.

Vanden Wymelenberg A, Minges P, et al. 2006. Computational analysis of the *Phanerochaete chrysosporium* v2.0 genome database and mass spectrometry identification of peptides in ligninolytic cultures reveals complex mixtures of secreted proteins. *Fungal Genet Biol.* 43: 343–356.

Varela E & Tien M. 2003. Effect of pH and oxalate on hydroquinone-derived hydroxyl radical formation during brown rot wood degradation. *Appl Environ Microbiol.* 69:6025–6031.

Watanabe T, Tsuda S, et al. 2010. Characterization of a Delta12-fatty acid desaturase gene from *Ceriporiopsis subvermispora*, a selective lignin-degrading fungus. *Appl Microbiol Biotechnol.* 87: 215–224.

Wei D, Houtman CJ, et al. 2009. Laccase and its role in production of extracellular reactive oxygen species during wood decay by the brown rot basidiomycete *Postia placenta*. *Appl Environ Microbiol.* 76: 2091–2097.

Westereng B, Ishida T, et al. 2011. The putative endoglucanase PcGH61D from *Phanerochaete chrysosporium* is a metal-dependent oxidative enzyme that cleaves cellulose. *PLoS One.* 6: e27807.

Whittaker J. 2002. Galactose oxidase. *Adv Protein Chem.* 60: 1–49.

Worrall JJ, Anagnost SE, et al. 1997. Comparison of wood decay among diverse lignicolous fungi. *Mycologia.* 89: 199–219.

Xu G & Goodell B. 2001. Mechanisms of wood degradation by brown-rot fungi: chelator-mediated cellulose degradation and binding of iron by cellulose. *J Biotechnol.* 87: 43–57.

Yadav JS & Loper JC. 2000. Cytochrome P450 oxidoreductase gene and its differentially terminated cDNAs from the white rot fungus *Phanerochaete chrysosporium*. *Curr Genet.* 37: 65–73.

Yadav JS, Soellner MB, et al. 2003. Tandem cytochrome P450 monooxygenase genes and splice variants in the white rot fungus *Phanerochaete chrysosporium*: Cloning, sequence analysis, and regulation of differential expression. *Fungal Genet Biol.* 38: 10–21.

Yelle DJ, Ralph J, et al. 2008. Evidence for cleavage of lignin by a brown rot basidiomycete. *Environ Microbiol.* 10: 1844–1849.

Zamocky M, Ludwig R, et al. 2006. Cellobiose dehydrogenase—a flavocytochrome from wood-degrading, phytopathogenic and saprotropic fungi. *Curr Protein Pept Sci.* 7: 255–280.

4 Aspergilli and Biomass-Degrading Fungi

Isabelle Benoit[1], Ronald P. de Vries[1], Scott E. Baker[2] and Sue A. Karagiosis[2]

[1] *CBS-KNAW Fungal Biodiversity Centre, Utrecht, The Netherlands*
[2] *Pacific Northwest National Laboratory, Richland, Washington*

Introduction

The distinctive nature of nutrient acquisition in fungi is central to the role that these microbes play in both industry and ecology. Fungi secrete digestive enzymes into their environment, depolymerizing the surrounding complex organic matter into simple biochemical building blocks that are then absorbed by the organism. This intrinsic mechanism used by fungi to secure nutrients also makes these organisms ideally suited for the industrial production of commercially valuable enzymes. Enzyme-manufacturing companies currently use a number of Ascomycetes as production hosts because of distinguishing degradative capacities and potent secretory systems. But there is an even longer "history" of fungal enzyme secretion that plays a role in the global carbon cycle; the deconstruction process carried out by fungi transforms the carbon of their environment and thereby replenishes carbon dioxide and other inorganic compounds.

Interestingly, the major industrial fungi are originally soil-borne fungi, namely species from the genus *Aspergillus* (Houbraken & Samson, 2011) as well as *Trichoderma reesei* (Samuels, 2006). The Aspergilli are widespread fungi that can be found globally in soils from forests, grassland, wetland, desert, and cultivated lands (reviewed in Klich, 2002). The most commonly reported species of this genus are *Aspergillus fumigatus*, *Aspergillus versicolor*, *Aspergillus terreus*, *Aspergillus flavus*, and *Aspergillus niger*. Their dispersion in these biotopes varies, with for instance a relatively higher abundance of species from the sections *Aspergillus*, *Nidulantes*, *Flavipeded*, and *Circumdati* in desert soil (Klich, 2002). The distribution of *T. reesei* in soil is limited to a narrow equatorial band, and therefore more specific than other *Trichoderma* species (Samuels, 2006). The third main fungal species addressed in this chapter, *Neurospora crassa*, has its main habitat on burned vegetation, although some reports from soil exist (Turner, Perkins, et al., 2001).

The Ecological Genomics of Fungi, First Edition. Edited by Francis Martin.

At the crossroads of the industrial relevance of fungi to enzyme production and these microbes' essential role in the environment is the compelling case for the molecular genetic and genomic analysis of a variety of ascomycetes. *A. niger* and *T. reesei* are important industrial enzyme production hosts, and as such, the mechanisms of enzyme induction have been well studied in these organisms. *N. crassa*, with its long history of use as an experimental model system, until recently, was relatively unexplored with regard to the mechanisms by which its degradative enzymes are induced. Genomic analyses of all three fungi shed light on the carbon assimilation capacities of these organisms. Not surprisingly, the ability of these organisms to grow on particular carbon sources is reflected, for the most part, in the catalog of enzymes encoded by their genomes. However, the regulatory circuits that govern the expression of these genes are less obvious. Genetic, biochemical, and cell biological analysis have all contributed to the current understanding of these transcriptional regulatory networks. And although genomic analysis readily shows shared control switches, each organism also contains some distinct regulatory mechanisms.

Regulatory Pathways for Induction of Biomass-Degrading Enzymes

Many fungi are highly efficient degraders of plant biomass. This is in part as a result of the broad range of plant biomass degrading enzymes they can produce (de Vries & Visser, 2001; Stricker, Grosstessner-Hain, et al. 2006; Stricker, Steiger, et al., 2007), but also to their ability to respond to changes in the composition of the plant biomass and to ability to access microhabitats thanks to their minute hyphae. Depending on the biotope they inhabit, the vegetation, and therefore the biomass composition, can vary significantly. For instance, fungi in grass soil will largely have access to monocot biomass, whereas fungi in forest soils will deal mainly with wood-based substrates. In addition, the composition of the biomass also varies depending on climate and seasons. For cosmopolitan fungi, such as *Aspergillus*, it is therefore of critical importance to be able to respond rapidly to these changes to ensure an efficient and competitive strategy in obtaining carbon sources. These responses are mediated by a number of transcriptional regulators that activate specific gene sets required to liberate and convert the available carbon sources (de Vries, 2003). Although some of these regulators are commonly present in fungi, others are specific for subsets of fungi.

In this chapter the main regulatory systems involved in plant biomass utilization are discussed. As most studies into regulation of gene expression related to plant biomass degradation have been performed in a relatively small number of Ascomycete fungi, the focus will be on these species.

Aspergilli

The genus *Aspergillus* consists of a large number of species (less than 300) that can be divided into several sections (Houbraken & Samson, 2011). *Aspergillus* is found globally in all natural and man-made environments but is particularly common in soil and indoor environments. In soil they are important players in degradation of organic biomass and the global carbon cycle. Several species, in particular *A. niger* and other members of the black Aspergilli (*Aspergillus* section *nigri*) and *Aspergillus oryzae*, have a long history of industrial use as producers of enzymes and metabolites (Baker & Bennett, 2008).

Among the main industrial enzymes of *Aspergillus* are phytases (Lei & Porres, 2003) and a large variety of plant polysaccharide-degrading or modifying enzymes (de Vries & Visser, 2001). This latter group has already been applied for several decades in many industrial sectors, such as the production of food, feed, beverages, textiles, paper and pulp, wine, and detergents (de Vries, 2003). More recently, they have been applied for the production of biofuels and biochemicals from plant-based substrates (Sorensen, Lubeck, et al., 2011; Sorensen, Teller, et al., 2011).

This long and broad industrial interest has also stimulated a wide range of research programs on Aspergilli and has resulted in one of the largest fungal research communities. It has also made this genus one of the best-studied groups of fungi with respect to genomics. Public genome sequences are available for 12 species, whereas at least 10 more are in progress and a comparative genome database (www.aspgd.org) has been set-up.

Aspergillus species are not among the fungal species with the highest number of genes encoding plant biomass-degrading enzymes in their genome, but they are among the species with the broadest range of enzymes (Coutinho, Andersen, et al., 2009). As a result, most Aspergilli can grow on nearly all plant polysaccharides. This is accompanied by a high number of transcriptional regulators, which is among the highest found in fungi (Pel, deWinde, et al. 2007; Andersen, Salazar, et al., 2011). In the next sections, the different identified regulators related to plant biomass utilization will be discussed.

The Amylolytic Regulator Amylotlytic regulator (AmyR) was the first plant polysaccharide-related regulator identified in *Aspergillus* and was initially studied in *Aspergillus nidulans* and *A. oryzae* (Petersen, Lehmbeck, et al. 1999; Gomi, Akeno, et al. 2000; Tani, Katsuyama, et al., 2001). It regulates the expression of genes-encoding enzymes involved in starch degradation, such as glucoamylase, α-amylase, and α-glucosidase. AmyR contains a Zn_2Cys_6 DNA-binding domain that binds to sequences in the promoters of its target genes that contain the consensus $CGGN_8CGG$ or CGGAAATTTAA (Ito, 2004). This type of DNA-binding domain was first described for

Saccharomyces cerevisiae GAL4, and this class of regulators is therefore also referred to as GAL4-like regulators (Pan & Coleman, 1990).

AmyR responds to the presence of starch or one of its components and then activates the expression of amylolytic genes. However, the actual inducer appears to vary in different *Aspergillus* species. In *A. oryzae*, isomaltose, a transglycosylation product of maltose, appears to be the strongest inducer (Ito, 2004), whereas maltose was suggested to be the inducer in *A. nidulans* and *A. niger* (Nakamura, Maeda, et al., 2006; Yuan, van der Kaaij, et al., 2008; Makita, Katsuyama, et al., 2009). A recent study in *A. niger* demonstrated that glucose itself can induce the expression of amylolytic genes through the action of AmyR (Vankuyk, Benen, et al., 2012). This effect was likely overlooked in previous studies, because glucose also causes repression of these genes through the carbon catabolite repressor CreA. At higher levels of glucose, the inducing effect is masked by the repressing effect, suggesting that it is maltose itself that induces AmyR. However, these effects can now be explained by the presence of a constant low level of glucose during growth on maltose, which would enable induction without causing significant CreA effects.

A recent study in *A. niger* demonstrated that the role of AmyR extends beyond starch degradation (Vankuyk, Benen, et al., 2012). AmyR also regulates the expression of α- and β-galactosidases and β-glucosidases. The physiological relevance of this was demonstrated by the reduced growth of an *amyR* disruptant strain on oligo- and polysaccharides containing α- and β-linked galactose and β-linked glucose residues.

The (Hemi-) Cellulolytic Regulator (Hemi-)cellulolytic regulator (XlnR) was first identified in *A. niger* and described as an activator of xylanolytic genes (van Peij, Visser, et al. 1998). Subsequently, it was demonstrated that XlnR also regulates cellulolytic genes in *Aspergillus* (van Peij, Gielkens, et al. 1998; Gielkens, Dekkers, et al., 1999) and later also a gene involved in xyloglucan degradation (Hasper, Dekkers, et al., 2002). This indicates a central role for XlnR in plant biomass degradation, which reflects its presence in all filamentous Ascomycetes for which a genome is available (Battaglia, Visser, et al., 2011). In addition, XlnR regulates an enzyme of the pentose catabolic pathway (D-xylose reductase) in *Aspergillus* and other fungi (Hasper, Visser, et al., 2000; Seiboth, Gamauf, 2007) and also affects the expression of two other genes of this pathway (xylitol dehydrogenase and D-xylulose kinase) that are under main control of AraR (de Groot, van den Dool, et al., 2007; Battaglia, Hansen, et al., 2011; Battaglia, Visser, et al., 2011). Studies of XlnR have also been performed with *A. oryzae* and *A. nidulans* (Marui, Tanaka, et al., 2002; Battaglia, Hansen, et al., 2011) as well as several other fungal species (e.g., *T. reesei*, *Fusarium oxysporum*) (Stricker, Grosstessner-Hain, et al., 2006; Brunner, Lichtenauer, et al., 2007).

In the presence of xylose, proteolytic cleavage of the c-terminal part of XlnR results in transport into the nucleus and activation of the expression of its target genes (Hasper, Trindade, et al., 2004). The expression of these target genes and *xlnR* itself are also under control of CreA and is dependent on the xylose concentration (de Vries, Visser, et al., 1999) as was described for AmyR and glucose previously.

XlnR contains a Zn_2Cys_6 DNA-binding domain, but its binding site has not yet been studied in detail. Based on analysis of the promoters of XlnR target genes in *A. niger*, the consensus binding site was first suggested to be GGCTAAA (van Peij, Visser, et al., 2008), but was later changed to GGCTAR (de Vries, van de Vondervoort, et al., 2002). Small variations on this have also been described for other Aspergilli (Noguchi, Sano, et al., 2009).

The Arabinanolytic Regulator Studies using ultraviolet mutants of *A. niger* indicated the presence of an arabinose/arabitol responsive transcriptional activator that controls the expression of genes encoding arabinose-releasing enzymes and genes of the L-arabinose catabolic pathway (de Groot, van de Vondervoort, et al., 2003). However, attempts to identify this regulator by complementation of these mutants were not successful. After the *A. niger* genome was sequenced (Pel, DeWinde, et al., 2007; Andersen, Salazar, et al. 2011), analysis of this genome identified three putative regulator-encoding genes with amino acid homology to XlnR. Analysis of these genes demonstrated that the closest *xlnR* homolog, arabinanolytic regulator (*araR*), controls the expression of the arabinolytic genes and genes of the L-arabinose catabolic pathway (Battaglia, Visser, et al., 2011). AraR contains a Zn_2Cys_6 DNA-binding domain, but so far the consensus sequence in the promoter of its target genes to which it binds has not been determined.

AraR controls the expression of several genes encoding α-L-arabinofuranosidase and endoarabinanase and also affects the expression of a β-galactosidase (*lacA*) and an arabinoxylan arabinofuranohydrolase (*axhA*) (Battaglia, Hansen, et al., 2011; Battaglia, Visser, et al., 2011). It also controls the L-arabinose–specific genes of pentose catabolism and has a stronger effect than XlnR on the common genes of L-arabinose and D-xylose catabolism.

An antagonistic relationship between AraR and XlnR has been described (de Groot, van de Vondervoort, et al., 2003; Battaglia, Visser, et al., 201). In an *xlnR* disruption strain, the AraR-regulated genes are also expressed on xylose, whereas in an *araR* disruption strain, the XlnR-regulated genes are also expressed on arabinose. This includes both genes encoding extracellular enzymes and the genes of the pentose catabolic pathway. This antagonistic effect is believed to be responsible for the small difference in growth on arabinose and xylose of the *araR* and *xlnR* disruptant strain, respectively, whereas the *araR*/*xlnR* double disruptant is not able to grow on these sugars (Battaglia, Visser, et al., 2011). The mechanism of this interaction is unclear at this time.

The overall function of AraR is conserved among the Aspergilli, although there are differences in the expression of specific genes (Battaglia, Hansen, et al., 2011). For instance, no expression could be observed for the L-arabitol dehydrogenase encoding gene (*ladA*) in the *A. niger araR* disruptant, whereas reduced, but detectable, expression levels were observed for this gene in the *A. nidulans araR* disruptant.

Although XlnR is commonly found in nearly all filamentous Ascomycetes, AraR seems restricted to the order of the Eurotiales that consists of the Aspergilli, Penicillia, and several other genera (e.g., *Talaromyces*) (Battaglia, Visser, et al., 2011). In light of this and because of the high homology between XlnR and AraR throughout their amino acid sequence, it was suggested that *araR* originated from an *xlnR* gene duplication just before the Eurotiales split from the other Ascomycetes.

The Galactose-Related Regulators Galactose-related regulator (GalR) was first detected in *A. nidulans* as a second regulator with homology to XlnR (in addition to AraR) (Christensen, Gruben, et al., 2011). Analysis of a *galR* disruption strain demonstrated that GalR affects galactose catabolism in *A. nidulans* similar to what was described for the *A. nidulans galA* mutant. Sequencing of the *galR* gene and its promoter region of this mutant did not reveal any mutations, but a mutation was found in a neighboring gene that also encodes a putative regulator. This second galactose-regulated regulator was called GalX and was shown to complement the *galA* mutant (Christensen, Gruben, et al., 2011). Expression analysis of both regulator genes and their putative targets was performed in a wild type, the *galR* disruptant and the *galA* mutant. This revealed that GalR regulates the expression of most genes of the Leloir pathway and some genes of the reductive D-galactose catabolic pathway. *A. nidulans* GalX controls the expression of GalR and one of the genes of the reductive pathway (Christensen, Gruben, et al., 2011). When the presence of these regulators was evaluated in other fungi, it was found that GalR is unique to *A. nidulans*, whereas GalX is commonly found in other Aspergilli and in other fungal species (Christensen, Gruben, et al., 2011). This suggests a specific and recent modification of this regulatory system in *A. nidulans*. Because the homology between GalX and GalR is relatively low, it is unlikely that GalR originates from a local gene duplication event of GalX. Whether all *A. nidulans* isolates contain GalR has not been reported.

In *A. niger*, GalX regulates the reductive galactose catabolic pathway, which appears to be the main pathway for galactose conversion in this species. This is in contrast with *A. nidulans*, *T. reesei*, and *S. cerevisiae* where the Leloir pathway is largely responsible for galactose catabolism (Fekete, Karaffa, et al., 2004; Flipphi, Sun, et al., 2009). Neither GalR nor GalX are homologs of the *S. cerevisiae* galactose regulator, Gal4 (Johnston & Hopper, 1982). Several other genes of the galactose regulatory system of *S. cerevisiae* are absent in *Aspergillus*, suggesting that regulation of galactose catabolism developed independently in *Aspergillus* and *Saccharomyces* (Christensen, Gruben, et al., 2011).

GalR and GalX have only a minor influence on the degradation of plant biomass because control of only a single α-galactosidase-encoding gene was detected for these regulators in *A. nidulans* (Christensen, Gruben, et al., 2011). Both regulators contain a Zn_2Cys_6 DNA-binding domain, but the binding sites in the promoters of their target genes have not yet been determined.

The Inulinolytic Regulator Inulinolytic regulator (InuR) was described in *A. niger* as the regulator controlling inulin degradation (Yuan, Roubos, et al., 2008). It was identified by its proximity to genes-encoding inulin degrading enzymes in the *A. niger* genome, and disruption of *inuR* resulted in a strongly reduced growth phenotype on inulin. InuR has homology to AmyR. It also contains a Zn_2Cys_6 DNA-binding site and analysis of the promoters of its target genes resulted in the consensus binding site $CGGN_8CGG$ (Yuan, Roubos, et al., 2008), which is identical to one of the AmyR binding sites.

The homology of AmyR and InuR suggests that the regulators related to storage polysaccharides (starch and inulin) may have a common ancestor as is likely also the case for XlnR, GalR, and AraR.

The Rhamnose-Related Regulator A large-scale transcriptomics study using *A. niger* microarrays and RNA of *A. niger* grown on a range of plant biomass-related carbon sources, revealed a gene-encoding putative regulator that was specifically induced on rhamnose (Gruben, de Vries, unpublished results). This gene was located on the genome next to three genes that are close homologs of recently identified genes of the rhamnose catabolic pathway of *Pichia stipitis* (Watanabe, Saimura, et al., 2008). Disruption of *rhaR* resulted in a strain that was strongly reduced in growth on rhamnose compared to the wild type and also had a small growth reduction on pectin (Gruben, de Vries, unpublished results). Transcriptome comparison of the wild type and the *rhaR* disruptant on rhamnose demonstrated that RhaR activates the expression of the genes of the rhamnose catabolic pathway and several genes involved in pectin degradation. These genes mainly encoded enzymes involved in the degradation of the rhamnogalacturonan I part of pectin, such as endo- and exo-rhamnogalacturonases, α-rhamnosidases and rhamnogalacturonan acetyl esterases. The expression of most of the genes-encoding homogalacturonan-related enzymes was not effected in the *rhaR* disruptant strain, suggesting the presence of at least one more regulator involved in pectin degradation.

CreA Although all of the regulators described are activating the expression of their target genes in response to the presence of specific compounds, the expression of nearly all plant biomass-related genes is also affected by a negatively acting regulator. This protein, CreA, is a member of the Cys_2His_2 family of transcriptional regulators and is the major factor controlling carbon catabolite repression in most filamentous fungi (Dowzer & Kelly, 1991). Unlike the regulators mentioned, it is not only identified in Ascomycete fungi, but also in Basidiomycetes (Todd & de Vries, unpublished results). The presence of this regulator in such a wide range of fungi indicates its central role in regulation of gene expression.

CreA responds to the presence of high amounts of mainly monosaccharides and, under these conditions, represses the expression of genes involved in monosaccharide release from polysaccharides and genes related to use of alternative carbon sources (Ruijter & Visser, 1997). This is an elegant system to prevent waste of energy on non-essential enzymes, when sufficient easily metabolizable carbon sources are available.

The strength of CreA repression depends on the nature and concentration of the monosaccharide that is present. A study into the expression of two feruloyl esterase-encoding genes from *A. niger* demonstrated the difference in CreA repression of a number of sugars (de Vries, Kester, et al., 2002). These genes are expressed in the presence of ferulic acid, and their expression was studied using combinations of ferulic acid with different sugars in a wild type and a *creA*-derepressed mutant. This demonstrated that glucose and xylose gave the strongest CreA effect, whereas the effect of other sugars (e.g., fructose and rhamnose) was much weaker. The effect of the concentration of the carbon source was identified by analyzing the expression of four xylanolytic genes in the presence of a range of xylose concentrations (from 1 to 200 mM) in a wild type and a *creA* mutant (de Vries, Visser, et al., 1999). Increasing the xylose concentration resulted in a strong decrease in expression levels of the genes in the wild type, whereas these remained constant in the *creA* mutant.

CreA not only affects genes involved in plant biomass utilization, but also many other gene systems (e.g., proteases, fatty acid metabolism) (Ruijter & Visser, 1997). It has been suggested that CreA may respond to imbalances of the co-factor pools, but its precise mechanism has not yet been unraveled. CreA bind to the consensus sequence SYGGRT in the promoters of its target genes and a reverted repeat of this sequence has been shown to result in particularly strong CreA effects (Kulmburg, Mathieu, et al., 1993).

Other Not Yet Identified Positively Acting Regulators Although much progress has been made over the last years in the identification of the various transcriptional activators, expression studies in *Aspergillus* have provided strong indications for additional regulators related to plant biomass utilization.

A detailed expression study of 28 pectinolytic genes under 48 conditions in a wild type and a *creA* mutant provided indications for three regulatory systems related to pectin degradation (de Vries, Jansen, et al., 2002). One responded to the presence of arabinose and was specific for genes-encoding enzymes related to arabinose and galactose release from the side chains of pectin. This regulator was later identified as AraR. A second system responded to the presence of rhamnose and seemed to be mainly involved in genes related to rhamnogalacturonan I degradation. This regulator was also recently identified as RhaR. However, the main system governing pectin degradation seemed to respond to the presence of galacturonic acid and galacturonic acid-containing polysaccharides (de Vries, Jansen, et al., 2002). Nearly all pectinolytic genes responded to these inducers, suggesting that this putative regulator provides the general control of pectin degradation. The gene encoding this regulator is currently unknown, but indications for a similar system in other fungi have been reported (Wubben, Mulder, et al., 1999).

Expression studies of genes related to galacto(gluco)mannan degradation suggested a general regulator for this process. Genes-encoding endomannanases, α-galactosidases, and β-mannosidases are coordinately expressed in response to the presence of mannan but to a less extend mannose (de Vries, van den Broeck, et al., 1999; Ademark, de Vries, et al., 2001).

Expression profiles of two feruloyl esterases of *A. niger* suggested the presence of a specific transcriptional activator responding to the presence of ferulic acid, but this gene remains unidentified at this time (de Vries & Visser, 1999; de Vries, Kester, et al., 2002).

The available data from *Aspergillus* demonstrates a complex system of regulation for plant biomass utilization involving many transcriptional activators and at least one repressor. It is likely that additional regulators play a role, for instance, in fine-tuning the expression levels of the different genes related to the same polysaccharide. This complex regulatory system will enable *Aspergillus* to respond rapidly to changes in the substrate composition, which gives it a competitive advantage in securing a steady supply of carbon.

Trichoderma

Trichoderma species are ubiquitous, robust colonizers of soil and root ecosystems. These filamentous Ascomycetes fungi secrete a broad spectrum of metabolites and are among the most prolific producers of plant cell wall-degrading enzymes. The cellulolytic system of the green-spored saprobe *T. reesei* (anamorph of *Hypocrea jecorina*) is well characterized (Harman, Herrera-Estrella, et al., 2012), and this microbe presents a paradigm for efficient depolymerization of plant cell wall polysaccharides. *T. reesei* was originally isolated from rotting tents of the US army in the Solomon Islands during World War II and identified as the culprit of a rampant infection of cotton-based army material (Reese, Levinsons, et al., 1950). The filamentous fungus was deposited in the Quartermaster collection at Natick, where its cellulolytic potential was realized in the late 1960s. Highly productive strains derived from this original isolate QM6a—selected for their potent secretion system and elevated enzyme expression level—are the workhorse organisms for industrial production of native cellulases and hemicellulases (Kumar, Singh, et al., 2008). These microbe-manufactured enzymes hydrolyze plant cell wall polysaccharides to mixed sugars and have applications in the pulp and paper, food, and textile industries and in the conversion of plant biomass materials into chemical intermediates and biofuels such as ethanol (Kubicek, Mikus, et al., 2009; Schuster & Schmoll, 2010).

Biosynthesis of *T. reesei* cellulase and hemicellulase is remarkably adaptive to the microbe's physiological conditions; carbon source availability promotes differential expression of overlapping but distinct sets of plant biomass-degrading enzyme-encoding genes. The sequencing and annotation of the 34 million base pairs of the *T. reesei* genome revealed a surprisingly streamlined repertoire of genes for hydrolysis of plant biomass (Martinez, Berka, et al., 2008). On average this microbe's genome encodes fewer glycoside hydrolases, carbohydrate esterases, polysaccharide lyases, and carbohydrate-binding module-containing proteins than other Ascomycete species with sequenced genomes (Martinez, Berka, et al., 2008). Previous reports have detailed the hydrolytic enzyme assemblages *T. reesei* generates when cultivated on cellulose, hemicellulose, or their respective degradation or transglycosylation products, for example, cellobiose, sophorose, D-xylose, and xylobiose (Aro, Pakula, et al., 2005; Stricker, Mach, et al., 2008; Kubicek, Mikus, et al., 2009). Sophorose, two β-1,2-linked D-glucose monomers, is derived from transglycosylation of cellobiose by β-glucosidase and is the putative natural inducer molecule of cellulases (Mandels & Reese, 1960; Mandels, Reese, et al., 1962). Interestingly, the most common soluble inducer used for industrial *T. reesei* cellulase biosynthesis is galactosyl-β-1,4-glucoside lactose; though this disaccharide is economical and a potent inducer molecule, it is not a constituent of plant cell wall polymers (Kubicek, Mikus, et al., 2009).

Elucidating the transcription regulatory networks coordinating carbohydrate-active enzyme-encoding gene expression in *T. reesei* has garnered much interest. Synthesis and secretion of large quantities of extracellular enzymes is an energy-expensive endeavor for the fungus. Thus, gene expression of biomass-degrading enzymes is tightly controlled at the transcriptional level (Ilmen, Saloheimo, et al., 1997; Gielkens, Dekkers, et al., 1999). The transcriptional activators XYR1, ACE2, and HAP2/3/5 complex as well as the repressors CRE1 and ACE1 have central roles in modulating the *T. reesei* hydrolytic enzyme system (Aro, Pakula, et al., 2005; Stricker, Mach, et al., 2008; Kubicek, Mikus, et al., 2009).

Xylanase Regulator 1 The central transcriptional activator xylanase regulator 1 (XYR1) is a zinc binuclear cluster protein essential for expression of the chief hydrolytic enzyme-encoding genes including *xyn1*, *xyn2*, *cbh1*, *cbh2*, and *egl1* (Stricker, Grosstessner-Hain, et al., 2006). The *xyr1* gene is an ortholog to *A. niger*'s *xlnR*. Deletion of *xyr1* abolished all cellulase gene expression regardless of the inducer molecule and impaired induction of hemicellulase genes necessary for xylan and arabinan degradation (Stricker, Grosstessner-Hain, et al., 2006; Stricker, Steiger, et al., 2007). DNA footprinting analysis identified functional XYR1-binding sites as a 5′GGCTAA motif arranged as an inverted repeat separated by either a 10-base pair spacer within the *xyn1* promoter or an intervening 12-base pair spacer within the *xyn2* promoter (Rauscher, Würleitner, et al., 2006; Stricker, Trefflinger, et al., 2008). Functional XYR1-binding sequences have also been demonstrated to include a solitary motif as well as 5′GGCTAA-like motifs, which have A or T substitutions in the 3' proximal three bases (Furukawa, Shida, et al., 2009). The transcriptional regulators ACE1 and ACE2 are believed to modulate XYR1 activity by several mechanisms including homo- and heterodimerization, competitive binding, and the recruitment of additional factors to the promoter region (Wurleitner, Pera, et al., 2003; Rauscher, Würleitner, et al., 2006; Stricker, Grosstessner-Hain, et al., 2006; Stricker, Trefflinger, et al., 2008;). For example, ACE1 antagonizes XYR1-dependent activation of *xyn1* (Rauscher, Würleitner, et al., 2006). XYR1 binds to inverted repeats within the *xyn1* promoter as either a homo- or heterodimer under repressing or inducing conditions, respectively (Rauscher, Würleitner, et al., 2006).

Activator of Cellulases 2 The second characterized *T. reesei* transcriptional activator was activator of cellulases 2 (ACE2), which also belongs to a class of zinc binuclear cluster proteins (Aro, Saloheimo, et al., 2001). The *ace2* gene was initially isolated in a yeast expression screen designed to identify factors binding to and activating *T. reesei*'s main cellulase promoter *cbh1* (Aro, Saloheimo, et al., 2001). Loss of *ace2* reduced the expression of all the main cellulase genes, lowered cellulase activity to 30 to 70 percent of wild type levels and reduced *xyn2* expression when the fungus was grown on cellulose (Aro, Saloheimo, et al., 2001; Stricker, Trefflinger, et al., 2008). However sophorose-induced cellulase expression in the *ace2* deletion mutant was comparable to wild type levels, alluding to a distinct ACE2-independent mechanism of transcriptional regulation (Aro, Saloheimo, et al., 2001). ACE2 was reported to bind in vitro to 5′GGCTAATAA sequences in the *cbh1* promoter. This motif also contains the XYR1 regulator binding sequence (Aro, Saloheimo, et al., 2001). One proposed mechanism of ACE2 activity requires phosphorylation and dimerization for binding to target promoters (Stricker, Trefflinger, et al., 2008). Notably, ACE2 has been identified solely in *Trichoderma* species thus far; the genome sequences of *A. niger, A. nidulans*, *N. crassa*, and *Magnaportha grisea* do not contain an *ace2* ortholog (Aro, Pakula, et al., 2005; Kubicek, Mikus, et al., 2009).

HAP2/3/5 Complex The HAP2/3/5 complex binds to a CCAAT motif and is key for complete transcriptional activation via the generation of an open chromatin structure. CCAAT motifs are found in the 5′-non-coding regions of about 30 percent of eukaryotic genes (Aro, Pakula, et al., 2005); this motif is present in the promoters of many fungal cellulase and hemicellulase genes. The original CCAAT-binding complex was identified in *S. cerevisiae* (Pinkham & Guarente, 1985; Olesen & Guarente, 1990; Mcnabb, Xing, et al., 1995), and homologs of the HAP-encoding genes have been cloned from several filamentous fungi including *T. reesei* (Zeilinger, Ebner, et al., 2001), *A. nidulans* (Papagiannopoulos, Andrianopoulos, et al., 1996; Steidl, Papagiannopoulos, et al., 1999), and *N. crassa* (Chen, Crabb, et al., 1998). In general, mutations within the 5′CCAAT sequence decrease the expression level of the gene of interest; this reduction occurs either at the basal expression level or in response to an inducer molecule (Aro, Pakula, et al., 2005). Expression of the *T. reesei cbh2* gene has been shown by promoter mutation and in vivo footprinting analysis to be dependent on a 5′CCAAT element bound by the HAP2/3/5 complex and a GTAATA motif bound to ACE2 (Zeilinger, Mach, et al., 1998). Mutations targeting either the CCAAT or GTAATA sequence reduced *cbh2* transcript levels whereas the double mutation abolished *cbh2* expression. Zeilinger, Schmoll, et al. (2003) detected a nucleosome-free area near the XYR1/ACE2/HAP2/3/5-binding region in the *cbh2* promoter, which is bordered by strictly positioned nucleosomes. Induction by sophorose resulted in a modification of nucleosome positioning downstream of this binding region, thereby promoting accessibility to the TATA box. A mutation in the CCAAT motif altered this positioning and reduced *chb2* transcription. Thus, the HAP2/3/5 complex may enhance accessibility of the promoter to other factors.

CRE1 Expression of a majority of the cellulase genes in *T. reesei* is inhibited in the presence of D-glucose by the transcriptional regulator CRE1. CRE1 is homologous to *A. niger's* CreA and related to *S. cerevisiae*'s Mig1/Mig2/Mig3 proteins and the mammalian Krox20/Egr and Wilms' tumor protein (Dowzer & Kelly, 1991; Westholm, Nordberg, et al., 2008). Mutations of the *cre1* gene generally lead to partial de-repression of biomass degradation enzyme gene expression when the fungus is cultivated in the presence of D-glucose. One of the best examples of this is the high-yielding industrial strain RUT C30, which contains a truncated *cre1* and thereby generates cellulases and most of the hemicellulases in the presence of D-glucose (Ilmen, Thrane, et al., 1996). The truncation consists of a 2,478-base pair fragment, starting downstream of the CRE1 zinc finger-encoded sequence and continuing into the 3′-non-coding region (Seidl, Gamauf, et al., 2008). Transformation of full-length *cre1* into RUT C30 restored carbon catabolite repression (Ilmen, Thrane, et al., 1996).

ACE1 The *ace1* gene was uncovered in a similar yeast based screen as the previously mentioned *ace2* to identify novel transcription factors binding to and activating the *T. reesei cbh1* promoter (Saloheimo, Aro, et al., 2000). ACE1 (activator of cellulases 1) contains three Cys_2His_2-type zinc fingers; this regulator was demonstrated to bind in vitro to eight sites in the 1.15-kb *cbh1* promoter, all of which contain the core 5′AGGCA followed by an A/T-rich sequence (Saloheimo, Aro, et al., 2000). Loss of *ace1* resulted in 2- to 30-fold increased expression of all major cellulase and hemicellulase genes in sophorose- and cellulose-induced cultures, indicating that ACEI acts as a transcriptional repressor (Aro, Ilmen, et al., 2003). The *ace1* and *ace2* double deletion mutant phenocopied the *ace1* knockout strain; the expressed cellulases and hemicellulases in this double knockout mutant are thought to be a result of the remaining XYR1 activity (Aro, Ilmen, et al., 2003). An ortholog of ACE1, the stress response factor-encoded *stzA* gene in *A. nidulans*, reveals a potential connection between intracellular amino acid availability and cellulase gene expression (Chilton, Delaney, et al., 2008).

Genomics and Metabolic Engineering The recent sequencing of the *T. reesei* genome paves the way for industrial strain development using targeted genetic engineering to boost enzyme production. Presently, the cost of cellulase production remains a considerable bottleneck to economic lignocellulose fuel ethanol. The main industrial production organisms are derived from classical mutagenesis using ultraviolet light or chemical mutagens, and the understanding of the molecular mechanisms behind their superior generation and secretion of hydrolytic enzymes is incomplete. One hypothesis is that highly productive strains may have undergone alterations in the transcription regulatory networks controlling cellulase gene expression. As evidence of this, significant differences of *xyr1*, *ace1*, and *ace2* expression patterns were observed during cellulase induction by lactose between the high producer strain RUT C30 and hyperproducer strain CL847 (Portnoy, Margeot, et al., 2011). Additionally, Zou, Shi, et al. (2012)recently demonstrated that modifications to the promoter regions have the capacity to significantly increase the expression efficiency of cellulase genes. The authors engineered a heterologous cellulase hyperexpression system in *T. reesei* by replacing the CRE1 binding sites within the *cbh1* promoter with binding sites for the transcriptional activators ACE2 and HAP2/3/5 complex. This modification resulted in a 5.5- and 7.4-fold increase of the green fluorescent protein-reporter expression level in inducing and repressing culture conditions, respectively. Transcriptional regulation is most likely only one piece to understanding the cellulase hyperproduction puzzle. Massively parallel sequencing and comparative high-density genome microarray analysis of the genomes of multiple cellulase high-producing mutants confirmed previously reported mutations and uncovered novel mutations in several genes (Le Crom, Schackwitz, et al., 2009; Vitikainen, Arvas, et al., 2010). The most abundant among them was encoded transcription factors, as well as components of nuclear transport, mRNA stability, secretion/vacuolar targeting, and metabolism. This heterogeneity of functional categories suggests that multiple changes may be necessary to improve cellulase production (Le Crom, Schackwitz, et al., 2009; Vitikainen, Arvas, et al., 2010).

Neurospora *and Other Ascomycetes*

Neurospora crassa, labeled "the fungal counterpart of *Drosophila*," has been used since the 1920s as a laboratory model organism; early studies with the fungus pioneered the use of microorganisms in genetics, biochemistry, and molecular biology (Davis & Perkins, 2002). This Ascomycete filamentous fungus is also a proficient degrader of plant biomass, although with a narrower substrate range than *Aspergillus* and *Trichoderma*. Natural isolates are commonly found as the earliest colonizers of burnt grasses and sugarcane. The *N. crassa* genome is predicted to contain 23 cellulase-encoding genes as well as 19 hemicellulase-encoding genes and additional genes with annotated functions associated with plant biomass degradation (Martinez, Berka, et al., 2008). The number of *N. crassa* cellulase-encoding genes is comparable to the predicted number in the *A. nidulans* genome (18) and twice as many as those in the *T. reesei* genome (10) (Martinez, Berka, et al., 2008). Current understanding of the molecular mechanisms behind *N. crassa* plant biomass degradation is fragmentary, although this model filamentous fungus was first reported to efficiently depolymerize cellulose in the 1970s (Eberhart, Beck, et al., 1977). Because *N. crassa* is a "domesticated" microbe with an extensive repertoire of genetic and molecular tools, this fungus is an attractive model system for investigating transcriptional regulation of cellulase- and hemicellulase-encoded genes in filamentous fungi. Additionally, *N. crassa* has an abundance of functional genomic resources, including whole genome microarrays and a near-full genome deletion strain set (Dunlap, Borkovich, et al., 2007) to further aide in unraveling the mechanisms underpinning biomass degradation.

N. crassa responds to a variety of inducer molecules and uses a broad range of carbon sources. And similar to what is observed in *T. reesei* and *Aspergillus* species, there is considerable cross-talk between inducers and regulatory networks that are involved in plant polysaccharide degradation. Znameroski, Coradetti, et al. (2012) teased apart the mechanism by which *N. crassa* senses cellulose. Insoluble cellulose is a potent inducer for many cellulolytic fungi but not an ideal substrate for industrial enzyme biosynthesis. The authors hypothesized that soluble cellobiose, the main by-product of cellulase activity when the fungus is exposed to cellulose, has the capacity to induce cellulase gene expression, but this action is quelled by β-glucosidase activity and carbon catabolite repression. In support of this, a *N. crassa* deletion strain lacking three key genes-encoding putative intracellular (*gh1-1*) and extracellular (*gh3-3* and *gh3-4*) β-glucosidase enzymes exhibited full induction of cellulase gene expression when grown on cellobiose. This induction was to the same level as that of the fungus cultivated on cellulose as the sole carbon source. The triple β-glucosidase gene deletion mutant grown on cellobiose showed a similar transcriptional and secretory profile as wild type cultivated on cellulose. Furthermore, mutants with a deletion of the carbon catabolite repressor *cre-1* in the triple β-glucosidase-deletion background secreted higher levels of active cellulases upon cellobiose induction, as compared to enzyme levels observed during cultivation on cellulose. Molecular mechanisms obtained from this model fungus may provide insights that can be applied to strain development of industrial cellulolytic fungi.

Transcription factors essential for *N. crassa* plant biomass degradation and utilization were revealed by exploiting the transcription factor deletion set for this filamentous fungus (Colot, Park, et al., 2006). *N. crassa* gene deletion strains of the *nit-2*, *pacC*, and *cre1* genes, which encode regulators with previously reported influence on cellulase production, were demonstrated to have aberrant growth on cellulose as the sole carbon source (Coradetti, Craig, et al., 2012). However, deletion strains of homologs to other known regulators in *A. niger* and *T. reesei* including *xlnR/xyr1*, *ace1*, and *hap2* exhibited near-normal growth on cellulose, suggesting that these transcription factors do not play a significant role in *N. crassa* cellulase gene expression. *N. crassa's xlnR/xyr1* ortholog, *xlr-1*, however, is indispensable for hemicellulose degradation; deletion of *xlr-1* abolished growth on xylan and xylose (Sun, Tian, et al., 2012). Transcriptome analysis of *N. crassa* cultivated on beechwood xylan showed that *xlr-1* is necessary for induction of hemicellulase and xylose metabolism genes. Induction of cellulase genes was not dependent on *xlr-1*, but *xlr-1* did modulate the expression levels of a subset of cellulase genes. Two novel transcription factors, CLR-1 and CLR-2, belonging to the zinc binuclear cluster superfamily, were identified in *N. crassa* as essential for cellulose depolymerization (Coradetti, Craig, et al., 2012). Strains with deletions of either *clr-1* or *clr-2* showed no cellulase activity and only trace levels of xylanase activity when cultured on cellulose; these mutants exhibited wild type growth when cultivated on sucrose or xylan. Homologs of *clr-1* and *clr-2* were identified in the genomes of a wide variety of filamentous Ascomycete species capable of degrading plant cell wall polysaccharides. In *A. nidulans*, induction of cellulase genes required the *clr-2* homolog but not *clr-1*, revealing both conserved and differing requirements of these regulators between fungi (Coradetti, Craig, et al., 2012). The diversity of transcriptional regulators between fungal species may be in part as a result of the independently evolved mechanisms for expression of cellulase and hemicellulase genes in response to specific inducers (Coradetti, Craig, et al., 2012).

Elucidating the transcriptional networks underlying thermophilic fungal regulation of genes-encoding lignocellulose-degrading enzymes has garnered increased attention. Thermostable enzymes are well suited for industrial conditions and may present a more economical avenue for the efficient hydrolysis of plant cell wall polymers to fermentable sugars. Industrial fermentation at elevated temperatures has several advantages, including increased growth rate of the production host, higher cellulase activity, and lower risk of contamination. Thermophilic fungi are commonly isolated from decaying plant material and present a potentially abundant reservoir of thermostable plant biomass depolymerizing enzymes. The regulatory networks of these microbes governing cellulase expression may share some similarities with those of mesophilic fungi (Li, Li, et al., 2011). Cellulase expression in thermophiles has been reported to use an inducer-repressor system (Maheshwari, Bharadwaj, et al., 2000). For examples, two thermophiles *Talaromyces emersonii* and *Thermoascus aurantiacus* use carbon catabolite repression via the CREI regulator (Li, Li, et al., 2011). And putative CREI and ACE1 binding sites have been identified in the promoter region of the *cbh2* gene in *T. emersonii* (Murray, Collins, et al., 2003). Identification of the full complement of thermophilic fungal transcription factors regulating cellulase gene expression is an ongoing endeavor.

Recently, the comparative genomic analysis of two thermophilic Ascomycete species, *Thielavia terrestris* and *Myceliophthora thermophila* was reported (Berka, Grigoriev, et al., 2011). The completed genomes for *T. terrestris* and *M. thermophila* are the first described for eukaryotic thermophiles. Among thermophilic fungi, *T. terrestris* and *M. thermophila* are well regarded for their cellulolytic capacities and have been reported as being suitable for large-scale production. The authors present evidence that both thermophiles hydrolyze all major polysaccharides found in plant biomass, and they have a similar complement of glycoside hydrolases as *T. reesei*. Transcriptome and secretome analyses suggest that *T. terrestris* and *M. thermophila* use similar mechanisms for cellulose and hemicellulose hydrolysis but distinct approaches for pectin degradation. Additionally, these fungi are amenable to genetic manipulation by classical mutagenesis and targeted engineering for strain development. These completed genomes lay the foundation for identifying the genetic variations underpinning differences between thermophile lignocellulose degradation capacities.

Physiology of Fungal Growth on Various Carbon Sources

The biological information required to support life of any living organism is reflected by the contents of its genome. Fungal growth on a particular type of plant biomass is the result of the production of enzymes capable of degrading the plant polysaccharides into sugar monomers, which are taken up by the fungal cells. The production of these enzymes depends on the presence of the corresponding genes in the genome and on their regulatory systems. Comparison of the ever-increasing number of sequenced and annotated fungal genomes enables the prediction of fungal specificities related to plant biomass utilization by using the carbohydrate active enzymes (CAZy) database.

Ascomycetes were the first group to be sequenced because they include several human and plant pathogens as well as relevant industrial and model organisms such as the Aspergilli. Currently, a limited number of highly varied Basidiomycetes genomes are available as well as a few genomes from Zygomycetes and Chytrids.

To evaluate whether the genomic potential reflects the ability to degrade plant cell wall polymers, fungal growth was monitored on a broad range of substrates from monosaccharides to crude plant biomass (www.Fung-Growth.org). CAZy annotation of genomes combined with growth profiling highlight diverse fungal behavior, from general to specialized life styles (Fig. 4.1; Table 4.1).

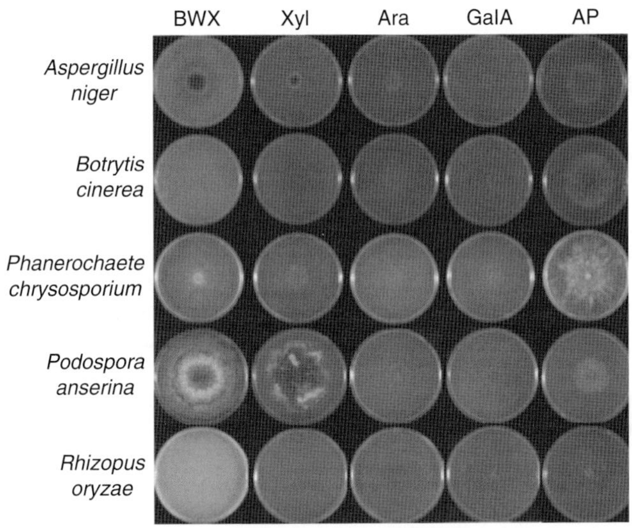

Figure 4.1 Growth profile of five fungi from general, such as *Aspergillus niger*, to more specialized life style such as *Podospora anserina* on plant cell wall polysaccharides. Ara, arabinose; AP, Apple pectin; BWX, beech wood xylan; GalA: galacturonic acid; Xyl, xylose. Corresponding CAZymes are displayed in Table 4.1. For color detail, please see color plate section.

Table 4.1 CAZy families of five fungi from general, such as *Aspergillus niger*, to more specialized life style, such as *Podospora anserina*. Corresponding growth profiles on plant cell polysaccharides are displayed in Figure 4.1.

		AN	BC	PA	PC	RO
Pectin	GH28,GH53, GH78, PL1, PL3, PL4, PL9, PL11, CE8, CE12	44	44	11	8	24
Xylan	GH10, GH11, GH62, CE1	9	9	31	12	0
Cellulose	GH6, GH7, GH45, GH61	11	14	45	24	5

AN, *Aspergillus niger*; BC, *Botrytis cinerea*; CAZy, carbohydrate active enzyme; PA, *Podospora anserina*; PC, *Phanerochaete chrysosoporium*; RO, *Rhizopus oryzae*.

Among Ascomycetes, genome analysis of the model organism *Podospora anserina* revealed genes potentially involved in lignin and cellulose degradation. This was supported by good growth on cellulose and more interestingly also on lignin (Espagne, Lespinet, et al., 2008). In natural environments, lignin degradation is thought to give access to (hemi-)cellulose. In contrast, no growth was observed on inulin or sucrose, which is in agreement with the absence of genes required to degrade those polysaccharides. *Sclerotinia sclerotiorum* and *Botrytis cinerea* are two closely related plant necrotrophs with wide host ranges. The CAZyme content in their genomes is smaller than in the other plant pathogens such as *Magnaporthe oryzae* or *Giberrella zeae* but equivalent in size to the saprobe *A. niger* and larger than *N. crassa* (Amselem, Cuomo, et al., 2011). Their growth profiles on monosaccharides as well as on plant cell-derived polysaccharides showed a clear preference for pectin whereas both grew poorly on xylan and cellulose (*see* Fig. 4.1). The wheat pathogen *Mycosphaerella graminicola* genome contains fewer genes for cellulose degradation and only about one-third as many genes for cell wall degradation in total compared to other plant pathogens such as *M. oryzae*, *G. zeae*, and *N. crassa* (Goodwin, M'barek, et al., 2011). This correlates well with reduced growth of *M. graminicola* on cellulose and xylan. In contrast, *M. graminicola* does contain α-amylases, which correlates with a good growth on starch. Two other interesting cases of specialization are the Oomycete *Pythium ultimum* (Levesque, Brouwer, et al., 2010) and the Zygomycete *Rhizopus oryzae* (Battaglia E, Benoit I, et al. 2011). CAZyme analysis of *P. ultimum* genome showed a limited, if not totally absent, capability of degrading xylan, confirmed by no growth on xylan or xylose. On the other hand, α-amylase, glucoamylase, and invertase candidates were found, suggesting that plant starch and sucrose are targeted. This was confirmed by good growth on these substrates. Growth on pectin was intermediate and could be explained by an incomplete set of pectinases (especially the lack of pectin methyl esterases) (Battaglia, Benoit, et al., 2011). In contrast to *P. ultimum*, in the genome of *R. oryzae*, pectin degradation appears to be the main focus with the highest number of putative CAZymes. The *R. oryzae* genome also contains a large number of chitinolytic and glucanolytic genes and showed a good growth on chitin, chitosan, and diverse fungal cell walls. Although some of these chitinolytic genes may be involved in the renewal or expansion of the cell wall of *R. oryzae*, both, genome content and growth profile suggest a non-plant–based nutritional ability (Battaglia, Benoit, et al., 2011).

Among the Basidiomycetes, *Ceriporiopsis subvermispora* is closely related to *Phanerochaete chrysporium*, and although the last one simultaneously degrades lignin and cellulose, *C. subvermispora* depolymerizes lignin but with relatively little cellulose degradation (Fernandez-Fueyo, Ruiz-Dueñas, et al., 2012). The *C. subvermispora* genome contains about half of the genes-encoding cellulose degrading enzymes compared to the *P. chrysporium* genome, which correlates well this lifestyle.

The examples cited previously clearly show a correlation between CAZymes genome contents and the ability to degrade the plant polysaccharides. However, this correlation does not apply to all cases. On crude substrates, several regulatory systems are involved to induce or repress enzyme production as mentioned before, and those regulatory systems are far from being fully elucidated. Furthermore the number of genes encoding one enzyme activity is not directly correlated to a higher activity (e.g., 10 putative α-amylases present in the genome instead of one does not lead to 10 times more efficient starch degradation). It could be that the different paralogs have differential tissue, or environmentally -regulated expression. For instance, although a relative increase in pectin-related genes in general leads to better growth on pectin, this correlation was not observed for some of the structural elements of pectin, such as methyl esterified pectin and pectin methyl esterases (Benoit, Coutinho, et al., 2012).

A factor that likely contributes to this is that approximately 40 percent of the genes of fungal genomes are still not associated with a known function (Galagan, Henn, et al., 2005). Some of these unspecified genes likely encode enzymes involved in polysaccharide degradation because novel enzyme families are still being discovered, such as recently the GH115 α-glucuronidases and CE15 glucuronoyl esterases (Duranova, Spanikova, et al., 2009; Chong, Battaglia, et al., 2011). This implies that the currently available data on fungal sets of plant polysaccharide degrading enzymes are incomplete, which prevents a perfect correlation between genome content and growth profile.

Conclusions and Future Perspectives

The availability of fungal genome sequences has significantly deepened the understanding of the utilization of plant biomass by fungi. The enzyme sets employed by fungi to degrade plant polysaccharides are significantly broader, and the regulatory network governing the production of these enzymes is more complex than was previously assumed. In addition, recent studies have demonstrated significant differences between fungi their strategy to degrade plant biomass (Table 4.2), both with respect to the enzymes encoded in the genome and the regulation of the expression of the corresponding genes. This topic has only been studied in a small number of

Table 4.2 Presence of transcriptional regulators related to plant biomass utilization in three filamentous fungi.

	AmyR	InuR	XlnR	AraR	RhaR	GalX	AceI	AceII	CreA
Aspergillus niger	+	+	+	+	+	+	+	−	+
Trichoderma reesei	+	−	+	−	+	−	+	+	+
Neurospora crassa	+	−	+	−	+	−	+	−	+

fungal species, most of them from the Ascomycota. Therefore it can be expected that detailed studies on a wider range of fungi, now that (post-) genomic studies are possible for them, will reveal an even bigger variety of fungal strategies for plant biomass degradation.

These studies will help in the understanding of the biotopes of the individual species because carbon source utilization is a critical factor to this. It will also enable the determination of evolutionary aspects of this topic and even the historical onset of a certain mechanism, as was recently done for lignin degradation in white rot fungi (Floudas, Binder, et al., 2012). In addition, it will provide leads for new strategies for industrial applications. For instance, current enzymatic pre-treatments of plant biomass for biofuel production do not release all the fermentable sugars. Combining the strategies of several fungi may result in a more efficient process.

References

Ademark P, de Vries RP, et al. 2001. Cloning and characterization of *Aspergillus niger* genes encoding an alpha-galactosidase and a beta-mannosidase involved in galactomannan degradation. *Eur J Biochem*. 268(10): 2982–2990.

Amselem J, Cuomo CA, et al. 2011. Genomic analysis of the necrotrophic fungal pathogens *Sclerotinia sclerotiorum* and *Botrytis cinerea*. *PLoS Genet*. 7(8): e1002230.

Andersen MR, Salazar MP, et al. 2011. Comparative genomics of citric-acid-producing *Aspergillus niger* ATCC 1015 versus enzyme-producing CBS 513.88. *Genome Res*. 21(6): 885–897.

Aro N, Ilmen M, et al. 2003. ACEI of *Trichoderma reesei* is a repressor of cellulase and xylanase expression. *Appl Environ Microbiol*. 69(1): 56–65.

Aro N, Pakula T, et al. 2005. Transcriptional regulation of plant cell wall degradation by filamentous fungi. *FEMS Microbiol Rev*. 29(4): 719–739.

Aro N, Saloheimo A, et al. 2001. ACEII, a novel transcriptional activator involved in regulation of cellulase and xylanase genes of *Trichoderma reesei*. *J Biol Chem*. 276(26): 24309–24314.

Baker SE & Bennett JW. 2008. An overview of the genus Aspergillus. In *The Aspergilli: Genomics, Medicine, Biotechnology and Research Methods* (eds. GH Goldman & S Osmani), 3–14. Boca Raton, FL: CRC Press.

Battaglia E, Benoit I, et al. 2011. Carbohydrate-active enzymes from the zygomycete fungus *Rhizopus oryzae*: A highly specialized approach to carbohydrate degradation depicted at genome level. *BMC Genomics*. 12:38.

Battaglia E, Hansen SF, et al. 2011. Regulation of pentose utilisation by AraR, but not XlnR, differs in *Aspergillus nidulans* and *Aspergillus niger*. *Appl Microbiol Biotechnol*. 91(2): 387–397.

Battaglia E, Visser L., et al. 2011. Analysis of regulation of pentose utilisation in *Aspergillus niger* reveals evolutionary adaptations in the Eurotiales. *Stud Mycol*. 69: 31–38.

Benoit I, Coutinho PM, et al. 2012. Degradation of different pectins by fungi: Correlations and contrasts between the pectinolytic enzyme sets identified in genomes and the growth on pectins of different origin. *BMC Genomics*. 13: 321.

Berka RM, Grigoriev IV, et al. 2011. Comparative genomic analysis of the thermophilic biomass-degrading fungi *Myceliophthora thermophila* and *Thielavia terrestris*. *Nat Biotechnol*. 29(10): 922–927.

Brunner K, Lichtenauer AM, et al. 2007. Xyr1 regulates xylanase but not cellulase formation in the head blight fungus *Fusarium graminearum*. *Curr Genet*. 52(5–6): 213–220.

Chen H, Crabb JW, et al. 1998. The *Neurospora* aab-1 gene encodes a CCAAT binding protein homologous to yeast HAP5. *Genetics*. 148(1): 123–130.
Chilton IJ, Delaney CE, et al. 2008. The *Aspergillus nidulans* stress response transcription factor StzA is ascomycete-specific and shows species-specific polymorphisms in the C-terminal region. *Mycol Res*. 112: 1435–1446.
Chong SL, Battaglia E, et al. 2011. The alpha-glucuronidase Agu1 from *Schizophyllum commune* is a member of a novel glycoside hydrolase family (GH115). *Appl Microbiol Biotechnol*. 90(4): 1323–1332.
Christensen U, Gruben BS, et al. 2011. Unique regulatory mechanism for D-galactose utilization in *Aspergillus nidulans*. *Appl Environ Microbiol*. 77(19): 7084–7087.
Colot HV, Park G, et al. 2006. A high-throughput gene knockout procedure for *Neurospora* reveals functions for multiple transcription factors. *Proc Natl Acad Sci USA*. 103(44): 16614–16614.
Coradetti ST, Craig JP, et al. 2012. Conserved and essential transcription factors for cellulase gene expression in ascomycete fungi. *Proc Natl Acad Sci USA*. 109(19): 7397–7402.
Coutinho PM, Andersen MR, et al. 2009. Post-genomic insights into the plant polysaccharide degradation potential of *Aspergillus nidulans* and comparison to *Aspergillus niger* and *Aspergillus oryzae*. *Fungal Genet Biol*. 46 Suppl 1: S161–S169.
Davis RH & Perkins DD. 2002. *Neurospora:* A model of model microbes. *Nat Rev Genet*. 3(5): 397–403.
de Groot MJL, van den Dool C, et al. 2007. Regulation of pentose catabolic pathway genes of *Aspergillus niger*. *Food Technol Biotechnol*. 45: 134–138.
de Groot MJL, van de Vondervoort PJ, et al. 2003. Isolation and characterization of two specific regulatory *Aspergillus niger* mutants shows antagonistic regulation of arabinan and xylan metabolism. *Microbiology*. 149: 1183–1191.
de Vries RP. 2003. Regulation of *Aspergillus* genes encoding plant cell wall polysaccharide degrading enzymes; relevance for industrial production. *Appl Microbiol Biotechnol*. 61:10–20.
de Vries RP, Jansen J, et al. 2002. Expression profiling of pectinolytic genes from *Aspergillus niger*. *FEBS Lett*. 530: 41–47.
de Vries RP, Kester HCM, et al. 2002. The *Aspergillus niger faeB* gene encodes a second feruloyl esterase involved in pectin and xylan degradation, and is specifically induced on aromatic compounds. *Biochem J*. 363: 377–386.
de Vries RP & Visser J. 1999. Regulation of the feruloyl esterase (*faeA*) gene from *Aspergillus niger*. *Appl Environ Microbiol*. 65(12): 5500–5503.
de Vries RP & Visser J. 2001. *Aspergillus* enzymes involved in degradation of plant cell wall polysaccharides. *Microb Mol Biol Rev*. 65: 497–522.
de Vries RP, Visser J, et al. 1999. CreA modulates the XlnR-induced expression on xylose of *Aspergillus niger* genes involved in xylan degradation. *Res Microbiol*. 150(4): 281–285.
de Vries RP, van den Broeck HC, et al. 1999. Differential expression of three a-galactosidase genes and a single b-galactosidase gene from *Apergillus niger*. *Appl Environ Microbiol*. 65: 2453–2460.
de Vries RP, van de Vondervoort PJI, et al. 2002. Regulation of the a-glucuronidase encoding gene (*aguA*) from *Aspergillus niger*. *Mol Gen Genet*. 268: 96–102.
Dowzer CEA & Kelly JM. 1991. Analysis of the *cre*A gene, a regulator of carbon catabolite repression in *Aspergillus nidulans*. *Mol Cell Biol*. 11: 5701–5709.
Dunlap JC, Borkovich KA, et al. 2007. Enabling a community to dissect an organism: overview of the *Neurospora* functional genomics project. *Adv Genet*. 57: 49–96.
Duranova M, Spanikova S, et al. 2009. Two glucuronoyl esterases of *Phanerochaete chrysosporium*. *Arch Microbiol*. 191(2): 133–140.
Eberhart BM, Beck RS, et al. 1977. Cellulase of *Neurospora crassa*. *J Bacteriol*. 130(1): 181–186.
Espagne E, Lespinet O, et al. 2008. The genome sequence of the model ascomycete fungus *Podospora anserina*. *Genome Biol*. 9: R77.

Fekete E, Karaffa L, et al. 2004. The alternative D-galactose degrading pathway of *Aspergillus nidulans* proceeds via L-sorbose. *Arch Microbiol*. 181: 35–44.

Fernandez-Fueyo E, Ruiz-Dueñas FJ, et al. 2012. Comparative genomics of *Ceriporiopsis subvermispora* and *Phanerochaete chrysosporium* provide insight into selective ligninolysis. *Proc Natl Acad Sci USA*. 109(14): 5458–5463.

Flipphi M, Sun J, et al. 2009. Biodiversity and evolution of primary carbon metabolism in *Aspergillus nidulans* and other *Aspergillus* spp. *Fungal Genet Biol*. 46 Suppl 1: S19–S44.

Floudas D, Binder M, et al. 2012. The Paleozoic origin of enzymatic lignin decomposition reconstructed from 31 fungal genomes. *Science*. 336(6089): 1715–1719.

Furukawa T, Shida Y, et al. 2009. Identification of specific binding sites for XYR1, a transcriptional activator of cellulolytic and xylanolytic genes in *Trichoderma reesei*. *Fungal Genet Biol*. 46(8): 564–574.

Galagan JE, Henn MR, et al. 2005. Genomics of the fungal kingdom: Insights into eukaryotic biology. *Genome Res*. 15(12): 1620–1631.

Gielkens MM, Dekkers E, et al. 1999. Two cellobiohydrolase-encoding genes from *Aspergillus niger* require D-xylose and the xylanolytic transcriptional activator XlnR for their expression. *Appl Environ Microbiol*. 65(10): 4340–4345.

Gomi K, Akeno T, et al. 2000. Molecular cloning and characterization of a transcriptional activator gene, *amyR*, involved in the amylolytic gene expression in *Aspergillus oryzae*. *Biosci Biotechnol Biochem*. 64: 816–827.

Goodwin SB, M'barek SB, et al. 2011. Finished genome of the fungal wheat pathogen *Mycosphaerella graminicola* reveals dispensome structure, chromosome plasticity, and stealth pathogenesis. *PLoS Genet*. 7(6): e1002070.

Harman GE, Herrera-Estrella AH, et al. 2012. Special issue: *Trichoderma*—from basic biology to biotechnology. *Microbiol-Sgm* 158: 1–2.

Hasper AA, Dekkers E, et al. 2002. EglC, a new endoglucanase from *Aspergillus niger* with major activity towards xyloglucan. *Appl Environ Microbiol*. 68(4): 1556–1560.

Hasper AA, Trindade LM, et al. 2004. Functional analysis of the transcriptional activator XlnR from *Aspergillus niger*. *Microbiology*. 150: 1367–1375.

Hasper AA, Visser J, et al. 2000. The *Aspergillus niger* transcriptional activator XlnR, which is involved in the degradation of the polysaccharides xylan and cellulose, also regulates D-xylose reductase gene expression. *Mol Microbiol*. 36: 193–200.

Houbraken J & Samson RA. 2011. Phylogeny of *Penicillium* and the segregation of Trichocomaceae into three families. *Stud Mycol*. 70(1): 1–51.

Ilmen M, Saloheimo A, et al. 1997. Regulation of cellulase gene expression in the filamentous fungus *Trichoderma reesei*. *Appl Environ Microbiol*. 63(4): 1298–1306.

Ilmen M, Thrane C, et al. 1996. The glucose repressor gene cre1 of *Trichoderma:* isolation and expression of a full-length and a truncated mutant form. *Mol Gen Genet*. 251(4): 451–460.

Ito T. 2004. Mode of AmyR binding to the $CGGN_8AGG$ sequence in the *Aspergillus oryzae taaG2* promoter. *Biosci Biotechnol Biochem*. 68(9): 1906–1911.

Johnston SA & Hopper JE. 1982. Isolation of the yeast regulatory gene GAL4 and analysis of its dosage effects on the galactose/melibiose regulon. *Proc Natl Acad Sci USA*. 79(22)6971–6975.

Klich MA. 2002. Biogeography of *Aspergillus* species in soil and litter. *Mycologia*. 94(1): 21–27.

Kubicek CP, Mikus M, et al. 2009. Metabolic engineering strategies for the improvement of cellulase production by *Hypocrea jecorina*. *Biotechnol Biofuels*. 2: 19.

Kulmburg P, Mathieu M, et al. 1993. Specific binding sites in the *alcR* and *alcA* promoters of the ethanol regulon for the CreA repressor mediating carbon catabolite repression in *Aspergillus nidulans*. *Mol Microbiol*. 7: 847–857.

Kumar R, Singh S, et al. 2008. Bioconversion of lignocellulosic biomass: Biochemical and molecular perspectives. *J Ind Microbiol Biot*. 35(5): 377–391.

Le Crom S, Schackwitz W, et al. 2009. Tracking the roots of cellulase hyperproduction by the fungus *Trichoderma reesei* using massively parallel DNA sequencing. *Proc Natl Acad Sci USA*. 106(38): 16151–16156.

Lei XG & Porres JM. 2003. Phytase enzymology, applications, and biotechnology. *Biotechnol Lett.* 25(21): 1787–1794.

Levesque CA, Brouwer H, et al. 2010. Genome sequence of the necrotrophic plant pathogen *Pythium ultimum* reveals original pathogenicity mechanisms and effector repertoire. *Genome Biol.* 11(7): R73.

Li DC, Li AN, et al. 2011. Cellulases from thermophilic fungi: recent insights and biotechnological potential. *Enzyme Res.* 2011: 308730.

Maheshwari R, Bharadwaj G, et al. 2000. Thermophilic fungi: Their physiology and enzymes. *Microbiol Mol Biol Rev.* 64(3):461–488.

Makita T, Katsuyama Y, et al. 2009. Inducer-dependent nuclear localization of a Zn(II)(2)Cys(6) transcriptional activator, AmyR, in *Aspergillus nidulans. Biosci Biotechnol Biochem.* 73(2): 391–399.

Mandels M & Reese ET. 1960. Induction of cellulase in fungi by cellobiose. *J Bacteriol.* 79(6): 816–826.

Mandels M, Reese ET, et al. 1962. Sophorose as an inducer of cellulase in *Trichoderma viride. J Bacteriol.* 83(2): 400–408.

Martinez D, Berka RM, et al. 2008. Genome sequencing and analysis of the biomass-degrading fungus *Trichoderma reesei* (syn. Hypocrea jecorina). *Nat Biotechnol.* 26(5): 553–560.

Marui J, Tanaka A, et al. 2002. A transcriptional activator, AoXlnR, controls the expression of genes encoding xylanolytic enzymes in *Aspergillus oryzae. Fungal Genet Biol.* 35(2): 157–169.

Mcnabb DS, Xing YY, et al. 1995. Cloning of yeast Hap5—a novel subunit of a heterotrimeric complex required for Ccaat binding. *Genes Dev.* 9(1): 47–58.

Murray PG, Collins CM, et al. 2003. Molecular cloning, transcriptional, and expression analysis of the first cellulase gene (cbh2), encoding cellobiohydrolase II, from the moderately thermophilic fungus *Talaromyces emersonii* and structure prediction of the gene product. *Biochem Biophys Res Commun.* 301(2): 280–286.

Nakamura T, Maeda Y, et al. 2006. Expression profile of amylolytic genes in *Aspergillus nidulans. Biosci Biotechnol Biochem.* 70(10): 2363–2370.

Noguchi Y, Sano M, et al. 2009. Genes regulated by AoXlnR, the xylanolytic and cellulolytic transcriptional regulator, in *Aspergillus oryzae. Appl Microbiol Biotechnol.* 85(1): 141–154.

Olesen JT & Guarente L. 1990. The Hap2 subunit of yeast Ccaat transcriptional activator contains adjacent domains for subunit association and DNA recognition—Model for the Hap2/3/4 complex. *Genes Dev.* 4(10): 1714–1729.

Pan T & Coleman JE. 1990. GAL4 transcription factor is not a "zinc finger" but forms a Zn(II)2Cys6 binuclear cluster. *Proc Natl Acad Sci USA.* 87(6): 2077–2081.

Papagiannopoulos P, Andrianopoulos A, et al. 1996. The *hapC* gene of *Aspergillus nidulans* is involved in the expression of CCAAT-containing promoters. *Mol Gen Genet.* 251(4): 412–421.

Pel HJ, DeWinde JH, et al. 2007. Genome sequencing and analysis of the versatile cell factory *Aspergillus niger* CBS 513.88. *Nat Biotechnol.* 25(2): 221–231.

Petersen KL, Lehmbeck J, et al. 1999. A new transcriptional activator for amylase genes in *Aspergillus. Mol Gen Genet.* 262(4–5): 668–676.

Pinkham JL & Guarente L. 1985. Cloning and molecular analysis of the Hap2 Locus—a global regulator of respiratory genes in *Saccharomyces cerevisiae. Mol Cell Biol.* 5(12): 3410–3416.

Portnoy T, Margeot A, et al. 2011. Differential regulation of the cellulase transcription factors XYR1, ACE2, and ACE1 in *Trichoderma reesei* strains producing high and low levels of cellulase. *Eukaryot Cell.* 10(2): 262–271.

Rauscher R, Würleitner E, et al. 2006. Transcriptional regulation of xyn1, encoding xylanase I, in *Hypocrea jecorina. Eukaryot Cell.* 5(3): 447–456.

Reese ET, Levinsons HS, et al. 1950. Quartermaster culture collection. *Farlowia.* 4: 45–86.

Ruijter GJG & Visser J. 1997. Carbon repression in Aspergilli. *FEMS Microbiol Lett.* 151: 103–114.

Saloheimo A, Aro N, et al. 2000. Isolation of the ace1 gene encoding a Cys(2)-His(2) transcription factor involved in regulation of activity of the cellulase promoter cbh1 of *Trichoderma reesei. J Biol Chem.* 275(8): 5817–5825.

Samuels GJ. 2006. *Trichoderma:* Systematics, the sexual state, and ecology. *Phytopathol.* 96(2): 195–206.

Schuster A & Schmoll M. 2010. Biology and biotechnology of *Trichoderma. Appl Microbio Biotechnol.* 87(3): 787–799.

Seiboth B, Gamauf C, et al. 2007. The D-xylose reductase of *Hypocrea jecorina* is the major aldose reductase in pentose and D-galactose catabolism and necessary for beta-galactosidase and cellulase induction by lactose. *Mol Microbiol.* 66(4): 890–900.

Seidl V, Gamauf C, et al. 2008. The *Hypocrea jecorina (Trichoderma reesei)* hypercellulolytic mutant RUT C30 lacks a 85 kb (29 gene-encoding) region of the wild-type genome. *BMC Genomics.* 9: 327.

Sorensen A, Lubeck PS, et al. 2011. β-glucosidases from a new *Aspergillus* species can substitute commercial beta-glucosidases for saccharification of lignocellulosic biomass. *Can J Microbiol.* 57(8): 638–650.

Sorensen A, Teller PJ, et al. 2011. Onsite enzyme production during bioethanol production from biomass: screening for suitable fungal strains. *Appl Biochem Biotechnol.* 164(7): 1058–1070.

Steidl S, Papagiannopoulos P, et al. 1999. AnCF, the CCAAT binding complex of *Aspergillus nidulans,* contains products of the *hapB, hapC,* and *hapE* genes and is required for activation by the pathway-specific regulatory gene amdR. *Mol Cell Biol.* 19(1): 99–106.

Stricker AR, Grosstessner-Hain K, et al. 2006. Xyr1 (xylanase regulator 1) regulates both the hydrolytic enzyme system and D-xylose metabolism in *Hypocrea jecorina. Eukaryot Cell.* 5(12): 2128–2137.

Stricker AR, Mach RL, et al. 2008. Regulation of transcription of cellulases- and hemicellulases-encoding genes in *Aspergillus niger* and *Hypocrea jecorina* (Trichoderma reesei). *Appl Microbiol Biotechnol.* 78(2): 211–220.

Stricker AR, Steiger MG, et al. 2007. Xyr1 receives the lactose induction signal and regulates lactose metabolism in *Hypocrea jecorina. FEBS Lett.* 581(21): 3915–3920.

Stricker AR, Trefflinger P, et al. 2008. Role of Ace2 (Activator of Cellulases 2) within the xyn2 transcriptosome of *Hypocrea jecorina. Fungal Genet Biol.* 45(4): 436–445.

Sun JP, Tian CG, et al. 2012. Deciphering transcriptional regulatory mechanisms associated with hemicellulose degradation in *Neurospora crassa. Eukaryot Cell.* 11(4): 482–493.

Tani S, Katsuyama Y, et al. 2001. Characterisation of the *amyR* gene encoding a transcriptional activator for the amylase genes in *Aspergillus nidulans. Curr Genet.* 39: 10–15.

Turner BC, Perkins DD, et al. 2001. *Neurospora* from natural populations: A global study. *Fungal Genet Biol.* 32(2): 67–92.

van Peij N, Gielkens MMC, et al. 1998. The transcriptional activator XlnR regulates both xylanolytic and endoglucanase gene expression in *Aspergillus niger. Appl Environ Microbiol.* 64(10): 3615–3619.

van Peij NN, Visser J, et al. 1998. Isolation and analysis of *xlnR*, encoding a transcriptional activator co-ordinating xylanolytic expression in *Aspergillus niger. Mol Microbiol.* 27(1): 131–142.

Vankuyk PA, Benen JA, et al. 2012. A broader role for AmyR in *Aspergillus niger*: regulation of the utilisation of D-glucose or D-galactose containing oligo- and polysaccharides. *Appl Microbiol Biotechnol.* 93(1): 285–293.

Vitikainen M, Arvas M, et al. 2010. Array comparative genomic hybridization analysis of *Trichoderma reesei* strains with enhanced cellulase production properties. *BMC Genomics.* 11: 441.

Watanabe S, Saimura M, et al. 2008. Eukaryotic and bacterial gene clusters related to an alternative pathway of nonphosphorylated L-rhamnose metabolism. *J Biol Chem.* 283(29): 20372–20382.

Westholm JO, Nordberg N, et al. 2008. Combinatorial control of gene expression by the three yeast repressors Mig1, Mig2 and Mig3. *BMC Genomics.* 9: 601.

Wubben JP, Mulder W, et al. 1999. Cloning and partial characterization of endopolygalacturonase genes from *Botrytis cinerea. Appl Environ Microbiol.* 65(4): 1596–1602.

Wurleitner E, Pera L, et al. 2003. Transcriptional regulation of *xyn2* in *Hypocrea jecorina. Eukaryot Cell.* 2(1): 150–158.

Yuan XL, Roubos JA, et al. 2008. Identification of InuR, a new Zn(II)2Cys6 transcriptional activator involved in the regulation of inulinolytic genes in *Aspergillus niger*. *Mol Genet Genomics*. 279(1): 11–26.

Yuan XL, van der Kaaij RM, et al. 2008. *Aspergillus niger* genome-wide analysis reveals a large number of novel alpha-glucan acting enzymes with unexpected expression profiles. *Mol Genet Genomics*. 279(6): 545–561.

Zeilinger S, Mach RL, et al. 1998. Two adjacent protein binding motifs in the *cbh2* (cellobiohydrolase II-encoding) promoter of the fungus *Hypocrea jecorina* (*Trichoderma reesei*) cooperate in the induction by cellulose. *J Biolog Chem*. 273(51): 34463–34471.

Zeilinger S, Ebner A, et al. 2001. The *Hypocrea jecorina* HAP 2/3/5 protein complex binds to the inverted CCAAT-box (ATTGG) within the *cbh2* (cellobiohydrolase II-gene) activating element. *Mol Genet Genomics*. 266(1): 56–63.

Zeilinger S, Schmoll M, et al. 2003. Nucleosome transactions on the *Hypocrea jecorina* (*Trichoderma reesei*) cellulase promoter *cbh2* associated with cellulase induction. *Mol Genet Genomics*. 270(1): 46–55.

Znameroski EA, Coradetti ST, et al. 2012. Induction of lignocellulose-degrading enzymes in *Neurospora crassa* by cellodextrins. *Proc Natl Acad Sci USA*. 109(16): 6012–6017.

Zou G, Shi S, et al. 2012. Construction of a cellulase hyper-expression system in *Trichoderma reesei* by promoter and enzyme engineering. *Microb Cell Fact*. 11:21.

5 Ecological Genomics of *Trichoderma*

Irina S. Druzhinina[1,2] and Christian P. Kubicek[1,2]

[1]*Research Area Biotechnology and Microbiology, Institute of Chemical Engineering, Vienna University of Technology, Vienna, Austria*
[2]*Austrian Center of Industrial Biotechnology, Institute of Chemical Engineering, Vienna University of Technology, Vienna, Austria*

Introduction: Domestication of *Trichoderma* and impact on man kind

Species of *Trichoderma* (teleomorph *Hypocrea*[1], Hypocreales, Ascomycota, Dikarya) are among the most frequent mitosporic fungi commonly detected in cultivation-based surveys. They have been isolated from an innumerable diversity of natural and artificial substrata that demonstrates their high opportunistic potential and adaptability to various ecological conditions. Among hundreds of fungal genera, *Trichoderma* is one of those with the broadest impact on mankind; some *Trichoderma* species are applied in contemporary biotechnology because of their ability to produce enzymes for conversion of plant biomass into soluble sugars that can be used for biofuel production and other biorefinery processes. This has mainly been studied in the domesticated and commercially exploited *Trichoderma reesei* (teleomorph *Hypocrea jecorina*). Mutants of the isolate QM 6a have been used for years both for production of polysaccharide hydrolytic enzymes and heterologous proteins (Kubicek & Penttilä, 1998; Kumar, Singh, et al. 2008; Kubicek, Mikus, et al. 2009). Yet another trait of *Trichoderma* is more broadly distributed within the genus: the profound ability of *Trichoderma* to parasitize or even prey on other fungi (necrotrophic hyperparasitism or mycoparasitism or mycotrophy) is widely used to combat phytopathogenic fungi (biological control of pests, biocontrol) (Hjeljord & Tronsmo, 1998; Kubicek & Penttilä, 1998; Sivasithamparam & Ghisalberti, 1998). At the moment strains of the species *T.* cf. *harzianum*, *T. atroviride* (teleomorph *H. atroviridis*), *T. virens* (teleomorph *H. virens*), and *T. asperellum* are applied as biocontrol agents against plant pathogenic fungi, such as *Rhizoctonia (Thanatephorus)*, *Botrytis (Botryotinia)*, *Sclerotinia*, and *Fusarium (Gibberella)* or fungi-like organisms *Phytophthora*, *Pythium* (Hjeljord & Tronsmo, 1998) for a wide variety of diseases, crops, and climates. More

The Ecological Genomics of Fungi, First Edition. Edited by Francis Martin.

recently, this spectrum was expanded to include the biocontrol of nematodes (Dababat, Sikora, et al. 2006; Kyalo, Affokpon, et al. 2007; Goswami, Pandey, et al. 2008; Druzhinina, Atanasova, et al. unpublished data). We have performed a genus-wide survey of *Trichoderma* antagonistic potential in three plant pathogenic Ascomycetes and one fungus-like organism and detected that, despite considerable infraspecific variability, nearly all species are able to reduce the development of tested prey fungi in the average range of 70 percent (Fig. 5.1).

The mycoparasitic activity of *Trichoderma* can also produce negative impacts. Strains of *T. aggressivum* on one hand and *T. pleuroticola*, *T. pleurotum*, and *T. mienum* on the other, are antagonistic to the commercial mushrooms *Agaricus* and *Pleurotus* respectively (Seaby, 1998; Samuels, Dodd, et al. 2002; Komon-Zelazowska, Bissett, et al. 2007; Kim, Shirouzu, et al., 2012). For more than a decade, it was believed that the infection of mushroom farms was the result of *T. harzianum* (Muthumeenakshi, Mills, et al. 1994; Castle, Speranzini, et al., 1998; Ospina-Giraldo, Royse, et al. 1999). However, Samuels, Dodd, et al. (2002) clearly showed that it was the result of a new species of *Trichoderma* (i.e., *T. aggressivum*). A similar disease outbreak occurred a few years ago on the oyster mushroom *Pleurotus ostreatus*, which was caused by two different although genetically closely related species, *T. pleurotum* and *T. pleuroticola* (Park, Bae, et al., 2006; Hatvani, Antal, et al., 2007; Komon-Zelazowska, Bissett, et al., 2007; Kredics, Kocsubé, et al., 2009). *Pleurotus* "green mould disease" is known from South Korea and Taiwan, as well as from Central (Poland, Hungary, Romania) and Southern Europe (Italy). During the time of this writing, the causal agent of *Pleurotus* green mould disease in Japan was identified to be a new species *T. mienum* (Semiorbis clade), which is unrelated to *T. pleurotum* and *T. pleuroticola* (Kim, Shirouzu, et al., 2012).

Most recently, *Trichoderma* spp. have increasingly been described as symptomless associates of plants or endophytes, a phenomenon that is common among bacteria and fungi. These microorganisms offer a wide range of benefits to the host, including stimulation of plant growth, delaying onset of drought stress and preventing attacks of pathogens (Aly, Debbab, et al. 2011). So far endophytic *Trichoderma* species were mainly isolated from tropical and subtropical ecosystems. Although some species such as *T. hamatum* are detected both as endophytes and common soil and rhizosphere inhabitants, it is still unclear whether obligate endophytic *Trichoderma* species do exist.

In a clinical context, a pair of genetically related species, a strictly clonal *T. longibrachiatum* and *H. orientalis* have been shown to occur as opportunistic pathogens of immunocompromised humans (Kredics, Antal, et al., 2003; Druzhinina, Komon-Zelazowska, et al. 2008). Although mycoparasitism is a common trait for a wide variety of species of the genus, the opportunistic attack of immunocompromised mammals seem to be restricted to section *Longibrachiatum* only.

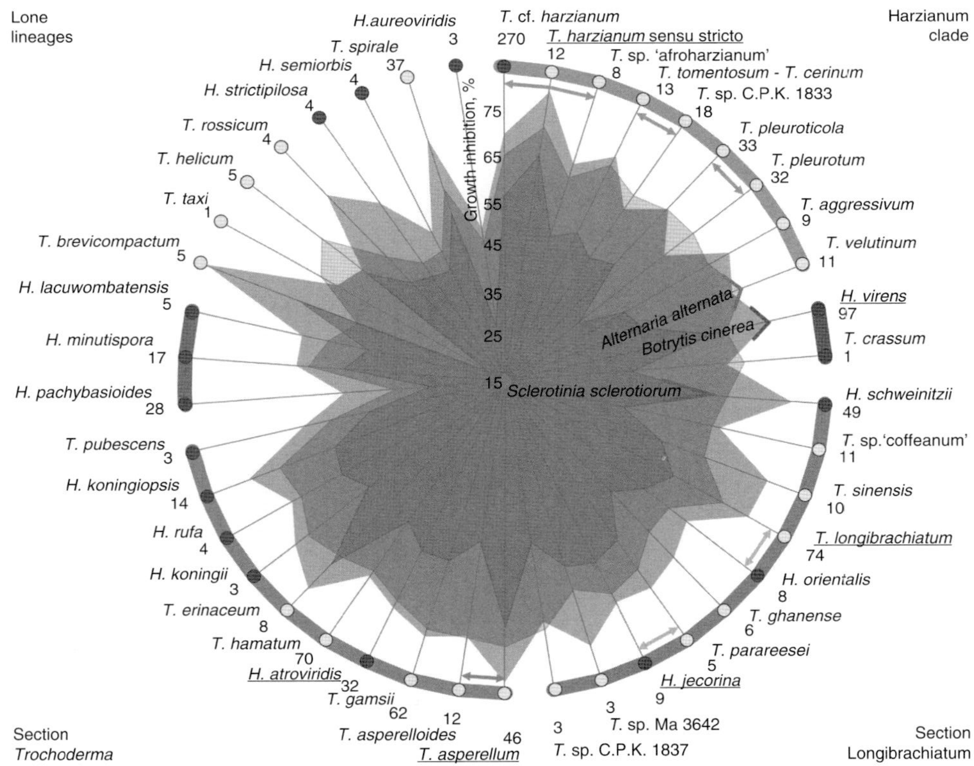

Figure 5.1 Antagonistic potential of the most common *Trichoderma* species estimated in dual confrontations after 10 days of incubation on potato dextrose agar at 77° F (25° C) in 12 hour-long illumination cycle. The scale represents the reduction of diameter of the prey colony (corrected for the growth rate in competition with itself) attributed to the mycoparasitic activity of *Trichoderma*. The three shadow areas correspond to confrontations with *Botrytis cinerea*, *Alternaria alternata*, and *Sclerotinia sclerotiorum*, respectively. All strains have been molecularly identified; species are grouped in phylogenetic clades as established on www.isth.info/biodiversity. Filled and open circles mark holomorphic and putatively clonal species, respectively. Arrows indicate sibling phylogenetic species. Digits below taxon names show number of strains analyzed per species. Underlined species indicate completed genome sequencing initiatives For color detail, please see color plate section.

The ecology of the genus and an overview the genetic background that allows *Trichoderma* to thrive in its natural habitats and serve multiple applications is presented.

Trichoderma in its Environment

Life Cycle and Surviving Strategies

As other Ascomycete fungi, species of *Trichoderma* are haploid during the vegetative stage of the life cycle and—so far identified—have a heterothallic mode of sexual reproduction. This means that mating is only possible between individuals that contain different mating types genes, *mat1-1* and *mat1-2*, which occupy the same chromosomal location but lack sequence similarity (termed *bipolar heterothallism*). Bipolar heterothallism has recently been genetically characterized in *T. reesei* (Seidl, Seibel, et al. 2009), and its occurrence is also evident in *T. virens* and *T. atroviride* from the findings of only one of the two mating types (*mat1-2* for both) in their genomes of the respective strains.

Numerous field observations in Central Europe indicate that the *Trichoderma* anamorphs develop before the *Hypocrea* teleomorph is formed, with some overlap in time (Jaklitsch 2009; 2011). Less commonly the conidiophores are found on overmature stromata, suggesting that anamorph to teleomorph to anamorph life cycle takes place only under optimal environmental conditions (Jaklitsch, 2009).

A considerable number of *Trichoderma* species still have no connections to a teleomorph and are therefore considered to be clonal (*T. hamatum, T. pleuroticola*, and *T. aggressivum, T. tomentosum, T. cerinum, T. spirale, T. gamsii*, and others). Yet clonality has so far been confirmed in silico in only a few species by the application of population analytic methods (Druzhinina, Komon-Zelazowska, et al., 2008; Atanasova, Jaklitsch, et al. 2010; Druzhinina, Kubicek, et al., 2010).

Importantly, nearly all confirmed clonal species of *Trichoderma* are cosmopolitan, being only biased toward a certain climatic conditions (such as tropical/subtropical *T. parareesei* and temperate *T. harzianum*), which suggests that its outstanding opportunistic ability is linked to the mode of reproduction. Moreover most of the confirmed agamospecies are extremely antagonistic toward other fungi and consequently have already found their applications in biocontrol.

Species Recognition and Taxonomy

The history of *Trichoderma* taxonomy before and after the introduction of molecular-based methods has been reviewed previously (see Druzhinina, Koptchinskiy, et al. 2006; Samuels, 2006; Jaklitsch 2009; 2011). Today the genus *Trichoderma* is exceptionally well documented by DNA barcoding and molecular evolutionary analyses through the use of universal DNA barcode markers, ITS1 and 2 of the rRNA gene cluster (Schoch, Seifert, et al., 2012); the fourth and fifth introns of translation elongation factor 1-alpha (*tef1*); a partial exon of endochitinase *chi18-5* (formerly *ech42*); partial intron containing sequences of calmodulin (*cal1*) and actin (*act*) genes; the coding fragment of the RNA polymerase subunit B II gene (*rpb2*); and some other markers. Public databases of DNA sequence data now contain at least two DNA loci for virtually every one of the 200 *Trichoderma* species, which makes molecular identification of strains feasible for researchers from different disciplines (National Center for Biotechnology Information, 2012). Moreover several dedicated tools for molecular identification of the most common *Trichoderma* species have been developed and are at www.isth.info (*TrichO*Key, *Tricho*BLAST; Kopchinskiy, Komon, et al. 2005; Druzhinina, Kopchinskiy, et al., 2006). It is widely accepted now that the phenotypic approach for identification of *Trichoderma* is severely impaired by the homoplasy or insufficient variability of characters, which makes morphological species recognition impossible. Thus, most of the studies on the ecology (Danielson & Davey, 1973), enzyme production (Wey, Hseu, et al., 1994; Kovacs, Szakacs, et al., 2004), biocontrol (Kullnig, Krupica, et al., 2001), human infection (Gautheret, Dromer, et al., 1995), and secondary metabolite formation (Cutler, Cutler, et al. 1999; Humphris, Bruce, et al., 2002) within *Trichoderma* that were performed before the availability of DNA barcoding are difficult to interpret. The detailed infrageneric taxonomy of *Trichoderma* is available in monographs of Jaklitsch (2009, 2011) and previously published reviews of Samuels (2006) and Druzhinina, Kopchinskiy, et al. (2006).

In Situ *Diversity of* Trichoderma

The biodiversity of higher Fungi is considered to be largely unknown (Hawksworth, 1991). Therefore, studies using cultivation-independent methods should result in the identification of a high percentage of still unknown taxa. In situ diversity of *Trichoderma* has so far been only studied in soils (Hagn, Wallisch, et al., 2006; Zachow, Berg, et al., 2009; Meincke, Weinert, et al., 2010; Friedl & Druzhinina, 2012). These pioneering studies, however, detected almost exclusively already known species of *Trichoderma*.

Friedl and Druzhinina (2012) found no hidden diversity of *Trichoderma* in primeval undisturbed soils (Austria). Among 411 ITS1 and 2 molecular operational taxonomic units (MOTUs), 407 were safely attributed to 15 existing species or to putatively new taxa that have previously been sampled. In contrast, the known diversity of *Trichoderma* in Europe consists of at least 75 holomorphic species (Jaklitsch, 2009; 2011) and 10 to 20 anamorphic species (see Friedl & Druzhinina, 2012, for references); in summary, their finds approaching 100 taxa. The finding of only a minor portion of potentially expected diversity (roughly 15 percent) in soil is in agreement with the previous hypothesis that soil itself is not the primary ecological niche for the genus (Druzhinina, Seidl-Seiboth, et al., 2011). A similar outcome was also obtained by Hagn, Wallisch, et al. (2007) for arable soil and Meincke, Weinert, et al. (2010) for rhizosphere of *Solanum tuberosum*. These results are also in agreement with findings of the in situ diversity of Pezizomycotina in soil; in these studies, *Trichoderma* MOTUs were found only at minor portions compared to other groups of Ascomycota (Buée, Reich, et al., 2009; Lim, Kim, et al., 2010).

Infrageneric Communities of Trichoderma

Irrespectively of whether cultivation-based or large-scale genotyping (rRNA ITS) approaches were used, all studies performed so far demonstrated the dominance of communities of highly opportunistic *Trichoderma* species (Migheli, Balmas, et al., 2009; Hagn, Wallisch, et al., 2007; Zachow, Berg, et al., 2009; Friedl & Druzhinina, 2012)

The interaction between coexisting *Trichoderma* species in a single habitat are largely unknown. Friedl and Druzhinina (2012) used an in vitro system to show that different *Trichoderma* species exhibit versatile effects on presence of tribal relatives ranging from inhibition to stimulation of both mycelial growth and conidiation. They concluded that many *Trichoderma* species inhabiting the same microecological niche not only compete with one another but are also able to act synergistically by accelerating the sensing of abiotic factors and thus facilitate the distribution of each other. It demonstrates that the knowledge about infrageneric communities and interactions are required for the screening of *Trichoderma* strains to be used for the biological control of soil-borne plant pathogenic fungi.

Habitats of Trichoderma

Trichoderma was considered to be a soil fungus for a long time. This perception was based on abundant isolations from soil samples worldwide. The general strong antifungal activity of *Trichoderma* spp. favors their detection in cultivation-based surveys because they are able to suppress other fungi and thrive on a petri plate. Qualitative analysis of the diversity revealed in such samples shows the dominance by the same 15 to 20 highly opportunistic species such as *T. asperellum*, *T.* cf. *harzianum*, *T. pleuroticola*, *T. hamatum*, *T. atroviride*, *T. virens*, *T. longibrachiatum*, *T. gamsii*, *T. spirale*, *T. asperelloides*, *T. alni* (teleomorph *Hypocrea alni*), *T. strignosum*, *T. brevicompactum*, *T. citrinoviride* (*H. schweinitzii*), *T. koningiopsis, T. koningii* complex, etc. It is likely that these species obtained the ability to saprotrophic growth in soil because of their general opportunistic potential as suggested based on genomes of *T. atroviride and T. virens* (Druzhinina, Seidl-Seiboth, et al., 2011; Kubicek, Herrera-Estrella, et al. 2011). Consequently, the general belief that *Trichoderma* is a "soil fungus" is not supported.

Although it is now known that some species are closely associated with higher plants (endophytes, Bailey, Bae, et al., 2006, and plant growth promotors), Basidiomycetes (mushroom green mold disease, see Komon-Zelazowska, Bissett, et al., 2007), invertebrates (marine sponge *Psamocinia* sp.: Paz, Komon-Zelazowska, et al., 2010; Gal-Hemed et al. 2011; terrestrial (soil) nematodes: Mennan & Erper, personal communication), and mammals (opportunistic pathogens of humans: Kredics, Antal, et al. 2003; Druzhinina, Komon-Zelazowska, et al., 2008), most of the taxa have been recovered from dead wood and fruiting bodies of other fungi, suggesting that these are the original ecological niches of the fungus (Table 5.1).

It appears that the most notable role of *Trichoderma* in microbial community is likely its ability to prey (or more generally to feed) on other fungi or to inhibit their growth by production of antifungal metabolites. Although it has been mentioned, the mycoparasitism of *Trichoderma* is widely exploited in agriculture, and therefore it is a focus of many geneticists and molecular biologists; however, the ecological role of this habit and its evolutionary significance are not understood.

Table 5.1 Summary on *Trichoderma* ecology.

Ecological niche	Nutritional strategy	Stage of the life cycle	Frequency
Dead wood and plant debris Fruiting bodies of fungi	Saprotrophy/biotrophy	Holomorph	Major
Soil Rhizosphere Indoor habitats	Saprotrophy Saprotrophy/ biotrophy Saprotrophy	Anamorph	Common
In plants as endophytes Marine sponges Immunocompromised humans	Biotrophy Unknown Biotrophy	Anamorph	Rare
Dead herbaceous materials Living herbaceous materials Soil filamentous fungi	Saprotrophy Biotrophy Unknown	Holomorph Teleomorph Anamorph	Putative
Mycorrhizae Soil nematodes	Unknown Biotrophy		

Genomic Attributes of *Trichoderma*

General Genomic Features

So far, the genomes of the industrial cellulase producer *T. reesei* and two mycoparasitic *Trichoderma* spp. have been sequenced and analyzed (Martinez, Berka, et al., 2008; Kubicek, Herrera-Estrella, et al., 2011). They are 34.1 (*T. reesei*), 36.1 (*T. atroviride*), and 38.8 Mbp (*T. virens*) in size and comprise 9,143, 11,865, and 12,518 gene models, respectively, which places the latter two into the average shown by other Ascomycota (see Department of Energy Joint Genome Institute program, MycoCosm, at http://genome.jgi.doe.gov/programs/fungi/index.jsf; Grigoriev, Nordberg, et al., 2011). Because the genome of *T. reesei* has recently been reviewed (Kubicek, 2013), *TT. atroviride* and *virens* will be the concentration of this discussion.

Genomes of *T. atroviride* and *T. virens* share 1,273 orthologues that are not present in the ecologically specialized and weaker mycoparasite, *T. reesei*, which could thus be important for mycotrophy or opportunistic behavior. They were particularly rich in protein families (PFAM, http://pfam.sanger.ac.uk), domains for fungal-specific Zn(2)Cys(6) transcription factors (PF00172, PF04082), and solute transporters (PF07690, PF00083). In addition, they encoded proteins with PFAM groups for oxidoreductases, monooxygenases, AMP activation of acids, phosphopantetheine attachment, and synthesis of isoquinoline alkaloids (Kubicek, Herrera-Estrella, et al., 2011). Thus, *T. atroviride* and *T. virens* may contain an as yet undiscovered reservoir of secondary metabolites.

Gene expansion in Mycoparasitic Trichoderma *spp.*

Markov cluster algorithm (MCL) analysis of the two *Trichoderma* mycoparasitic and opportunistic species, together with *T. reesei* and 10 other Ascomycetes, identified 46 such gene families that are expanded in all *Trichoderma* spp., of which 26 were expanded only in *T. atroviride* and *T. virens*. Zn(2)Cys(6) transcription factors, solute transporters of the major facilitator superfamily, short chain alcohol dehydrogenases, S8 peptidases, and proteins bearing ankyrin domains were expanded in all three *Trichoderma* spp. In addition, *T. atroviride* and *T. virens* contained even more expanded gene families comprising ankyrin proteins, proteins with CCHC zinc finger domains, with WD40 domains, heteroincompatibility (HET) and NACHT domains, and NmrA-type NAD-dependent epimerases (Kubicek, Herrera-Estrella, et al., 2011). An even more detailed analysis based on 44 Pezizomycotina genomes that were available on April 1, 2012, in the Joint Genome Institute database (at http://genome.jgi.doe.gov/programs/fungi/index.jsf) further refined these data and shows that mycoparasitic and opportunistic *Trichoderma* species (*T. atroviride* and *T. virens*) have a unique genomic architecture among all Peziziomycotina (Druzhinina & Kubicek, unpublished data). In a single linkage clustering (a method that is based on the determination of the distance between the two closest objects), the moderately mycoparasitic and not opportunistic *T. reesei* appears as the nearest neighbor to the two other *Trichoderma* species, being however attributed to a different cluster (Fig. 5.2). In the complete linkage method (that clusters the objects based of their differences) *T. reesei* belongs to a more remote cluster that contains no members from the Sordariomycetes. This analysis reflected that indeed *T. atroviride* and *T. virens* harbor the highest number of genes that encode proteins with ankyrin and HET/ankyrin/NACHT domains among all Pezizomycotina (Fig. 5.3). A preliminary phylogenetic analysis of the randomly chosen Trire2:30084 ankyrin protein and 99 homologous fungal sequences-encoding proteins with ankyrin domain (retrieved by blastp withTrire2:30084 as a query) reveals a supported clade dominated by *Trichoderma* genes with a few Peziziomycotina, mainly Sordariomycetes. Consistent data were also obtained for several other ankyrin proteins of *Trichoderma* (Kubicek, unpublished data), indicating that ankyrin-domain proteins may evolve by extensive gene duplication. These findings suggest that only the strongly opportunistic and cosmopolitan species, *T. atroviride* and *T. virens*, harbor enlarged numbers of ankyrin-encoding genes, a claim also supported by a preliminary analysis of the just recently available genome sequences of *T. harzianum* and *T. asperellum*. It appears, therefore, justified to speculate that these genes contribute in an as yet unknown way to the unique opportunistic success of *Trichoderma*.

The ankyrin repeat is a 33-residue motif that mediates protein–protein interactions, and proteins with ankyrin domains are involved in several cellular functions in higher eukaryotes, such as transcriptional regulation, cell cycle, signal transduction, and tumor development (Mosavi, Cammett, et al., 2004). In bacteria, some of them play important roles in microbial pathogenesis. Interestingly, the gram-negative obligate endosymbiont *Wolbachia pipientis* (Proteobacteria), which infects 20 to 75 percent of insect species and also some spiders, mites, and nematodes (Breeuwer& Jacons, 1996; Bouchon, Rigaud, et al., 1998; Jeyaprakash & Hoy, 2000) contains 60 ankyrin genes, the highest number reported in a prokaryote; they are believed to play an important biological role in endosymbiosis of *Wolbachia*. Proteins containing the ankyrin domain have not been studied systematically in Pezizomycotina. The expansion of genes-encoding ankyrin domains in mycotrophic *Trichoderma* and the findings in *Wolbachia* described leads to speculation that they may be involved in the interaction of *Trichoderma* with other organisms such as prey or host fungi or plants.

Also the amplification of genes-encoding proteins with HET/NACHT domains is worth some comments: the heterokaryon incompatibility (HET)

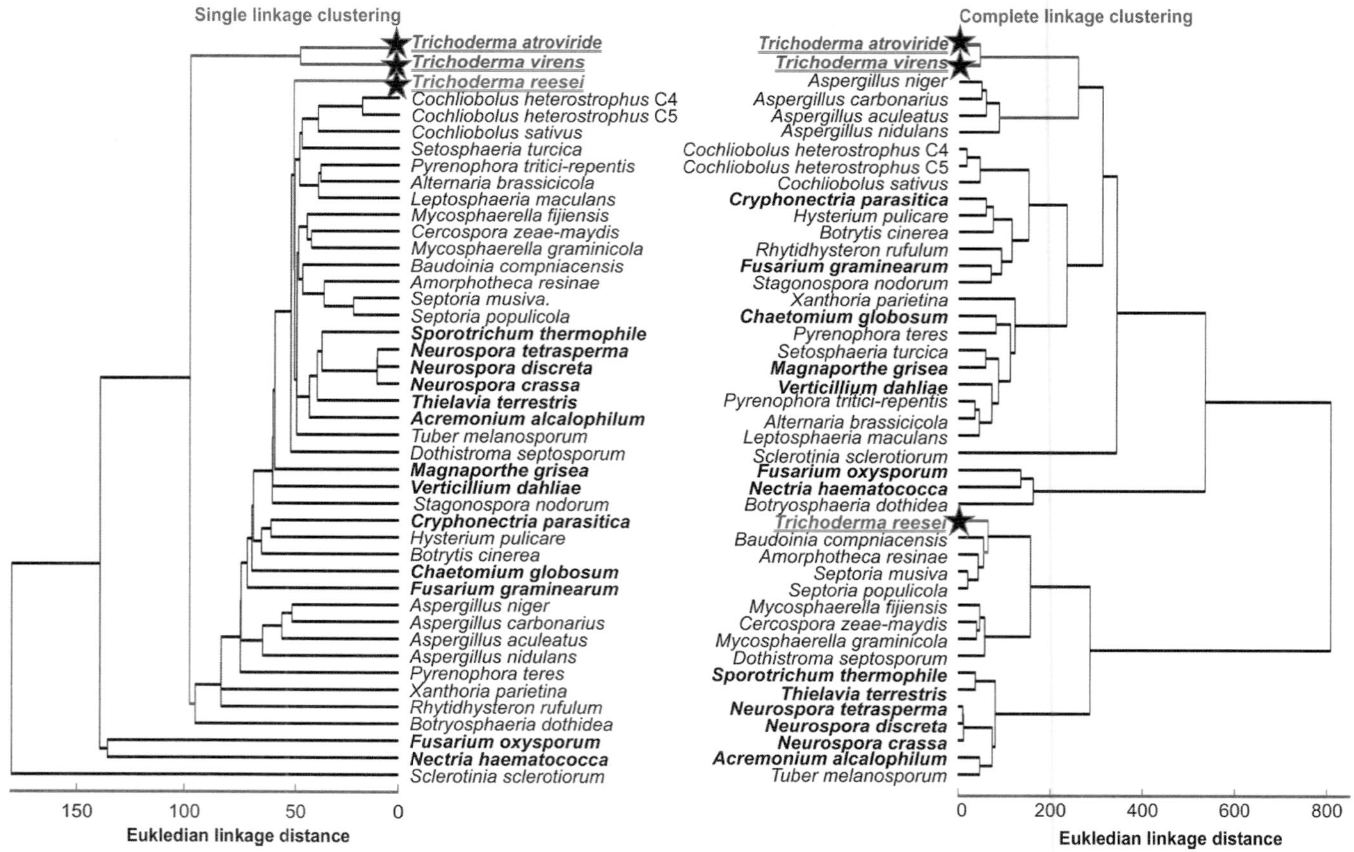

Figure 5.2 Cladograms based on the gene number per cluster matrix resulted from the MCL analysis of orthologous sequences between the three *Trichoderma* and 41 other Peziziomycotina fungi. Members of the class Sordariomycetes are in bold. Single linkage method distance between two clusters is determined by the distance of the two closest objects ("nearest neighbors") in the different clusters. Complete linkage method the distances between clusters are determined by the greatest distance between any two objects in the different clusters (i.e., by the "furthest neighbors"). For color detail, please see color plate section.

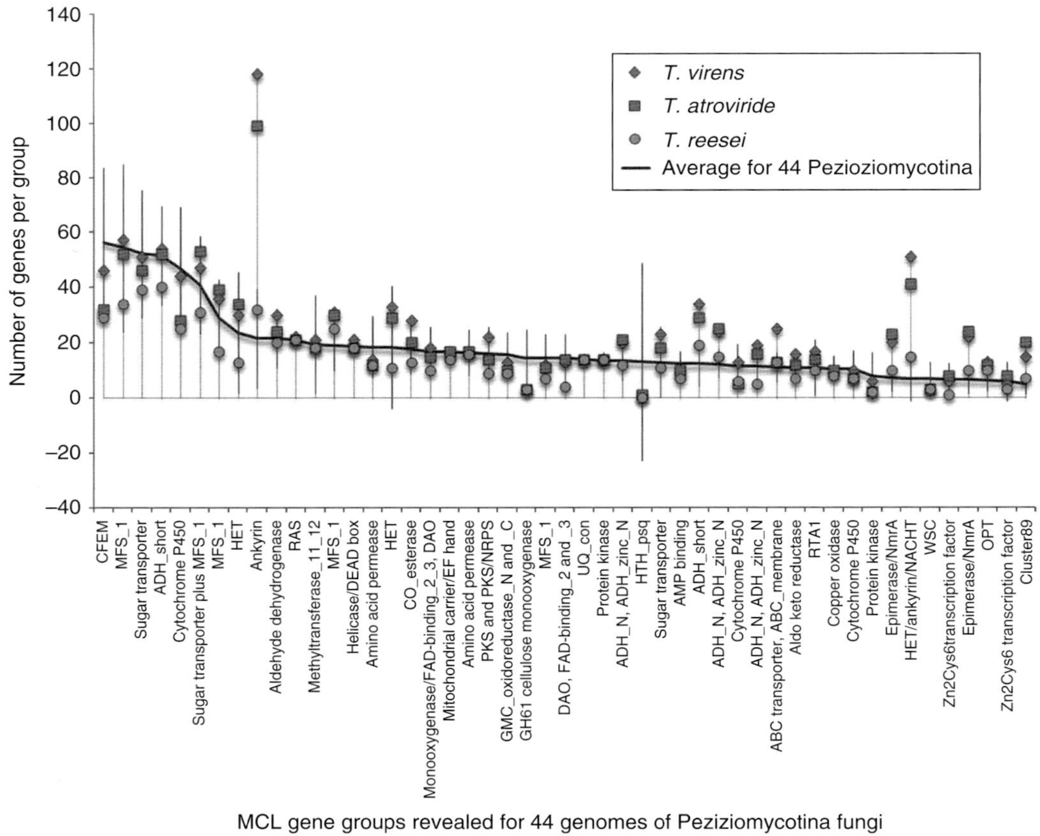

Figure 5.3 Gene numbers per cluster of orthologous genes revealed by MCL analysis for 44 Peziziomycotina fungi. Black solid vertical bars indicate the standard deviation value for the average for Peziziomycotina, $N=44$. For color detail, please see color plate section.

domain genes are part of the genetic systems that lead to recognition of and response to non-self during cell fusion between different individuals belonging to the same species. In case of sexual incompatibility (no mating possible) *het* genes specifically lead to the rejection of cospecific non-self: when fungi grow, their hyphae often fuse with each other (=anastomosis), which can occur also between genetically different isolates of the same species, leading to the formation of a heterokaryon. Fedorova, Badger, et al. (2005), however, have expanded the role of HET proteins to claim that the HET domain may represent a niche adaptation strategy of filamentous Ascomycetes to process a large number of similar stimuli associated with defense against pathogens, self/non-self recognition, differentiation, or analogous roles. In line with this, Paoletti and Saupe (2009) proposed that the *het* genes might also have a function in the recognition and response to non-self pathogens. Interestingly, bacterial pathogens of fungi have been shown to make use of this system, too; the necrotrophic bacterium *Pseudomonas syringae* (Gamma Proteobacteria) harbors a gene homologous to the *het-C* VI gene, which expression in *N. crassa* (Sordariales, Ascomycota) triggers—like *het-C*—a cell death reaction and is apparently used by *P. syringae* to induce cell death in the fungus to feed on it. It will be interesting to test whether some of the HET proteins are indeed involved in a similar way in mycoparasitism in *Trichoderma*.

The Secretome Reveals Strategies for Interaction of Trichoderma *with Its Environment*

Fungi, and any other organisms whose cells are surrounded by a rigid cell wall, have to secrete enzymes and proteins, which aid in the breakdown of nutrients and in the interaction with the biogenic and non-biogenic environment. The efficacy of this process has a strong link to successful competition with other (micro)organisms. Therefore, the inventory of secreted proteins may be informative about potential habitat adaptation. Ideally, a thorough and complete proteomic analysis of *Trichoderma* grown under various conditions relevant to its competition in nature (saprotrophic growth, interaction with other organisms) would form the basis for such an interpretation. Proteomic studies under some of these conditions have been published (Grinyer, Hunt, et al., 2005; Marra, Ambrosino, et al., 2006; Suarez, Sanz, et al., 2006), but they all were performed before the genome sequence of the investigated species was available, and peptide sequences were thus aligned by cross-species identification. Unfortunately, the peptide sequences have not been deposited, and therefore it is impossible to verify these analyses with the available genome sequences.

We have recently exploited an alternative approach, that is in silico identification of secreted proteins and analysis of their occurrence in the transcriptome during mycoparasitism (Atanasova, Le Crom, et al., 2012;

Druzhinina, Shelest, et al., 2012). This revealed 781 and 865 putative secreted proteins in *T. atroviride* and *T. virens*, respectively, of which 71 (from a total of 346 significantly upregulated genes) and 7 (from 75) were shown to be induced during confrontation with *Rhizoctonia solani*. The low number in *T. virens* is probably related to the fact that this fungus mainly responds by activating genes for secondary metabolism (Atanasova, Le Crom, et al., 2012). Proteases belonging to various families represented the largest group of induced proteins in *T. atroviride*, followed by small secreted cysteine-rich proteins. Interestingly, only 6 carbohydrate active enzymes were induced, and they were dominated by 4 GH16 β-1,3/1,4-glucanases. Enzymes with hydrolytic activity on other polymers (nucleases, lipases, and phosphatases) and with bound FAD were also present in the *T. atroviride* secretome but in lower numbers.

Interestingly, *T. atroviride* also induced a ferrooxidoreductase when confronted with *R. solani*, which may reflect an enhanced ability to compete for iron. In addition, both *T. atroviride* and *T. virens* formed an oxalate decarboxylase, which may aid *Trichoderma* in removing this toxic acid, produced by many fungi, and also aid iron acquisition that can easily be trapped as Fe^{3+}-trioxalate chelate.

Hydrolytic enzymes have traditionally been regarded as key features in mycoparasitism, because—whatever the mechanism of competition is—the necessity to (at least partially) degrade the cell wall of the prey is mandatory for feeding on it. This is indeed also reflected by an increased abundance of chitinolytic enzymes (composing most of the carbohydrate active enzymes (CAZyme) glycoside hydrolase (GH) family GH18 fungal proteins along with more rare endo-β-N-acetylglucosaminidases), GH75 chitosanases, and various β-1,3-glucanases (families GH17, GH55, GH64, and GH81) in *Trichoderma* relative to other fungi. The properties of these hydrolases have been described in detail in several recent papers and reviews (Seidl, 2008; Kubicek, Herrera-Estrella, et al., 2011).

Similar to plant pathogenic fungi (Gibson, King, et al. 2011), we have also observed an expansion of some plant cell wall degrading enzyme gene families (for review see Druzhinina, Shelest, et al., 2012). Overall, the CAZyme machinery also of the mycoparasitic species is compatible with a saprotrophic behavior. Of interest is the reduction in the set of enzymes involved in the degradation of pectin. An endopolygalacturonase gene from *T.* cf. *harzianum* T34 is required for root colonization, but it does not induce plant defense reactions (Moran-Diez, Hermosa, et al., 2009). A reduced activity on pectin may minimize plant defense reactions, and thereby aid to the interaction of *Trichoderma* with the plants.

As for the proteolytic enzymes, *Trichoderma* seems to possess one of the largest sets of proteases among fungi (as predicted by the MEROPS Batch Blast tool; Rawlings & Morton, 2008). Indeed, the numbers of predicted proteases are 3.75 percent and 3.85 percent of all predicted genes in *T. atroviride* and *T. virens*, of which about 20 percent are secreted. The proteases dominating this secretome have been recently reviewed (Druzhinina, Shelest, et al., 2012). The transcriptomic studies (Seidl, Seibel, et al., 2009; Atanasova, Le Crom, et al., 2012) showed that the attack of *R. solani* (as a model prey) not only leads to the expression of protease genes but also of genes-encoding transporters for oligopeptides and amino acids.

The small secreted cysteine-rich proteins (SSCPs)—together with unknown but conserved proteins—actually make up the largest group of proteins secreted by *Trichoderma* spp. Their definition is an Mr less than or equal to 300 amino acids in length and containing four or more cysteine residues (Kubicek, Herrera-Estrella, et al., 2011). An in-depth analysis in *Trichoderma* secreted shows that they fall into four groups: (1) hydrophobins and hydrophobin-like proteins; (2) elicitor-like proteins; (3) proteins with similarity to *T. virens* MRSP1 (Mukherjee, Hadar, et al., 2006); and (4) SSCPs for which no member with a known function has as yet been identified (Kubicek, Herrera-Estrella, et al., 2011). The latter contain a large number of orphan genes, and most of them are present only in a single *Trichoderma* species. Their properties have recently been described in detail (Druzhinina, Shelest, et al., 2012).

Trichoderma *Genes for Secondary Metabolites*

With respect to gene families commonly associated with secondary metabolite biosynthetic pathways, the three *Trichoderma* spp. contain a varying assortment of non-ribosomal peptide synthetases (NRPS) and polyketide synthases (PKS); *T. virens* comprises the highest number (50) because of the abundance of NRPS genes (28). A phylogenetic analysis showed that this was as a result of recent duplications of cyclodipeptide synthases, cyclosporin/enniatin synthase-like proteins, and NRPS-hybrid proteins (Kubicek, Herrera-Estrella, et al., 2011). Half of the genes present in *T. atroviride* and *T. virens* are unique for the respective species and occur within non-syntenic islands of the genome, indicating their origin by recent genome rearrangements, which is also reflected in a higher nucleotide dissimilarity (about 30 percent) than the average of genes between *T. atroviride* and *T. virens*. The genes-encoding enzymes that synthesize NRPS, PKS, and isoprenoid secondary metabolites have recently been reviewed (Mukherjee, Horwitz, et al., 2012).

The Molecular Biology of *Trichoderma* Mycotrophy

Mycoparasitism is a directed process, which can be divided into several stages: waiting for a prey ("ambushing"); recognition of the presence of a potential prey ("sensing"); induction of the biochemical tools to besiege the prey ("hunting"); and actual attack and eventual "killing" and feeding on

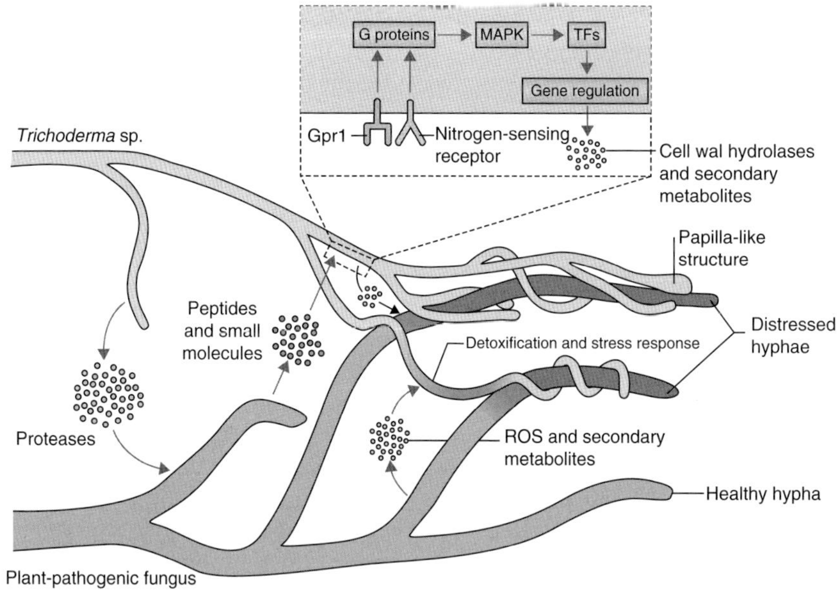

Figure 5.4 Mycoparasitism of *Trichoderma* spp. within the soil community. *Trichoderma* spp. recognize a plant-pathogenic fungus (a prey) via small molecules that are released by the pathogen; some of these molecules may be peptides that are released by the action of proteases secreted by the *Trichoderma* sp. before contact. These molecules may bind to G protein-coupled receptors (such as Gpr1) or nitrogen-sensing receptors on the surface of the *Trichoderma* sp. hyphae, thereby eliciting a signaling cascade, comprising G proteins and mitogen-activated protein kinases (MAPKs), which may ultimately modulate the activities of as-yet-unknown transcription factors (TFs). These factors then enhance the constitutive expression of genes that encode enzymes for the biosynthesis of secondary metabolites and for cell wall lysis. Lectins from the pathogenic fungus and proteins harboring cellulose-binding modules from hyphae of *Trichoderma* spp. may collaborate in the attachment of the predator to the prey. At the same time, the prey responds by forming secondary metabolites and reactive oxygen species (ROS) that elicit a stress response and detoxification in *Trichoderma* spp. (Figure from Druzhinina IS, Seidl-Seiboth V, et al. 2011. *Trichoderma*: The genomics of opportunistic success. *Nat Rev Microbiol.* 16:749–759.) For color detail, please see color plate section.

the prey. It is, therefore, useful to group the description according to these lines. The generalized summary of all these processes is shown in Figure 5.4.

Ambushing

The transition from the commensalism to parasitic state necessitates the molecular dissection of traits responsible for both interactions. The availability of the genome sequence of *T. atroviride* and *T. virens* has enabled full genome arrays to be used to study the sequential events occurring during confrontation of these *Trichoderma* spp. with *R. solani* at a genome-wide transcriptomic level (Atanasova, Le Crom, et al., 2012). This study revealed that both *T. atroviride* and *T. virens* reacted to the presence of *R. solani* already before physical contact. Yet they showed essentially different behavior: *T. virens* only overexpressed 78 genes, of which those involved in gliotoxin biosynthesis and its precursor metabolites accounted for the largest group. *T. atroviride*, in contrast, overexpressed 400 genes, which were enriched PTH11-G-protein coupled receptors, lectins and β-glucanases, small secreted cysteine-rich proteins, and secondary metabolite synthases.

However, there were also common responses shared by *T. atroviride* and *T. virens*: one was the overexpression of a high number of genes for proteolytic enzymes and oligopeptide transporters, which is consistent with the findings that the overexpression of the alkaline protease gene *prb1* enhances the mycoparasitic ability (Flores, Chet, et al., 1997). Seidl, Seibel, et al., (2009) hypothesized that the receptors, which sense the nitrogen status of the medium, are modulated by components derived from the host fungus and thereby mimic nitrogen limitation. Such a mechanism would be reminiscent of nematophagous fungi, where trapping of the prey is induced by oligopeptides from the nematodes (Dijksterhuis, Veenhuis, et al., 1994).

Another event, common to both *T. atroviride* and *T. virens*, is the induction of genes of the heat shock response such as HSP23, HSP70, HSP90, and HSP104, genes of oxidative stress response (cytochrome C peroxidase, proline oxidase, and ER-bound glutathione-S-transferases), and genes for detoxification processes (ABC efflux transporters, the pleiotropic drug resistance (PDR) transporters, and the multidrug resistance MDR-type transporters). *R. solani* has been shown to use radical oxygen species as signaling molecules during sclerotia formation (Papapostolou & Georgiou, 2010) and excrete antifungal components (Aliferis & Jabaji, 2010), both of which may have elicited this response. An ABC-transporter from *T. atroviride* (TAABC2) has already been shown to be involved in biocontrol of *R. solani* (Ruocco, Lanzuise, et al., 2009).

Sensing

The notion of a prey-specific expression pattern in *Trichoderma* and the observation that species such as *T. atroviride* display directed growth toward the prey suggests an efficient sensing mechanism. There is some evidence for the involvement of G-protein coupled receptors (GPCRs) in the process. The *T. atroviride* GPR1 (ID 160995) that belongs to the class of cAMP receptor-like proteins, is involved in coiling and expression of some chitinases (Omann, Lehner, et al. 2012).

T. atroviride and *T. virens* also contain a large number of PTH11-like G-protein coupled receptors, which were first described in *Magnaporthe grisea*

(Magnaporthales, Ascomycota) to be required for appressorium development and pathogenicity (De Zwaan, et al. 1999), and which are restricted to Pezizomycotina and represent the largest group of GPCRs there (Kulkarni, Thon, et al., 2005). As mentioned previously, these receptor genes were enriched among the genes that were upregulated during confrontation of *T. atroviride* with *R. solani* (Atanasova, Le Crom, et al., 2012). As for the G-proteins, *Trichoderma* contain the conserved signaling cascades comprising three G-protein α-subunits, one G β-subunit, one G γ-subunit, an adenylate cyclase, and three MAP kinases (Fig. 5.4, Kubicek, Herrera-Estrella, et al., 2011). The role of G-protein signaling has been reviewed by Omann and Zeilinger (2010).

Hunting

Recognition and attachment to the host hyphae is the first essential step in the contact with the prey, although the observed morphological changes depend strongly on the host fungus. Lu, Tombolini, et al. (2004) used a *T. atroviride* strain carrying a green fluorescent protein under a constitutive promoter to study the necrotrophic parasitic interaction between the Oomycete *Pythium ultimum* (Heterokontophyta) and the Basidiomycete *R. solani*. Growing alongside the host hyphae and formation of papillae-like structures were observed as the most common events. These authors further showed that the hyphae of *T. atroviride* also frequently branched toward the host, suggesting an active, probably chemotactic response to its presence. The formation of helix-shaped hyphae ("coiling")—a morphological response that has most frequently been associated with mycoparasitic attack—was however not only observed during contact with the host but also in its absence. Coiling around the host has been linked to lectin-type interactions between *Trichoderma* and the prey fungus (Inbar & Chet, 1996). All three *Trichoderma* spp. contain an arsenal of genes encoding proteins with lectin-domains, and the C-type lectins are particularly abundant and overexpressed before and at contact in *T. atroviride* (Atanasova, Le Crom, et al., 2012). However, *T. atroviride* and *T. virens* also contain proteins consisting of carbohydrate-binding (CBM13) modules that resemble those of plant lectins such as ricin (Notenboom, Boraston, et al., 2002). Interestingly, *T. atroviride* also induced a gene consisting of a cyanovirin domain, a mannose-binding lectin (Xiong, Fan, et al., 2010) during contact with *R. solani*.

Killing

Despite the wealth of information on genes that contribute to the mycoparasitic activity of *Trichoderma*, only a little is known about the molecules that are actually used to kill the prey. This is likely also the result of the fact that different strains and species use several strategies for this purpose. Secondary metabolites of *Trichoderma* are generally believed to play a role in this process. However, functional genetic evidence is still lacking, and in vitro data may be misleading; as an example, the peptaibols can act synergistically with secreted hydrolytic enzymes to promote ingress into pathogen structures, suggesting a role in antagonistic actions against plant pathogens (Schirmboeck, Lorito, et al., 1994). However, both NRPS and PKS encoding genes are down-regulated during confrontation of *T. atroviride* and *R. solani* (Atanasova, Le Crom, et al., 2012) and knock-out mutants are not affected in their mycoparasitic abilities (Seidl-Seiboth & Kubicek, unpublished data). Reverse genetic data for an involvement in antagonism have so far only been obtained for trichodermin (in *T. brevicompactum*; Tijerino et al. 2011) and gliotoxin (Atanasova, Le Crom, et al., 2012; Mukherjee, Buensanteai, et al., 2012).

***Trichoderma* Species in Rhizosphere: Opportunists and Commensals**

Recognition of the Plant

Trichoderma spp. have been known for decades to be "rhizosphere competent," that is, they grow and develop within the plant rhizosphere, thereby eventually antagonizing other pathogenic microorganisms (Lewis & Papavizas, 1984). This interaction of *Trichoderma* spp. with living plants usually does not cause disease, which led to the consideration of *Trichoderma* as an opportunistic symbiont (Harman, Howell, et al., 2004).

The ability of *Trichoderma* to be rhizosphere-competent likely depends both on an appropriate gene inventory for attaching and eventually penetrating the roots. *Trichoderma*'s affinity to the rhizosphere can be explained by two of its nutritional preferences. First, the roots of 92 percent of land plants are establishing a mutualistic symbiosis with mycorrhizal fungi, which represents an attractive ground for a mycotroph. In fact, mycoparasitic attack of arbuscular mycorrhizal fungi by *Trichoderma* and inhibition of the proliferation of mycorrhizal fungi have been reported (Werner, Zadworny, et al., 2003). Second, the roots and especially root tips are covered by a gel-like slimy capsule ("mucigel"), which is composed of highly hydrated polysaccharides (pectins and hemicelluloses, particularly rhamnogalacturonans and arabinoxylans). *Trichoderma* spp. have an expanded arsenal of genes-encoding secreted enzymes for degradation of the latter (Druzhinina, Shelest, et al., 2012). In addition, a successful establishment of *T.* cf. *harzianum* CECT 2413 in the tomato (*Solanum lycopersicum*) rhizosphere was shown to require the expression of an endopolygalacturonase gene (Moran-Diez, Hermosa, et al., 2009).

In addition, mono- and disaccharides excreted by plant roots into the rhizosphere are known to provide an important carbon substrate for mycorrhizae

(Nehls, Göhringer, et al., 2010). A similar role for sucrose has recently been demonstrated for the establishment of *T. virens* in the rhizosphere (Vargas, Mandawe, et al., 2009). As the genomes of *T. atroviride, T. virens* and *T. reesei* contain genes for intracellular invertases but not for extracellular invertases, sucrose must be taken up by a sucrose permease before being hydrolyzed. *T. virens* contains a highly specific sucrose transporter that is induced in the early stages of root colonization with biochemical properties similar to plant-encoded sucrose carriers (Vargas, Crutcher, et al., 2011), which suggests an active sucrose transfer from plant to fungus. However, sucrose hydrolysis is not an essential trait for rhizosphere competence because a *T. virens* knock-out mutant in the respective invertase gene could still colonize roots (Vargas, Crutcher, et al., 2011). In addition, the genomes of *T. atroviride* and *T. virens* encode a great number of major facilitator solute transporters (Kubicek, Herrera-Estrella, et al., 2011), whose role in acquisition of other root exudates still waits testing.

Other molecules that contribute to rhizosphere competence could be hydrophobins: a hydrophobin from *T. asperelloides* is essential for penetration of roots (Viterbo & Chet, 2006). Also, the availability of mechanisms to detoxify inhibitory chemicals produced by the plant is important for establishment in the rhizosphere. To this end, the *Trichoderma* genomes possess a large number of ABC transporters (maximum being in *T. virens*) that might aid in establishment of these species in the rhizosphere.

Plant Response to Trichoderma

Plants, as all other organisms, have developed mechanisms to monitor potential hazard in their environment. Contact with a pathogen causes the so-called systemic acquired resistance (SAR), which renders non-infected plant tissues more resistant to subsequent pathogen attack and is characterized by increased levels of salicylic acid and the coordinate activation of a specific set of pathogenesis-related genes, many of which encode PR proteins with antimicrobial activity (Van Loon, Rep, et al., 2006). The presence of non-pathogenic organisms, such as rhizobacteria or *Trichoderma*, however, triggers an induced systemic resistance (ISR) in the plants. This is different from SAR and is mediated by the jasmonate signaling pathway (Van der Ent, Verhagen, et al., 2008). Induction of the ISR response starts with the recognition of pathogen- or microbe-associated molecular patterns (PAMPs or MAMPs, respectively; Schwessinger & Zipfel, 2008), by pattern recognition receptor of the plant. This subsequently activates a primary defense (Van der Ent, Van Wees, et al., 2009). In accordance with these concepts, *Trichoderma* induces the jasmonic acid pathway of plant defense. Using *Arabidopsis thaliana* microarrays, Mathys, De Cremer, et al. (2012) further showed that *T. hamatum* mainly induces the phenylpropanoid pathway in ISR.

Whether the ISR by *Trichoderma* indicates a kind of symbiotic relationship with the plant or is simply the consequence of the likely situation that the MAMP receptors of plants also recognize orthologous proteins from nonpathogenic microbes is in debate (cf. Harman, Howell, et al., 2004; Druzhinina, Seidl-Seiboth, et al., 2011). *Trichoderma* molecules that have been shown to trigger ISR include secreted xylanases, cellulases, and the cellulose-binding protein swollenin (see Shoresh, Harman, et al. 2010), small cysteine-rich secreted protein (Djonovic, Pozo, et al. 2006; Djonovic, Vargas, et al., 2007), peptaibols like in *T. reesei* (erroneously published as "*T. viride*") alamethicin and TEX1 (*T. virens*) (Viterbo, Wiest, et al., 2007; Leitgeb, Szekeres, et al., 2007), and an unknown PKS-NRPS product (Mukherjee, Horwitz, et al., 2012b). In all these cases, knock-out in the respective genes did not impair the ability of *Trichoderma* to colonize the roots, although the ISR was abolished in most cases. Thus, *Trichoderma* does not seem to benefit from the plants response on a first glance.

However, *T. virens* produced the auxin-related compounds indole-3-acetic acid, indole-3-acetaldehyde, and indole-3-ethanol (Contreras-Cornejo, Macias-Rodriguez, et al., 2009). These compounds increased biomass production and growth of lateral roots of *A. thaliana*, and the effect was not observed in plant mutants in the auxin response pathways. Similarly, *T. asperellum* expresses a 1-aminocyclopropane-1-carboxylate (ACC) deaminase during interaction with roots of canola (*Brassica napus*), and strains in which this gene have been inactivated showed decreased ability to promote root elongation (Viterbo, Landau, et al., 2010). This gene is also present in *T. atroviride* and *T. virens* and can form the precursor of biosynthesis of the plant growth regulator ethylene. It is speculated that the stimulation of root growth aids some *Trichoderma* species to develop in the rhizosphere.

Endophytism

In addition to being rhizosphere colonizers, several *Trichoderma* taxa (including some novel species) are reported to live inside the plants as endophytes, offering a wide range of benefits to plants, such as growth promotion, delaying onset of drought stress, and inhibition of pathogens (Bailey, Bae, et al., 2006; Jaklitsch, Samuels, et al., 2006; Samuels, Suarez, et al., 2006; Tejesvi, Mahesh, et al. 2006; Hanada, de Jorge Souza, et al., 2008; Bae, Sicher, et al., 2009; Hanada, Pomella, et al., 2010; Samuels & Ismaiel, 2009). Almost all of the isolated endophytes comprise new taxa and—with the exception of *Hypocrea stilbohypoxili* and *Hypocrea stromatica*—have no known teleomorphs. It has recently been discussed (Druzhinina, Seidl-Seiboth, et al., 2011) that mycotrophs may have become endophytes by entering the plant roots by parasitizing hyphae of mycorrhizal fungi colonizing the plant roots as described by de Jaeger et al. (2010). No genomes from *Trichoderma* strains that were isolated as endophytes have yet been sequenced.

Trichoderma Interaction with Other Organisms

Trichoderma spp. have also been reported to undergo various types of interactions with other organisms, including invertebrates and mammals. Some species of *Trichoderma* spp. are also known to successfully antagonize and kill plant parasitic nematodes, which offers new not yet fully explored possibilities to combat these agricultural pests. *Trichoderma* recently also joined the emerging list of such opportunistic pathogens that cause invasive mycoses of mammals, including humans with impaired immune systems. So far only two species—*T. longibrachiatum* and *H. orientalis*—infect patients who are immunocompromised, but these two closely related species share identical multilocus haplotypes with isolates from soil and plant materials (Druzhinina, Komon-Zelazowska, et al., 2008). *Trichoderma* invasive mycoses may therefore be potentially nosocomial. Both subjects have recently been discussed in some detail (Druzhinina, Seidl-Seiboth, et al., 2011) and shall not be repeated here. The mechanisms of these interactions have not been studied yet, but the enhanced arsenal of proteases (as described previously) may play an important role in this trait.

Tasks and Questions for the Future

As has been shown, *Trichoderma* serves mankind in various ways, by acting as one of the most important organisms for biotechnology and a versatile biofungicide and biofertilizing agent. It may also affect mushroom cultivations and be a pathogen of humans. Yet the understanding of this important genus is limited by the small number of species for which genome sequences have become available. In fact, the three species that have been sequenced and annotated (*T. atroviride* and *T. virens* and *T. reesei*) and the two whose genome sequence was just recently released (*TT. asperellum* and *harzianum*) reflect the commercial interest by biotechnology and agriculture only. Analysis of the sequence of *T. longibrachiatum* (http://genome.jgi-psf.org/Trilo1/Trilo1.home.html) will provide an interesting complement to the other five. It will help to answer the question whether the ability to interact with mammals is due to the gain or loss of genes or a change in their regulation. In addition, *T. longibrachiatum* is one of the few *Trichoderma* spp. capable of growing at moderately high temperatures (104° F [40° C]), and its comparison to the phylogenetically close *T. reesei* may reveal genes involved in tolerance to increased temperatures.

Besides this, the sequencing of the genomes of some of the species that have so far only been detected as endophytes would help to understand the mechanisms that have driven *Trichoderma* to adapt to this lifestyle.

However, even the potential of the existing genome sequences has not been fully exhausted: the genomes of *T. atroviride* and *T. virens* contain a large number of genes putatively encoding oxidative enzymes (cytochrome P450 monooxygenases, FAD-linked oxidases/monooxygenases) methyltransferases, esterases, and transcription factors that occur in clusters in the genome. They likely encode the machinery for synthesis of unknown secondary metabolites. In addition, the high number of unknown and orphan genes requests a systematic approach to investigate their function, which, however, needs a sufficiently large community to collaborate.

Finally, the *Trichoderma* genomes may still bear secrets that could tell a story of *Trichoderma* evolution. When annotating the presently sequenced *Trichoderma* genomes, a number of genes have their closest neighbor only in certain soil bacteria. This raises the possibility of operation of horizontal gene transfer, whose investigation appears a challenging topic for the future.

Acknowledgments

Genome sequencing and analysis was supported by the Office of Science of the US Department of Energy under contract number DE-AC02-05CH11231. The authors' own work on this topic was supported by grants from the Austrian Science Fund P-17895 to I. S. D.

Note

1 In this review, we accommodate the changes proposed at the International Botanical Congress in July 2011 for the International Code of Botanical Nomenclature and the ongoing discussion on the future single taxon name for *Hypocrea/Trichoderma* that may be followed at the website of the IUMS International Subcommission on *Trichoderma* taxonomy at http://www.isth.info/content.php?page_id=102. Therefore we use the single generic name *Trichoderma* not only for asexual species but also for holomorphs when the sexual stage is described. However at first mention of holomorphic species both teleomorph (*Hypocrea*) and anamorph (*Trichoderma*) names are given. When the whole genus of *Trichoderma* and *Hypocrea* spp. is considered, the term *Trichoderma* is applied.

References

Aliferis KA & Jabaji S. 2010. Metabolite composition and bioactivity of *Rhizoctonia solani* sclerotial exudates. *J Agric Food Chem*. 58: 7604–7615.

Aly AH, Debbab A, et al. 2011. Fungal endophytes: Unique plant inhabitants with great promises. *Appl Microbiol Biotechnol*. 90:1829–1845.

Atanasova L, Jaklitsch WM, et al. 2010. The clonal species *Trichoderma parareesei* sp. nov., likely resembles the ancestor of the cellulase producer *Hypocrea jecorina/T. reesei*. *Appl Environ Microbiol*. 76: 7259–7267.

Atanasova L, Le Crom, et al. 2013. Comparative transcriptomics reveals versatile strategies of *Trichoderma* mycoparasitism. *BMC Genomics*. 14:121.

Bae H, Sicher RC, et al. 2009. The beneficial endophyte *Trichoderma hamatum* isolate DIS 219b promotes growth and delays the onset of the drought response in *Theobroma cacao*. *J Experiment Bot*. 60: 3279–3295.

Bailey BA, Bae H, et al. 2006. Fungal and plant gene expression during the colonization of cacao seedlings by endophytic isolates of four *Trichoderma* species. *Planta*. 224: 1449–1464.

Bouchon D, Rigaud T, et al. 1998. Evidence for widespread *Wolbachia* infection in isopod crustaceans: Molecular identification and host feminization. *Proc Biol Sci*. 265: 1081–1090.

Breeuwer JA & Jacobs G. 1996. *Wolbachia*: Intracellular manipulators of mite reproduction. *Exp Appl Acarol*. 20: 421–434

Buée M, Reich M, et al. 2009. 454 Pyrosequencing analyses of forest soils reveal an unexpectedly high fungal diversity. *New Phytol*. 184: 449–456.

Castle A, Speranzini D, et al. 1998. Morphological and molecular identification of *Trichoderma* isolates on North American mushroom farms. *Appl Environ Microbiol*. 64: 133–137.

Contreras-Cornejo HA, Macias-Rodriguez L, et al. 2009. *Trichoderma virens*, a plant beneficial fungus, enhances biomass production and promotes lateral root growth through an auxin-dependent mechanism in *Arabidopsis*. *Plant Physiol*. 149: 1579–1592.

Cutler HG, Cutler SJ, et al. 1999. Koninginin G, a new metabolite from *Trichoderma aureoviride*. *J Nat Prod*. 62: 137–139.

Dababat AA, Sikora RA, et al. 2006. Use of *Trichoderma harzianum* and *Trichoderma viride* for the biological control of Meloidogyne incognita on tomato. *Commun Agric Appl Biol Sci*. 71: 953–961.

Danielson R & Davey C. 1973. The abundance of *Trichoderma* propagules and the distribution of species in forest soils. *Soil Biol Biochem*. 5: 485–494.

De Jaeger N, Declerck S, et al. 2010. Mycoparasitism of arbuscular mycorrhizal fungi: a pathway for the entry of saprotrophic fungi into roots. *FEMS Microb Ecol*. 73: 312–322.

De Zwaan TM, Carroll AM, et al. 1999. *Magnaporthe grisea* pth11p is a novel plasma membrane protein that mediates appressorium differentiation in response to inductive substrate cues. *Plant Cell*. 11:2013–2030.

Dijksterhuis J, Veenhuis M, et al. 1994. Nematophagous fungi: Physiological aspects and structure-function relationships. *Adv Microb Physiol*. 36: 111–143.

Djonovic S, Pozo MJ, et al. 2006. Sm1, a proteinaceous elicitor secreted by the biocontrol fungus *Trichoderma virens* induces plant defense responses and systemic resistance. *Mol Plant Microbe Interact*. 19: 838–853.

Djonovic S, Vargas WA, et al. 2007. A proteinaceous elicitor Sm1 from the beneficial fungus *Trichoderma virens* is required for induced systemic resistance in maize. *Plant Physiol*. 145: 875–889.

Druzhinina IS, Komon-Zelazowska M, et al. 2008. Different reproductive strategies of *Hypocrea orientalis* and genetically close but clonal *Trichoderma longibrachiatum*, both capable to cause invasive mycoses of humans. *Microbiology*. 154: 3447–3459.

Druzhinina I, Koptchinski A, et al. 2005. An oligonucleotide barcode for species identification in *Trichoderma* and *Hypocrea*. *Fungal Genet Biol*. 42: 813–828.

Druzhinina IS, Koptchinskiy A, et al. 2006. The first one hundred *Trichoderma* species characterized by molecular data. *Mycoscience*. 47: 55–64.

Druzhinina IS, Kubicek CP, et al. 2010. The *Trichoderma harzianum* demon: Complex speciation history resulting in coexistence of hypothetical biological species, recent agamospecies and numerous relict lineages. *BMC Evolution Biol*. 10: 94.

Druzhinina IS, Seidl-Seiboth V, et al. 2011. *Trichoderma*: The genomics of opportunistic success. *Nat Rev Microbiol.* 16:749–759.

Druzhinina IS, Shelest E, et al. 2012. Novel traits of Trichoderma predicted through the analysis of its secretome (invited minireview). *FEMS Microbiol Lett.* 337(1): 1–9.

Fedorova ND, Badger JH, et al. 2005. Comparative analysis of programmed cell death pathways in filamentous fungi. *BMC Genomics.* 6: 177.

Flores A, Chet I., et al. 1997. Improved biocontrol activity of *Trichoderma harzianum* by over-expression of the proteinase-encoding gene *prb1*. *Curr Genet.* 31: 30–37.

Friedl MA & Druzhinina IS. 2012. Taxon-specific metagenomics of *Trichoderma* reveals a narrow community of opportunistic species that regulate each other's development. *Microbiology.* 158: 69–83.

Gal-Hemed I, Atanasova L, et al. 2011. Marine isolates of *Trichoderma* spp. as potential halotolerant agents of biological control for arid-zone agriculture. *Appl Environ Microbiol.* 77: 5100–5109.

Gautheret A, Dromer F, et al. 1995. *Trichoderma pseudokoningii* as a cause of fatal infection in a bone marrow transplant recipient. *Clinic Infect Dis.* 20: 1063–1064.

Gibson DM, King BC, et al. 2011. Plant pathogens as a source of diverse enzymes for lignocellulose digestion. *Curr Opin Microbiol.* 14: 264–270.

Goswami J, Pandey RK, et al. 2008. Management of root knot nematode on tomato through application of fungal antagonists, *Acremonium strictum* and *Trichoderma harzianum. J Environ Sci Health B.* 43: 237–240.

Grigoriev IV, Nordberg H, et al. 2011. The Genome Portal of the Department of Energy Joint Genome Institute. *Nucl Acids Res.* 40(Database issue): D26–D32.

Grinyer J, Hunt S, et al. 2005. Proteomic response of the biological control fungus *Trichoderma atroviride* to growth on the cell walls of *Rhizoctonia solani. Curr Genet.* 47:381–388.

Hagn A, Wallisch S, et al. 2007. A new cultivation independent approach to detect and monitor common *Trichoderma* species in soils. *J Microbiologic Methods.* 69: 86–92.

Hanada RE, de Jorge Souza T, et al. 2008. *Trichoderma martiale* sp. nov., a new endophyte from sapwood of *Theobroma cacao* with a potential for biological control. *Mycol Res.* 112: 1335–1343.

Hanada RE, Pomella AW, et al. 2010. Endophytic fungal diversity in *Theobroma cacao (cacao)* and *T. grandiflorum* (cupuaçu) trees and their potential for growth promotion and biocontrol of black-pod disease. *Fungal Biol.* 114: 901–910.

Harman GE, Howell CR, et al. 2004. *Trichoderma* species—opportunistic, avirulent plant symbionts. *Nat Rev Microbiol.* 2: 43–56.

Hatvani L, Antal Z, et al. 2007. Green Mmold diseases of *Agaricus* and *Pleurotus* are caused by related but phylogenetically different *Trichoderma* species. *Phytopathology.* 97: 532–537.

Hawksworth DL 1991. The fungal dimension of biodiversity: magnitude, significance, and conservation. *Mycol Res.* 95: 641–655.

Hjeljord L & Tronsmo A. 1998. *Trichoderma* and *Gliocladium* in biological control: An overview. In Trichoderma and Gliocladium, vol 2. *Enzymes, biological control and commercial applications* (eds. GE Harman & CP Kubicek), 131–151. London: Taylor and Francis.

Humphris SN, Bruce A, et al. 2002. The effects of volatile microbial secondary metabolites on protein synthesis in *Serpula lacrymans. FEMS Microbiol Lett.* 210:215–219.

Inbar J & Chet I. 1996. The role of lectins in recognition and adhesion of the mycoparasitic fungus *Trichoderma* spp. to its host. *Adv Experiment Med Biol.* 408: 229–231.

Jaklitsch WM. 2009. European species of *Hypocrea* Part I. *The green-spored species. Stud Mycol.* 63: 1–91.

Jaklitsch WM. 2011. European species of *Hypocrea* part II: Species with hyaline ascospores. *Fungal Divers.* 48: 1–250.

Jaklitsch WM, Samuels GJ, et al. 2006a. *Hypocrea rufa/Trichoderma viride*: a reassessment, and description of five closely related species with and without warted conidia. *Stud Mycol.* 56: 135–177.

Jaklitsch WM, Samuels GJ, et al. 2006b. *Hypocrea rufa/Trichoderma viride*: a reassessment, and description of five closely related species with and without warted conidia. *Stud Mycol.* 56: 135–177.

Jeyaprakash A & Hoy MA. 2000. Long PCR improves *Wolbachia* DNA amplification: *wsp* sequences found in 76% of sixty-three arthropod species. *Insect Mol Biol.* 9: 393–405.

Kim CS, Shirouzu T, et al. 2012. *Trichoderma mienum* sp. nov., isolated from mushroom farms in Japan. *Antonie Van Leeuwenhoek.* 102(4): 629–641.

Komon-Zelazowska M, Bissett J, et al. 2007. Genetically closely related but phenotypically divergent *Trichoderma* species cause world-wide green mould disease in oyster mushroom farms. *Appl Environment Microbiol.* 73: 7415–7426.

Koptchinski A, Komon M, et al. 2005. *Tricho*BLAST: A multiloci database of phylogenetic markers for *Trichoderma* and *Hypocrea* powered by sequence diagnosis and similarity search tools. *Mycol Res.* 109: 658–660.

Kovacs K, Szakacs G, et al. 2004. Production of chitinolytic enzymes with *Trichoderma longibrachiatum* IMI 92027 in solid substrate fermentation. *Appl Biochem Biotechnol.* 118: 189–204.

Kredics L, Antal Z, et al. 2003. Clinical importance of the genus *Trichoderma. A review. Acta Microbiologica et Immunologica Hungarum.* 50:105–117

Kredics L, Kocsubé S, et al. 2009. Molecular diagnosis of *Trichoderma* species associated with *Pleurotus ostreatus* and its natural substrate. *FEMS Microbiol Lett.* 300: 58–67.

Kubicek CP. 2013. Systems biological approaches towards understanding cellulase production by *Trichoderma reesei. J Biotechnol.* 163(2):133–142.

Kubicek CP, Herrera-Estrella A, et al. 2011. Comparative genome sequence analysis underscores mycoparasitism as the ancestral life style of *Trichoderma. Genome Biol.* 12: R40.

Kubicek CP, Mikus M, et al. 2009. Metabolic engineering strategies for the improvement of cellulase production by *Hypocrea jecorina. Biotechnol Biofuels.* 2: 19.

Kubicek CP & Penttilä ME. 1998. Regulation of production of plant polysaccharide degrading enzymes by *Trichoderma.* In *Trichoderma* and *Gliocladium*, Vol. 2: *Enzymes, Biological Control and Commercial Applications* (eds. GE Harman & CP Kubicek), 49–71. London: Taylor and Francis Ltd.

Kulkarni RD, Thon MR, et al. 2005. Novel G-protein-coupled receptor-like proteins in the plant pathogenic fungus *Magnaporthe grisea. Genome Biol.* 6: R24.

Kullnig CM, Krupica T, et al. 2001. Confusion abounds over identity of *Trichoderma* biocontrol isolates. *Mycol Res.* 105: 769–772.

Kumar R, Singh S, et al. 2008. Bioconversion of lignocellulosic biomass: Biochemical and molecular perspectives. *J Ind Microbiol Biotechnol.* 35: 377–391.

Kyalo G, Affokpon A, et al. 2007. Biological control effects of *Pochonia chlamydosporia* and *Trichoderma* isolates from Benin (West-Africa) on root-knot nematodes. *Commun Agric Appl Biol Sci.* 72: 219–223.

Leitgeb B, Szekeres A, et al. 2007. The history of alamethicin: A review of the most extensively studied peptaibol. *Chem Biodivers.* 4: 1027–1051.

Lewis JA & Papavizas GC. 1984. A new approach to stimulate population proliferation of *Trichoderma* species and other potential biocontrol fungi introduced into natural soils. *Phytopathology.* 74: 1240–1244.

Lim YW, Kim BK, et al. 2010. Assessment of soil fungal communities using pyrosequencing. *J Microbiol.* 48: 284–289.

Lu Z, Tombolini R, et al. 2004. In vivo study of *Trichoderma*-pathogen-plant interactions, using constitutive and inducible green fluorescent protein reporter systems. *Appl Environ Microbiol.* 70: 3073–3081.

Marra R, Ambrosino P, et al. 2006. Study of the three-way interaction between *Trichoderma atroviride*, plant and fungal pathogens by using a proteomic approach. *Curr Genet.* 50:307–321

Martinez D, Berka RM, et al. 2008. Genome sequence analysis of the cellulolytic fungus *Trichoderma reesei* (syn. *Hypocrea jecorina*) reveals a surprisingly limited inventory of carbohydrate active enzymes. *Nat Biotechnol.* 26: 553–560.

Mathys J, De Cremer K, et al. 2012. Genome-wide characterization of ISR induced in arabidopsis thaliana by *Trichoderma hamatum* T382 against *Botrytis cinerea* infection. *Front Plant Sci.* 3:108.

Meincke R, Weinert N, et al. 2010.Development of a molecular approach to describe the composition of *Trichoderma* communities. *J Microbiol Method.* 80: 63–69.

Migheli Q, Balmas V, et al. 2009. Soils of a Mediterranean hotspot of biodiversity and endemism (Sardinia, Tyrrhenian Islands) are inhabited by pan-European and likely invasive species of *Hypocrea/Trichoderma*. *Environ Microbiol.* 11: 35–46.

Moran-Diez E, Hermosa R, et al. 2009. The ThPG1 endopolygalacturonase is required for the *Trichoderma harzianum*-plant beneficial interaction. *Mol Plant Microbe Inter.* 22: 1021–1031.

Mosavi LK, Cammett TJ, et al. 2004. The ankyrin repeat as molecular architecture for protein recognition. *Protein Sci.* 13: 1435–1448.

Mukherjee PK, Buensanteai N, et al. 2012. Functional analysis of non-ribosomal peptide synthetases (NRPSs) in *Trichoderma virens* reveals a polyketide synthase (PKS)/NRPS hybrid enzyme involved in the induced systemic resistance response in maize. *Microbiology.* 158:155–165.

Mukherjee PK, Hadar R, et al. 2006. MRSP1, encoding a novel *Trichoderma* secreted protein, is negatively regulated by MAPK. *Biochem Biophysic Res Commun.* 350: 716–722.

Mukherjee PK, Horwitz BA, et al. 2012b. Secondary metabolism in *Trichoderma*—a genomic perspective. *Microbiology.* 158: 35–45.

Muthumeenakshi S, Mills PR, et al. 1994. Intraspecific molecular variation among *Tricho-derma harzianum* isolates colonizing mushroom compost in the British Isles. *Microbiology.* 140: 769–777.

National Center for Biotechnology Information. 2012. Taxonomy site guide. Accessed June 22, 2012, at http://www.ncbi.nlm.nih.gov/guide/taxonomy/.

Nehls U, Göhringer F, et al. 2010. Fungal carbohydrate support in the ectomycorrhizal symbiosis: A review. *Plant Biol (Stuttg).* 12: 292–301.

Notenboom V, Boraston AB, 2002. High-resolution crystal structures of the lectin-like xylan binding domain from *Streptomyces lividans* xylanase 10A with bound substrates reveal a novel mode of xylan binding. *Biochemistry.* 41: 4246–4254.

Omann M & Zeilinger S. 2010. How a mycoparasite employs g-protein signaling: Using the example of *Trichoderma*. *J Signal Transduct.* 2010: 123126.

Omann MR, Lehner S, et al. 2012. The seven-transmembrane receptor Gpr1 governs processes relevant for the antagonistic interaction of *Trichoderma atroviride* with its host. *Microbiology.* 158: 107–118.

Ospina-Giraldo MD, Royse DJ, et al. 1999. Molecular phylogenetic analysis of biological control strains of *Trichoderma harzianum* and other biotypes of *Trichoderma* spp. associated with mushroom green mold. *Phytopathology.* 89: 308–313.

Paoletti M & Saupe SJ. 2009. Fungal incompatibility: Evolutionary origin in pathogen defense? *Bioessays.* 31: 1201–1210.

Papapostolou I & Georgiou CD. 2010. Superoxide radical induces sclerotial differentiation in filamentous phytopathogenic fungi: A superoxide dismutase mimetics study. *Microbiology.* 156: 960–966.

Park MS, Bae KS, et al. 2006. Two new species of *Trichoderma* associated with green mold of oyster mushroom cultivation in Korea. *Mycobiology.* 34: 11–113.

Paz Z, Komon-Zelazowska M, et al. 2010. Diversity and potential antifungal properties of fungi associated with a Mediterranean sponge. *Fungal Divers.* 42: 17–26.

Rawlings ND & Morton FR. 2008. The MEROPS batch BLAST: A tool to detect peptidases and their non-peptidase homologues in a genome. *Biochimie.* 90: 243–259.

Ruocco M, Lanzuise S, et al. 2009. Identification of a new biocontrol gene in *Trichoderma atroviride*: The role of an ABC transporter membrane pump in the interaction with different plant-pathogenic fungi. *Mol Plant Microbe Inter.* 22: 291–301.

Samuels GJ. 2006 *Trichoderma*: Systematics, the sexual state, and ecology. *Phytopathology.* 96: 195–206.

Samuels GJ & Ismaiel A. 2009. *Trichoderma evansii* and *T. lieckfeldtiae*: two new *T. hamatum*-like species. *Mycologia.* 101: 142–152.

Samuels GJ, Suarez C, et al. 2006. *Trichoderma theobromicola* and *T. paucisporum*: Two new species isolated from cacao in South America. *Mycol Res.* 110: 381–392.

Samuels GJ, Dodd SL, et al. 2002. *Trichoderma* species associated with the green mold epidemic of commercially grown *Agaricus bisporus. Mycologia.* 94: 146–170.

Schirmböck M, Lorito M, et al. 1994. Parallel formation and synergism of hydrolytic enzymes and peptaibol antibiotics, molecular mechanisms involved in the antagonistic action of *Trichoderma harzianum* against phytopathogenic fungi. *Appl Environ Microbiol.* 60: 4364–4370.

Schoch CL, Seifert KA, et al. 2012. Nuclear ribosomal internal transcribed spacer (ITS) region as a universal DNA barcode marker for fungi. *Proc Natl Acad Sci USA.* 109: 6241–6246.

Schwessinger B & Zipfel C. 2008. News from the frontline: Recent insights into PAMP-triggered immunity in plants. *Cur Opin Plant Biol.* 11: 389–395.

Seaby D. 1998. *Trichoderma* as a weed and pathogen in mushroom cultivation. In *Trichoderma* and *Gliocladium.* Vol. 2, *Enzymes, Biological Control and Commercial Applications* (eds. GE Harman & CP Kubicek), 267–283. London: Taylor and Francis Ltd.

Seidl V. 2008. Chitinases of filamentous fungi: A large group of diverse proteins with multiple physiological functions. *Fungal Biol Rev.* 22: 36–42.

Seidl V, Seibel C, et al. 2009. Sexual development in the industrial workhorse *Trichoderma reesei. Proc Natl Acad Sci USA.* 106: 13909–13914.

Shoresh M, Harman GE, et al. 2010. Induced systemic resistance and plant responses to fungal biocontrol agents. *Annu Rev Phytopathol.* 48: 21–43.

Sivasithamparam K & Ghisalberti EL. 1998. Secondary metabolism in *Trichoderma* and *Gliocladium.* In *Trichoderma* and *Gliocladium.* Vol 1: *Basic Biology, Taxonomy and Genetics* (eds. GE Harman & CP Kubicek), 139–191. London: Taylor and Francis, Ltd.

Suarez MB, Sanz L, et al. 2006. Proteomic analysis of secreted proteins from *Trichoderma harzianum* identification of a fungal cell wall-induced aspartic protease. *Fungal Genet Biol.* 42: 924–934.

Tejesvi MV, Mahesh B, et al. 2006. Fungal endophyte assemblages from ethnopharmaceutically important medicinal trees. *Can J Microbiol.* 52: 427–435.

Tijerino A, Cardoza RE, et al. 2011. Overexpression of the trichodiene synthase gene *tri5* increases trichodermin production and antimicrobial activity in *Trichoderma brevicompactum. Fungal Genet Biol.* 48: 285–296.

Van der Ent, S., Van Wees, S.C., et al. 2009. Jasmonate signaling in plant interactions with resistance-inducing beneficial microbes. *Phytochemistry.* 70:1581–1588.

Van der Ent S, Verhagen BWM, et al. 2008. MYB72 is required in early signaling steps of rhizobacteria-induced systemic resistance in *Arabidopsis. Plant Physiol.* 146:1293–1304.

Van Loon LC, Rep M, et al. 2006. Significance of inducible defense-related proteins in infected plants. *Annu Rev Phytopathol.* 44: 135–162.

Vargas WA, Crutcher FK, et al. 2011. Functional characterization of a plant-like sucrose transporter from the beneficial fungus *Trichoderma virens. Regulation of the symbiotic association with plants by sucrose metabolism inside the fungal cells. New Phytol.* 189: 777–789.

Vargas WA, Mandawe JC, et al. 2009. Plant-derived sucrose is a key element in the symbiotic association between *Trichoderma virens* and maize plants. *Plant Physiol.* 151: 792–808.

Viterbo A & Chet I. 2006. TasHyd1, a new hydrophobin gene from the biocontrol agent *Trichoderma asperellum*, is involved in plant root colonization. *Mol Plant Pathol.* 7: 249–258.

Viterbo A, Landau U, et al. 2010. Characterization of ACC deaminase from the biocontrol and plant growth-promoting agent *Trichoderma asperellum* T203. *FEMS Microbiol Lett.* 305: 42–48.
Viterbo A, Wiest A, et al. 2007. The 18mer peptaibols from *Trichoderma virens* elicit plant defence responses. *Mol Plant Pathol.* 8: 737–746.
Werner A & Zadworny M. 2003. In vitro evidence of mycoparasitism of the ectomycorrhizal fungus *Laccaria laccata* against *Mucor hiemalis* in the rhizosphere of *Pinus sylvestris*. *Mycorrhiza.* 13: 41–47.
Wey TT, Hseu TH, et al. 1994. Molecular cloning and sequence analysis of the cellobiohydrolase I gene from *Trichoderma koningii* G-39. *Curr Microbiol.* 28:31–39.
Xiong S, Fan J, et al. 2010. The antiviral protein cyanovirin-N: The current state of its production and applications. *Appl Microbiol Biotechnol.* 86: 805–812.
Zachow C, Berg C, et al. 2008. Fungal biodiversity in the soils/rhizospheres of Tenerife (Canary Islands): Relationship to vegetation zones and environmental factors. *ISME J.* 3: 79–92.

Section 3
Plant-Interacting Fungi

6 Dothideomycetes: Plant Pathogens, Saprobes, and Extremophiles

Stephen B. Goodwin

USDA-ARS, Crop Production and Pest Control Research Unit, Department of Botany and Plant Pathology, Purdue University, West Lafayette, Indiana

Introduction

The Dothideomycetes is the largest class of fungi, both for number of species and for ecological and biological diversity. As currently defined, the class contains more than 19,000 species in 1,300 genera, 90 families, and 11 (Lumbsch & Huhndorf, 2010) or 12 (Zhang, Crous, et al., 2011) orders. As with many fungi, the number of species described is only a small fraction of those that occur in nature. DNA sequence data have revealed that many taxa actually are complexes of numerous, morphologically indistinguishable sibling species. Many plants that were thought to be hosts for one or at most a few species of Dothideomycetes, on closer examination can contain dozens of species (Arzanlou, Groenewald, et al., 2008; Crous, Wingfield, et al., 2006), yet most hosts have not been analyzed thoroughly. With this in mind, the total number of Dothideomycetes fungi is huge, and 20,000 species is likely to be a conservative estimate.

This huge abundance of species is matched by a correspondingly high ecological and biological diversity. Some Dothideomycetes are lichenized (Nelsen, Lücking et al., 2011), and there is some speculation that the ancestor of all Dothideomycetes may have been a lichen (Schoch, Crous, et al., 2009). Saprotrophs range from passive degraders of dead plant biomass to extremophiles that exploit harsh environmental niches. The latter include the meristematic black yeasts that often are tolerant of high solar radiation, desiccation, and extremes of temperature, both high and low. For example, the rock-inhabiting Dothideomycete *Taeniolella fagina* grows on marble surfaces and can survive temperatures of 248° F (120° C) at 0 percent relative humidity (Sterflinger, 1998). At the other extreme, several Dothideomycetes such as

The Ecological Genomics of Fungi, First Edition. Edited by Francis Martin.

Cryomyces antarcticus survive the high solar radiation, extreme desiccation, and intense cold experienced on exposed rock faces in Antarctica (Onofri, Selbmann, et al., 2007; Onofri, Barreca, et al., 2008). Thermal tolerance of the whiskey warehouse-staining fungus, *Baudoinea compniacensis*, is induced by low levels of ethanol vapor, giving rise to hard black crusts covering many exposed surfaces downwind of brandy maturation warehouses and bakeries (Scott, Untereiner, et al., 2007).

The most well-known Dothideomycetes are those that are associated with plants, particularly as pathogens. Virtually every major crop and almost all wild plant hosts are infected by multiple species of Dothideomycetes. These pathogens greatly increase costs and can be major impediments to sustainable agricultural production. For example, widespread planting of hybrid corn with a genetic susceptibility to a toxin produced by a previously minor race of *Cochliobolus heterostrophus* led to rampant proliferation of toxin-producing strains that caused the southern corn leaf blight epidemic during the early 1970s (Tatum, 1971). This epidemic reduced US corn harvests by 15 percent (Ullstrup, 1972) and could have been much worse if the cause of the susceptibility had not been identified rapidly and eliminated. Other important corn diseases caused by Dothideomycetes include northern leaf blight (caused by *Setosphaeria turcica*) and gray leaf spot (*Cercospora zeae-maydis*). Black Sigatoka, caused by *Mycosphaerella fijiensis*, is one of the most pressing constraints to banana production worldwide and a major expense for fungicides. This disease was first discovered in the Sigatoka Valley of Fiji during the 1960s, but during the past 50 years has spread to most banana-production areas worldwide. Similar to the southern corn leaf blight epidemic, the black Sigatoka problem is exacerbated by the high susceptibility of the Cavendish group of banana clones that dominate world production (Marín, Romero, et al., 2003).

Plant diseases caused by Dothideomycetes usually are not fatal or do not cause complete crop loss, but they are almost universally present and exact a huge toll for disease management. The world's most widely planted food crop, wheat, is affected by numerous Dothideomycetes. Among the most important are those that cause *Stagonospora nodorum* blotch (*Phaeosphaeria nodorum*), Septoria tritici blotch (*Mycosphaerella graminicola*), and tan spot (*Pyrenophora tritici-repentis*). Similar lists can be made for almost every crop plant, whether temperate or tropical, herbaceous or woody. Even rice, which has relatively few fungal pathogens including no rusts or powdery mildews, is infected by multiple Dothideomycetes, including species of *Alternaria*, *Cercospora*, and *Cochliobolus*, although they usually are not major pathogens. Tree pathogens within the Dothideomycetes include the Septoria leaf and canker pathogen *Mycosphaerella populorum* and the red band needle blight of pine pathogen, *Dothistroma septosporum*. Both of these pathogens have spread to new areas recently, and epidemics of Dothistroma may become more common as the climate warms (Woods, Coates, et al., 2005). Other tree nibblers include species of *Botryosphaeria*, which cause cankers and damage populations of tree crops worldwide.

Associations of Dothideomycetes with plants are not limited to pathogenesis. Some Dothideomycetes are endophytes (Guo, Xu, et al., 2004; Rhoden, Garcia, et al., 2012), which can live within plant hosts without causing disease or even may be beneficial. A few Dothideomycetes such as *Cenococcum geophilum* have mycorrhizal associations with plants, presumably helping with nutrient uptake in return for photosynthate.

Another major role of Dothideomycetes is as saprobes degrading dead plant biomass. Two of the most common saprobes are *Alternaria alternata* and *Cladosporium herbarum* (*Davidiella tassiana*). As ubiquitous colonizers of dead plant biomass, these fungi play a major role in global nutrient cycling.

Fortunately, most Dothideomycetes are not pathogenic to animals, although a few exceptions are known. *Hortaea werneckii* is the cause of tinea nigra, a superficial skin infection of humans in tropical climates primarily on the palms of hands and the soles of feet (Bonifaz, Badali, et al., 2008). The main effect on human health of Dothideomycetes is as allergens. Species of *Alternaria* and *Cladosporium* living on dead organic matter produce huge quantities of airborne conidia. These asexually produced spores make these Dothideomycetes among the most common allergenic molds.

The high economic importance of the plant pathogens, biological diversity of the extremophiles, and huge impact on human allergies have led to much interest in sequencing the genomes of Dothideomycetes. Many species have been sequenced or are in progress (Table 6.1), however only a few of these have been published. The sequenced species span a wide section of the Dothideomycetes evolutionary tree, particularly for members of the orders *Capnodiales* and *Pleosporales* (Fig. 6.1). This chapter will describe some of the evolutionary history and biological diversity of the Dothideomycetes with a particular focus on some of the plant pathogens with published genome sequences. It will close with a discussion of future sequencing needs and some significant unanswered questions.

Taxonomy, Origin, and Early Evolution of the Dothideomycetes

Taxonomy of the Dothideomycetes has been in flux over the past 15 years. Placement of many taxa had been problematic based on morphological characters, but recent multigene phylogenies of DNA sequence data are starting to provide resolution. The largest and most recent analysis of 356 isolates representing 10 orders of Dothideomycetes (Schoch, Crous, et al., 2009) revealed that the common ancestor of all Dothideomycetes most likely was a terrestrial saprobe. Lichenized species in the Trypetheliales diverged early and possibly

Table 6.1 A noncomprehensive list of species of Dothideomycetes that have been completely or partially sequenced.

Order	Species	Lifestyle	Genome size	No. of genes	Status
Incertae cedis	*Acidomyces richmondensis*	Saprobe/extremophile	29.88	11,202	http://genome.jgi-psf.org/Aciri1_meta/Aciri1_meta.home.html
Pleosporales	*Alternaria alternata*	Saprobe/Necrotrophic plant pathogen	33.2	—	Not yet available
Pleosporales	*Alternaria brassicicola*	Necrotrophic plant pathogen	31.97	10,688	http://genome.jgi-psf.org/Altbr1/Altbr1.home.html
Capnodiales	*Baudoinia compniacensis*	Saprobe/extremophile	21.88	10,513	http://genome.jgi-psf.org/Bauco1/Bauco1.home.html
Botryosphaeriales	*Botryosphaeria dothidea*	Plant pathogen	43.50	14,998	http://genome.jgi-psf.org/Botdo1/Botdo1.home.html
Incertae cedis	*Cenococcum geophilum*	Mycorrhizal	—	—	Not available
Capnodiales	*Cercospora zeae-maydis*	Hemibiotrophic plant pathogen	46.61	12,020	http://genome.jgi.doe.gov/Cerzm1/Cerzm1.home.html
Capnodiales	*Cladosporium fulvum*	Biotrophic plant pathogen	61.1	14,127	Not yet available
Pleosporales	*Cochliobolus carbonum*	Necrotrophic plant pathogen	31.27	12,857	http://genome.jgi-psf.org/Cocca1/Cocca1.home.html
Pleosporales	*Cochliobolus heterostrophus*	Necrotrophic plant pathogen	32.93-36.46	12,720-13,336	http://genome.jgi-psf.org/CocheC4_1/CocheC4_1.home.html
Pleosporales	*Cochliobolus lunatus*	Necrotrophic plant pathogen	31.17	12,131	http://genome.jgi.doe.gov/Coclu2/Coclu2.home.html
Pleosporales	*Cochliobolus miyabeanus*	Necrotrophic plant pathogen	31.36	12,007	http://genome.jgi-psf.org/Cocmi1/Cocmi1.home.html
Pleosporales	*Cochliobolus sativus*	Necrotrophic plant pathogen	34.42	12,250	http://genome.jgi.doe.gov/Cocsa1/Cocsa1.home.html
Pleosporales	*Cochliobolus victoriae*	Necrotrophic plant pathogen	32.83	12,894	http://genome.jgi.doe.gov/Cocvi1/Cocvi1.home.html
Capnodiales	*Dothistroma septosporum*	Hemibiotrophic plant pathogen	30.21	12,580	http://genome.jgi-psf.org/Dotse1/Dotse1.home.html

Hysteriales	*Hysterium pulicare*	Saprobe	38.43	12,352	http://genome.jgi-psf.org/Hyspu1/Hyspu1.home.html
Pleosporales	*Leptosphaeria maculans*	Hemibiotrophic plant pathogen	44.89	12,469	http://genome.jgi-psf.org/Lepmu1/Lepmu1.info.html
Capnodiales	*Mycosphaerella fijiensis*	Hemibiotrophic plant pathogen	73.7	13,107	http://genome.jgi-psf.org/Mycfi2/Mycfi2.home.html
Capnodiales	*Mycosphaerella graminicola*	Hemibiotrophic plant pathogen	39.67	10,933	http://genome.jgi-psf.org/Mycgr3/Mycgr3.home.html
Pleosporales	*Pyrenophora teres* f. *teres*	Necrotrophic plant pathogen	33.58	11,799	http://genome.jgi-psf.org/Pyrtt1/Pyrtt1.home.html
Pleosporales	*Pyrenophora tritici-repentis*	Necrotrophic plant pathogen	37.84	12,171	http://www.broadinstitute.org/annotation/genome/pyrenophora_tritici_repentis.3/Info.html
Hysteriales	*Rhytidhysteron rufulum*	Saprobe	40.18	12,117	http://genome.jgi-psf.org/Rhyru1/Rhyru1.home.html
Capnodiales	*Septoria musiva*	Hemibiotrophic plant pathogen	29.35	10,233	http://genome.jgi-psf.org/Sepmu1/Sepmu1.home.html
Capnodiales	*Septoria populicola*	Hemibiotrophic plant pathogen	33.19	9,739	http://genome.jgi-psf.org/Seppo1/Seppo1.home.html
Pleosporales	*Setosphaeria turcica*	Hemibiotrophic plant pathogen	43.01	11,702	http://genome.jgi-psf.org/Settu1/Settu1.home.html
Pleosporales	*Stagonospora nodorum*	Necrotrophic plant pathogen	37.21	12,380	http://www.broadinstitute.org/annotation/genome/stagonospora_nodorum/MultiHome.html
Venturiales	*Venturia inaequalis*	Hemibiotrophic plant pathogen	—	—	Not available
Incertae cedis	*Zasmidium cellare*	Saprobe	38.25	16,015	http://genome.jgi.doe.gov/Zasce1/Zasce1.home.html

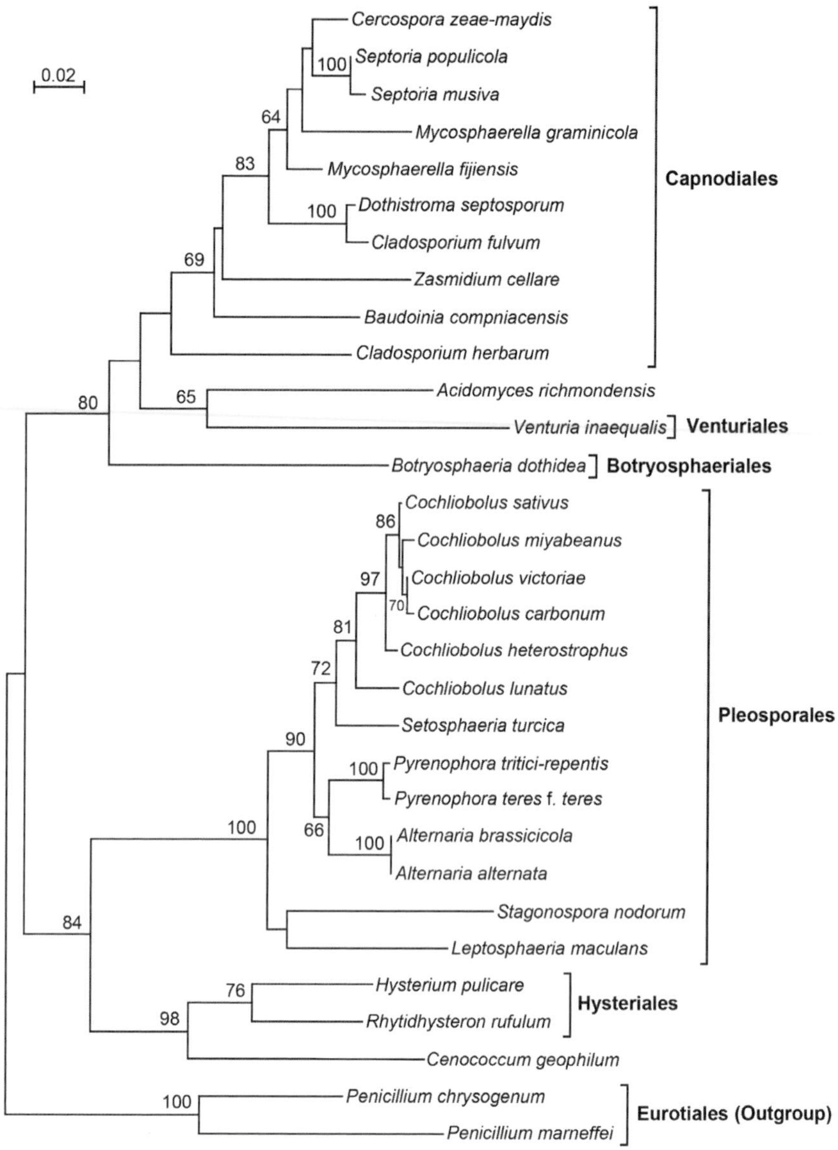

Figure 6.1 Phylogenetic tree of sequenced Dothideomycetes made from sequences of the Internal Transcribed Spacer (ITS) region of the ribosomal DNA. The tree is not comprehensive but includes most of those that have been sequenced at least partly. Several others are in process but were not included because the sequencing has not progressed. The tree was constructed with the neighbor-joining method from ITS sequences obtained from individual genomic sequences or from submissions to GenBank if the ITS region was not recovered in the genomic assembly. Sequences were aligned with Clustalx and corrected manually. Positions with gaps were excluded and a correction was made for multiple substitutions. Bootstrap values of 60 percent or higher (1,000 bootstrap replications) are indicated at the appropriate nodes. The scale bar indicates genetic distance. Orders are indicated with brackets to the right; no order was indicated for taxa that are incertae sedis. The sequences from two species of *Penicillium* (*Eurotiomycetes*) were included as an outgroup.

indicate that Dothideomycetes could have evolved from a lichen, but this is not yet certain. Other lichenized species are scattered throughout the Dothideomycetes phylogenetic tree (Schoch, Crous, et al., 2009), indicating either multiple origins of this lifestyle or representing remnants from an original lichenized ancestor.

Adaptation to freshwater or saltwater habitats appears to have occurred multiple times independently (Schoch, Crous, et al., 2009). Interestingly, almost every clade that contains aquatic species was limited to fresh or salt water, not both. The only exception was the Morosphaeriaceae, which did contain species from both types of aquatic habitat, although on separate branches (Schoch, Crous, et al., 2009). Whether this difference between freshwater or saltwater species is biologically significant or simply a sampling phenomenon is not known.

Unfortunately, sequencing of taxonomically interesting but economically unimportant Dothideomycetes has been limited and genomic resources are rare. Most of these species grow slowly and obtaining sufficient DNA for sequencing from highly melanized cultures has been difficult. One that has been sequenced recently is *C. geophilum*. This species forms mycorrhizal associations with plants and is the dominant ectomycorrhizal fungus in some ecosystems, such as coastal pine forests in Japan (Matsuda, Hayakawa, et al., 2009). As such, it is important for ecosystem health and for helping forests grow in nutrient-poor environments. The genome of this species has recently been sequenced and yielded an assembly of 268 scaffolds totaling ~177 Mbp (*see* Chapter 9). It will be important to help understand the changes in genomic content and architecture that occur during evolution from a free-living saprobe to plant-associated symbiont.

Other Dothideomycetes targeted for sequencing based on taxonomic interest and biological diversity include the lichen *Trypethelium virens* and the mangrove fungus *Aigialus grandis*. The genome of *T. virens* will help explain the early evolution of this class and to test the hypothesis that all Dothideomycetes are descended from a lichenized ancestor. *A. grandis* grows on prop roots of mangroves and so is alternately exposed to high solar radiation and desiccation at low tide followed by the relative anoxia and osmotic stress of saltwater submersion. Comparing the genomes of saltwater-adapted fungi to those from terrestrial and freshwater relatives should help to identify some of the genes that allow these fungi to survive these alternating cycles of environmental extremes.

Unfortunately, both species grow slowly in culture and produce highly pigmented mycelia that thwart efforts to obtain high-quality DNA. Therefore, although they have been chosen for genome projects, progress to date has been slow.

Better sampling of these evolutionarily interesting fungi is needed for a complete understanding of the genetic changes that occurred during their adaptation to diverse niches. A mycorrhizal symbiont can obtain much of its nutrition from its host, so might be expected to have lost many of the genes used by saprobes, such as those for degrading plant cell walls and other complex carbohydrates, or any genes that might trigger a defense response from the host of a pathogen (*see* Chapter 9). A lichenized fungus may show similar adaptations to coexist with its symbiotic alga (*see* Chapter 10). Little is known about how marine species such as *A. grandis* adapt to the high osmotic stress of their saltwater habitats. These species may have developed mechanisms to export sodium from their cells or to prevent it from accumulating to lethal levels. Sequencing of additional salt-adapted Dothideomycetes will indicate whether the same or similar changes occurred during the independent evolution of this trait within each lineage. For example, it would be interesting to compare those species adapted to temporary saltwater immersion with some of the extreme halophiles within the Dothideomycetes, such as *H. werneckii*, which is one of the most commonly found fungi in hypersaline environments (Plemenitaš, Vaupotič, et al. 2008), yet also occurs terrestrially and can cause human disease of the ear. However, these goals are blocked temporarily by the more immediate problems of culturing the fungi and extracting sufficient quantities of high-quality nucleic acids.

Saprobes and Extremophiles

Most Dothidomycetes are saprobes, usually found in association with dead or dying plants. Several have been chosen for sequencing, either because they are extremely common decomposers of dead plant biomass or to increase the phylogenetic diversity of sequenced organisms. One of these fungi is *C. herbarum*, recently renamed as *D. tassiana* (Braun, Crous, et al., 2003). This species was chosen for sequencing because it is extremely common on dead plants, yet cannot infect those that are living. The isolate chosen for sequencing is a recently described epitype from dead barley leaves (Schubert, Groenewald, et al., 2007). Although it cannot cause disease, this species clearly has the ability to catabolize leaf tissue of a cereal crop. Comparisons of the genes present in this species with those from related cereal pathogens such as *Septoria passerinii* from barley or *M. graminicola* from wheat could help to identify the genomic changes that occurred during the evolution of pathogens from saprobes.

As a saprobe, *C. herbarum* does not need the genes required for interaction with a host. Therefore, it might be expected to have a greatly reduced set of genes involved in signal transduction and for the production of effectors that are required for pathogenicity. Conversely, it might be expected to have an increased arsenal of genes involved in interactions with other microorganisms. By adapting to living hosts, plant pathogens access a niche that is free of most other microbes that otherwise might compete for the same substrates. In contrast, *C. herbarum* has to compete with bacteria and other fungi that can quickly colonize dead plants once their defense mechanisms are inactive. Therefore, the genome of *C. herbarum* might be expected to be expanded for genes that produce antibiotics or other toxic secondary metabolites that might reduce the growth of competing organisms. It also might have genes to protect itself from toxic metabolites produced by other organisms and a means for exchanging genes with other members of the same or different species to rapidly evolve its arsenal of offensive and defensive weapons. Unlike the relatively constant environment of a host, the substrate for a saprobe may be variable and ephemeral, may vary from dry to wet, and may require genes for adaptations to thermal or osmotic stresses above those endured by pathogens in their more limited, relatively protected host niches.

These basic questions about genome content can be addressed easily with a genome sequence. Unfortunately, sequencing of the *C. herbarum* genome has been delayed as a result of difficulties in producing sufficient quantities of high-quality nucleic acids. Some sequencing has been done, but there is no assembly and nothing has been published. Hopefully these difficulties can be overcome quickly so the sequencing of this genome can be completed.

Another common sabrobic Dothideomycete is *A. alternata*, which together with *C. herbarum* are two of the most commonly detected fungi in air samples worldwide (Gioulekas, Damialis, et al., 2004). This species is of interest because it has a plastic biology. It occurs most commonly as a saprobe in soil or on dead or dying plant tissues (Thomma, 2003), but it also can be pathogenic on a variety of plant species. Pathogenicity appears to evolve through horizontal transfer of conditionally dispensable chromosomes (CDCs) that contain genes for biosynthesis of host-selective toxins. Transfer of a specific set of toxin genes can allow a previously nonpathogenic individual to infect a host (Izumi, Kamei, et al., 2012). Combining CDCs that contain genes with different host specificities can allow a strain of *A. alternata* to infect both hosts (Akagi, Akamatsu, et al., 2009). Because most strains of *A. alternata* are not pathogenic, it is believed that the CDCs must confer a fitness cost during saprobic growth and are preferentially lost when not needed to infect a host (Akagi, Akamatsu, et al., 2009).

This ability to switch between pathogenic and nonpathogenic lifestyles in a predominantly saprobic organism makes the *A. alternata* genome extremely interesting for comparative analyses. The CDCs contain genes for host-selective toxins (Akagi, Akamatsu et al., 2009), but whether they possess other genes, for example to facilitate their transfer, is not known. It is possible that they contain other genes for interacting with the host or they may possess regulatory genes that alter the expression of genes on the essential chromosomes. Global analyses of gene expression during pathogenic versus saprobic growth would help to answer many of the questions about the roles and effects of the CDCs on growth, development, and niche specificity.

Another reason for interest in these saprobic Dothideomycetes is because of their effects on human health. Together *A. alternata* and *C. herbarum* represent two of the four most commonly detected allergenic fungi (Gioulekas, Damialis, et al., 2004). This is most likely because of their ubiquity and huge production of airborne spores. The primary allergen produced by *A. alternata*, Alt a1, is specific to that species (Achatz, Oberkofler, et al., 1995), but homologs of other allergenic proteins are found in both *A. alternata* and *C. herbarum* (Hong, Cramer, et al., 2005). The huge effects of these fungi on human allergies and asthma have led to additional genomics efforts. A recently funded National Science Foundation (NSF) project will lead to sequencing of 14 strains of *Alternaria* species and additional sequencing of allergenic fungi is underway in Europe (C. Lawrence, personal communication). A 20× coverage of the *A. alternata* genome yielded an assembly of 325 contigs totaling ~33 Mbp (C. Lawrence, personal communication). Related species are expected to have similarly small genomes. These additional genomes should help identify the proteins that trigger allergenic responses in mammals in addition to providing an excellent base for comparative genomics with strains isolated from plant hosts.

Comparative analyses between the genomes of *C. herbarum* and *Alternaria* species should help to identify the genes involved in adaptations to saprobic growth. Competing for ephemeral niches requires genes that facilitate rapid dispersal and growth on newly available resources, such as fallen leaves. Genes for rapid production of airborne conidia and fast growth may have a selective advantage. As discussed for *C. herbarum*, this also may include production of secondary metabolites that are inhibitory or toxic to competing organisms. It is known that pathogenic isolates of *A. alternata* possess CDCs containing genes for production of host-selective toxins that are absent from nonpathogenic strains. However, it is not known whether the saprobes have CDCs that help with that lifestyle or whether they exchange genes on a more limited basis. Nothing is known about the presence of CDCs in *C. herbarum* or other aspects of its genomic architecture. As a ubiquitous colonizer of dead plant tissue with a cosmopolitan distribution, there is a great need for sequencing additional genomes of *C. herbarum* from other substrates and parts of the world. There also is a great need for sequencing other species of *Cladosporium*, a large genus with many important, widely distributed species, including human allergens, plant pathogens, and endophytes as well as saprobes (Schubert, Groenewald, et al., 2007).

In addition to the common saprobes, the Dothideomycetes contains more extremophiles from more severe environments than any other class. The Eurotiomycetes, particularly those in the genus *Aspergillus*, also contains many extremophiles, but they are not as numerous as those in the Dothideomycetes. Several extremophiles have been targeted for sequencing because of the possibility of obtaining enzymes with interesting biological properties and for understanding the genomic changes that occur during adaptations to extreme environments. These include the Antarctic rock-inhabiting fungus *C. antarcticus*, the mangrove fungus *A. grandis*, the lichen *T. virens*, and the whiskey cask or warehouse staining fungus *B. compniacensis*. The habitat of *C. antarcticus* is exposed rock faces in Antarctica (Onofri, Selbmann, et al., 2007). In addition to extreme cold, this fungus must endure high solar radiation during the brief Antarctic summer and the desiccating conditions brought about by the strong winds and extremely low humidities of winter; relative humidity in Antarctica often is less than 1 percent (Selbmann, de Hoog, et al., 2005), making it a cold, polar desert. To adapt to this environment, *C. antarcticus* grows endolithically, possibly assisted by enzymes that help break down the hard rock surfaces. This organism also has a meristematic type of growth that is unusual for fungi (Onofri, Barreca, et al., 2008).

The genome of this species should be missing the genes required by plant pathogens during interactions with their hosts, although it may have genes involved in symbiotic associations because there is some evidence that many of these meristematic, rock-inhabiting fungi derive some of their nutrition through associations with algae in a primitive lichen-like stage (Selbmann, de Hoog, et al., 2005). Presumably the number of organisms competing for its cold, rocky substrate is greatly reduced compared to what it would be for organisms from more temperate climates, so there may be little need for production of or protection from toxic secondary metabolites. It most likely will contain genes to facilitate its endolithic lifestyle and for production of melanin or other pigments for protection against solar radiation. Overall, the genome of *C. antarcticus* is expected to be greatly reduced relative to those for plant pathogens or other saprobes and to have modifications that permit growth under extreme cold. Unfortunately, producing enough high-quality DNA from this slow-growing, heavily pigmented organism has been problematic so its sequencing has begun but has not been completed.

The mangrove fungus *A. grandis* also must survive environmental extremes, but in a different niche from *C. antarcticus*. The primary substrate for *A. grandis* is prop roots of mangrove trees. These are alternately exposed to high solar radiation and desiccation at low tides followed by several hours of saltwater immersion and concomitant osmotic stress when the tide is high. The species has a cosmopolitan distribution in tropical coasts worldwide (Kohlmeyer & Schatz, 1985; Chinnaraj 1993) and often is one of the most commonly detected fungi on the intertidal regions of mangrove prop roots (Alias & Jones, 2000; Nambiar & Raveendran, 2009; Pang, Sharuddin, et al., 2010). The phylogenetic placement of *A. grandis* is not certain but it appears to be related to fungi in the Pleosporales (Tam, Pang et al., 2003). It is of interest for sequencing because of its unusual biology and for phylogenetic diversity to better understand the evolution of fungi in the Dothideomycetes. Sequencing of this species has been delayed as a result of a natural disaster in

Thailand, where the DNA was going to be prepared initially, and subsequently because of its slow growth and concerns about the identity of an isolate obtained from a culture collection. These issues are being addressed, so hopefully the sequencing will commence quickly.

Dothideomycetous extremophiles that have been sequenced successfully include *Acidomyces richmondensis* and *B. compniacensis*. The habitat of *A. richmondensis* is acid mine drainages, which are characterized by extremely low pH (<1), high temperatures of 104–122°F (40–50°C), and high concentrations of dissolved metals (Baker, Lutz, et al., 2004) or sulfur (Selbmann et al., 2008). It often is a major component of the biofilms that form in these environments (Baker, Lutz, et al., 2004). The species was rediscovered and named invalidly recently (Baker, Lutz, et al., 2004), but sequencing of the ribosomal DNA revealed that it is a Dothideomycete and is identical to the previously named *Scytalidium acidophilum* (Selbmann et al., 2008). The correct name for both species now has been validly published as *Acidomyces acidophilus* (Selbmann et al., 2008).

Two genome sequences are available for this fungus, one from a pure culture (http://genome.jgi-psf.org/Aciri1_iso/Aciri1_iso.home.html) and the other a metagenomic assembly derived from an environmental sample of an acid mine biofilm (http://genome.jgi-psf.org/Aciri1_meta/Aciri1_meta.home.html). Neither assembly is particularly good, with from 1,683 to 3,164 scaffolds totaling 26.8 to 29.9 Mbp for the metagenomic and pure-culture assemblies, respectively. However, the number of predicted genes ranged from 10,352 to 11,202 depending on assembly, so most of the gene space appears to have been recovered. Because neither assembly has been published, little is known about the genome characteristics. Comparative analyses of these genomes with other Dothideomycetes should help identify genes involved in adaptations to limited, extreme environments. A reduction of genes involved in host-pathogen interactions should be expected, as well as those for producing secondary metabolites involved in defense against competing organisms because those are already selected against in these extreme environments. However, the total number of genes does not appear to be drastically reduced in either genome, so either *A. acidophilus* survives periodically in more permissive environments or there may have been selection for different genes that are required for life in their extreme environments.

The second extremophile with a recently sequenced genome is the whiskey cask or warehouse staining fungus, *B. compniacensis*. This fungus has been known since the 1880s as *Torula compniacensis* (Richon & Petit, 1881), but received little scientific attention until it was rediscovered and renamed *B. compniacensis* in 2007 (Scott, Untereiner, et al., 2007). So far it has only been found near bakeries and distilleries where it grows on exposed, external surfaces causing a characteristic black staining on buildings. Two aspects of this organism's biology make it interesting. The first is that its growth is stimulated by low levels of ethanol vapor, which it can metabolize as a carbon source (Ewaze et al., 2007) even though high concentrations of ethanol are toxic (Ewaze et al., 2008). This growth stimulation by ethanol evidently confers a fitness advantage and allows *B. compniacensis* to outcompete other fungi when the vapor is present. The second interesting aspect of the biology of *B. compniacensis* is that it can be extremely thermotolerant; when dry or after exposure to ethanol, the mycelia of this fungus can survive temperatures higher than 158° F (70° C) and still remain viable. The mechanism for this extraordinary thermal tolerance is not known completely but appears to involve the generation of heat-shock proteins in response to prior thermal or ethanol stress (Ewaze et al., 2008); direct exposure to 158° F(70° C) without preconditioning is fatal.

Because of these extraordinary properties, the genome of *B. compniacensis* is of great interest and was sequenced recently by the Joint Genome Insitute (http://genome.jgi.doe.gov/Bauco1/Bauco1.home.html). The genome is an excellent assembly with only 19 scaffolds totaling 21.9 Mbp. This is the smallest genome reported among Dothideomycetes, and it also has the most reduced mitochondrial genome at only 26 kb with a reduced set of mitochondrial genes (S. B. Goodwin, unpublished). The total number of nuclear genes, at approximately 10,500, is reduced relative to other Dothideomycetes but not drastically. As expected, the genome of *B. compniacensis* had the lowest numbers of genes for small secreted proteins, polyketide synthases, and nonribosomal peptide synthetases compared to related plant pathogens (Ohm, Feau, et al., 2012). So far this genome has only been analyzed superficially as part of an overall comparative analysis (Ohm, Feau, et al., 2012); additional details await publication in a comprehensive genome paper.

Plant Pathogens

Pathogenicity to plants appears to be a derived character in the Dothideomycetes that arose multiple times in different lineages (Schoch, Crous, et al., 2009). Species of Dothideomycetes infect almost every plant family, from primitive to advanced, aquatic to terrestrial, from those exhibiting herbaceous growth to trees including both Gymnosperms and Dicots. Complementing this huge diversity of species and growth forms is a similarly large variation in the methods of pathogenesis. Although many Dothideomycetes pathogens are necrotrophs, killing their hosts with toxins and extracting nutrients from dead tissue, many others are hemibiotrophs which infect living hosts and survive for some time without causing symptoms before switching to a necrotrophic growth phase. A few are biotrophic, growing on living cells without causing extensive host necrosis. No other fungal group contains such a wide range of pathogenic lifestyles or adaptations to such a huge diversity of hosts. So far three Dothideomycetes plant pathogens have been the subjects of comprehensive genome papers, but many others have been included in a large-scale comparative analysis or are the subjects of genome papers in progress.

The first dothideomycetous genome to be sequenced and published was that of the wheat pathogen *S.* (aka *Phaeosphaeria*) *nodorum*. This pathogen occurs in most wheat-growing regions worldwide. A characteristic this species shares with other members of the order Pleosporales is that it uses host-selective toxins or necrotrophic effectors to cause disease (Friesen, Meinhardt, et al., 2007; Friesen & Faris, 2010). To cause disease, a toxin must interact with a specific sensitivity gene in the host; plants with the corresponding host gene will be susceptible to the toxin whereas those lacking the sensitivity gene are resistant (Friesen & Faris, 2010). A fungal individual may produce many different toxins and will be able to infect a host carrying any of the corresponding sensitivity genes.

To better understand the pathogenicity of *S. nodorum*, its genome was sequenced with funding from Australian wheat growers. It is a nice assembly with 107 scaffolds totaling 37.2 Mbp at approximately 10× coverage (Hane, Lowe, et al., 2007). The complete mitochondrial genome of approximately 50 kb also was obtained. Because this was the first Dothideomycetes genome to be published, comparative analyses were limited to distantly related fungi or confined to only short regions of the genome. Genes for nonribosomal peptide synthetases and polyketide synthases were similar to those seen in other fungi (Hane, Lowe, et al., 2007), consistent with prior observations of toxin production by *S. nodorum*. The genome of *S. nodorum* contained an extra G-alpha protein compared to other fungi (Hane, Lowe, et al., 2007), although its function has not been determined.

There was little conservation of gene order or orientation between the genomes of *S. nodorum* and those of other fungi. This lack of synteny was attributed to the great evolutionary distances between the species used for comparison (Hane, Lowe, et al., 2007). However, some regions of microsynteny were identified, particularly around the mating type region between *S. nodorum* and the related Dothideomycete *Leptosphaeria maculans* and for a cluster of genes involved in quinate catabolism across many fungal genomes. The mitochondrial genome was similar to those in many other fungi, but only had 12 of the 14 protein-coding genes found in many other fungal mitochondrial genomes; the *Atp8* and *Atp9* genes were missing (Hane, Lowe, et al., 2007). Although the genome of *S. nodorum* had relatively few distinctive features, it was the first for a Dothideomycete and so it provided a baseline for the next species to be sequenced.

The second Dothideomycete genome to be published was that of *L. maculans*. This fungus causes blackleg or stem canker of *Brassica napus* and related brassicaceous plants. Like *S. nodorum* it is in the order Pleosporales, but its interaction biology is different. Unlike many pathogens in this order, which employ toxins and have a predominantly necrotrophic lifestyle, *L. maculans* can live asymptomatically in an endophytic, biotrophic phase before switching to necrotrophic growth so is considered to be a hemibiotroph (Rouxel, Grandaubert, et al., 2011). Pathogenicity of *L. maculans* is conditioned at least in part by effectors, which are small, secreted proteins that interact with the host to cause disease (Parlange, Daverdin, et al., 2009). A role in pathogenicity has been shown for some but not all effector proteins (Rouxel, Grandaubert, et al., 2011). Some of the same effectors that facilitate disease can be recognized by host resistance genes to trigger a defense response so they also function as avirulence genes.

The genome of *L. maculans* assembled well with 76 scaffolds totaling 45.1 Mbp and approximately 12,400 genes (Rouxel, Grandaubert, et al., 2011). The most striking feature of the genome was that GC content varied segmentally with alternating regions of high or low GC. Each segment ranged in size from approximately 1 to 320 or 500 kb for the low- or high-GC regions, respectively (Rouxel, Grandaubert, et al., 2011). Average GC content of the high-GC segments was 51 percent, consistent with coding regions in many fungi. The low-GC segments were different, with GC contents averaging only 34 percent. This difference appears to be caused by multiple insertions of transposable elements (TEs) into the low-GC regions (Rouxel, Grandaubert, et al., 2011). These TEs are recognized as repeats by the fungus and subjected to repeat-induced point (RIP) mutation (Selker, Cambareri, et al., 1987), a process unique to fungi that is thought to protect their genomes from invasion by TEs (although apparently the process was not efficient in the case of *L. maculans*). RIP works by causing CG to TA transitions in repeated sequences during meiosis (Watters, Randall, et al., 1999). This rapidly decreases the GC content of affected sequences and alone could explain the differences in GC content between the gene-rich, high-GC regions and the TE-enriched, low-GC segments of the *L. maculans* genome.

An interesting discovery was that the low-GC regions of the *L. maculans* genome are relatively gene poor, containing only 5 percent of the genes even though they comprise 36 percent of the genome (Rouxel, Grandaubert, et al., 2011). In addition to TEs, these regions are enriched for genes that might function in pathogenicity including numerous small, secreted proteins, nonribosomal peptide synthetases and polyketide synthases, which might produce toxic secondary metabolites, and genes that are predicted to respond to biotic or abiotic stress (Rouxel, Grandaubert, et al., 2011). Although these genes are not repeats, the RIP machinery is known to sometimes read through into single-copy genes (Irelan, Hagemann, et al., 1994), increasing their rates of mutation and thus facilitating a faster rate of evolution. Therefore, the segmented genome of *L. maculans*, with pathogenicity-related genes in fast-evolving, low-GC regions may have been selected for as an adaptive mechanism to speed up the rate of evolution in response to resistance genes in evolving host populations or other aspects of a variable environment (Rouxel, Grandaubert, et al., 2011).

In contrast to *S. nodorum* and *L. maculans*, which are in the order Pleosporales, the third Dothideomycete sequenced, the wheat pathogen *M. graminicola*, represents the order Capnodiales. This order contains numerous plant pathogens on diverse hosts, but most are hemibiotrophs rather than necrotrophs and a few are biotrophic, although of a different type from that seen in obligate biotrophs in that it does not involve haustoria or other specialized feeding structures. Another difference from the Pleosporales is that toxins produced by fungi in the Capnodiales are not known to be host specific and may or may not be required for pathogenicity (Schwelm, Barron, et al., 2009; Shim & Dunkle, 2003).

The genome of *M. graminicola* was targeted for sequencing because this pathogen causes a common and economically significant disease of wheat so has been the subject of numerous prior investigations. Genetic analyses of this organism had shown that it had large linkage groups that were missing in progeny isolates from sexual crosses. Loss of these linkage groups had no obvious effects on fitness, so they were assumed to represent dispensable chromosomes and this was confirmed by analyses of pulsed-field gels (Wittenberg, van der Lee, et al., 2008). However, the size and gene content of these chromosomes, their function (if any), and where they originated were not known.

Unlike the shotgun sequences of other fungi, which contain many gaps and incomplete assemblies, the genome of *M. graminicola* was finished so that all of the gaps within and between scaffolds were filled. The final product is the most completely finished genome of any filamentous Ascomycete, missing only two internal gaps of unclonable DNA and the telomere of one scaffold (Goodwin, M'Barek, et al., 2011). All remaining scaffolds are complete from telomere to telomere so represent finished chromosomes. The genome consists of 21 chromosomes in total. Eight of the chromosomes were different from the other 13 for every parameter measured; they were smaller, contained fewer genes at a much lower density, and had a lower GC content and different codon usage compared to the core chromosomes, among many other statistics (Goodwin, M'Barek, et al., 2011). Comparisons of the genome sequence with sequenced markers on the genetic linkage maps confirmed that these eight smallest chromosomes were those that were shown to be dispensable genetically. Most of the content on the dispensable chromosomes also was present in the genome of a close relative from wild grasses (Stukenbrock, Jørgensen, et al., 2010), which is believed to have diverged from *M. graminicola* more than 10,000 years ago (Stukenbrock, Banke, et al., 2007). Therefore, the set of eight dispensable chromosomes, collectively called the *dispensome*, must have been acquired by a common ancestor of both species and has been maintained since their speciation event onto different hosts. This was surprising because these chromosomes are lost readily during culture. Although no function has yet been identified, it seems likely that the dispensome must confer a fitness advantage at some time during the life cycle of the pathogen to have been maintained for such a long time against the rapid loss seen in culture.

The origin of the dispensome is not known. Because the chromosomes making up the dispensome are so different from those of the core set, the most likely origin was proposed to have been by horizontal transfer from an unknown donor species (Goodwin, M'Barek, et al., 2011). The mechanism of transfer is not known but most likely involved a sexual hybrid or asexual fusion of complete genomes from two species, followed by a reduction primarily of one genome. Follow-up work and additional comparisons with closely related species have recently confirmed an interspecific combination as the origin of the dispensome of *M. graminicola* and its close relatives (Stukenbrock, Christiansen, et al., 2012).

The second major discovery from the *M. graminicola* genome paper was a reduction in the number of genes for cell wall-degrading enzymes compared to most other plant pathogens (Goodwin, M'Barek, et al., 2011). This was thought to be an adaptation to avoid detection by the host during the relatively long phase of biotrophic growth. The low number of genes for cell wall-degrading enzymes in *M. graminicola* was most similar to those seen in the genomes of non-pathogens and endophytes, which was interpreted as possibly indicating that fungi in the genus *Mycosphaerella* may have been derived originally from an endophytic ancestor (Goodwin, M'Barek, et al., 2011). However, this hypothesis has not been tested directly, so the true origin of the reduced number of cell wall-degrading enzymes in the genome of *M. graminicola* remains obscure.

A New Type of Chromosomal Evolution

Comparisons among these first three sequenced Dothideomycetes genomes revealed an interesting pattern of conservation of gene content between scaffolds or chromosomes of different species. This phenomenon, mentioned in the *L. maculans* and *M. graminicola* genome papers, was analyzed completely and developed more fully by Hane, Rouxel, et al. (2011). Genes that are found together in the same order and orientation are said to be syntenic. Synteny can extend to whole chromosomes or large chromosomal segments (macrosynteny) or may be restricted to only small groups of 2 to 10 genes (microsynteny), depending on the evolutionary distance between the species being compared and their rates of chromosome structural evolution. In plants and animals, dot plots of gene content between species typically show macrosynteny, visible as strong diagonal lines on the graphs (Cannon, Sterck, et al., 2006; Shultz, Ray, et al., 2007). When organisms are too divergent, the dots are distributed randomly on the plots indicating no synteny.

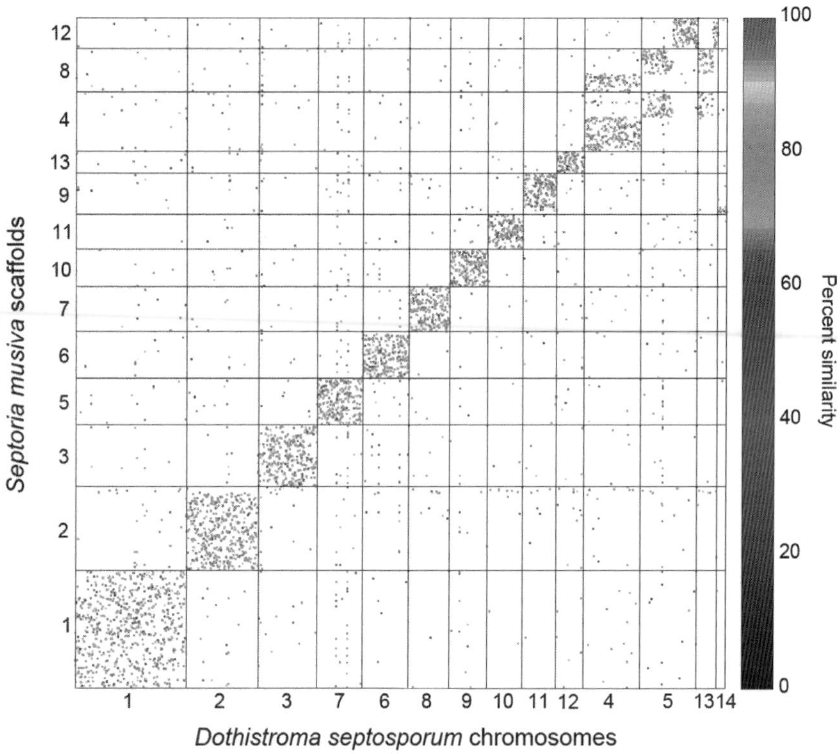

Figure 6.2 Dot plot of the 14 chromosomes of the *Dothistroma septosporum* genome assembly versus the 13 largest scaffolds of *Septoria musiva*. The content of each chromosome or scaffold is plotted from left to right or bottom to top, respectively, for the indicated chromosomes and scaffolds. Macrosynteny would be indicated as diagonal lines within each box defined by a particular chromosome-scaffold comparison. No synteny would be indicated if the contents of a particular chromosome are present over all of the scaffolds of the other genome. Instead, there is a clear pattern of mesosynteny where the contents of each chromosome of *D. septorum* are strongly conserved in a corresponding scaffold of *S. musiva*. However, there is no conservation of gene order or orientation so that the dots fill the box defined by each chromosome-scaffold combination. The results show an almost 1:1 correspondence between gene content on the chromosomes of *D. septosporum* and the largest scaffolds of *S. musiva*. For color detail, please see color plate section.

When Hane, Rouxel, et al. (2011) looked at scaffold-by-scaffold comparisons of Dothideomycetes genomes they saw a different pattern that did not conform to the previously described types of synteny. Instead, there was a strong conservation of gene content, but the order and orientation of the genes were randomized, giving a pattern of dots that filled the space defined by the two scaffolds or chromosomes being compared (Fig. 6.2); they did not see the diagonal lines of macrosynteny nor the random dispersal of the dots from a scaffold or chromosome in one species over the whole genome of the second that is expected when there is no synteny. For example, the genomes of the pine pathogen *D. septosporum* and the poplar pathogen *Septoria musiva* showed an almost 1:1 correspondence between the gene contents for 10 of the 13 largest scaffolds of each species (*see* Fig. 6.2). The exceptions indicate translocations, duplications, or possible misassemblies. For example, the content of *D. septosporum* chromosome 4 occurs on the first halves of *S. musiva* scaffolds 4 and 8, most likely indicating a translocation or possibly a misassembly. Similarly, the gene content on the first half of *D. septosporum* chromosome 5 is found on the second halves of *S. musiva* scaffolds 4 and 8 (*see* Fig. 6.2), most likely as a result of a translocation or duplication, whereas the second half of *D. septosporum* chromosome 5 corresponds to *S. musiva* scaffold 12, most likely indicating a translocation or that some scaffolds of *S. musiva* should be joined. By analyzing these patterns of mesosynteny between scaffolds or pieces of scaffolds, it is possible to deduce past chromosomal rearrangements. It also may be possible to use this information to aid the assembly of genomic sequences (Hane, Rouxel, et al., 2011); many of the scaffolds that were joined from version 1 to version 2 of the *M. graminicola* genomic assembly could have been predicted from their mesosyntenic relationships (Goodwin, M'Barek, et al., 2011).

A surprising result of these analyses was that the phenomenon of mesosynteny was confined almost exclusively to Dothideomycetes. Mesosynteny or degraded mesosynteny was seen in comparisons between Dothideomycetes and fungi in certain other classes, particularly the Eurotiomycetes and Leotiomycetes, but no synteny was observed in comparisons between Dothideomycetes and other fungal classes such as the Sordariomycetes (Hane, Rouxel, et al., 2011). Some comparisons between fungi within other fungal orders occasionally showed evidence of mesosynteny, but the level of conservation of gene content usually was not as strong as that seen in almost all comparisons between species in the Dothideomycetes (Hane, Rouxel, et al., 2011). Comparisons involving yeasts or non-Ascomycete fungi showed either macrosynteny or no synteny within classes and no synteny in comparisons between fungi from different groups. Thus, the phenomenon of mesosynteny appears to be confined almost exclusively to fungi in the Pezizomycotina and is particularly strong between species in the class Dothideomycetes.

The mechanism by which mesosynteny is generated is not known but was hypothesized to involve frequent inversions (Hane, Rouxel, et al., 2011). To test this hypothesis, Ohm, Feau, et al. (2012) performed a simulation analysis by taking two identical sequences, randomly generating inversions, plotting the results, and comparing them to those seen in dot plots between chromosomes from different Dothideomycetes genomes. After 500 random inversions, the simulated data looked similar to the mesosynteny seen in most comparisons between Dothideomycetes (Ohm, Feau, et al., 2012), indicating that intrachromosomal inversions alone are sufficient to explain the observed patterns of mesosynteny. Analyses of the genome sequences identified many possibly inverted chromosomal segments that were bounded by repetitive elements, providing a possible mechanism for generating the required high rate of inversions. Whether the rate of intrachromosomal inversions is much higher in the Dothideomycetes compared to other fungal groups is not known, but it is a hypothesis that can be tested in the future.

Comparisons among Multiple Fungal Genomes

Since the publication of the first three Dothideomycetes genomes in 2007 and 2011, many others have been sequenced, mostly through the Community Sequencing Program (CSP) of the US Department of Energy's Joint Genome Institute. For this effort, the Dothideomycetes research community came together to propose a unified wish list of species and strains to be sequenced. The first several rounds of CSP sequencing led to numerous additional Dothideomycetes sequenced with many others in progress. One goal of this project was to have coordinated publication of multiple genomes, and this came to fruition during 2012. Manuscripts describing at least eight new sequences representing five species are now in press and several others are in the works.

To provide an overview of these projects, including those with individual genome papers in press and those in process, a comparative analysis of 18 Dothideomycetes genomes representing 17 species was performed (Ohm, Feau, et al., 2012). This analysis revealed a striking variation in genome size among Dothideomycetes, ranging from a low of 21.9 Mbp for the extremophile *B. compniacensis* to a high of 74.1 Mbp for the banana pathogen *M. fijiensis*. Most of the differences in genome size could be attributed to an accumulation of TEs in species with large genomes and their near absence in others (Ohm, Feau, et al., 2012). For *B. compniacensis*, the reduced genome was achieved also by a lower number of genes compared to most other species, by having fewer genes with an intron and by reduced distances between genes (Ohm, Feau, et al., 2012).

The cause of the expansion of the large-genome species is not known. In the biotrophic pathogen *Blumeria graminis* (class Leotiomycetes), a greatly expanded genome (to ~120 Mbp) also was attributed to accumulation of TEs (Spanu, Abbott, et al., 2010). In that species, the genes involved in RIP were missing, and there was no evidence for RIP in the sequences. Therefore, loss of the RIP machinery allowing unmitigated replication of TEs was proposed as the most likely cause of the genome expansion (Spanu, Abbott, et al., 2010). In the three Dothideomycetes with the largest genomes (*L. maculans*, ~45 Mbp; *Cladosporium fulvum*, 61 Mbp; *M. fijiensis*, 74 Mbp), the genes involved in RIP are present and appear to be functional because most repetitive elements contained high numbers of the C/G to T/A transitions that are characteristic of

RIP (Rouxel, Grandaubert, et al., 2011; de Wit, van der Burgt, et al., 2012, Ohm, Feau, et al., 2012). This active RIP machinery should have inactivated the transposons to limit their spread and the reason it was not effective is not known. RIP is only active during meiosis (Watters, Randall, et al., 1999), so species with extended asexual stages may provide an opportunity for TE proliferation if the sexual cycle occurs only rarely. This was proposed as a possible explanation for the genome expansion of *C. fulvum* (de Wit, van der Burgt, et al., 2012) but is unlikely for the other two species in which sexual reproduction remains an important part of the life cycle. Thus, unlike *B. graminis*, the reason for the apparently independent genome expansions among species of Dothideomycetes remains mysterious.

The dispensome of *M. graminicola* was different from the core set of chromosomes (Goodwin, M'Barek, et al., 2011), and this provided a means to search for possible dispensable chromosomes in other species. Based on criteria developed from analyses of the *M. graminicola* dispensome (lower GC content, gene density, percent of predicted proteins with a PFAM domain, higher repetitive content), the other 17 Dothideomycetes genomes were tested for possible dispensable chromosomes (Ohm, Feau, et al., 2012). The analysis identified 14 potential dispensable chromosomes in the genome of *M. fijiensis*, two in *L. maculans*, and one each in *C. heterostrophus*, *S. turcica*, and *S. nodorum*. This likely overestimates the true number of possible dispensable chromosomes as a result of difficulties in assembling scaffolds with high amounts of repetitive DNA; a single dispensable chromosome might be represented by more than one scaffold. Lack of any kind of cross-species synteny among potentially dispensable chromosomes in the dot plots indicates that they all had separate origins. Whether any of these represent true dispensable chromosomes is not known for certain and must be tested explicitly. However, it may indicate that Dothideomycetes are capable of exchanging large amounts of genetic materials among species, making their genomes much more plastic than believed previously.

The overall comparative analysis also indicated that RIP, in addition to its role inactivating TEs, may have been harnessed by some of these fungi to speed up their rates of adaptive evolution. The RIP process overall is not precise and can read through from repetitive into single-copy genes (Irelan, Hagemann, et al., 1994), increasing their rates of evolution as noted previously for *L. maculans* (Rouxel, Grandaubert, et al., 2011). This phenomenon appeared to be common in the plant-pathogenic Dothideomycetes with high repetitive sequence content, particularly for genes that may have a role in pathogenicity (Ohm, Feau, et al., 2012). Therefore, it is possible that these fungi have harnessed the RIP machinery to facilitate a rapid rate of evolution of genes involved in adaptation to biotic or abiotic stresses.

Searches for microsynteny blocks revealed only two with from 5 to 10 genes each that were present in at least 14 of the 18 genomes. There was no obvious functional relationship among the genes in the first block, although half of them were down regulated in the host during infection based on microarray data from *L. maculans* (Ohm, Feau, et al., 2012). In contrast, the genes in the second conserved block contained two dehydrogenases and two oxidoreductases and so may be functionally related. All of the genes in block 2 also were upregulated during infection by *L. maculans* and may play a role in pathogenicity that is derived from the common ancestor of all Dothideomycetes (Ohm, Feau, et al., 2012). Such a low level of microsynteny is surprising given that most of the species analyzed are plant pathogens; there could be an advantage in keeping genes required for pathogenicity together but it is consistent with the idea of frequent inversions causing the observed pattern of mesosynteny. From this analysis, it appears that the randomizing force driving mesosynteny is stronger than any selective benefits from keeping potentially co-adapted gene complexes together.

A final significant result from the overall comparative genomics analysis was the discovery of a major difference between fungi in the Capnodiales versus those in the Hysteriales and Pleosporales. Analysis of the *M. graminicola* genome had identified reduced numbers of genes for cell wall-degrading enzymes and for production of potentially toxic secondary metabolites (Goodwin, M'Barek, et al., 2011), but it was not known whether this phenomenon was common among fungi in the Capnodiales or peculiar to *M. graminicola*. The overall comparative analysis identified a clear split along phylogenetic lines, where the Capnodiales genomes contained far fewer genes for cell wall-degrading enzymes and for secondary metabolites, especially polyketide synthases (Ohm, Feau, et al., 2012). The separation was particularly striking for genes involved in the catabolism of cellulose but also included those for degrading xylan and pectin. For example, the number of genes for enzymes in glucoside hydrolase family GH61 (oxidative degradation of cellulose) varied from 20 to 30 with a mean of 24 for fungi in the Hysteriales and Pleosporales. In contrast, fungi in the Capnodiales contained only from one to three genes in the GH61 family (Ohm, Feau, et al., 2012). Similar results were obtained for many other gene families. Taken together, these results identify a significant difference between the ancestors that gave rise to fungi in the Capnodiales versus the Hysteriales and Pleosporales, reflecting different modes of nutrition and giving rise to the diverse mechanisms of pathogenicity seen between fungi in these orders.

In addition to the overall comparative analysis, several interesting results have come from more in-depth analyses of individual genomes or genome sets. One of these involved the plant pathogens *D. septosporum* and *C. fulvum*. These species were analyzed together because they are relatively closely related (Goodwin, Dunkle, et al., 2001b), but their hosts are different: pine trees for *D. septosporum* and tomato for *C. fulvum*. They also have divergent lifestyles, with *D. septosporum* a typical Mycosphaerella-like hemibiotroph, whereas *C. fulvum* is one of the few Dothideomycetes that is a biotroph. An additional advantage of *C. fulvum* is that it has been studied for many years and its interaction with its tomato host is well understood. Previous work has shown that *C. fulvum* produces effector proteins that help cause disease in susceptible host genotypes but can be recognized by corresponding resistance genes in resistant genotypes to trigger a defense response (de Wit, Joosten, et al., 2009). An interesting discovery from the genome comparison was that homologs for many of the *C. fulvum* effector proteins were found in the *D. septosporum* genome and could trigger a response by the corresponding tomato resistance genes (de Wit, van der Burgt, et al., 2012). Therefore, they appear to have retained their function since the divergence of these species from a common (presumably plant pathogenic) ancestor and are likely to play a role in the pathogenicity of *D. septosporum* to pine. Specific adaptations of each species to its respective host also were identified. For example, the *C. fulvum* genome contained a gene for α-tomatinase, thought to be important for detoxification of the toxic glycoalkaloid tomatine, which is absent from *D. septosporum* (de Wit, van der Burgt, et al., 2012). Genes for production of the *D. septosporum* toxin dothistromin were present in the genome of *C. fulvum* but showed little if any expression *in planta* so probably are not used as pathogenicity factors. Overall, these analyses suggest that these species have evolved specific gene sets and differences in regulation of gene expression during their adaptation to different hosts since their divergence from a common ancestor.

Analyses of individual Dothideomycetes genomes also have provided insight into some specific mechanisms of genome evolution, particularly regarding TEs. The wheat pathogen *P. tritici-repentis* is interesting because it causes the economically important disease tan spot, and similar to *S. nodorum*, it uses host-selective toxins as necrotrophic effectors during pathogenesis. Three genome sequences were obtained for this organism: a reference genome of ~40 Mbp from a wheat-pathogenic strain that is known to produce two host-selective toxins and resequenced genomes including a wheat pathogen that produces a third host-selective toxin and a grass-pathogenic strain that is non-pathogenic to wheat (Manning, Pandelova, et al., 2012). One interesting feature of these genomes is that there was little to no evidence of RIP, despite apparently intact copies of the genes involved in the RIP machinery. This led to multiple copies of TEs with unmutated reading frames that were 95 percent or more identical (Manning, Pandelova, et al., 2012). Therefore, if RIP occurs at all in *P. tritici-repentis* its efficiency must be low. This is in stark contrast to other Dothideomycetes, including other members of the Pleosporales, which show strong evidence for RIP.

Another interesting aspect of the *P. tritici-repentis* genome is that some genes appear to have been picked up and amplified as part of class II (DNA-based) tranposons (Manning, Pandelova, et al., 2012). This phenomenon, known as *transduplication* (Juretic, Hoen, et al., 2005), is well known in plants, in which it can alter gene expression as well as increasing copy number (Jiang, Bao, et al., 2004; Hoen, Park, et al., 2006; Hanada, Vallejo, et al., 2009), and it recently has been reported in animals (Sela, Stern, et al., 2008) but was not known previously in fungi. The first evidence for possible transduplication in *P. tritici-repentis* came from identification of an unexpectedly large number of histone genes in the reference genome sequence (Manning, Pandelova, et al., 2012). These near-identical, multiple copies were associated with a transposase gene and had other characteristics of class II transposons. It appears that an original copy of the histone gene was replicated into a transposon, then amplified and moved to multiple locations within the genome (Manning, Pandelova, et al., 2012). Follow-up scans of the *P. tritici-repentis* genome identified many other genes, including one with an osmosensing transporter coiled-coil domain and numerous domains associated with nonribosomal peptide synthesatases, that were present in multiple copies and looked to be parts of DNA transposons. Thus, it appears that transduplication occurs commonly in *P. tritici-repentis* and may facilitate rapid evolution of the transduplicated genes (Manning, Pandelova, et al., 2012). This process is possible because of the near absence of RIP, suggesting that suppression of RIP activity might be advantageous for *P. tritici-repentis*.

Comparisons with different species of Dothideomycetes revealed variable fates for sequences that are duplicated within genomes. In *M. graminicola*, a DNA methyl transferase gene that occurred in multiple copies also appeared initially to be part of a TE (Dhillon, Cavaletto, et al., 2010). Closer inspection revealed only a small piece of a TE at one end of the sequence, which was not enough to permit transduplication. Instead, it appeared that the multiple copies of the DNA methyl transferase gene began as a single copy on chromosome 6, which was duplicated to one of the telomeres and subsequently replicated by an unknown mechanism, most likely by exchanges among the telomeres (Dhillon, Cavaletto, et al., 2010). All of the copies were telomeric or subtelomeric except for the original on chromosome 6. Unlike *P. tritici-repentis*, in which multicopy genes remain intact and unaffected by RIP, in *M. graminicola* every copy of the DNA methyl transferase gene was heavily RIPped, generating multiple stop codons that would prevent translation, including the original copy on chromosome 6 (Dhillon, et al., 2010). The result of this amplification and RIPping should be to eliminate cytosine methylation from *M. graminicola*. A direct test for cytosine methylation proved that it does not occur in *M. graminicola* (Dhillon, Cavaletto, et al., 2010), verifying that all copies of the DNA methyl transferase gene were inactivated and confirming a previous limited test with methylation-sensitive and -insensitive isoschizomers of restriction enzymes (Goodwin, Cavaletto, et al., 2001a). Thus, repeated sequences had completely different origins and fates in the two species, from how they replicated to whether they were affected by RIP. This great diversity of genomic processes underscores the need for sequencing multiple species because generalizations cannot be made safely from one species to another.

Future Sequencing Needs and Major Questions

Despite the large number of genomes available, sequencing of Dothideomycetes is still extremely limited relative to the number of species and has only scratched the surface of the existing biological diversity. The current sequences primarily represent only three of the 12 Dothideomycetes orders; sequenced representatives of the remaining orders are needed for a more complete understanding of the evolution of this group. Much broader sequencing for biological diversity is also needed. For example, aquatic species are underrepresented among those that have been sequenced and need to include both freshwater and marine species. It also would be helpful to include saprobes that are closely related to the pathogenic species to understand the genetic differences and origins of pathogenicity.

One of the major unanswered questions remaining is to identify the likely ancestor that gave rise to the Dothideomycetes. Was it a lichen? An endophyte? What was the major force driving the split between the Capnodiales, with their reduced sets of genes for cell wall-degrading enzymes and secondary metabolites, from the Hysteriales/Pleosporales? Where do the other orders fit relative to this split? How frequently and by what mechanism does horizontal transfer of genes or genomes occur still needs to be understood. Is this a major force for adaptive evolution of the Dothideomycetes? Do the Dothideomycetes really have a higher rate of inversions compared to other organisms? If so, how is it generated? Has it been selected for to drive more rapid evolution? Does rapid chromosomal change generate reproductive isolation to speed up the speciation process? Is mesosynteny one of the reasons there are so many species of Dothideomycetes? What adaptations have occurred that allow the Dothideomycetes to exist in so many extreme environments? The number of questions is endless. Hopefully many more will be addressed during the next few years.

Conclusions

The Dothideomycetes is the largest class of fungi and also has the greatest biological diversity. This provides a huge opportunity both for understanding evolutionary adaptations to stress and extreme environments, but it also provides a possibility to identify enzymes with novel properties that will be useful for industry. The relatively limited sampling of sequenced Dothideomycetes has already revealed new types of genome evolution and helped to identify the mechanisms of plant pathogenesis. The future of Dothideomycetes genomic research is bright and likely to yield new insights into biological processes for many years into the future.

References

Achatz G, Oberkofler H, et al. 1995. Molecular cloning of major and minor allergens of *Alternaria alternata* and *Cladosporium herbarum*. *Mol Immunol.* 32: 213–227.

Akagi Y, Akamatsu H, et al. 2009. Horizontal chromosome transfer, a mechanism for the evolution and differentiation of a plant-pathogenic fungus. *Eukaryotic Cell.* 8: 1732–1738.

Arzanlou M, Groenewald JZ, et al. 2008. Multiple gene genealogies and phenotypic characters differentiate several novel species of *Mycosphaerella* and related anamorphs on banana. *Persoonia.* 20: 19–37.

Alias SA & Jones EBG. 2000. Vertical distribution of marine fungi on Rhizophora apiculata at Morib mangrove, Selangor, *Malaysia. Mycoscience.* 41:431–436.

Baker BJ, Lutz MA, et al. 2004. Metabolically active eukaryotic communities in extremely acidic mine drainage. *Appl Environ Microbiol.* 70: 6264–6271.

Bonifaz A, Badali H, et al. 2008. Tinea nigra by *Hortaea werneckii*, a report of 22 cases from Mexico. *Stud Mycol.* 61: 77–82.

Braun U, Crous PW, et al. 2003. Phylogeny and taxonomy of cladosporium-like hyphomycetes, including *Davidiella* gen. nov., the teleo-morph of *Cladosporium* s. str. *Mycological Progress.* 2: 3–18.

Cannon SB, Sterck L, et al. 2006. Legume genome evolution viewed through the *Medicago truncatula* and *Lotus japonicus* genomes. *Proc Natl Acad Sci USA.* 103: 14959–14964.

Chinnaraj S. 1993. Higher marine fungi from mangroves of Andaman and Nicobar Islands. *Sydowia.* 45: 109–115.

Crous PW, Wingfield MJ, et al. 2006. Phylogenetic reassessment of *Mycosphaerella* spp. and their anamorphs occurring on Eucalyptus. II. *Stud Mycol.* 55: 99–131.

de Wit PJGM, van der Burgt A, et al. 2012. The genomes of the fungal plant pathogens *Cladosporium fulvum* and *Dothistroma septosporum* reveal adaptation to different hosts and lifestyles but also signatures of common ancestry. *PLoS Genet.* 8(11): e1003088.

de Wit PJGM, Joosten MHAJ, et al. 2009. Gene-for-gene models and beyond: The *Cladosporium fulvum*-tomato pathosystem. *The Mycota. V*: 135–156.

Dhillon B, Cavaletto JR, et al. 2010. Accidental amplification and inactivation of a methyltransferase gene eliminates cytosine methylation in *Mycosphaerella graminicola*. *Genetics.* 186: 67–77.

Ewaze JO, Summerbell RC, et al. 2007. Physiological studies of the warehouse staining fungus, *Baudoinia compniacensis. Mycol Res.* 111:1422–1430.

Ewaze JO, Summerbell RC, et al. 2008. Ethanol physiology in the warehouse-staining fungus, *Baudoinia compniacensis. Mycol Res.* 112:1373–1380.

Friesen TL & Faris JD. 2010. Characterization of the wheat-*Stagonospora nodorum* disease system: what is the molecular basis of this quantitative necrotrophic disease interaction? *Can J Plant Pathol.* 32: 20–28.

Friesen TL, Meinhardt SW, et al. 2007. The *Stagonospora nodorum*–wheat pathosystem involves multiple proteinaceous hostselective toxins and corresponding host sensitivity genes that interact in an inverse gene-for-gene manner. *Plant J.* 51: 681–692.

Gioulekas D, Damialis A, et al. 2004. Allergenic fungi spore records (15 years) and sensitization in patients with respiratory allergy in Thessaloniki-Greece. *J Investigl Allergol Clin Immunol.* 14: 225–231.

Goodwin SB, M'Barek SB, et al. 2011. Finished genome of the fungal wheat pathogen *Mycosphaerella graminicola* reveals dispensome structure, chromosome plasticity, and stealth pathogenesis. *PLoS Genet.* 7(6): e1002070. doi:10.1371/journal.pgen.1002070.

Goodwin SB, Cavaletto JR, et al. 2001a. DNA fingerprint probe from *Mycosphaerella graminicola* identifies an active transposable element. *Phytopathology.* 91: 1181–1188.
Goodwin SB, Dunkle LD, et al. 2001b. Phylogenetic analysis of *Cercospora* and *Mycosphaerella* based on the internal transcribed spacer region of ribosomal DNA. *Phytopathology.* 91: 648–658.
Guo LD, Xu L, et al. 2004. Genetic variation of *Alternaria alternata*, an endophytic fungus isolated from *Pinus tabulaeformis* as determined by random amplified microsatellites (RAMS). *Fungal Divers.* 16: 53–65.
Hanada K, Vallejo V, et al. 2009. The functional role of pack-MULEs in rice inferred from purifying selection and expression profile. *Plant Cell.* 21: 25–38.
Hane JK, Lowe RG, et al. 2007. Dothideomycete plant interactions illuminated by genome sequencing and EST analysis of the wheat pathogen *Stagonospora nodorum*. *Plant Cell.* 19: 3347–3368.
Hane JK, Rouxel T, et al. 2011. Mesosynteny: A novel mode of chromosomal evolution peculiar to filamentous Ascomycete fungi. *Genome Biol.* 12: R45. doi:10.1186/gb-2011-12-5-r45.
Hoen DR, Park KC, et al. 2006. Transposon-mediated expansion and diversification of a family of ULP-like genes. *Mol Biol Evol.* 23: 1254–1268.
Hong SG, Cramer RA, et al. 2005. Alt a 1 allergen homologs from *Alternaria* and related taxa: analysis of phylogenetic content and secondary structure. *Fungal Genet Biol.* 42: 119–129.
Irelan JT, Hagemann AT, et al. 1994. High frequency repeat-induced point mutation (RIP) is not associated with efficient recombination in Neurospora. *Genetics.* 138: 1093–1103.
Izumi Y, Kamei E, et al. 2012. Role of pathotype-specific *ACRTS1* gene encoding a hydroxylase involved in the biosynthesis of host-selective ACR-toxin in the rough lemon pathotype of *Alternaria alternata*. *Phytopathology.* 102: 741–748.
Jiang N, Bao Z, et al. 2004. Pack-MULE transposable elements mediate gene evolution in plants. *Nature.* 431: 569–573.
Juretic N, Hoen DR, et al. 2005. The evolutionary fate of MULE-mediated duplications of host gene fragments in rice. *Genome Res.* 15: 1292–1297.
Kohlmeyer J & Schatz S. 1985. *Aigialus gen. nov.* (Ascomycetes) with two new marine species from mangroves. *Transactions of the British Mycological Society.* 85:699–707.
Lumbsch HT & Huhndorf SM. 2010. Outline of Ascomycota—2009. *Fieldiana Life and Earth Sciences.* 1: 1–60
Manning VA, Pandelova I, et al. 2013. Comparative genomics of a pathogenic fungus reveals trans-duplication and the impact of repeat elements on pathogenicity and population divergence. *G3 (Bethesda).* 3(1): 41–63.
Marín DH, Romero RA, et al. 2003. Black Sigatoka: An increasing threat to banana cultivation. *Plant Dis.* 87: 208–222.
Matsuda Y, Hayakawa N, et al. 2009. Local and microscale distributions of *Cenococcum geophilum* in soils of coastal pine forests. *Fungal Ecol.* 2: 31–35.
Nambiar GR & Raveendran K. 2009. Manglicolous marine fungi on *Avicennia* and *Rhizophora* along Kerala Coast (India). *Middle-East J Sci Res.* 4: 48–51.
Nelsen MP, Lücking R, et al. 2011. New insights into relationships of lichen-forming Dothideomycetes. *Fungal Divers.* 51: 155–162.
Ohm RA, Feau N, et al. 2012. Diverse lifestyles and strategies of plant pathogenesis encoded in the genomes of eighteen *Dothideomycetes* fungi. *PLoS Pathog.* 8(12): e1003037.
Onofri S, Barreca D. et al. 2008. Resistance of Antarctic black fungi and cryptoendolithic communities to simulated space and Martian conditions. *Stud Mycol.* 61: 99–109.
Onofri S, Selbmann L, et al. 2007. Evolution and adaptation of fungi at boundaries of life. *Adv Space Res.* 40: 1657–1664.
Pang K-L, Sharuddin SS, et al. 2010. Diversity and abundance of lignicolous marine fungi from the east and west coasts of Peninsular Malaysia and Sabah (Borneo Island). *Botanica Marina.* 53: 515–523.

Parlange F, Daverdin G, et al. 2009. *Leptosphaeria maculans* avirulence gene *AvrLm4-7* confers a dual recognition specificity by the *Rlm4* and *Rlm7* resistance genes of oilseed rape, and circumvents *Rlm4*-mediated recognition through a single amino acid change. *Mol Microbiol.* 71: 851–863.

Plemenitaš A, Vaupoti⊠ T, et al. 2008. Adaptation of extremely halotolerant black yeast *Hortaea werneckii* to increased osmolarity: a molecular perspective at a glance. *Stud Mycol.* 61: 67–75.

Rhoden SA, Garcia A, et al. 2012. Phylogenetic diversity of endophytic leaf fungus isolates from the medicinal tree *Trichilia elegans* (Meliaceae). *Genet Mol Res.* 11: 2513–2522.

Richon MM & Petit P. 1881. Note sur la plante cryptogame des murs do Cognac (*Torula compniacensis* sp. n.). *Brebissonia.* 3: 113–116.

Rouxel T, Grandaubert J, et al. 2011. Effector diversification within compartments of the *Leptosphaeria maculans* genome affected by repeat induced point mutations. *Nature. Commun.* 2: 202. doi: 10.1038/ncomms1189.

Schoch CL, Crous PW, et al. 2009. A class-wide phylogenetic assessment of *Dothideomycetes*. *Stud Mycol.* 64: 1–15.

Schubert K, Groenewald JZ, et al. 2007. Biodiversity in the *Cladosporium herbarum* complex (*Davidiellaceae*, *Capnodiales*), with standardisation of methods for *Cladosporium* taxonomy and diagnostics. *Stud Mycol.* 58: 105–156.

Schwelm A, Barron NJ, et al. 2009. Dothistromin toxin is not required for dothistroma needle blight in *Pinus radiata*. *Plant Pathol.* 58: 293–304.

Scott JA, Untereiner WA, et al. 2007. *Baudoinia*, a new genus to accommodate *Torula compniacensis*. *Mycologia.* 99: 592–601.

Sela N, Stern A, et al. 2008. Transduplication resulted in the incorporation of two protein-coding sequences into the *Turmoil*-1 transposable element of *C. elegans*. *Biol Direct.* 3: 41. doi:10.1186/1745-6150-3-41.

Selbmann L, de Hoog GS, et al. 2005. Fungi at the edge of life: Cryptoendolithic black fungi from Antarctic Desert. *Stud Mycol.* 51: 1–32.

Selbmann L, de Hoog GS, et al. 2008 Drought meets acid: Three new genera in a dothidealean clade of extremotolerant fungi. *Stud Mycol.* 61:1–20.

Selker EU, Cambareri EB, et al. 1987. Rearrangement of duplicated DNA in specialized cells of *Neurospora*. *Cell.* 51: 741–752.

Shim WB & Dunkle LD. 2003. CZK3, a MAP kinase kinase kinase homolog in *Cercospora zeae-maydis*, regulates cercosporin biosynthesis, fungal development, and pathogenesis. *Mol Plant-Microbe Interact.* 16: 760–768.

Shultz JL, Ray JD, et al. 2007. A sequence based synteny map between soybean and *Arabidopsis thaliana*. *BMC Genomics.* 8: 8. doi:10.1186/1471-2164-8-8.

Spanu PD, Abbott JC, et al. 2010. Genome expansion and gene loss in powdery mildew fungi reveal tradeoffs in extreme parasitism. *Science.* 330: 1543–1546.

Sterflinger K. 1998. Temperature and NaCl- tolerance of rock-inhabiting meristematic fungi. *Antonie van Leeuwenhoek.* 74: 271–281.

Stukenbrock EH, Banke S, et al. 2007. Origin and domestication of the fungal wheat pathogen *Mycosphaerella graminicola* via sympatric speciation. *Mol Biol Evol.* 24: 398–411.

Stukenbrock EH, Christiansen FB, et al. 2012. Fusion of two divergent fungal individuals led to the recent emergence of a new widespread pathogen species. *Proc Natl Acad Sci USA*. doi: 10.1073/pnas.1201403109.

Stukenbrock EH, Jørgensen FG, et al. 2010. Whole-genome and chromosome evolution associated with host adaptation and speciation of the wheat pathogen *Mycosphaerella graminicola*. *PLoS Genet.* 6(12): e1001189. doi:10.1371/journal.pgen.1001189.

Tam WY, Pang K-L, et al. 2003. Ordinal placement of selected marine *Dothideomycetes* inferred from small subunit ribosomal DNA sequence analysis. *Botanica Marina.* 46: 487–494.

Tatum LA. 1971. The southern corn leaf blight epidemic. *Science.* 171: 1113–1116.

Thomma BPHJ. 2003. *Alternaria* spp.: From general saprophyte to specific parasite. *Mol Plant Pathol.* 4: 225–236.

Ullstrup AJ. 1972. The impacts of the southern corn leaf blight epidemics of 1970–1971. *Annu Rev Phytopathol.* 10: 37–50.

Watters MK, Randall TA, et al. 1999. Action of repeat-induced point mutation on both strands of a duplex and on tandem duplications of various sizes in *Neurospora*. *Genetics.* 153: 705–714.

Wittenberg AHJ, van der Lee TAJ, et al. 2009. Meiosis drives extraordinary genome plasticity in the haploid fungal plant pathogen *Mycosphaerella graminicola*. *PLoS One.* 4: e5863. doi:10.1371/journal.pone.0005863.

Woods A, Coates KD, et al. 2005. Is an unprecedented Dothistroma needle blight epidemic related to climate change? *BioScience.* 55: 761–769.

Zhang Y, Crous PW, et al. 2011. A molecular, morphological and ecological re-appraisal of Venturiales—a new order of *Dothideomycetes*. *Fungal Divers.* 51: 249–277.

7 Biotrophic Fungi (Powdery Mildews, Rusts, and Smuts)

Sébastien Duplessis[1], Pietro D. Spanu[2], and Jan Schirawski[3]

[1] *Laboratory of Excellence ARBRE, UMR 1136 INRA-Université de Lorraine, Interactions Arbres-Microorganismes, INRA-Nancy, Champenoux, France*
[2] *Department of Life Sciences, Imperial College London, London, United Kingdom*
[3] *Microbial Genetics, Aachen Biology and Biotechnology, RWTH Aachen University, Aachen, Germany*

Powdery Mildews, Rusts, and Smuts: Biotrophy on Distant Arms of the Tree

Powdery mildews, rusts, and smuts are plant pathogenic fungi that belong to distant fungal lineages. Whereas the powdery mildews belong to the Erysiphales order of the Ascomycota, the rusts and smuts belong to the Pucciniales and the Ustilaginales orders of the Basidiomycota, respectively (Fig. 7.1). All three groups have independently developed the ability to colonize their host plants and to obtain nutrients for production of dissemination structures (i.e., sexual or asexual spores). Despite their microscopic size, these fungi can be easily spotted on their hosts because each leads to the development of distinct macroscopic symptoms: mildews develop white powdery patches on the infected organ surfaces, rusts form orange to brown/black pustules of various sizes and shapes on their colonized hosts, and smuts generally lead to black spore-filled tumorous plant organs (*see* Fig. 7.1; Piepenbring, 2001; Glawe, 2008; Voegele, Hahn, et al., 2009). All mildews and rusts are obligate biotrophs because they cannot be cultured to any significant degree in axenic media; the smuts can be cultured but may be considered ecologically obligate biotrophs because their sexual reproduction and mass offspring formation is dependent on a live host plant.

Biotrophy in plant pathogenic fungi has arisen independently and repeatedly in many fungal lineages, and it has been argued that biotrophic interactions may in fact be an ancestral state to nectrotrophic relationships (Spanu, 2012). Indeed, the arbuscular mycorrhizal biotrophic symbionts are among the most ancient plant-fungus associations: fossil evidence indicates that they were

The Ecological Genomics of Fungi, First Edition. Edited by Francis Martin.

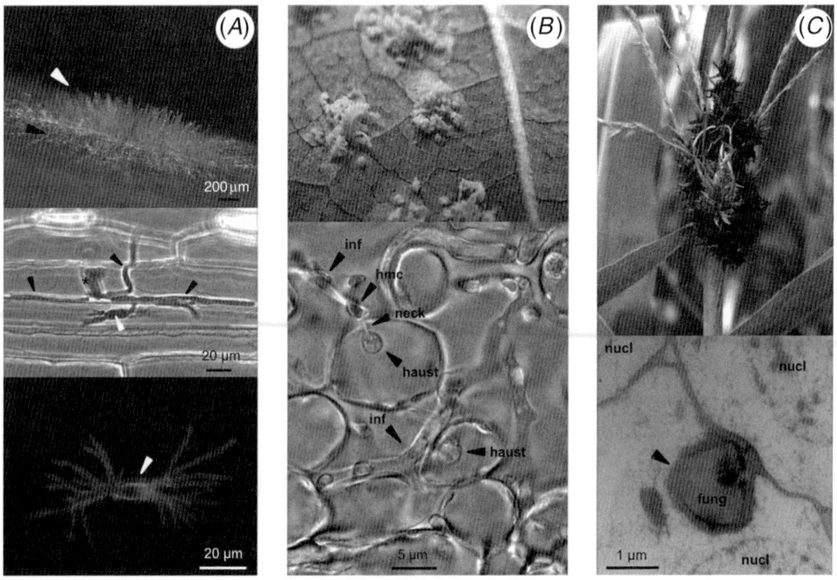

Figure 7.1 Illustration of symptoms and infection structures formed by (*A*) powdery mildew, (*B*) rust, and (*C*) smut fungi in their host plants. *A*, Mildews, top: Sporulating colony of *Blumeria graminis* f. sp. *hordei* on the surface of a barley leaf 5 days after inoculation. The surface hyphae (*black arrow*) extending the colony over the leaf are clearly differentiated from the conidiophores (*white arrow*), which emerge at right angles and produce the abundant asexual conidia that disseminate the fungus and cause epidemics. Bottom: *B. graminis* f. sp. *hordei* on the surface of a barley leaf 2 days after inoculation. The conidium from which the colony originated is still visible (*). The functional haustorium (*white arrow*) that feeds the fungus and delivers effectors to the host is contained within an epidermal cell. Nutrients taken up from the plant are delivered to the external hyphae that extend on the surface of the leaf (*black arrows*). *B*, Rusts, top: uredinia pustules formed on the surface of the lower epidermis of poplar leaves by *Melampsora larici-populina* 7 days after infection. These typical orange sporulating structures are filled with newly formed asexual urediniospores. Bottom: *M. larici-populina* infection structures formed in the spongy mesophyll of a poplar leaf. Infection hyphae (inf) between host cells; haustorial mother cell (hmc) at the contact of an infected cell and haustoria (haust) formed in the cavity of host cells can be distinguished. The extra-haustorial matrix is visible as a white halo around haustoria. Note the neck between the haustorial mother cell and the haustorium. *C*, Smuts, *top*: head smut on maize caused by *Sporisorium reilianum*. Bottom: electron microscopy image of a colonizing hypha of *S. reilianum* in the inflorescence. The intracellular fungal hypha (fung) is surrounded by a thick interaction zone (*arrow*). The nuclei (nucl) of the three adjacent plant cells are indicated. (*A*, Pietro Spanu; *B* (top), photograph by Benjamin Petre, INRA; (bottom), Sébastien Duplessis, INRA; *C* (top), photograph by Tilman Schirawski; (bottom), Martin Engelhaupt and Michael Hoppert, Göttingen University.)

extant in the Devonian 450 million years ago (Simon, Bousquet, et al., 1993). The genome sequence of pathogenic biotrophic fungi (and Oomycetes) published in recent years has revealed some unexpected similarities that indicate evolutionary convergence (Dodds, 2010).

In mildew and rust fungi, nutrient uptake during plant colonization is achieved by the haustoria (*see* Fig. 7.1), which can develop in different plant cell types (i.e., epidermal cells or parenchyma cells). The haustorium forms after penetration of the plant cell wall by an infection hypha that invaginates the plasma membrane of the plant cell. Each haustorium is surrounded by an extra-haustorial matrix that has a different composition specific to the particular fungal and plant species (Voegele & Mendgen, 2003). Haustoria have emerged several times in the course of evolution. In addition to biotrophic fungal pathogens, haustoria are also used by other plant-interacting microorganisms such as the downy mildews (Coates & Beynon, 2010) and the endomycorrhizal fungi that form arbuscules (Parniske, 2008). The role of haustoria in nutrient acquisition has been demonstrated (Voegele & Mendgen, 2003), and there are now many reports indicating that these structures are also involved in the delivery of pathogenic effectors into infected host cells (Catanzariti, Dodds, et al., 2007). Haustoria can also be formed by hemibiotrophic pathogens but are not thought to be maintained after the switch from biotrophic to necrotrophic growth.

Smut fungi do not form true haustoria. Instead, smuts form intercellular and intracellular invasive hyphae during the colonization of their hosts. Growth of intracellular hyphae leads to an invagination of the plant plasma membrane, and the hyphae are surrounded by a matrix through which the plant-fungus communication and nutrient uptake occur (*see* Fig. 7.1; Brefort, Doehlemann, et al., 2009; Doehlemann, van der Linde, et al., 2009); thus, arguably, the functional equivalent of haustoria. Beyond all the listed differences, powdery mildews, rusts, and smuts share a singular survival strategy, feeding on the host and keeping it alive from the earliest steps of colonization to the production of spores for dispersal and further infection. The colonized plant tissues are the habitats of these fungi for most, if not all, of their life cycles. The considerable adaptation processes to convert this hostile environment into an ecological niche is reflected in significant impacts on genome structure and composition: the genome "landscape."

The Impact of Fungal Genomics on Plant Pathology

In less than 20 years, genomics has revolutionized the understanding of plant pathology. Since the completion of the *Saccharomyces cerevisiae* genome—the first eukaryotic genome to be sequenced (Goffeau, Barrell, et al., 1996)—the development of a new generation of sequencing technologies has led to an exponential increase in the number of sequenced microbial genomes (http://genome.jgi.doe.gov/programs/fungi/index.jsf; http://1000.fungalgenomes.org/home/) (Grigoriev, Cullen, et al., 2011; Grigoriev, Nordberg, et al., 2012; Raffaele & Kamoun, 2012) and has had a great impact on our understanding of fungal biology. Genome sequencing of saprotrophs, followed by cataloging and characterizing all genes related to physiologic traits, revealed the underlying genomic adaptation that led to metabolic and physiologic typologies, such as survival in soil, growth on dead material, or dependency on living organisms (Stajich, Wilke, et al., 2010; Floudas, Binder, et al., 2012; Morin, Kohler, et al., 2012). Several recent comparative studies have pointed out the major commonalities and differences in the genomes of obligate and nonobligate biotrophs in fungi or plant mutualistic symbionts, including Oomycetes (Spanu, 2012; Plett & Martin, 2011; Kemen & Jones, 2012; Raffaele & Kamoun, 2012). These comparative genome analyses at the intra- or interspecific levels have led to defining hypotheses on the molecular bases of pathogenicity and on how pathogenic microbes could have arisen several times through the evolution of plant-microbe interactions.

Here, we summarize the salient genomic features of the three archetypal biotrophic fungal groups: the powdery mildew, the rust, and the smut fungi. We present the major similarities and differences of these plant biotroph genomes with a special emphasis on genomic features that relate to their ecology.

Genomics of Powdery Mildews, Rusts, and Smuts

Genomics of Powdery Mildews

The first report of mildew genomes was published a few years ago: this included a full genome sequence and annotation of the barley powdery mildew *Blumeria graminis* f. sp. *hordei* and its comparison to the mildews of pea (*Erysiphe pisi*) and Arabidopsis (*Golovinomyces orontii*) (Spanu, Abbott, et al., 2010). The genome of two more strains of *B. graminis* f. sp. *hordei* and of the wheat powdery mildew *Blumeria graminis* f. sp. *tritici* have now also been published (Hacquard, Kracher, et al. 2013; Wicker, Oberhaensli, et al. 2013).

Surprising Genome Size The majority of Ascomycete genomes known are about 30 to 40 Mb (Spanu, Abbott, et al., 2010). As soon as the first tentative assemblies of the *B. graminis* genome were obtained, it appeared that the initial sequence coverage was much lower than that predicted. It came as a surprise to learn that genomes of the powdery mildews exceed 130 Mb (Spanu, Abbott, et al., 2010). In the powdery mildew genomes of *B. graminis* f. sp. *hordei*, *B. graminis* f. sp. *tritici*, *G. orontii*, and *E. pisi* this is because of exceptionally large quantities of repetitive DNA originated mostly from retro-transposons. Interestingly, the retro-elements identified in the mildews appear to be mildew-specific, suggesting that they are derived from similar retro-elements present in the ancestral mildew.

Ascomycetes have three well-known and conserved mechanisms to limit the amount of repetitive DNA: quelling, meiotic silencing by unpaired DNA

(MSUD) (Shiu & Metzenberg, 2002), and repeat-induced point (RIP) mutations (Hane & Oliver, 2008). Remarkably, the highly conserved genes necessary for RIP are absent in the mildew genomes sequenced to date (Spanu, Abbott, et al., 2010). An in-depth comparative analysis of fungal genomes has conclusively shown that there is practically no evidence of RIP in any of the mildew repeats (J. Amselem et al., personal communication). This suggests that accumulation of retro-transposon–derived repetitive DNA is causally linked to the absence of RIP in the mildews.

Low Gene Density The genome size increase observed in powdery mildews does not lead to an increase in the number of encoded proteins (Table 7.1). Whereas other filamentous fungi with a genome size of 30 to 40 Mb contain about 10,000 to 15,000 genes, the large powdery mildew genomes encode less than 7,000 protein coding genes (http://www.blugen.org/). This low gene count relative to other filamentous fungi reflects the near absence of paralogs in genes-encoding enzymes of primary metabolism, an overall reduction of the size of gene families, and the absence of some highly conserved Ascomycete genes. The most notable reduction in gene family size is observed in the genes-encoding carbohydrate-active enzymes (CAZymes), such as glycoside hydrolases, glycosyltransferases, polysaccharide lyases, and carbohydrate esterases (Spanu, Abbott, et al., 2010). This was recently reiterated by the publication of two *Colletotrichum* genomes in which a comparative analysis indicated that *B. graminis* has the lowest number of genes encoding CAZymes of all the fungi surveyed there (O'Connell, Thon, et al., 2012). The second biggest reduction in gene family size of powdery mildews was observed for secondary metabolic enzymes: Only one polyketide synthase and one nonribosomal peptide synthetase are encoded in the *B. graminis* f. sp. *hordei* genome (Spanu, Abbott, et al., 2010). Of the genes normally conserved in Ascomycetes, 99 are missing in the *B. graminis* genome—a large portion of which are also missing in other obligate biotrophic plant pathogens. However, these genes are present in the hemibiotrophic fungus *Colletotrichum higginsianum*, demonstrating that the missing genes are not *per se* detrimental to biotrophy and suggesting that they are dispensable for the specialized powdery mildew lifestyle (Spanu, Abbott, et al., 2010).

A few additional genes missing from the powdery mildew genomes are interesting because they are also lacking in other plant pathogenic fungi and Oomycetes: genes-encoding enzymes for the reduction of inorganic nitrogen and sulfur, as well as anaerobic fermentation enzymes such as alcohol dehydrogenase. The fact that these primary metabolic genes are missing probably reflects the nutritional environment of the fungi in their ecological niche: inorganic nitrates, nitrites, and sulfates are likely to be in short supply in the host plant, in which organic nitrogen and sulfur are abundantly present in the form of amino acids. Anaerobic fermentation may not be necessary for the powdery

Table 7.1 Genome characteristics of sequenced powdery mildews, rusts, and smuts illustrating the contrasted genomics features of fungal biotrophy.

	Powdery Mildews			Rust Fungi			Smut Fungi		
Fungal species	*Blumeria graminis f. sp. hordei*	*Erisiphe pisi*	*Golovinomyces orontii*	*Melampsora larici-populina*	*Puccinia graminis f. sp. tritici*	*Puccinia striiformis f. sp. tritici*	*Sporisorium reilianum*	*Ustilago hordei*	*Ustilago maydis*
Genome size (Mb)	>120	151	160	101	89	68	19	26	21
Sequencing approach	Whole-genome Sanger shotgun sequencing, 454-sequencing, ABI SOLiD-sequencing	454-sequencing	454-sequencing	Whole-genome Sanger shotgun sequencing	Whole-genome Sanger shotgun sequencing	Illumina (draft genome)	454-sequencing, Paired-end 454-sequencing, Optical Mapping	454-sequencing, Paired-end 454-sequencing BAC-library Sanger-sequencing, Optical Mapping	Whole-genome shotgun Sanger-sequencing
Sequencing depth (x-fold)	140	8.4	8.9	6.9	12	>50	29	25	13
Transposable elements and repetitive DNA content (%)	64%	nd	nd	45%	43.7%	17.8%	0.8%	8%	2%
Gene numbers	6,470	—	—	16,399	17,773	22,815	6,648	7,113	6,786
Number of CSEPs	491[a]	—	—	1,848 secreted proteins, 1184 CSEPs[b]	1,386 secreted proteins, 1106 CSEPs	1,088 secreted proteins	767 secreted proteins, >319 CSEPs	623 secreted proteins, >39 CSEPs	739 secreted proteins, >265 CSEPs

Number of CAZymes[c]	141[d] (GH:61; GT:56; PL:0; CE:10; CBM:14)	—	—	305 (GH:173; GT:84; PL:6; CE:37; CBM:5)	283 (GH:158; GT:85; PL:4; CE:28; CBM:8)	—	188[c] (GH:103; GT:66; PL:3; CE:13; CBM:8)	nd	73[d] (GH:94; GT:58; PL:1; CE:13; CBM:7)
References for genome sequences	Spanu, Abbott, et al., 2010	Spanu, Abbott, et al., 2010	Spanu, Abbott, et al., 2010	Duplessis,Cuomo, et al., 2011	Duplessis, Cuomo, et al., 2011	Cantu, Govindarajulu, et al., 2011	Schirawski, Mannhaupt, et al., 2010	Laurie, Ali, et al., 2012	Kämper, Kahmann, et al., 2006

CBM, carbohydrate-binding domain; CE, carbohydrate esterase; GH, glycoside hydrolase; GT, glycosyltransferase; PL, polysaccharide lyase.
[a]Pedersen, Ver Loren van Themaat et al., 2012.
[b]Hacquard, Joly, et al., 2012
[c]CAZy.org.
[d]O'Connell, Thon, et al., 2012.

mildews because the surface of an infected leaf, stem, or fruit is always fully oxygenated. Similarly, there are no hydrophobins in the mildews. Hydrophobins are well-known proteins otherwise ubiquitous in filamentous fungi that are important in the crossing of water-air interfaces (Wösten, 2001). The unusual absence of hydrophobins in the mildews may thus be a result of the fact that mildews, possibly uniquely among filamentous fungi, live in an exclusively "dry" environment.

Mildew Effector Genes A comprehensive and targeted bioinformatic search for candidate secreted effector protein (CSEP) genes was completed after the initial publication of the *B. graminis* f. sp. *hordei* genome (Pedersen, Ver Loren van Themaat, et al., 2012). To achieve this, the initial set of CSEPs found by identifying open reading frames that encode small secreted proteins (i.e., with a predicted signal peptide but no identifiable transmembrane domain) with no distinct homologies in related fungi, was used in an iterated cycle of BLAST against the *B. graminis* f. sp. *hordei* genome. Using this approach, a superfamily of 491 CSEP-encoding genes was identified, which is equivalent to more than 7 percent of the gene coding capacity of the mildew genome (*see* Table 7.1). The existence of these predicted protein-coding genes was independently confirmed by EST sequences, RNAseq data, and peptide mass-spectra derived from infected plants. Overall, the majority of the CSEP genes are upregulated in haustoria.

Despite that CSEPs were initially identified as lineage-specific genes with no evident homologs in filamentous fungi, computational prediction of the CSEP protein structure appeared to indicate that many may have protein folds that bear a distant similarity to secreted RNAses. This CSEP subset, dubbed "RNase-like proteins expressed in haustoria" (RALPHs), consists of at least 71 RALPHs in *B. graminis* f. sp. *hordei* (Pedersen, Ver Loren van Themaat, et al., 2012). However, this is likely to be a significant underestimate because in many cases, the RALPH genes belong to small gene families in which some members have diversified to the extent that the methods used for assigning the "RNA-like" character fail to do so. The protein sequences of many of the families containing RALPHs can be aligned, and this revealed the conservation of a single intron at the same position: a strong indication that the RALPHs may have originated by a "proto-RALPH" RNAse. Interestingly, other powdery mildews, including the more distantly related *Podosphaera plantaginis*, *E. pisi* and *G. orontii* also have many CSEPs, and the majority of the ones that are common are in fact RALPHs (E. ver Loren van Themaat & C. Tollenaere, personal communication). This observation is consistent with the idea that many, if not all, of the CSEPs originated from a common proto-RALPH effector.

Many of the CSEP families characteristically show significant evidence of positive diversifying selection between family members; the observation that in many cases the diversification of the proteins did not disrupt the predicted protein fold, suggests that a common structure may be important for effector function, much like the situation observed in the Oomycete pathogens (Win, Krasileva, et al., 2012). Another important observation is that many of the paralogous CSEPs are flanked by the same retro-transposon/repetitive element, and that these paralogs are often relatively close together (Pedersen, Ver Loren van Themaat, et al., 2012). This suggests that in *B. graminis* f. sp. *hordei*, effectors have evolved by a series of gene duplications aided by transposon-derived repetitive elements, which was followed by differentiation of the paralogs. It is likely that a similar scenario also applies to the other powdery mildews.

The close association between effector genes and retro-transposons, and the role that repetitive elements may have played in effector gene duplication, offers one explanation for why, at some point early in the evolution of the mildews, a lineage with greater retro-transposon activity may have been at a selective advantage. Such lineage may have been more adaptable to the changes imposed by interactions with a host plant, in particular to challenges imposed by pathogen recognition and immunity in the host. These advantages have been "traded off" against an increased genome size, with the associated metabolic burden that necessarily accompanies this. The losses of genes observed in the mildews may have been a consequence of the retro-transposon activity, too, and as long as they consisted in loss of genes that were not essential for life on a plant, they were tolerated. When the losses were those of genes necessary for life on a nonliving substrate, the mildews became obligate biotrophs.

Genomics of Rust Fungi

Rust fungi belong to the basal and monophyletic order of Pucciniales, one of the largest within the Basidiomycota (Aime, Matheny, et al., 2006). All rusts are obligate biotrophic pathogens of plants and, as a group, can colonize a wide variety of hosts including monocots, dicots, annual, and perennial plant species. Rusts are important pathogens that cause enormous damage to plants, and unraveling the mechanisms underlying host infection at the molecular level is crucial for agriculture. Wheat rust fungi that are responsible for major diseases in wheat plantations, have recently become a major concern in Africa and Asia after the emergence of the new virulent Ug99 strains of *Puccinia graminis* f. sp. *tritici*: Ug99 strains are able to infect wheat cultivars that for decades had been resistant to the disease (Singh 2011; http://www.globalrust.org/). The poplar leaf rust *Melampsora larici-populina*, like other poplar rusts, is a major threat to poplar plantations worldwide. This rust species strongly affects plantations of poplar hybrid cultivars in Northern Europe (Frey, Gérard, et al., 2005; Duplessis, Major, et al., 2009) and in America, where a few epidemics have already been reported (Newcombe & Chastagner, 1993; Innes, Marchand, et al., 2004). Studying rust biology is particularly challenging because rusts are obligate biotrophs and because their complex life cycles commonly involve multiple hosts. Because of these challenging complexities, genomics has been a great step forward to unraveling host interaction determinants (Fernandez, Talhinhas, et al., 2013).

Large Rust Genomes The genomes of the poplar leaf rust *M. larici-populina* and the wheat stem rust *P. graminis* f. sp. *tritici* were sequenced by a Sanger sequencing-based shotgun strategy, which resulted in better genome assemblies than for those rusts sequenced with new generation sequencing technologies. Two large consortia in collaboration with the Joint Genome Institute (JGI; http://genome.jgi.doe.gov/programs/fungi/index.jsf) and the Broad Institute (http://www.broadinstitute.org/scientific-community/science/projects/fungal-genome-initiative/fungal-genome-initiative) joined forces for comprehensive genome analysis. *M. larici-populina* and *P. graminis* f. sp. *tritici* exhibit the largest genomes with a gene content among the highest reported so far in the Basidiomycota with 101 and 89 Mb, and 16,399 and 17,773 genes, respectively (Duplessis, Cuomo, et al., 2011) (*see* Table 7.1). The genomes of *P. striiformis* f. sp. *tritici* and the pine rust fungus *Cronartium quercuum* f. sp. *fusiforme* are only slightly smaller with 80 Mb and 77 Mb, respectively (JGI, http://genome.jgi.doe.gov/Croqu1/Croqu1.home.html; Cantu, Govindarajulu, et al., 2011). The large size of these genomes is the result of the high content of both protein coding sequences and transposable elements (TEs): almost half of the poplar and wheat rust genomes consists of TEs and repeats. In both rusts, the genes and TEs are evenly distributed along the genome, and they share this feature with the powdery mildews.

Sequencing of other rust genomes is underway. *Melampsora lini* (P. N. Dodds, CSIRO, Australia, personal communication) and other poplar-infecting *Melampsora* spp. (R. C. Hamelin, Canada Natural Resources, and British Columbia University, personal communication) have a similar or even greater genome size and complexity. The genome of the soybean rust fungus *Phakopsora pachyrhizi* that caused severe epidemics in soybean plantations in the Southern United States (Schneider, Hollier, et al., 2005) is predicted to be larger than 800 Mb (I. V. Grigoriev, JGI, personal communication). Altogether, these data suggest that some rust genomes might be far more complex and larger than any reported so far.

Rust-Specific Genes The annotation of *M. larici-populina* and *P. graminis* f. sp. *tritici* genomes revealed a large proportion (about 30 percent) of rust-specific genes, with more than half of the genes lacking homologs in international databases (Duplessis, Cuomo, et al., 2011). Most of the rust-specific genes are of unknown function and belong to large multigene families

(Duplessis, Cuomo, et al., 2011). These multigene families may have evolved with the help of TEs, but there is no systematic physical association between the largest TE and gene families (Duplessis, Cuomo, et al., 2011). For several members of the multigene families, matching EST sequences have been detected (either Sanger or 454 sequences, or both), confirming that they are transcribed entities rather than the result of false bioinformatic prediction. In a few cases, ESTs and tBLASTn-driven homology searches helped in identifying new genes or gene family members (Hacquard, Joly, et al., 2012), which highlights the importance of available transcript sequences for de novo gene prediction in genome annotation.

The rust genome annotation revealed several striking findings. (1) The loss or partial loss of essential genes in nitrogen and sulfur assimilation pathways, which mirror similar findings in the powdery mildews and other obligate biotrophic plant pathogens including Ascomycetes and Oomycetes (Baxter, Tripathy, et al., 2010; Spanu, Abbott, et al., 2010; Duplessis, Cuomo, et al., 2011; Raffaele & Kamoun, 2012). (2) The reduction of the CAZyme repertoire compared with those of hemibiotrophic or saprotrophic fungi, with a profile that is closer to the mutualistic biotrophic fungus *Laccaria bicolor* (Martin, Aerts, et al., 2008; Duplessis, Cuomo, et al., 2011). (3) The notable expansion of genes-encoding secreted proteins (Duplessis, Cuomo, et al., 2011): in total, 1,184 and 1,106 genes-encoding CSEPs lower than 300 amino acids were identified in the genomes of *M. larici-populina* and *P. graminis* f. sp. *tritici*, respectively (*see* Table 7.1). Most CSEPs of *M. larici-populina* (70 percent) belong to multigene families, the largest having 111 members. Transcriptome analyses with custom oligonucleotide arrays of the two rusts showed that more than 50 percent of the CSEPs are expressed *in planta*, and several of them are among the most highly expressed genes (Duplessis, Cuomo, et al., 2011; Duplessis, Hacquard, et al., 2011). The detected CSEPs contained several homologs of haustorially expressed secreted avirulence factors of the flax rust *M. lini* (Catanzariti, Dodds, et al., 2006) and of the rust transferred protein RTP1 of the bean rust *Uromyces fabae* (Kemen, Kemen, et al., 2005). Quite strikingly, the large majority of the CSEP genes uncovered in each of the rust fungi were species-specific (74 percent and 84 percent of the CESP genes in *M. larici-populina* and in *P. graminis* f. sp. *tritici*, respectively) (Duplessis, Cuomo, et al., 2011; Duplessis, Hacquard, et al., 2011). This strongly suggests a functional involvement of these genes in host adaptation and biotrophic growth. The diversification of CSEP genes in many expanded families in both the poplar and the wheat rust genomes may be explained by their ecology; the wheat and poplar rusts are heteroecious (i.e., they infect two different host plants in the course of their life cycle). It will be interesting to see if there are significant differences in the number and diversification of the CSEPs in the genomes of monoecious rusts that are yet to be sequenced.

Genomics of Smut Fungi

Smut fungi are biotrophic plant parasites with a narrow host range. Most smut fungi infect particular members of the Graminaceae, and the knowledge of the host plant usually serves as a confident identification criterion. The vast majority of the smut fungi lead to symptoms in the inflorescences of the host, even though it is seedlings that are the primary infection target. This implies that most smuts behave like endophytes during plant growth and only show their parasitic potential at flowering time, when particular floral organs or complete inflorescences are replaced by fungal sori. Opening of the sori releases millions of brown to black spores, which gives the plant a burnt and the eponymous smutty appearance. As a group, smut fungi cause disease on virtually all of the economically important crop plants. The most famous member, the maize smut pathogen *Ustilago maydis* received considerable attention because it is a facultative biotroph that can be grown in axenic culture and at that stage is easily accessible to molecular modifications and genetic investigations. It served as a model to genetically investigate the principles of biotrophy that cannot yet be achieved in obligate biotrophs. Its 20-Mb genome was sequenced by two different companies (Exelexis and Bayer Crop Science) before it was also sequenced by the Broad institute. When the genome sequence was finally published, it was still one of the first Basidiomycete genomes and the first genome of a biotrophic fungus available (Kämper, Kahmann, et al., 2006). Since then, two more smut fungal sequences have been published: the maize head smut fungus *Sporisorium reilianum* f. sp. *zeae* and the barley smut fungus *Ustilago hordei* (Schirawski, Mannhaupt, et al., 2010; Laurie, Ali, et al., 2012) (*see* Table 7.1).

Clustered Effector Genes The *U. maydis* genome sequence provided several surprises. One surprise came from analysis of the gene inventory that revealed a much smaller number of CAZymes compared to those in other fungi. It was speculated that reducing the number of CAZymes is beneficial for biotrophic fungi that need to live unrecognized within the host plant because cell wall degradation serves as a cue for the induction of plant defense (Kämper, Kahmann, et al., 2006). Later sequencing projects corroborated this view: the nectrotrophic plant pathogens *Sclerotinia sclerotiorum* and *Botrytis cinerea* contained a similar multitude of CAZymes as the necrotrophic Ascomycete *Fusarium graminearum* (Cuomo, Gueldener, et al., 2007; Amselem, Cuomo, et al., 2011), and the biotrophic mycorrhizal fungus *L. bicolor* as well as the obligate biotrophic rust fungi *M. larici-populina* and *P. graminis* f. sp. *tritici* lacked the multitude of CAZymes displayed by necrotrophic Ascomycetes (Martin, Aerts, et al., 2008; Duplessis, Cuomo, et al., 2011).

The second surprise came from analysis of gene organization that revealed the presence of clusters of adjacent genes-encoding potentially secreted proteins (Kämper, Kahmann, et al., 2006). Interestingly, these proteins are small and lack evident catalytic domains and homologs in other fungi. As shown by transcriptome analysis, the clustered genes are strikingly upregulated during plant colonization, whereas immediately neighboring genes outside this region were either expressed to the same level as in axenically grown fungi or were even downregulated (Kämper, Kahmann, et al., 2006). The hypothesis that these clustered CSEPs were determinants of the plant-fungus interaction was indeed confirmed by deletion of complete cluster regions in *U. maydis*; about half of the clusters contained compatibility genes with an effect on virulence of *U. maydis* (Kämper, Kahmann, et al., 2006).

The genome of *S. reilianum* was sequenced to learn why *U. maydis* behaves differently to other smuts after plant penetration. In terms of its biology, *U. maydis* is an atypical smut fungus that is not restricted to spore formation in the flowers. Within 1 week after penetration of the host tissue, *U. maydis* induces the formation of tumors in which the fungal spores develop. Plant infection by *S. reilianum* takes place at seedling stage and includes an extended endophytic phase. The 18-Mb genome of *S. reilianum* turned out to be extremely similar to that of *U. maydis* with a shared gene content of about 94 percent and a high conservation of gene synteny (*see* Table 7.1). Striking differences in gene conservation at particular loci were discovered through gene-by-gene comparison. These clusters of highly divergent homologous genes contained a high percentage of CSEP-encoding genes, and a deletion analysis confirmed their contribution to the full virulence potential of *U. maydis* (Schirawski, Mannhaupt, et al., 2010). It was speculated that the different infection strategies of *U. maydis* and *S. reilianum* led to evolution of effectors targeting different host molecules. This implies that relevant host proteins show a different temporal or spatial distribution that would foster the evolution of different sets of effectors in pathogens with different infection strategies.

Evolution through Transposable Elements The genomes of *U. maydis* and *S. reilianum* were also striking because they contained only few repetitive sequences. In contrast, the 26-Mb genome of the barley smut fungus *U. hordei* contained many more TEs, which made genome assembly a challenging task (Laurie, Ali, et al., 2012) (*see* Table 7.1). The repetitive elements in the *U. hordei* genome belong to an only small number of families, which suggests a recent introduction and expansion in the *U. hordei* lineage. All three smut genomes are organized in 23 chromosomes and show an overall conserved synteny. Only one large chromosome rearrangement can be observed between *U. hordei* and *S. reilianum*, and a different rearrangement is obvious between *S. reilianum* and *U. maydis*. This is a clear indication that the ancestral genome organization was close to that of *S. reilianum*, and that both *U. maydis* and *U. hordei* are in derived lineages where different chromosomal rearrangements took place

(Laurie, Ali, et al., 2012). In the case of *U. hordei*, this rearrangement had fundamental biological consequences because it physically linked the *a* and *b* mating type loci on the same chromosome. A striking accumulation of TEs in the intervening region of the mating type loci might have led to such significant sequence differences relative to the other mating type region that recombination was suppressed, thereby enforcing a bipolar mating behavior on *U. hordei*.

The repetitive elements in *U. hordei* also seem to have shaped other aspects of fungal biology. Genomic signatures of retro-transposition can be found at loci encoding-RNA-silencing components (RNA-dependent RNA polymerases, ARGONAUTE, and DICER) suggesting that these loci were active sites of recombination (Laurie, Ali, et al., 2012). Whereas the RNA-silencing components are found at syntenic positions in the *S. reilianum* genome, the genes seem to be cleanly excised from the genome of *U. maydis*. Remnants of numerous copies of small repeats suggest that the components were lost from the *U. maydis* genome by intrachromosomal recombination. Why would the RNA-silencing genes that are distributed at different sites in the chromosomes of *U. hordei* and *S. reilianum* be so cleanly deleted from the *U. maydis* genome? One possible answer stems from work in *S. cerevisiae*, where it was found that the presence of double-stranded RNA viruses is incompatible with RNAi (Drinnenberg, Fink, et al., 2011). *U. maydis* is known to contain double-stranded RNA viruses producing a killer toxin that would provide a competitive advantage to *U. maydis* in the presence of other microbes. And this advantage would be more of an advantage to *U. maydis* than to *U. hordei* or *S. reilianum* because *U. maydis* has a shorter endophytic phase and spends more time on the phylloplane where presence of a killer toxin will be an evolutionary advantage. *U. hordei* or *S. reilianum*, on the other hand, minimize exposure to competing microbes by an extended endophytic period in the competition-free zone of the interior of the host plant.

Signatures of TEs were also found in the vicinity of effector candidates located at syntenic positions to the diversity clusters detected by comparing the *S. reilianum* and the *U. maydis* genomes (Laurie, Ali, et al., 2012). In *U. hordei*, these clusters are present, but individual genes are more spread out over the genome. One class of cysteine-rich effectors that has 3 members in *U. maydis*, 8 members in *S. reilianum*, and 19 members in *U. hordei*. Whereas in both *U. maydis* and *S. reilianum* all members occur clustered at the same locus on chromosome 8, in *U. hordei* they are spread out over the whole genome. Interestingly, in *U. hordei* most of these are flanked by repetitive DNA, suggesting that in this fungus the activity of TEs contributed to rearrangements, including activation and inactivation of effector genes. This could increase their rate of evolution and create a selective advantage for the fungus to overcome host resistance (Laurie, Ali, et al., 2012).

Genome comparison of the three available smut fungal genomes allowed a deeper insight into the molecular understanding of the adaptation of the different species to their respective ecological niches. However, much is still unclear, and this includes the exact molecular reason for the specific adaptation to the different hosts. Genome analysis of other smut fungi infecting other hosts is required to understand how these fungi interact with, manipulate, and exploit their hosts. To this end, the genomes of the sorghum head smut fungus *Sporisorium reilianum* f. sp. *reilianum* and of the sugarcane smut *Sporisorium scitamineum* have been sequenced, and their analysis will be published in the future (R. Kahmann and J. Schirawski, personal communication).

Evolution of Powdery Mildew, Rust, and Smut Genomes

In the three groups of biotrophic fungi surveyed in this chapter, past TE activity has differently shaped the genome landscapes. On the one hand, the overall TE content in the powdery mildews and rust fungi is so preponderant that it makes up most of the genome sequence. On the other hand, TE content is limited in smut fungi. As a consequence, the powdery mildews and rust fungal genomes are large, whereas the smut fungi exhibit small and compact genomes, placing smuts and rusts at opposite ends of the Basidiomycete genome size spectrum. The striking difference in TE abundance and genome size may be functionally related to the differences in the frequency of sex during host infection. Whereas smut fungi undergo sexual reproduction at each round of infection on their hosts, powdery mildews and rust fungi rely on asexual reproduction during their epidemic phases. It is tempting to speculate that a more frequent sexual exchange offers a greater potential for variation of the CSEP repertoire needed for adaptation to the ever-changing plant environment and the dynamic evolutionary pressure from the host immune system. In the case of mildews and rusts, variations occur through TE activity, which might compensate for the rarer opportunities of sexual recombination.

Another clear difference between these groups of biotrophic fungi is the number of protein-coding genes. Both powdery mildew and smut genomes harbor a limited number of genes compared to other sequenced fungal genomes, with marked gene family contractions and losses in genes encoding CAZymes and secondary metabolic enzymes. In contrast, rust fungal genomes show extensive expansions of gene families (e.g., secreted proteases, oligopeptide transporters, and signaling components) with potential importance for host tissue colonization. This difference in the number of protein-coding genes might reflect the different infection strategies followed by these fungi; the sequenced mildews and smuts are specific to only one host plant, whereas the rusts are heteroeicous and can infect alternate hosts in different plant taxa (e.g., barberry and wheat for *P. graminis* f. sp. *tritici* and poplar and larch for *M. larici-populina*).

For obligate biotrophs convergent gene losses have been reported that may reflect their adaptation to, but also their dependence on, the phyllosphere. In powdery mildews, rusts and biotrophic Oomycetes, impaired capacity for nitrogen and sulfur assimilation pathways was observed (Baxter, Tripathy, et al., 2010; Spanu, Abbot, et al. 2010; Kemen, Gardiner, et al., 2011). In addition, the capacity to synthesize thiamin has been lost in different fungal and Oomycete biotrophic pathogens, including powdery mildews (Spanu, Abbot, et al. 2010; Kemen, Gardiner, et al., 2011). Interestingly, some of the common missing genes are still present in the poplar rust genome, and rust fungi still possess the thiamin biosynthesis genes, which are highly expressed in haustorial structures during the biotrophic growth (Hahn & Mendgen, 1997; Sohn, Voegele, et al., 2000; Wirsel, Voegele, et al., 2001; Duplessis, Cuomo, et al., 2011; Hacquard, Veneault-Fourrey, et al., 2011). This indicates that such an adaptation to the host niche is not representative for the whole Pucciniales order (Duplessis, Cuomo, et al., 2011). Despite obvious convergence between unrelated biotrophic plant pathogens, the latter example illustrates that definitive adaptation to the plant host niche did not reach the same level in all fungi surveyed.

Rust fungi have probably evolved different strategies to drain nutrients from their hosts as smut fungi and powdery mildews. In rust fungi, a striking expansion of genes-encoding oligopeptide transporters is observed, and these are highly expressed during biotrophic growth (Duplessis, Cuomo, et al., 2011). This strongly suggests that peptide uptake from the host is an important nutritional strategy for rusts. In contrast, a biotrophy-expressed high-affinity sucrose transporter (Srt1) is required for virulence of the smut fungus *U. maydis*, for which secreted invertases and sucrose-6-phosphate hydrolases are not necessary for infection (Wahl, Wippel, et al., 2010). The situation might be similar for *B. graminis* because it has two *srt1*-homologous genes but no secreted invertase or sucrose-6-phosphate hydrolase (Spanu & Kämper, 2010). It therefore seems likely that both mildew and smut fungi acquire the bulk of their fixed carbon by direct uptake of sucrose from the plant cell, highlighting a convergent adaptation to extract this sugar source without prior hydrolysis in the apoplast.

Conclusions and Perspectives

The elucidation and analysis of the genomes of powdery mildews, rust, and smut fungi has brought to light an incredible wealth of information. Even though only a few members of each group have been sequenced, it is already clear that the respective lifestyles have left their marks on the genome (e.g., loss of essential assimilation and biosynthesis genes in the obligate biotrophs), and that some genomic changes are responsible for the fungal lifestyle (e.g., the bipolar mating behavior of *U. hordei* that resulted from a genome rearrangement). The need to adapt to the changing plant environment has put the biotroph genomes under an unequaled ecological pressure for diversification, which has been achieved differently in the different systems (i.e., evolution with the help of repetitive elements versus recombination during sex).

The great challenge of the future is to prove the ideas generated from the current data set. Several approaches will need to be followed. To identify loci related to pathogenicity in the genomes of considered pathogens, a population genomics approach would be most informative. Depending on the ability to recover adequate populations with enough individuals and appropriate demographic features for the surveyed sets, accurate evolution metrics based on polymorphism could be calculated over virulent versus avirulent individual sets in populations. This would lead to the identification of target virulence factors that are responsible for overcoming resistance.

A different approach will be necessary to answer which genes are responsible for adaptation to a particular host. Did the CSEPs evolved to allow host adaptation or are the species-specific genes needed to achieve compatible interactions? To answer these questions, we will need genomic, transcriptomic, and functional data. Genome sequence comparisons will be particularly interesting between closely related pathogens of different hosts (i.e., comparison of the maize pathogen *S. reilianum* f. sp. *zeae* and the sorghum pathogen *S. reilianum* f. sp. *reilianum*, or comparison of the barley pathogen *B. graminis* f. sp. *hordei* and the wheat pathogen *B. graminis* f. sp. *tritici*) and will reveal which subset of genes is likely responsible for infection of a particular host and the inability to infect the other host. In addition to more genomes, more transcriptome analyses are needed. For example, the investigation of the expression profiles of CSEP genes of the same heteroecious rust fungus on its different host plants will show whether the same or different CSEP sets are used on the different hosts. Finally, a thorough functional analysis of identified candidates and their targets is needed to define how host adaptation and host selection are achieved in the different systems. A combination of these approaches is required to gain a comprehensive understanding of the complex interactions between powdery mildews, rusts, smuts, and their respective host plants.

Acknowledgments

Sébastien Duplessis would like to acknowledge the ANR for supporting rust fungi genomics projects (young scientist grant POPRUST ANR-2010-JCJC-1709-01 and Labex ARBRE ANR-12-LABXARBRE-01), as well as the US Department of Energy Joint Genome Institute (Office of Science of the US

Department of Energy under contract no. DE-AC02-05CH11231) and the Broad Institute of Harvard and MIT, for the sequencing of the poplar rust genome and the wheat stem rust genome, respectively. Pietro Spanu would like to acknowledge the BBSRC for supporting the powdery mildew sequencing project.

References

Aime MC, Matheny PB, et al. 2006. An overview of the higher level classification of Pucciniomycotina based on combined analyses of nuclear large and small subunit rDNA sequences. *Mycologia.* 98: 896–905.

Amselem J, Cuomo CA, et al. 2011. Genomic analysis of the necrotrophic fungal pathogens *Sclerotinia sclerotiorum* and *Botrytis cinerea*. *PLoS Genet.* 7(8): e1002230.

Baxter L, Tripathy S, et al. 2010. Signatures of adaptation to obligate biotrophy in the *Hyaloperonospora arabidopsidis* genome. *Science.* 330: 1549–1551.

Brefort T, Doehlemann G, et al. 2009. *Ustilago maydis* as a Pathogen. *Annu Rev Phytopathol.* 47: 423–445.

Cantu D, Govindarajulu M, et al. 2011. Next generation sequencing provides rapid access to the genome of *Puccinia striiformis* f. sp *tritici*, the causal agent of wheat stripe rust. *PLoS One*. 6: 8.

Catanzariti A-M, Dodds PN, et al. 2006. Haustorially expressed secreted proteins from flax rust are highly enriched for avirulence elicitors. *Plant Cell.* 18: 243–256.

Catanzariti AM, Dodds PN, et al. 2007. Avirulence proteins from haustoria-forming pathogens. *FEMS Microbiol Lett.* 269: 181–188.

Coates ME & Beynon JL 2010. Hyaloperonospora arabidopsidis as a Pathogen Model. *Annu Rev Phytopathol.* 48: 329–345.

Cuomo CA, Gueldener U, et al. 2007. The Fusarium graminearum genome reveals a link between localized polymorphism and pathogen specialization. *Science.* 317: 1400–1402.

Dodds PN 2010. Genome evolution in plant pathogens. *Science.* 330: 1486–1487.

Doehlemann G, van der Linde K, et al. 2009. Pep1, a secreted effector protein of *Ustilago maydis*, is required for successful invasion of plant cells. *PLoS Pathog.* 5(2): e1000290.

Drinnenberg IA, Fink GR, et al. 2011. Compatibility with killer explains the rise of RNAi-deficient fungi. *Science.* 333: 1592.

Duplessis S, Cuomo CA, et al. 2011. Obligate biotrophy features unraveled by the genomic analysis of rust fungi. *Proc Natl Acad Sci USA*. 108: 9166–9171.

Duplessis S, Hacquard S, et al. 2011. *Melampsora larici-populina* transcript profiling during germination and timecourse infection of poplar leaves reveals dynamic expression patterns associated with virulence and biotrophy. *Mol Plant-Microbe Interact.* 24: 808–818.

Duplessis S, Major I, et al. 2009. Poplar and pathogen interactions: Insights from *Populus* genome-wide analyses of resistance and defense gene families and gene expression profiling. *Crit Rev Plant Sci.* 28: 309–334.

Fernandez D, Talhinhas P, et al. 2013. Rust fungi: Achievements and future challenges on genomics and host-parasite interactions. In: The Mycota, Agricultural applications, Vol. 11 (ed. F Kempen). In press. Berlin: Springer.

Floudas D, Binder M, et al. 2012. The Paleozoic origin of enzymatic lignin decomposition reconstructed from 31 fungal genomes. *Science.* 336: 1715–1719.

Frey P, Gérard P, et al. 2005. Variability and population biology of Melampsora rusts on poplar. In: Rust Diseases of Willow and Poplar (eds. MH Pei & AR McCracken), 63–72. Cambridge: CABI Publishing.

Glawe DA 2008. The powdery mildews. A review of the world's most familiar (yet poorly known) plant pathogens. *Annu Rev Phytopathol.* 46: 27–51.

Goffeau A, Barrell BG, et al. 1996. Life with 6000 genes. *Science.* 274: 546, 563–567.

Grigoriev IV, Cullen D, et al. 2011. Fueling the future with fungal genomics. *Mycology Int J Fungal Biol.* 2: 192–209.

Grigoriev IV, Nordberg H, et al. 2012. The Genome Portal of the Department of Energy Joint Genome Institute. *Nucl Acids Res.* 40: D26–D32.

Hacquard S, Joly DL, et al. 2012. A comprehensive analysis of genes encoding small secreted proteins identifies candidate effectors in *Melampsora larici-populina* (Poplar Leaf Rust). *Mol Plant Microbe Interact.* 25: 279–293.

Hacquard S, Kracher B, et al. 2013. Mosaic genome structure of the barley powdery mildew pathogen and conservation of transcriptional programs in divergent hosts. *Proc Natl Acad Sci USA.* doi:10.1073/pnas.1306807110

Hacquard S, Veneault-Fourrey C, et al. 2011. Validation of *Melampsora larici-populina* reference genes for in planta RT-quantitative PCR expression profiling during time-course infection of poplar leaves. *Physiol Mol Plant Pathol.* 75: 106–112.

Hahn M & Mendgen K. 1997. Characterization of *in planta* induced rust genes isolated from a haustorium-specific cDNA. *Mol Plant Microbe Inter.* 10: 427–437.

Hane JK & Oliver RP. 2008. RIPCAL: A tool for alignment-based analysis of repeat-induced point mutations in fungal genomic sequences. *BMC Bioinformatics.* 9: 12.

Innes L, Marchand L, et al. 2004. First report of *Melampsora larici-populina* on *Populus* spp. in Eastern North America. *Plant Dis.* 88: 85.

Kämper J, Kahmann R, et al. 2006. Insights from the genome of the biotrophic fungal plant pathogen *Ustilago maydis. Nature.* 444: 97–101.

Kemen E, Gardiner A, et al. 2011. Gene gain and loss during evolution of obligate parasitism in the white rust pathogen of *Arabidopsis thaliana. PLoS Biol.* 9: e1001094.

Kemen E & Jones JDG. 2012. Obligate biotroph parasitism: Can we link genomes to lifestyles? *Trends Plant Sci.* 17: 448–457.

Kemen E, Kemen AC, et al. 2005. Identification of a protein from rust fungi transferred from haustoria into infected plant cells. *Mol Plant Microbe Inter.* 18: 1130–1139.

Laurie JD, Ali S, et al. 2012. Genome comparison of barley and maize smut fungi reveals targeted loss of RNA silencing components and species-specific presence of transposable elements. *Plant Cell.* 24: 1733–1745.

Martin F, Aerts A, et al. 2008. The genome of Laccaria bicolor provides insights into mycorrhizal symbiosis. *Nature.* 452: 88–92.

Morin E, Kohler A, et al. 2012. Genome sequence of the button mushroom *Agaricus bisporus* reveals mechanisms governing adaptation to a humic-rich ecological niche. *Proc Natl Acad Sci USA.* 109: 17501–17506.

Newcombe G, Chastagner GA. 1993. 1st Report of the Eurasian poplar leaf rust fungus, *Melampsora-larici-populina*, in North-America. *Plant Dis.* 77: 532–535.

O'Connell RJ, Thon MR, et al. 2012. Lifestyle transitions in plant pathogenic *Colletotrichum* fungi deciphered by genome and transcriptome analyses. *Nat Genet.* 44: 1060–1065.

Parniske M. 2008. Arbuscular mycorrhiza: The mother of plant root endosymbioses. *Nat Rev Microbiol.* 6: 763–775.

Pedersen C, Ver Loren van Themaat E, et al. 2012. Structure and evolution of barley powdery mildew effector candidates. *BMC Genomics.* 13: 694

Piepenbring M. 2001 Smut fungi (*Ustilaginomycetes* and *Microbotryales, Basidiomycota*) in Panama. *Revista de Biologia Tropical* 49: 411–428.

Plett JM & Martin F. 2011. Blurred boundaries: lifestyle lessons from ectomycorrhizal fungal genomes. *Trends Genet.* 27: 14–22.

Raffaele S & Kamoun S. 2012. Genome evolution in filamentous plant pathogens: Why bigger can be better. *Nat Rev Microbiol.* 10: 417–430.

Schirawski J, Mannhaupt G, et al. 2010 Pathogenicity determinants in smut fungi revealed by genome comparison. *Science.* 330: 1546–1548.

Schneider RW, Hollier CA, et al. 2005. First report of soybean rust caused by *Phakopsora pachyrhizi* in the continental United States. *Plant Dis.* 89: 774–774.

Shiu PKT & Metzenberg RL. 2002. Meiotic silencing by unpaired DNA: Properties, regulation and suppression. *Genetics.* 161: 1483–1495.

Simon L, Bousquet J, et al. 1993. Origin and diversification of endomycorrhizal fungi and conicidence with vascular land plants. *Nature.* 363: 67–69.

Singh RP, Hodson DP, et al. 2011. The emergence of Ug99 races of the stem rust fungus is a threat to world wheat production. *Annu Rev Phytopathol.* 49:465–481.

Sohn J, Voegele RT, et al. 2000. High level activation of vitamin B1 biosynthesis genes in haustoria of the rust fungus *Uromyces fabae*. *Mole Plant Microbe Interact.* 13: 629–636.

Spanu P. 2012. The genomics of obligate (and non-obligate) biotrophs. *Annu Rev Phytopathol.* 50: 91–109.

Spanu P & Kämper J. 2010. Genomics of biotrophy in fungi and oomycetes—emerging patterns. *Curr Opin Plant Biol.* 13: 409–414.

Spanu PD, Abbott JC, et al. 2010. Genome expansion and gene loss in powdery mildew fungi reveal tradeoffs in extreme parasitism. *Science.* 330: 1543–1546.

Stajich JE, Wilke SK, et al. 2010. Insights into evolution of multicellular fungi from the assembled chromosomes of the mushroom Coprinopsis cinerea (*Coprinus cinereus*). *Proc Natl Acad Sci USA.* 107: 11889–11894.

Voegele RT, Hahn M, et al. 2009. The Uredinales: Cytology, biochemistry, and molecular biology. In: *The Mycota Plant Relationships*, Vol. 5 (ed. HB Deising),69–98. Berlin: Springer.

Voegele RT & Mendgen K. 2003. Rust haustoria: Nutrient uptake and beyond. *New Phytol.* 159: 93–100.

Wahl R, Wippel K, et al. 2010. A novel high-affinity sucrose transporter is required for virulence of the plant pathogen Ustilago maydis. *PLoS Biol.* 8(2): e1000303.

Wicker T, Oberhaensli S, et al. 2013. The wheat powdery mildew genome shows the unique evolution of an obligate biotroph. *Nature Genet.* doi:10.1038/ng.2704

Win J, Krasileva KV, et al. 2012. Sequence divergent RXLR effectors share a structural fold conserved across plant pathogenic oomycete species. *PLoS Pathog.* 8: e1002400.

Wirsel SG, Voegele RT, et al. 2001. Differential regulation of gene expression in the obligate biotrophic interaction of Uromyces fabae with its host Vicia faba. *Mol Plant Microbe Interact* 14: 1319–1326.

Wösten HAB. 2001. Hydrophobins: Multipurpose proteins. *Annu Rev Microbiol.* 55: 625–646.

8 The Mycorrhizal Symbiosis Genomics

Francis Martin and Annegret Kohler

Laboratory of Excellence ARBRE, UMR 1136 INRA-Université de Lorraine, Interactions Arbres-Microorganismes, INRA-Nancy, Champenoux, France

Introduction

Covering large areas of land, forests have a large impact on global climate, biodiversity, and human activity. Forest health, productivity and sustainability depend on above- and belowground microbial interactions to exchange nutrients, recycle carbon, and sustain diseases and harsh environmental conditions. As people's understanding of biological systems improves, thanks in part as a result of rapid advancements in molecular biology and ecology techniques, it is clear that organisms do not function independently but rather that they are heavily influenced by their microbial environment and, in their turn, they influence the microbial communities (microbiome) around them. In the past decade, it has become apparent that trees are colonized by microbes that probably shape many of their most important physiological processes (Martin, 2007; Martin, Perotto, et al., 2007; Gottel, Castro, et al., 2011; Wullschleger, Weston, et al., 2012). For effective forest management, it is therefore critical to understand these entangled biological interactions as well as effects of soil fertilization and other treatments, such as forest management, on their balance.

Low concentration of soil bio-available nutrients has driven tree species into mutualistic relations, so-called mycorrhizal symbiosis, with taxonomically diverse clades of rhizospheric fungi. In this beneficial symbiosis, fungal hyphal networks colonize plant roots and use host photosynthate to support extensive extramatrical hyphal webs with high absorptive surface area for nutrient element mass transfer from the substrate (Martin, 2007; Martin & Nehls, 2009). Because exploratory hyphae of mycorrhizal fungi radiate out from plant root systems in the upper soil profile and leaf litter (Lindahl, Ihrmark, et al., 2007), this extensive web of hyphae has access to an impressive array of nutrients, including inorganic and organic nitrogen compounds and inorganic phosphate, not normally available to plants (Martin & Nehls, 2009). It is via the mycorrhizal interface that up to 20 percent of all the carbon fixed by plants enters soils (Högberg, Nordgren, et al., 2001), while as much as 80 percent of plant

The Ecological Genomics of Fungi, First Edition. Edited by Francis Martin.

nitrogen and phosphate can be supplied by their mycorrhizal fungal partners (Brandes, Godbold, et al., 1998). Mycorrhizal symbioses are at the core of linkages between soil microbial processes, vegetation, soil carbon storage and release, and movements of nutrients and water. The mechanisms controlling the off-loading of these nutrients in the mycorrhizal root tips, however, and uptake by the plant remain largely a mystery (Chalot, Blaudez, et al., 2006; Müller, Avolio, et al., 2007; Doidy, Grace, et al., 2012). This is as a result in large part to the complexity of the tissue as well as the size and quantity mycorrhizal tissues available for study. With advancements in comparative genomics, microscopy, tissue dissection, and other molecular biology tools it is hoped that studies in these tissues will be facilitated in the coming years.

The knowledge surrounding the mode by which mycorrhizal fungi cause symbiosis is insufficient and the nature and regulation of the genes defining symbiosis still elude us (Plett & Martin, 2011; Veneault-Fourrey & Martin, 2011). Understanding the biology, ecology, and evolution of mutualistic associations between trees and symbiotic fungi in forest soils requires a novel, community-driven, multidisciplinary approach based on genomics and ecological data and tools. A better understanding of the interactions between symbiotic fungi and their plant hosts, but also wood- and litter decayers present in forest environments, and their evolutionary adaptive history in the face of changing conditions will create tools to predict how they are likely to adapt to future climate change. From this, forest management will be able to anticipate the consequences of future global climate change on the forest microbiome and mitigate problems before they arise, allowing for the preservation of important forest resources.

The high ecological, evolutionary, and economic importance of mycorrhizal symbionts has led to much interest in sequencing the genomes of mycorrhizal fungi. Completion of the genome sequence of the model tree *Populus trichocarpa* (Tuskan, Difazio, et al., 2006) has been a flagship project for mycorrhizal research; in its wake, the United States Department of Energy Joint Genome Institute (JGI) has sequenced the first mycorrhizal genome from the basidiomycete *Laccaria bicolor*. A decade ago it was emphasized that with the sequencing of the *L. bicolor* genome a series of opportunities would arise (Lammers, Tuskan, et al., 2004). Among them, that having in hand the genetic blueprints for the mycobiont and its host tree would provide the ability to take a holistic approach in understanding how the symbionts interact with the tree host and a platform for detailed comparative genomic analysis across the fungal taxa, including a comparison of saprotrophic, pathogenic and mutualistic species. The challenge being to use these sequenced genomes to determine how mycorrhizal fungi evolve and function. Substantial progress has been made in this direction. Since then, many species have been sequenced or are in progress (Table 8.1); however only a few of these have been published (Martin, Aerts, et al., 2008; Martin, Kohler, et al., 2010). The sequenced species span a wide section of the evolutionary tree of

Table 8.1 Current sequencing status of the mycorrhizal species targeted for sequencing within the framework of the Mycorrhizal Genome Initiative (MGI). Public release means that the genome sequence is available at the JGI MycoCosm site: http://genome.jgi.doe.gov/programs/fungi/index.jsf.

Species	Genome	Transcriptome
Amanita muscaria (L.) Lam.	Public release	Complete
Boletus edulis Bull.	Public release	Complete
Cenococcum geophilum Fr.	Public release	Complete
Cortinarius glaucopus (Schaeff.) Fr.	Complete	Complete
Hebeloma cylindrosporum Romagn.	Public release	Complete
Laccaria amethystina Cooke	Public release	Complete
Laccaria bicolor (Maire) P.D. Orton	Public release	Public release
Meliniomyces bicolor Hambl. & Sigler	Public release	Complete
Meliniomyces variabilis Hambl. & Sigler	Public release	Complete
Oidiodendron maius G.L. Barron	Public release	Complete
Paxillus involutus P.D. Orton	Public release	Complete
Paxillus rubicundulus P.D. Orton	Public release	Complete
Piloderma croceum J. Erikss. & Hjortstam	Public release	Complete
Pisolithus microcarpus (Cooke & Massee) G. Cunn.	Public release	Complete
Pisolithus tinctorius (Pers.) Coker & Couch	Public release	Complete
Rhizopogon vinicolor A.H. Sm.	Complete	Complete
Rhizopogon vesiculosus A.H. Sm.	Complete	Complete
Rhizoscyphus ericae (D.J. Read) W.Y. Zhuang & Korf	Assembly	Sequencing
Scleroderma citrinum Pers.	Public release	Complete
Sebacina vermifera Oberw.	Public release	Complete
Suillus luteus (L.) Roussel	Public release	Complete
Terfezia boudieri Chatin	Assembly	Rna prep
Tricholoma matsutake Singer	Public release	Complete
Tulasnella calospora (Boud.) Juel	Public release	Complete
Tuber melanosporum Vittad.	Public release	Public release
Tuber magnatum Pico	Sequencing	Complete
Tuber aestivum Vittad.	Assembly	Complete
Cantharellus cibarius Fr.	DNA prep	RNA prep
Choiromyces venosus (Fr.) Th. Fr.	Assembly	Sequencing
Gyrodon lividus (Bull.) Sacc.	Sequencing	Complete
Lactarius quietus (Fr.) Fr.	DNA prep	RNA prep
Leccinum scabrum (Bull.) Gray	DNA prep	RNA prep
Russula sp.	DNA prep	RNA prep
Thelephora terrestris Ehrh.	DNA prep	RNA prep
Tomentella sublilacina (Ellis & Holw.) Wakef.	DNA prep	RNA prep

Ascomycota and Basidiomycota (Fig. 8.1). This chapter will describe some of the evolutionary history and biological and ecological diversity of various types of mycorrhizal fungi with a particular focus on the ectomycorrhizal species with publicly released genome sequences. Additionally, it will show how sequencing of the *L. bicolor* and *Tuber melanosporum* genomes has enabled research relevant to ecosystem-scale processes, thereby opening opportunities for ecological genomics. A discussion of some significant unanswered questions will close the chapter.

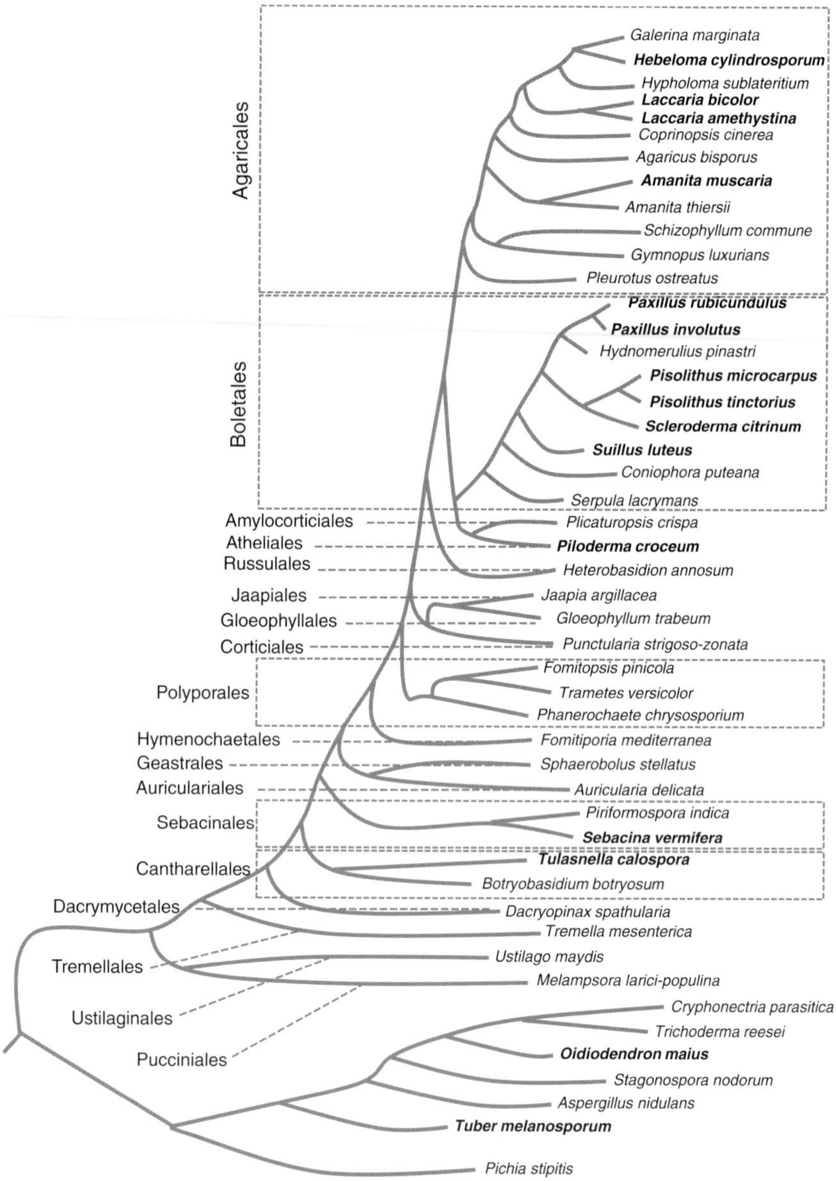

Figure 8.1 Phylogenetic distribution of mycorrhizal fungi sequenced within the Mycorrhizal Genomics Initiative. Bold font indicates ectomycorrhizal symbionts.

Ecological and Evolutionary Significance of Mycorrhizal Symbioses

Mycorrhizal symbioses are nearly universal in terrestrial plants. Based on host plant and characteristic symbiotic structures, several classes of mycorrhizal symbioses are currently recognized, with the two major types being the endocellular arbuscular mycorrhiza (AM) and the intercellular ectomycorrhiza (ECM). In AM association, the fungal hyphae penetrates host roots to form intracellular arbuscules and vesicles. In ECM, colonizing hyphae remain in the intercellular, apoplastic space forming the Hartig net. They do not penetrate the root cells. ECM are mostly formed by Basidiomycetes (e.g., *Amanita, Boletus, Sebacina*), but some are formed with ascomycetes (e.g., *Tuber*, *Terfezia*). Additionally, the ericoid mycorrhiza (ERM) has been regarded as the most specific of mycorrhizas because of its limitation to hosts belonging to a restricted number of families of the Ericales and the participation of a small group of ascomycetous fungi (e.g., Helotiales) as mycobionts in the association (Smith & Read 2008). Ericoid fungi form hyphal coils in outer cells of the narrow "hair roots" of plants in the family Ericaceae, such as *Vaccinium* and *Calluna*. Arbutoid mycorrhizal associations are variants of ECM found in certain plants in the Ericaceae in the genera *Arctostaphylos* and *Arbutus* characterized by hyphal coils in epidermal cells. The fungi of arbutoid mycorrhizas are Basidiomycetes, often the same fungal species that form ectomycorrhizal associations (Kennedy, Smith, et al., 2012). All orchids are myco-heterotrophic at some stage during their life cycle and form orchid mycorrhizas with a range of Basidiomycete fungi (e.g., *Tulasnella*). The mycobiont forms coils of hyphae within roots or stems of orchidaceous plants. This type of mycorrhiza is unique because the endophytic fungus supplies the plant with carbon during the heterotrophic seedling stage of orchidaceous plants. The mycorrhizal fungi are often Tulasnellales, a basidiomycetous order that contains plant parasites and saprobes capable of degrading complex carbohydrates, such as cellulose. Whether these different types of mycorrhizal fungi forming strikingly different anatomical structures and with contrasted biology and ecology differ in their gene repertoires and symbiosis-related gene networks is currently unknown. The genome of representatives of these various types of mycorrhizal symbioses are currently sequenced within the framework of the Mycorrhizal Genome Initiative (MGI), and this genomic resources should provide highlights on the biology, genetics, and ecology of these symbioses.

The invasion of the land by the ancestor of the vascular plants clearly seems to have been facilitated by the origin of symbiotic associations between these plants and certain soil fungi similar to those that are involved in AM symbiosis at the present time (Pirozynski & Malloch, 1975). Fossil stems of the prevascular land plant *Aglaophyton major* of the Lower Devonian found in the Rhinie chert (~ 400 million years ago) and early gymnosperms from the

Carboniferous contain AM symbiosis-related structures, and this type of widespread mycorrhizal symbiosis continues to be found in many tree species (Smith & Read, 2008). Independently evolving ECM fungi diversified as early as the late Jurassic or early Cretaceous period. Recent work would suggest that the origin of ECM fungi might in fact be from tropical climes. ECM fungi likely developed in present-day South America ~ 135 Mya, forming mycorrhizal associations with the Pinaceae and angiosperm trees in the Betulaceae, Myrtaceae, Fabaceae and Dipterocarpaceae, that can form dominant stands in warm temperate and tropical regions (Tedersoo, May, et al., 2010). Nowadays, trees hosting ECM form extensive forests in areas, such as much of Eurasia and North America, that now experience strongly seasonal climatic conditions or have poor soils. Multigene phylogenetic analyses and identification studies suggest that ECM symbiosis has arisen independently and persisted at least 66 times in fungi, in the Basidiomycota, Ascomycota, and Zygomycota (Tedersoo, May, et al., 2010). A recent evolutionary scenario suggests that ECM fungi have evolved from saprotrophic precursors (wood and litter decayer lineages) at multiple times through convergent evolution (Hibbett, Gilbert, et al., 2000; James, Kauff, et al., 2006; Hibbett & Matheny, 2009; Tedersoo, May, et al., 2010; Floudas, Binder, et al., 2012). The evolutionary path that led to emergence of ERM symbioses is less clear. ERM fungi appear less dependent on plants than other mycorrhizal types, because of their superior saprotrophic abilities; so much so that they have been hypothesized to be "facultative symbionts," representing recently recruited lineages of soil decomposer fungi. An alternative hypothesis is that they have evolved with their host from the ECM habit and have reacquired de novo the genetic information required to degrade complex organic compounds. Whether ancestor-derived or acquired de novo, ERM saprophytic capabilities must be subject to stringent levels of regulation.

Mycorrhizal Genomics

The First Ectomycorrhizal Genomes

A major step toward unlocking the similarities and differences between mycorrhizal fungi has been the cooperative effort of our group at INRA and the US Department of Energy JGI and Genoscope into genome sequencing of the ECM fungi, *L. bicolor* and *T. melanosporum* (Martin, Aerts, et al., 2008; Martin, Kohler, et al., 2010). Comparative analysis of these genomes set the stage for future genome-supported studies on the biology and ecology of this important group of plant-interacting fungi. This analysis shed light on the genetic similarities between ECM fungi and their saprotrophic cousins as well as identified key genes in the regulation of symbiosis (Plett & Martin, 2011).

L. bicolor belongs to the Hydnangiaceae in the Agaricales, a large order within the Basidiomycota. The genome of *L. bicolor* was found to be larger than most of its saprotrophic relatives—60 Mbp with ~23,000 predicted protein-coding genes, of which most have been verified by transcript profiling using NimbleGen arrays and RNA-Seq (Martin, Aerts, et al., 2008). In comparison to other sequenced Basidiomycetes that cover 550 million years of evolutionary history, *L. bicolor* has the largest complement of predicted proteins to date, suggesting that it was through expansion of gene coding space that symbiotic genes were acquired in Agaricales (Martin, Aerts, et al., 2008).

In contrast, the large 125 Mbp-genome of the ascomycete *T. melanosporum* (Tuberaccae, Pezizales) (Martin, Kohler, et al., 2010) has a low gene repertoire with only ~7,500 predicted protein-coding regions. These contrasting genomes, therefore, show that absolute coding space is not the prerequisite of a symbiotic organism, but rather it is likely in the high percentage of orphan genes that we may be able to define a "symbiotic toolbox"— the complement of genes used by ECM fungi to broker symbiosis with plants. The genome sequencing of these two ECM fungal species has given a number of insights into the molecular mechanisms in action during the symbiotic development (Plett & Martin, 2011).

Compared to their saprotrophic relatives, genomes of these ectomycorrhizal symbionts exhibit common genetic trends:

- Proliferation of repetitive elements, suggesting a fluid genome.
- Reduction of gene families coding for secreted degradative enzymes acting on plant cell wall polysaccharides; this may be a way to conceal the hyphae from the host defense sensing systems during infection.
- Lack of secondary metabolite biosynthetic gene clusters.

These genomic features appear to be necessary adaptations for mutualistic symbiosis (Plett & Martin, 2011). They have also been observed in obligate biotrophic pathogens (Spanu, Abbott, et al., 2010; Kemen, Gardiner, et al., 2011) and the AM fungus *Glomus irregulare* DAOM197198 (Tisserant, Da Silva, et al., 2011). It must be stressed, however, that such a conclusion is based on the availability of only two ECM genomes (Martin & Selosse, 2008; Plett & Martin, 2011), although new research into the saprotrophic and symbiotic *Amanita* species would suggest similar mechanisms are at work in the evolution of this genus (Wolfe, Tulloss, et al., 2012).

Despite the aforementioned shared genomic features. the Ascomycete *T. melanosporum* and the basidiomycete *L. bicolor* encode strikingly different proteomes—compact with few multigene families, versus large with many expanded multigene families. Effector-like proteins, such as the *L. bicolor*–induced mycorrhiza-induced small secreted protein (MiSSP) MiSSP7 (Plett, Kemppainen, et al., 2011), are not expressed in *T. melanosporum* ectomycorrhizas. Differences between the enzyme repertoires of *T. melanosporum* and *L. bicolor* also suggest differences in the mode of metabolic interaction of the two symbionts with their respective host plants. A striking difference is the presence of an invertase gene in *T. melanosporum*, whereas *L. bicolor* has none and is therefore completely dependent on its host for its provision of glucose (Martin, Kohler, et al., 2010). In contrast, *T. melanosporum* could access and hydrolyse the plant-derived sucrose. This would suggest that although both fungi develop symbiotic relationships with plants, *T. melanosporum* is probably less dependent than *L. bicolor*.

Sequencing of the *L. bicolor* and *T. melanosporum* genomes and the subsequent development of genetic, transcriptomic, and proteomic resources have solidified the role of *Laccaria* and *Tuber* as a model taxa for molecular studies in ectomycorrhizal biology research (Deveau, Kohler, et al., 2008; Fajardo López, Dietz, et al., 2008; Labbé, Zhang, et al., 2008; Lucic, Fourrey, et al., 2008; Niculita-Hirzel, Labbé, et al., 2008; Courty, Hoegger, et al., 2009; Kemppainen, Duplessis, et al., 2009a; Kemppainen, Pardo, et al., 2009b; Rajashekar, Kohler, et al., 2009; Reich, Göbel, et al., 2009; Larsen, Trivedi, et al., 2010; Plett, Kemppainen, et al., 2011; Lackner, Misiek, et al., 2012; Vincent, Kohler, et al., 2011). These studies have revealed the identity of numerous symbiosis-related enzymes and membrane transporters and the role of MiSSP7 in *L. bicolor* mycorrhizal ability. In addition, they have led to population genetic analysis studies of these organisms (Rubini, Belfiori, et al., 2011a; Rubini, Belfiori, et al. 2011b; Hortal, Troch, et al., 2012). Available genome sequences and transcript profiles, together with other "-omics" methods, such as metabolomics, glycomics, and lipidomics, will be more powerful, and accordingly, will strengthen the understanding and characterization of mycorrhizal symbiosis.

The Mycorrhizal Genomics Initiative

The findings obtained on *L. bicolor* and *T. melanosporum* genomes and symbiosis-related transcriptomes suggest that the ECM condition represents a syndrome of variable traits and that ECM fungi share fewer functional similarities in their molecular toolboxes than anticipated (Plett & Martin, 2011). This contention emphasizes the importance of having sequence data for more than one representative of each phylum of mycorrhizal fungi. In addition to *L. bicolor*, the species targeted for genome sequencing at JGI were *Paxillus involutus* by the Community Sequencing Program (CSP) in 2008, *Rhizopogon salebrosus* by CSP in 2009, and *Pisolithus tinctorius* and *P. microcarpus* by CSP in 2010 (http://www.jgi.doe.gov/CSP/overview.html). The latter taxa belong to the Boletales, a large phylum of symbiotic basidiomycetes. This overall lack of broad phylogenetic considerations in the selection of mycorrhizal genomes for sequencing has led to a strongly biased representation of symbiotic phylogenetic diversity. To evaluate the potential benefits of a more systematic effort we proposed to embark on a large-scale project, the so-called Mycorrhizal Genomics Initiative (MGI), to sequence 30 additional genomes of mycorrhizal species selected for (1) their phylogenetic novelty, (2) their ability to establish different types of mycorrhizal symbiosis (ECM, ERM, and orchid mycorrhizas), (3) their prominence in ecological settings, (4) their host specificity, (5) their ability to promote growth of trees with sequenced genomes (*Populus, Eucalyptus, Quercus*), (6) their use in bioremediation, and (7) their taxonomic relationships with already sequenced mycorrhizal genomes to explore the intraclade variability in symbiosis gene repertoire. The availability of mycelial cultures and the feasibility of producing high-quality DNA were also key factors in the selection. As of this writing, 16 mycorrhizal genomes have been publicly released on the JGI MycoCosm web portal (Grigoriev, Nordberg, et al., 2012; http://genome.jgi.doe.gov/programs/fungi/index.jsf; *see* Table 8.1) and 20 additional genomes will be publicly released by the end of 2013 (see also the MGI web portal: http://mycor.nancy.inra.fr/IMGC/MycoGenomes/index.html).

The MGI taxa include representatives of the major clades (orders or subclasses) of culturable Fungi that contain mycorrhizal taxa (*see* Fig. 8.1). The fact that mycorrhizal fungi appear to be independently derived from multiple saprobic lineages means that genomic data will provide independent assessments of what is required to become mycorrhizal and the retained saprotrophic ability of the selected species. The ECM Basidiomycota selected for sequencing represent 9 of the approximately 18 major clades (orders and subclasses) of Agaricomycotina (*see* Fig. 8.1). The 9 clades that are not targeted in the MGI contain only wood decayers as far as has been demonstrated. This set of target species includes the first ECM genomes of five of the major groups of Agaricomycotina, including the Atheliales, Russulales, Thelephorales, Cantharellales, and Sebacinales. Three of these clades—the Russulales, Thelephorales, and Cantharellales—contain some of the most diverse and abundant ECM formers. Other groups targeted are significant largely because of their phylogenetic position. In particular, the Atheliales (such as *Piloderma croceum*) is the sister group of the Boletales, which contains a major concentration of ECM forms (e.g., the porcini mushroom, *Boletus edulis*, and the gasteromycetes *P. tinctorius* and *P. microcarpus*) and could provide insight into the origins of ECM in this important assemblage. The Sebacinales is noteworthy because it is the sister group of all other Agaricomycetes (the clade of Agaricomycotina that excludes Dacrymycetales and Tremellomycetidae, which both lack ECM species). Thus, the sample of species selected in the MGI span the root node of the Agaricomycetes and will provide an opportunity to estimate the gene content, and therefore, the ECM potential of the common ancestor of the Agaricomycetes. Genome sequences of this suite of taxa will also enable resolution of the backbone of the phylogeny of Agaricomycotina using phylogenomic approaches, which has remained poorly resolved, despite analysis of data sets with five or six genes constructed through polymerase chain reaction–based methods (James, Kauff, et al. 2006).

Seven of the groups from which ECM species are targeted also contain saprotrophic species, which have been sequenced (Floudas, Binder, et al., 2012). Comparison of the genomes of closely related ECM and non-ECM taxa will provide clues to the genetic bases of transitions between ECM and non-ECM lifestyles, the expansion or contraction of polysaccharide-degrading enzymes. Such comparisons may also shed light on the possibility of reversals from ECM to decayer ecologies, which was suggested based on early phylogenetic studies but which remains controversial (Hibbett & Matheny, 2009).

The Ascomycota selected for sequencing in the MGI represent two distantly related orders, the Pezizales (Pezizomycetes) and Helotiales (Leotiomycetes). The mycorrhizal condition in these two groups is almost certainly independently evolved. The recently sequenced genome of the black truffle (*T. melanosporum*) provides an example of ECM in the Pezizales and will be a useful comparison to that of *Terfezia boudieri* and *Choiromyces venosus* (also Pezizales), especially in relation to acquisition or loss of capability for dual endo- or ectomycorrhizal colonization. The Helotiales species selected for sequencing (*Meliniomyces bicolor, Meliniomyces variabilis, Rhizoscyphus ericae*) are all closely related, yet represent a range of contrasted abilities for colonization of their host intracellularly (ERM) or formation of ectomycorrhizal structures. A comparison of these genomes and symbiosis-related transcriptomes may shed more light on the evolution and regulation of gene families involved in the degradation of host cell walls. These helotialean taxa have also some of the highest saprotrophic capabilities known to mycorrhizal fungi (Straker, 1996). A function-driven comparison of their genomes with that of basidiomycetes fungal wood and litter decayers (Eastwood, Floudas, et al., 2011; Floudas, Binder, et al., 2012; Morin, Kohler, et al., 2012) will provide some interesting insights in the evolution of gene families involved in organic matter decomposition. Further benefit could be gained in understanding the genetic basis of resistance to heavy metal contamination and potential for bioremediation, from the exploration of the genome of the helotialean *Rhizoscyphus ericae*, in tandem with that of *Oidiodendrion maius*, a non-helotialean Leotiomycete.

What follows is the summary of the main biological and ecological features of the mycorrhizal taxa having their genome publicly available on the JGI MycoCosm portal (http://genome.jgi-psf.org/programs/fungi/index.jsf).

Hebeloma cylindrosporum: A Model Species for Ectomycorrhizal Research The Basidiomycete Agaric *H. cylindrosporum* Romagnesi (Agaricales, Cortinariaceae) has only been reported to occur in Europe. It is frequently found in forest stands developing on sand dunes with little organic matter along the Atlantic or Mediterranean coasts (Marmeisse, Guidot, et al., 2004). *H. cylindrosporum* thrives in newly established forests or in disturbed areas where it is frequently associated with different pine trees such as *Pinus pinaster*. The most remarkable feature of *H. cylindrosporum* as an ectomycorrhizal fungal species is that its entire life cycle, including fruit body formation, can be obtained under axenic conditions in the laboratory using defined culture media (Marmeisse, Guidot, et al., 2004). It can be routinely transformed using *Agrobacterium tumefaciens*, and insertional mutant libraries are available making it possible to reverse genetic approach to identify fungal functions essential for symbiosis establishment. A 126× coverage of the genome yielded an assembly v2.0 of 176 scaffolds totaling ~38 Mbp (http://genome.jgi.doe.gov/Hebcy2/Hebcy2.home.html). The total number of nuclear genes, at approximately 15,380, is reduced relative to *L. bicolor* but not drastically.

Amanita muscaria: The Charismatic Mushroom The charismatic *A. muscaria* (Agaricales, Amanitaceae) may be the most widely recognized fungus in the world; drawings of its mushroom are featured in books for children and fairy tales. The morphological species encompasses a complex of eight undescribed biological species, and the sequenced strain belongs to "clade 1" in the phylogeny of Geml, Laursen, et al. (2006). The species is a geographically widespread symbiont of conifers and hardwoods in boreal and temperate ecosystems. *Amanita* species have been introduced to America, Australia, and New Zealand, where they spread in association with local tree species; these invasive species are of increasing concern to foresters (Pringle, Adams, et al., 2009). Unlike most ectomycorrhizal species, *A. muscaria* can be cultured and its symbiosis synthesized in vitro. For this reason, it was an early target of research focused on the molecular underpinnings of ectomycorrhizal symbioses (Nehls, Mikolajewsk, et al. 2001, Nehls, Grunze, et al., 2007). A 125× coverage of the genome yielded an assembly of 1,101 scaffolds totaling ~41 Mbp (http://genome.jgi.doe.gov/Amamu1/Amamu1.home.html). The total number of predicted nuclear genes is ~18,000.

Laccaria amethystina: The Amethyst Deceiver *Laccaria* Berkeley and Broome is a cosmopolitan genus of mushrooms (Agaricales, Hydnangiaceae) collected frequently throughout North America and Eurasia. Its taxa make up a sizeable part of the basidiomycetous ECM species and have been reported from every continent except Antarctica. *L. amethystina* (Bull. ex Mérat) Murr, commonly known as the Amethyst Deceiver, is ectomycorrhizal, forming symbiotic associations with hardwoods or conifers. It produces deep purple, edible mushrooms, that grow among moss and leaf litter under deciduous as well as coniferous trees. A 156× coverage of the genome yielded an assembly of 1,299 scaffolds totaling ~52 Mbp (http://genome.jgi.doe.gov/Lacam1/Lacam1.

info.html). The total number of nuclear genes, at approximately 21,000, is similar to *L. bicolor*. The comparison of the genomes of *L. bicolor* and *L. amethystina*, which split from the outgroup species, *Laccaria laccata* and *Laccaria proxima* ~15 million years ago (Ryberg & Matheny, 2011), will provide insights on the various types of genomic variations in this species. The analysis of sequence polymorphism in MiSSPs, such as the effector protein MiSSP7 (Plett, Kemppainen, et al., 2011), will highlight the evolution of key symbiosis-related effectors in the *Laccaria* clade. This would allow mycologists to extract the most information about adaptive mutations that are most likely to be important to symbiosis in a well-investigated symbiotic clade.

Paxillus rubicundulus: An Alder-specific ECM Symbiont *P. rubicundulus* P.D.Orton (Basidiomycota, Agaromycetideae, Boletales Paxillaceae) is an ectomycorrhizal basidiomycete specifically associated to alders. It is found associated to *Alnus glutinosa* and *Alnus incana* in wetlands or along rivers. This species belongs to the Paxillaceae family, in which some members are hygrophilic and highly specialized on alders, such as *P. rubicundulus* or *Gyrodon lividus*, whereas some other members have a large ecological range and are generalist, such as *Paxillus involutus*. A 80× coverage of the genome yielded an assembly v2.0 of 1,671 scaffolds totaling ~60.7 Mbp (http://www.jgi.doe.gov/Paxillus_rubicundulus/). The total number of predicted nuclear genes is ~22,000, a repertoire slightly higher than *P. involutus* (17,968 genes for 58 Mbp; http://www.jgi.doe.gov/Paxillus/).

Piloderma croceum *P. croceum* (synonym *P. fallax*) is a Basidiomycete (Agaromycetideae, Atheliales, Atheliaceae), which forms ECM symbiosis. *P. croceum* is a broad host range fungus and a common mutualist of both conifer and hardwood species in the North America, Europe, and Australasia. It is an established model fungus for ecological and for physiological studies. Established plant models for interaction studies with *P. croceum* include the broad-leaved trees *Quercus robur* and *Betula pendula* and the conifer *Picea abies*. The fungus has been detected from the mineral soil horizons as well as on granitic rocks and has been shown to scavenge ions as a result of organic acid extraction and efficient mineral uptake. A 134× coverage of the genome yielded an assembly of 715 scaffolds totaling ~59 Mbp (http://www.jgi.doe.gov/Piloderma/). The total number of predicted nuclear genes is ~21,580.

Pisolithus tinctorius, Pisolithus microcarpus, and Scleroderma citrinum: Common and Widespread Ectomycorrhizal Boletales The three ectomycorrhizal taxa, *P. tinctorius (P. arhizus)*, *P. microcarpus*, and *S. citrinum* are classified within the Sclerodermataceae (Boletales). They are a common and widespread powdery-spored Gasteromycetes, which produces sporocarps termed *earthballs* in different forest environments or adjacent to forest area. These sporocarps appear early in the fruiting succession of ECM fungi. There are primary colonizers of mining waste, which enables it to spread rapidly and to colonize young root systems of numerous tree species. These Sclerodermataceae have a worldwide range. *P. tinctorius* appears to be mainly associated to conifer in the Northern Hemisphere, whereas *P. microcarpus* is found on *Eucalyptus* roots in Australasia (Martin, Diez, et al., 2002). A range of host genera has been reported for *S. citrinum*, including *Populus* and *Eucalyptus*, two sequenced host trees. A 81× coverage of the *S. citrinum* genome yielded an assembly of 938 scaffolds totaling ~56 Mbp (http://jgi.doe.gov/Scleroderma/). The total number of predicted nuclear genes is ~21,000. A 74× coverage of the *P. tinctorius* genome yielded an assembly of 610 scaffolds totaling ~71 Mbp with ~22,700 predicted nuclear genes (http://jgi.doe.gov/Pisolithus_tinctorius/). The *P. microcarpus* genome with 53 Mbp (http://jgi.doe.gov/Pisolithus_microcarpus/) is significantly smaller than *P. tinctorius* genome but with a similar set of nuclear genes (~21,000). This size difference is mainly related to the abundance of transposable elements in *P. tinctorius*. The comparison of the pine-associated *P. tinctorius* and eucalypt-associated *P. microcarpus* should facilitate the identification of gene networks involved in host specificity.

Suillus luteus: A Cosmopolitan Ectomycorrhizal Fungus Commonly referred to as slippery Jack, *S. luteus*, also belongs to the Boletales (Suillaceae). The large genus *Suillus* is a sister group of the genus *Rhizopogon*. *Rhizopogon vinicolor* has been sequenced within the AFTOL project (J. Spatafora, personal communication). Both genera are differentiated by a distinct ontogeny of the reproductive organs. Comparative analyses of these genomes will contribute to the exploration of the intraclade variability in symbiosis gene repertoire. *S. luteus* is a cosmopolitan ECM fungus whose natural range of distribution matches the range of distribution of its host plants, the *Pinus* species. It is particularly abundant in young pine forest or planted stands, from the Andes to the boreal forests. The species is a pioneer species, which quickly starts sexual reproduction from large edible sporocarps that produce massive quantities of basidiospores, spread by wind and mammals. It forms conspicuous, though relatively few mycorrhizas from which an extensive external mycelium develops into the mineral soil. It appears frequently on man-disturbed sites wherever pines are planted or start primary succession. A 81× coverage of the *S. luteus* genome yielded an assembly of 1,944 scaffolds totaling ~37 Mbp (http://jgi.doe.gov/SlipperyJack/). The total number of predicted nuclear genes is ~18,300. A significant limitation of the mycorrhizal genomics project to date was the inclusion of species belonging to distantly related lineages in a single analysis. The approach makes it difficult to identify genetic changes caused by any single evolutionary force. The availability of several saprotrophic and symbiotic Boletales genomes (*Serpula lacrymans, Coniophora puteana, S. citrinum, P. involutus, P. microcarpus, P. tinctorius,*

S. luteus, Boletus edulis) should facilitate the identification of symbiosis-related gene networks at the order level.

Sebacina vermifera: The Orchid Mycorrhizal Fungus The selected Australian orchid mycorrhiza isolate (MAFF 305830) belongs to the basidiomycetous order Sebacinales (subgroup B). This order encompasses ubiquitously distributed taxa that are basal in the Agaricomycetes with diverse mycorrhizal abilities, ranging from ECM to ERM, orchid mycorrhiza, and root endophytes. Because of their inconspicuous or even absent basidiomes, this group of fungi has been often overlooked and underestimated in its ecological and potential economic importance. The orchid mycorrhiza represent the most basal group with known mycorrhizal capabilities. Because of the relative ease with which some species of the subgroup B can be grown and manipulated in the laboratory, several taxa including *Piriformospora indica* and various *S. vermifera* isolates are now widely used in basic research of plant and fungus interaction. Genome sequencing of a *S. vermifera* isolate is therefore useful in evolutionary genomics and in dissecting the molecular mechanisms of symbiosis. The MAFF 305830 strain was isolated from the terrestrial orchid, *Cyrtostylis reniformis*, from South Australia. It is able to stimulate germination of seed of species of *Microtis* (Orchidaceae) and to colonize by inter- and intracellular hyphae roots of barley, tomato, and switchgrass. A 117× coverage of the *S. vermifera* genome yielded an assembly of 546 scaffolds totaling ~38 Mbp (http://genome.jgi.doe.gov/Sebve1/Sebve1.home.html) The total number of predicted nuclear genes is ~ 15,312.

Tulasnella calospora: The Orchid Mycorrhizal Symbiont *T. calospora* belongs to Tulasnellaceae, a taxonomic group of Basidiomycetes that is currently nested in the Cantharellales but that may represent a sister group, the Tulasnellales. The effuse, inconspicuous fruiting bodies found in nature are most often overlooked in field surveys. Fungi in the genus *Tulasnella* are the major symbionts of terrestrial and epiphytic orchids, but they have also been reported to form ECM on different plant hosts. *Tulasnella* species have been found in the mycorrhizal roots of orchid species growing in forest as well as in open habitats. As ECM symbionts have been found in the genus *Tulasnella*, sequencing of this fungus will open the possibility to compare the genetic background of an endomycorrhizal and an ectomycorrhizal behavior in Basidiomycetes. Given the obligate nature of the mycorrhizal relationship for the germinating orchid embryos, the genome analysis will not only allow considerable advances in the understanding of the genetic and functional basis of the orchid symbiosis, but it may also have implications for the conservation of these endangered plant species. A 100× coverage of the *T. calospora* genome yielded an assembly of 1,335 scaffolds totaling ~62

Mbp (http://genome.jgi.doc.gov/Tulca1/Tulca1.home.html). The total number of predicted nuclear genes is ~19,600.

Cenococcum geophilum: The Pan-Global Fungus *C. geophilum* is an ascomycetous fungus placed into the Dothideomycetes, where it represents the only known ectomycorrhizal species within this large and ecologically diverse class of Ascomycota (*see* Chapter 6). *C. geophilum* is one of the most common and globally abundant genera of mycorrhizal fungi, forming black ECM with darkly pigmented, stringy hyphae emanating from root tips. It has broad host and habitat ranges and is often the dominant mycorrhizal fungus on the tree root systems in forests of arctic, temperate, and subtropical environments. Therefore, understanding its ecological role in forest ecosystems is of great significance. *C. geophilum* is highly resistant to desiccation and its ectomycorrhizas are abundant during drought when other mycorrhizal species decline. Although ubiquitous, the biology of *C. geophilum* is poorly understood. It forms sclerotia as resistant propagules, but no definitive sexual or asexual spore-producing structures are known. Studies of fine-scale diversity of *C. geophilum* populations, however, revealed a high level of genetic polymorphism among individuals consistent with the occurrence of recombination mechanisms and suggesting that the fungus is reproducing sexually in nature.

The genome assembly was hindered by a high content in repeated sequences. A 75× coverage of the *C. geophilum* genome yielded an assembly of 268 contigs, totaling ~177 Mbp (http://genome.jgi.doe.gov/Cenge1/Cenge1.home.html). As of this writing, an improved draft assembly has been completed, and nearly 2,300 gaps were closed using PBJelly (English, Richards, et al., 2012) and Pacific Biosciences RS long-read sequencing technology. The total number of predicted nuclear genes is ~27,500. However, the gene predictions are short (on average) and have poor transcript (EST) support compared to a typical Ascomycete. In addition, this genome has an astonishing number of gene predictions compared to other JGI Dothideomycete genomes (http://jgi.doe.gov/Dothideomycetes/; Ohm, Feau, et al., 2012). The predicted gene repertoire has a huge spike of short proteins (<100 amino acids), which are probably artefactual. Version 2 of the annotation should confirm this contention. The genomic sequence of *C. geophilum* will provide insights in the evolution of the mycorrhizal symbiosis in Dothideomycetes by comparing the presence of symbiosis-related genes found in other ECM ascomycetes, such as *Tuber* and *Terfezia* species.

Oidiodendron maius: An Endomycorrhizal Ascomycete *O. maius* belongs to the Leotiomycetes (Ascomycotina). *O. maius* is an interesting experimental organism, being both an endomycorrhizal fungus (with ericaceous plants, i.e. *Vaccinium myrtillus, Calluna vulgaris*) and a metal-tolerant fungus. The metal tolerance is particularly evident for strains isolated from the roots of

V. myrtillus growing on heavy-metal–polluted soils (Martino,Turnau, et al., 2000). *O. maius* can be easily grown in vitro, where it reproduces asexually by forming conidia with a single haploid nucleus, that can germinate and produce a homokaryotic mycelium. The *O. maius* genome was sequenced using the Illumina and Roche (454) platforms and this resulted in a 29× coverage assembly with 100 scaffolds totaling ~46 Mbp (http://genome.jgi.doe.gov/Oidma1/Oidma1.home.html). The total number of predicted nuclear genes is ~ 16,700. The ericaceous *O. maius* contains the largest set (>250) of glycosyl hydrolases acting on plant cell wall polysaccharides identified in any sequenced fungus so far.

Beyond the Genomes

In summary, the genome of mycorrhizal species released over the last few years, combined with previous studies of the *L. bicolor* and *T. melanosporum* genomes, provides a rich foundation for future studies to elucidate the unique features of these ubiquitous plant symbionts. However, many long-term challenges remain for the application of genomics to enhance understanding of the evolution, development, and functioning of mycorrhizal symbioses. The genome sequenced within the framework of the MGI will be used in comparative studies that illustrate the diversity and evolution of the mycorrhizal symbioses. Comparisons of multiple genomes should enable determination of the essential components of symbiosis mechanisms and genome-enable transcriptome and secretome analyses should allow within- and between-species analyses of the transcriptomes of the symbiotic interactions, in addition to the gene expression of key carbohydrate-cleaving enzymes (CAZymes), if any, during the saprotrophic phase preceding the interaction with the plant. Analyses of transposon distribution, synteny, and other higher level genomic features should provide clues to processes of genome evolution. Mycologists are increasingly aware that some fungal species have dual or multiple ecological abilities, and a potentially exciting avenue for future research would target the sequencing of species with this kind of trophic complexity, for example, the endophytic and mycorrhizal Sebacinales (Zuccaro, Lahrmann, et al., 2011), or the saprotrophic and mycorrhizal fungi colonizing Ericaceae and Orchids (Martos, Dulormne, et al., 2009). Moreover, the multiple apparent transitions in the saprotrophism-mutualism lifestyles suggest that much remains to be learned about how, when, and where symbiosis molecular toolboxes have been acquired.

Concurrent with the sequencing of the genome of *L. bicolor* and *T. melanosporum* (Martin, Aerts, et al., 2008; Martin, Kohler, et al., 2010), microarray- and RNA-Seq based analysis of the transcriptome of free-living mycelium and ectomycorrhizal root tips were carried out (Martin, Aerts, et al.,

2008; Martin, Kohler, et al., 2010; Tisserant, Da Silva, et al., 2011). These studies were foundational in establishing that *L. bicolor* encodes symbiotic specific effector-like genes, that it expresses few CAZymes during interaction with the plant and that, although there were similarities between the gene sets induced in *L. bicolor* by two different plant hosts, that the host plant also had an impact on the transcriptome of the colonizing fungus. Combined to laser capture microdissection system, they also revealed that the fungal mantle (known as the preferential storage compartment) and the Hartig net (known as the metabolically active tissue) have relatively similar metabolic activity because most genes-encoding enzymes of the nitrogen and carbon metabolism pathways are constitutively expressed in both compartments (Hacquard, Tisserant, et al., 2013). This finding suggests that there is no clear metabolic zonation between the fungal mantle and the Hartig net and supports the idea that only slight or targeted alteration of gene expression might be sufficient to regulate the functions specific to each compartment (e.g., membrane transporters). Genome-based RNA-Seq transcript profiles from different parts of a mycorrhizal individual (symbiotic tissues, mycelial mats, rhizomorphs, extraradical hyphal webs) could be used as a proxy to infer either physiological specialization or uniformity between these fungal compartments. These kinds of transcriptomic experiments, currently conducted in *Paxillus-Betula* microcosms (Wright, Johansson, et al., 2005), should be transposed in the near future to more natural settings, including perhaps the edges and center of a network growing on a forest floor.

Despite significant advances in recent years, gaps remain in the understanding of basic biological processes that underlie how mycorrhizal individuals, populations, communities, and ecosystems respond to the environment. Besides the genomics of single species, metagenomics or ecological genomics emerged as a rapidly expanding research field, and this new research will likely help in filling these gaps. Whole-genome shotgun analyses begin with sequences sampled from the entire community metagenome (*see* Chapters 13 and 14). These sequences can mapped or BLASTed to reference sequence databases and the frequencies of enzymes and other gene products so determined can be assigned to pathways, allowing inference of the overall metabolic potential of the community and inference of potentially explanatory functional biomarkers (Baldrian, Kolařík, et al., 2012). In the case of bacteria, metagenomics has benefited from the large number of available genomes for precise taxonomic annotation of anonymous environmental sequences (*see* Chapter 13). Within fungi, the current taxonomic distribution of sequenced species is a severe limitation to the precise taxonomic identification of soil fungal sequences. For this reason, ongoing sequencing efforts, such as the CSP project "*Metatranscriptomics of Forest Soil Ecosystems*" (http://mycor.nancy.inra.fr/blogGenomes/?page_id=3262) are targeting ecologically relevant and abundant "keystone" fungal species found in soil or other substrates in which fungi play critical roles. In complex systems involving many interacting fungal and tree species, it is believed that the a genes-to-ecosystem approach will provide a better understanding of the role of a mycorrhizal fungal species' impact on the plant and soil microbial communities and ecosystems.

Acknowledgments

The manuscript benefited from discussions in the "Mycorrhizal Genomics Initiative" workshops. In particular, we thank Igor Grigoriev, David Hibbett, Claude Murat, Emmanuelle Morin, and Kerrie Barry. The Mycorrhizal Genome Initiative is supported by INRA, the Region Lorraine Research Council, the US Department of Energy (DOE)—Oak Ridge National Laboratory Scientific Focus Area for Genomics Foundational Sciences, the US Department of Energy Joint Genome Institute (Office of Science of the US Department of Energy under contract no. DE-AC02-05CH11231) and the European Commission. FM's lab is part of the Laboratory of Excellence ARBRE (ANR-12-LABX-ARBRE-01).

References

Baldrian P, Kolařík M, et al. 2012. Active and total microbial communities in forest soil are largely different and highly stratified during decomposition. *ISME J.* 6: 248–58.

Brandes B, Godbold DL, et al. 1998. Nitrogen and phosphorus acquisition by the mycelium of the ectomycorrhizal fungus *Paxillus involutus* and its effect on host nutrition. *New Phytol.* 140:735–743.

Chalot M, Blaudez D, et al. 2006. Ammonia: A candidate for nitrogen transfer at the mycorrhizal interface. *Trends Plant Sci.* 11: 263–266.

Courty PE, Hoegger PJ, et al. 2009. Phylogenetic analysis, genomic organization, and expression analysis of multi-copper oxidases in the ectomycorrhizal basidiomycete *Laccaria bicolor. New Phytolog.* 182: 736–750.

Doidy J, Grace E, et al. 2012. Sugar transporters in plants and in their interactions with fungi. *Trends Plant Sci.* 17(7): 413–422.

Deveau A, Kohler A, et al. 2008. The major pathways of carbohydrate metabolism in the ectomycorrhizal basidiomycete *Laccaria bicolor* S238N. *New Phytol.* 180: 379–390.

Eastwood DC, Floudas D, et al. 2011. The plant cell wall- decomposing machinery underlies the functional diversity of forest fungi. *Science.* 333(6043): 762–765.

English AC, Richards S, et al. 2012. Mind the gap: Upgrading genomes with Pacific Biosciences RS long-read sequencing technology. *PLoS One.* 7: e47768.

Fajardo López M, Dietz S, et al. 2008. The sugar porter gene family of *Laccaria bicolor*: function in ectomycorrhizal symbiosis and soil-growing hyphae. *New Phytol.* 180: 365–378.

Floudas D, Binder M, et al. 2012. The Paleozoic origin of enzymatic lignin decomposition reconstructed from 31 fungal genomes. *Science.* 336 (6089): 1715–1719.

Geml J, Laursen GA, et al. 2006. Beringian origins and cryptic speciation events in the fly agaric (*Amanita muscaria*). *Mol Ecol.* 15: 225–239.

Gottel NR, Castro HF, et al. 2011. Distinct microbial communities within the endosphere and rhizosphere of *Populus deltoides* roots across contrasting soil types. *Appl Environ Microbiol.* 77: 5934–5944.
Grigoriev IV, Nordberg H, et al. 2012. The Genome Portal of the Department of Energy Joint Genome Institute. *Nucl Acids Res.* 40: D26–D32
Hacquard S, Tisserant E, et al. 2013. Laser microdissection and microarray analysis of *Tuber melanosporum* ectomycorrhizas reveal functional heterogeneity between mantle and Hartig net compartments. *Environ Microbiol.* doi:10.1111/1462-2920.12080.
Hibbett DS, Gilbert L-B, et al. 2000. Evolutionary instability of ectomycorrhizal symbioses in basidiomycetes. *Nature.* 407: 506–508.
Hibbett DS & Matheny PB. 2009. The relative ages of ectomycorrhizal mushrooms and their plant hosts estimated using Bayesian relaxed molecular clock analyses. *BMC Biol.* 7: 13.
Högberg P, Nordgren A, et al. 2001. Large-scale forest girdling shows that current photosynthesis drives soil respiration. *Nature.* 411: 789–792.
Hortal S, Trocha LK, et al. 2012. Beech roots are simultaneously colonized by multiple genets of the ectomycorrhizal fungus *Laccaria amethystina* clustered in two genetic groups. *Mol Ecol.* 21: 2116–2129.
James TY, Kauff F, et al. 2006. Reconstructing the early evolution of Fungi using a six-gene phylogeny. *Nature.* 443: 812–822.
Kemen E, Gardiner A, et al. 2011. Gene gain and loss during evolution of obligate parasitism in the white rust pathogen of *Arabidopsis thaliana*. *PLoS Biol.* 9: e1001094.
Kemppainen M, Duplessis S, et al. 2009a. RNA silencing in the model mycorrhizal fungus *Laccaria bicolor*: Gene knock-down of nitrate reductase results in inhibition of symbiosis with *Populus*. *Environ Microbiol.* 11: 1878–1896.
Kemppainen MJ, Pardo AG. 2009b. pHg/pSILBAgamma vector system for efficient gene silencing in homobasidiomycetes: optimization of ihpRNA-triggering in the mycorrhizal fungus *Laccaria bicolor*. *Microb Biotechnol.* 3(2):178–200.
Kennedy PG, Smith DP, et al. 2012. *Arbutus menziesii* (Ericaceae) facilitates regeneration dynamics in mixed evergreen forests by promoting mycorrhizal fungal diversity and host connectivity. *Am J Bot.* 99: 1691–1701.
Labbé J, Zhang X, et al. 2008. A genetic linkage map for the ectomycorrhizal fungus *Laccaria bicolor* and its alignment to the whole-genome sequence assemblies. *New Phytol.* 180: 316–328.
Lackner G, Misiek M, et al. 2012. Genome mining reveals the evolutionary origin and biosynthetic potential of basidiomycete polyketide synthases. *Fungal Genet Biol.* 49: 996–1003
Lammers P, Tuskan GA, et al. 2004. Mycorrhizal symbionts of populus to be sequenced by the United States Department of Energy's Joint Genome Institute. *Mycorrhiza* 14(1): 63–64.
Larsen PE, Trivedi G, et al. 2010. Using deep RNA sequencing for the structural annotation of the *Laccaria bicolor* mycorrhizal transcriptome. *PLoS One.* 5: e9780.
Lindahl BJ, Ihrmark K, et al. 2007. Spatial separation of litter decomposition and mycorrhizal nitrogen uptake in a boreal forest. *New Phytol.* 173: 611–620.
Lucic E, Fourrey C, et al. 2008. A gene repertoire for nitrogen transporters in *Laccaria bicolor*. *New Phytol.* 180: 343–364.
Marmeisse R, Guidot A, et al. 2004. *Hebeloma cylindrosporum*—a model species to study ectomycorrhizal symbiosis from gene to ecosystem. *New Phytol.* 163: 481–498.
Martin F, Diez J, et al. 2002. Phylogeography of the ectomycorrhizal *Pisolithus* species as inferred from nuclear ribosomal DNA ITS sequences. *New Phytol.* 153: 345–357.
Martin F. 2007. Fair trade in the underworld: The ectomycorrhizal symbiosis. In: *The Mycota*, Vol. 8: *Biology of the Fungal Cell*, 2nd ed. (eds. RJ Howard, NAR Gow), 291–308. Berlin: Springer.
Martin F & Nehls U. 2009. Harnessing ectomycorrhizal genomics for ecological insights. *Curr Opin Plant Biol.* 12: 1–8.
Martin F, Aerts A, et al. 2008. The genome of *Laccaria bicolor* provides insights into mycorrhizal symbiosis. *Nature.* 452: 88–92.

Martin F, Kohler A, et al. 2010. Périgord black truffle genomeuncovers evolutionary origins and mechanisms of symbiosis. *Nature.* 464: 1033–1038.

Martin F, Perotto S, et al. 2007. Mycorrhizal fungi: A fungal community at the interface between soil and roots. In: *The Rhizosphere: Biochemistry and Organic Substances at the Soil-Plant Interface* (eds. R Pinton, Z Varanini, et al.), 263–296. New York: Marcel Dekker.

Martin F & Selosse MA. 2008. The *Laccaria* genome: A symbiont blueprint decoded. *New Phytol.* 180: 296–310.

Martino E, Turnau K, et al. 2000. Ericoid mycorrhizal fungi from heavy metal polluted soils: their identification and growth in the presence of zinc ions. *Mycol Res.* 104: 338–344.

Martos F, Dulormne M, et al. 2009. Independent recruitment of saprotrophic fungi as mycorrhizal partners by tropical achlorophyllous orchids. *New Phytol.* 184: 668–681.

Morin E, Kohler A, et al. 2012. Genome sequence of the button mushroom *Agaricus bisporus* reveals mechanisms governing adaptation to a humic-rich ecological niche. *Proc Natl Acad Sci USA.* 109(43): 17501–17506.

Müller T, Avolio M, et al. 2007. Nitrogen transport in the ectomycorrhiza association: The *Hebeloma cylindrosporum—Pinus pinaster* model. *Phytochemistry.* 68: 41–51.

Nehls U, Grunze N, et al. 2007. Sugar for my honey: Carbohydrate partitioning in ectomycorrhizal symbiosis. *Phytochemistry.* 68: 82–91.

Nehls U, Mikolajewski S, et al. 2001. Carbohydrate metabolism in ectomycorrhizas: Gene expression, monosaccharide transport and metabolic control. *New Phytol.* 150: 533–541.

Niculita-Hirzel H, Labbé J, et al. 2008. Gene organization of the mating type regions in the ectomycorrhizal fungus *Laccaria bicolor* reveals distinct evolution between the two mating type loci. *New Phytol.* 180: 329–342.

Ohm RA, Feau N, et al. 2012. Diverse lifestyles and strategies of plant pathogenesis encoded in the genomes of eighteen dothideomycetes fungi. *PLoS Pathog.* 8: e1003037.

Pirozynski, KA & Malloch DW. 1975. The origin of land plants: A matter of mycotrophism. *Biosystems.* 5:153–164.

Plett JM, Kemppainen M, et al. 2011. A secreted effector protein of *Laccaria bicolor* is required for symbiosis development. *Curr Biol.* 21: 1197–1203.

Plett JM & Martin F. 2011. Blurred boundaries: Lifestyle lessons from ectomycorrhizal fungal genomes. *Trends Genet.* 27:14–22.

Pringle A, Adams RI, et al. 2009. The ectomycorrhizal fungus *Amanita phalloides* was introduced and is expanding its range on the west coast of North America. *Mol Ecol.* 18: 817–833.

Rajashekar B, Kohler A, et al. 2009. Expansion of signal pathways in the ectomycorrhizal fungus *Laccaria bicolor*-evolution of nucleotide sequences and expression patterns in families of protein kinases and RAS small GTPases. *New Phytol.* 183: 365–379.

Reich M, Göbel C, et al. 2009. Fatty acid metabolism in the ectomycorrhizal fungus Laccaria bicolor. *New Phytol.* 182: 950–964.

Rubini A, Belfiori B, et al. 2011a. *Tuber melanosporum*: mating type distribution in a natural plantation and dynamics of strains of different mating types on the roots of nursery-inoculated host plants. *New Phytol.* 189: 723–735.

Rubini A, Belfiori B, et al. 2011b. Isolation and characterization of *MAT* genes in the symbiotic ascomycete *Tuber melanosporum. New Phytol.* 189: 710–722.

Ryberg M & Matheny PB. 2011. Asynchronous origins of ectomycorrhizal clades of Agaricales. *Proc R Soc B.* doi:10.1098/rspb.2011.

Smith SE & Read JR. 2008. *Mycorrhizal Symbiosis.* San Diego: Academic Press.

Spanu PD, Abbott JC et al. 2010. Genome expansion and gene loss in powdery mildew fungi reveal tradeoffs in extreme parasitism. *Science.* 330: 1543–1546.

Straker CJ.1996. Ericoid mycorrhiza: ecological and host specificity. *Mycorrhiza.* 6: 215–225

Tedersoo L, May TW, et al. 2010. Ectomycorrhizal lifestyle in fungi: global diversity, distribution, and evolution of phylogenetic lineages. *Mycorrhiza.* 20: 217–263.

Tisserant E, Da Silva C, et al. 2011. Deep RNA sequencing improved the structural annotation of the *Tuber melanosporum* transcriptome. *New Phytol.* 189: 883–891.

Tuskan GA, Difazio S, et al. 2006. The genome of black cottonwood, *Populus trichocarpa. Science.* 313: 1596–1604.

Veneault-Fourrey C & Martin F. 2011. Mutualistic interactions on a knife-edge between saprotrophy and pathogenesis. *Curr Opin Plant Biol.* 14: 444–450.

Vincent D, Kohler A, et al. 2011. Secretome of the Free-living Mycelium from the Ectomycorrhizal Basidiomycete *Laccaria bicolor. J Proteome Res.* 11(1): 157–171

Wolfe BE, Tulloss RR, et al. 2012. The irreversible loss of a decomposition pathway marks the single origin of an ectomycorrhizal symbiosis. *PLoS One.* 7: e39597.

Wright DP, Johansson T, et al. 2005. Spatial patterns of gene expression in the extramatrical mycelium and mycorrhizal root tips formed by the ectomycorrhizal fungus *Paxillus involutus* in association with birch (*Betula pendula*) seedlings in soil microcosms. *New Phytol.* 167: 579–596.

Wullschleger SD, Weston DJ, et al. 2012. Revisiting the sequencing of the first tree genome: *Populus trichocarpa. Tree Physiol.* doi:10.1093/treephys/tps081.

Zuccaro A, Lahrmann U, et al. 2011. Endophytic life strategies decoded by genome and transcriptome analyses of the mutualistic root symbiont *Piriformospora indica. PLoS Pathog.* 7: e1002290.

9 Lichen Genomics: Prospects and Progress

Martin Grube[1], Gabriele Berg[2], Ólafur S. Andrésson[3], Oddur Vilhelmsson[4], Paul S. Dyer[5], and Vivian P.W. Miao[6]

[1] *Institut für Pflanzenwissenschaften, Karl-Franzens-Universität Graz, Graz, Austria*
[2] *Institute for Environmental Biotechnology, Graz University of Technology, Graz, Austria*
[3] *Institute of Life and Environmental Sciences, University of Iceland, Reykjavik, Iceland*
[4] *Department of Natural Resource Sciences, University of Akureyri, Borgir vid Nordurslod, Akureyri, Iceland*
[5] *School of Biology, University of Nottingham, Nottingham, United Kingdom*
[6] *Department of Microbiology and Immunology, University of British Columbia, Vancouver, Canada*

Introduction

Lichens are distinctive, symbiotic life forms that are present in terrestrial environments worldwide. They dominate the landscape of some parts of the planet, and it has been estimated that they may cover up to 8 percent of the total land surface (Ahmadjian, 1995). In contrast to most other fungal symbioses that remain hidden within substrata or other organisms, lichens form vegetative thalli (singular: thallus) which are conspicuous, easily recognized macroscopic structures. They present a variety of often colorful forms and diverse morphologies on surfaces exposed to light (Fig. 9.1; Nash, 2008; Lumbsch, Ahti, et al., 2011). Lichens are found in an extremely wide range of habitats, including many that are generally characterized as being subject to some form of environmental stress, such as low nutrient or water availability or extremes of temperature (Boddy, Dyer, et al., 2010). For example, lichens occur on rocks and soils in harsh and hostile polar habitats, form belts of vegetation in intertidal zones of rocky coastlines, grow on trees in all climatic zones, and even colonize living leaves in tropical rain forests. Lichen structures are perennial and where ecological conditions and substrates are stable

The Ecological Genomics of Fungi, First Edition. Edited by Francis Martin.

Figure 9.1 Lichens used in ecological genomics studies. Upper row: left, *Peltigera membranacea*; right, *Xanthoria parietina*. Middle row: left, *Cladonia grayi*; right, *Lobaria pulmonaria*. Lower row: left, *Solorina crocea* with lichenicolous infection by *Rhagadostoma lichenicola*; right, *Cetraria aculeata* (Ch. Printzen). For color detail, please see color plate section.

(e.g., in Antarctica, Øvstedal & Lewis Smith 2001), can persist for many to thousands of years.

Lichens are traditionally characterized as associations of a fungus (mycobiont) and a photoautotrophic organism (primary photobiont) that is either a green alga or a cyanobacterium. In the symbiotic stage, the partners come together to form a self-sustaining thallus that is more complex and robust than the individual organisms, able to tolerate and sustain growth in stressful environmental conditions where neither alone could survive (Nash, 2008). The lichen thallus is primarily shaped by the fungal mycelia, with the mycobiont being the dominant partner in biomass (the Latin binomial of the mycobiont is the name of the association as a whole, and the basis for lichen taxonomy; Tehler & Wedin, 2008). Most mycobionts belong to the Ascomycota, whereas only few species of Basidiomycota form lichens. Lichenization is typical of species in the classes Lecanoromycetes, Arthoniomycetes, and Verrucariales, but it is also observed occasionally in others (e.g., the Dothideomycetes) consistent with multiple independent origins of lichenization (Gargas, DePriest et al., 1995; Schoch, Sung, et al., 2009). The class Lecanoromycetes is almost exclusively lichenized, and with about 13,500 species, it is the most species-rich lineage in Ascomycota. Within this class, preference for specific photobiont lineages is observed in major clades (Miadlikowska, Kauff, et al., 2006). Lichen photobionts are predominantly eukaryotic algae, primarily coccal algae of the class Trebouxiophyceae, or filamentous representatives of the Trentepohliales (Ulvophyceae), but approximately 10 percent of the lichenized fungi associate with coccal or filamentous cyanobacteria (Friedl & Büdel, 2008).

Once the mycobiont has encountered an appropriate photobiont(s), a developmental process occurs, leading to the eventual formation of the lichen thallus. Depending on the species concerned, such lichen thalli can have diverse growth forms described as crustlike, leaflike (foliose), or shrublike (fruticose). In the thalli, fungal hyphae surround the photobionts, forming a biological growth chamber for the photobionts. The main function of the photobiont is to provide fixed carbon to the fungal partner, which is likely to supply mineral elements in return and protect the photobiont (Nash, 2008). Lichens reproduce and disperse by various methods, including production of vegetative fragments containing both partners (e.g., soredia and isidia) and release of sexual spores by the mycobiont (Murtagh, Dyer, et al., 2000; Honegger & Scherrer, 2008). Sexual reproduction has only been reported from mature thalli; thus the fungus appears to require the formation of a thallus for sexual reproduction. By contrast, sexuality of the algal partner is suppressed in most lichen symbioses.

The lichen symbiosis often involves more organisms than the two typically considered functional partners. It has long been known that tripartite lichens can acquire cyanobacteria as part of their thalli, in addition to their more common green algal photobionts. These cyanobacteria form structures (cephalodia) inside the thallus in some species, and outside in others. Whereas the cyanobacteria that are primary photobionts supply both fixed carbon and nitrogen for the lichen, those in tripartite lichens are focused specifically on nitrogen fixation. In addition, other so-called "lichenicolous fungi" can colonize lichens and complete their life cycles as commensals or parasites (Hawksworth, 2003; Lawrey & Diederich, 2003). Lichenicolous fungi have been phenotypically well described, but there have been relatively few molecular studies concerning these fungi (e.g., Ruibal, Millanes, et al., 2011). Most recently, the presence of bacterial communities has been highlighted as an important part of the composition of many lichen symbioses. Although their presence has been noted before (Cardinale, Puglia, et al., 2006 and references therein), the large number and diversity of bacteria present have only been adequately revealed and appreciated as a result of investigations with modern, culture-independent analytical approaches (Grube, Berg, et al., 2009; Hodkinson & Lutzoni, 2009; Bates, Cropsey, et al., 2011). Therefore, the classical paradigm of the lichen symbiosis is evolving from that based on a primarily myco-centric view to a larger concept whereby in some instances lichens might arguably be better considered as a microbial community regularly comprised of a large number of diverse associated taxa, in addition to the main symbionts. These potentially interact and affect each other. This notion is being fueled by the continuing influx of new knowledge concerning all elements of the lichen symbiosis whether on a genomics, transcriptomics, proteomics, or metabolomics level.

Studies of lichen genomics began a few years ago, initially with the larger and more complex genomes of the primary mycobionts and photobionts, which when completed will result in detailed annotation of individual symbiont strains (whether cultured or in situ). Three of these projects will be described in this chapter. Investigations of lichen-associated and intrathalline bacteria began later, addressing different types of questions and using different forms of analysis, but these studies have proceeded quickly and are leading the way in terms of implementing new technologies, such as proteomics and metabolomics for studying lichen biology; some of these projects will also be reviewed in this chapter. The anticipated and welcomed challenge for lichenologists and mycologists studying lichen fungi will be to use genomic and other new methodological tools to consider all the biological entities and their contributions, and thereby arrive at a better understanding of the symbiotic biology and ecology of lichens.

Experimental Demands of Work with Lichens and Lichen Symbionts

Lichens offer particular advantages, but also obstacles, for experimental work. Samples can be collected relatively easily from nature given the prominence and long-lived nature of thalli, and analysis of the functioning of the whole thallus is possible. However, it is often of interest to study the functioning of the individual lichen symbionts, which provides challenges both in the isolation and in the maintenance of the mycobiont and photobiont partners. Lichen fungi in general are considered nearly obligate symbionts and are notoriously difficult to isolate, establish, and sustain in vitro (Crittenden, David, et al.,

1995; Stocker-Wörgötter & Hager, 2008). Ascospores often germinate only after a prolonged dormancy (melanized spores of some species seem to do so only after drastic pretreatment such as exposure to "outer space conditions," S. Ott, personal communication). Once established, mycobionts grow slowly, with low metabolic turnover, and great care must be taken to avoid contamination of cultures. By contrast, many algal symbionts grow readily in axenic culture, but much care is still demanded for their isolation (Friedl & Büdel, 2008).

In vitro resynthesis of lichen thalli from the independent symbiotic partners is one of the holy grails of lichenology, but even when lichens are resynthesized with their photobionts, development into the characteristic thallus morphology usually fails under standard laboratory conditions. Thallus regeneration is possible in some instances under oligotrophic conditions, such as culture on sterilized soil substrates, but it may take years to develop (e.g., Stocker-Wörgötter & Türk, 1991). These constraints, as well as the lack of opportunities for genetic manipulation (e.g., making mutant strains), make cultured lichen mycobionts challenging as study objects for many types of standard experimentation. For example, there can even be difficulties in generating adequate biomass for isolation of DNA and RNA. Ideally for genome sequencing, all DNA submitted should be of the same genotype because the presence of polymorphisms can complicate genome assembly. This means that in vitro cultures for DNA isolation should be established from either a single genetic source or genetically identical ascospores. This was one reason for the selection of *Xanthoria parietina* (*see* Fig. 9.1) as a model for genome studies because this has a homothallic (self-fertile) breeding system (Honegger, Zippler, et al., 2004; Itten & Honegger, 2010), and therefore multiple axenic cultures could be established from ascospores from the same thallus, allowing bulking up of cultures for DNA extraction (uniformity of cultures can be confirmed by DNA fingerprinting; Murtagh, Archer, et al., unpublished results). Similar, painstaking work has led to development of multiple cultures of *Cladonia grayi* (*see* Fig. 9.1), and DNA and RNA extraction has been facilitated given that the mycobiont is relatively fast growing in vitro. This latter system also has the benefit that compatible mycobiont and photobionts are capable of forming "lichenoids," a callus tissue, when cocultured.

Given the difficulties of axenic culture, much work on lichens has relied on the use of samples taken directly from nature, which then need to be analyzed by culture-independent approaches (e.g., a great deal of genetic work on lichens has relied on DNA extracted directly from natural thalli [metagenomic DNA]). Care is required to ensure that once collected, lichen thalli are processed quickly and appropriately, to minimize postsampling change of biological conditions. This is particularly important for analyses of gene expression and transcriptional analyses of genes, as well as for general microbiome profiling. Because lichens are adapted to poikilohydry (i.e., variation in hydration condition) the analysis of gene expression must carefully consider the hydration status at the time of sampling. A full understanding of patterns of expression will only be possible after sampling over the full range of hydration conditions. Even when this is achieved, the slow metabolic rates may pose problems for obtaining sufficient RNA by extraction. Also, many natural lichens, especially the foliose forms, are highly structured with respect to their morphology, and differential gene expression is expected in the different strata of a lichen thallus, and in the fruiting organs, in addition to age gradients that are often seen in fully grown thalli (e.g., Miao, Manoharan, et al., 2012). Although these caveats may make sampling more complicated, on the other hand, awareness and incorporation of these factors into experimental design can lead to more relevant and productive studies.

Previous Molecular Approaches and Status of Exploration

Molecular genetic studies on lichen fungi have used genomic DNA from pure cultures of mycobionts as well as natural lichens. Although cultured strains are highly desirable, because of the daunting task involved in their establishment and maintenance, only a limited number of genetic studies have been conducted using in vitro propagated mycobionts. These include investigation of DNA methylation status in *Cladonia grayi* (Armaleo & Miao, 1999), analysis of breeding systems in *Graphis scripta* and *Ochrolechia parella* (Murtagh, Dyer, et al., 2000), and studies of mycobiont hydrophobin and mating-type (*MAT*) encoding genes in *Xanthoria parietina* and relatives (Scherrer, Haisch, et al., 2002; Scherrer, Zippler, et al., 2005). Most studies have instead relied on metagenomic DNA extracted from whole lichens and analysis using taxon-specific primers or probes to recover informative amplicons or hybridization patterns. For example, one of the earliest genetic studies in lichens (Armaleo & Clerc, 1991) used Southern hybridization of DNA from whole thalli to identify symbionts in lichen chimeras. In general, use of metagenomic DNA and taxon-specific primers or probes circumvent issues of establishing and propagating pure cultures, allowing certain research questions to be addressed quickly. By contrast, establishment of pure cultures is often more important for long-term programs addressing questions relating to symbiont recognition, mycobiont differentiation, and lichen thallus development.

Most previous molecular research on lichen symbioses has been conducted within a phylogenetic context, aiming to establish an evolutionary framework, and also to some extent, to gain information on symbiont specificity (De Priest, 2004). There has been relatively little research on genes other than those that primarily serve as phylogenetic markers. However, progress has been made in characterization of some genes. One particular group of interest concerns genes involved in polyketide biosynthesis because lichens are well known for production of a tremendous diversity of secondary metabolites (Huneck & Yoshimura, 1996; Huneck, 1999; Boustie, Tamasi, et al., 2011). A polymerase chain reaction (PCR)-based approach has been used to survey lichens for polyketide synthase (PKS) genes that are responsible for the production of depsides and depsidones, polyketide-derived secondary metabolites which are typical of lichens and lichen mycobionts, but uncommon elsewhere. The approach has proved productive, with amplified fragments of PKS genes being obtained from numerous species of lichens. This has allowed their analysis in a phylogenetic context that has revealed both diversity and evidence of purifying selection (e.g., Muggia, Schmitt, et al., 2008). Although the results indicated that genomes harbor many such genes, likely to have resulted from ancient gene duplications, a PCR-based approach alone cannot provide a comprehensive picture. In complementary work, Sinnemann, Andrésson, et al. (2000) pioneered one aspect of molecular work in lichenology when they cloned the mycobiont *pyrG* gene (encoding orotidine 5′-phosphate carboxylase, the essential terminal enzyme in uridine 5′-phosphate biosynthesis) from a phage library of *Solorina crocea*, the "chocolate chip lichen," and expressed it in a heterologous fungal host. It was hoped that an approach based on creation of phage and cosmid metagenomic libraries and heterologous expression could pave the way to understanding functions of lichen genes as well as making their technological exploitation possible (Miao, Coëffet-LeGal, et al., 2001).

Even though methodology for constructing gene libraries has much improved since then, there remain few lichens studies using clone libraries (Kim, Hong, et al., 2012), and none have been used for de novo genome sequencing. One reason for the low number of large insert libraries has to do with the problem of obtaining high-quality, high-molecular–weight DNA. Some mycobionts have thick cell walls and attempts to open the cells result in shearing the DNA to fragments usually well below 20 kb. Another reason is related to library size; although the mycobiont is thought to contribute the bulk of the biomass, the photobiont(s) and associated organisms may in fact contribute more DNA, thereby greatly increasing the number of clones needed for adequate coverage. In addition, there has been no consensus in the lichenological community on a standardized model system for further genomic exploration, with competing systems all having different advantages.

Next-Generation Sequencing Family Platforms

High-throughput sequencing (*see* Chapter 1) has made it feasible to perform whole-genome sequencing using genomic DNA from cultured symbionts, or metagenomic DNA from lichen thalli. The 454-pyrosequencing platform generates reads of up to 600 nt (av. 350–400 nt) and can easily produce 10-fold coverage of a hypothetical 40-Mb mycobiont genome that is sufficient for a good working database covering more than 98 percent of the genes. A lower cost method, offered by Illumina, provides improvements (e.g., reduced homopolymer uncertainty) and offers 2–600 Gb per run capacity that can be subdivided into lanes as well as multiplexed; one 2-Gb run can yield more than 50-fold mean coverage of a mycobiont genome, or more than 10-fold mean coverage of the main symbionts in a lichen metagenome. Improvements in methodology as well as development of related applications have significantly improved sequence quality and expanded the breadth of associated studies (e.g., "mate-pair" sequencing can bridge regions difficult to sequence or assemble and help build scaffolds approaching the size of full-length chromosomes). In addition, gene expression studies ("RNA-Seq")—in which cDNA, reverse transcribed from cellular RNAs recovered from symbionts or lichens are sequenced—can elucidate not only which genes are active (and by inference the proteins produced) but also their relative levels of expression in different contexts (e.g., Miao, Manoharan, et al., 2012). Also, epigenetic modifications such as the presence of 5-methycytosine in lichen genomes can be determined in conjunction with next-generation sequencing platforms. For example, methylation in the *Peltigera membranacea* mycobiont, appears mainly in transposons and repeat elements (Manoharan & Andrésson, unpublished).

Model Systems and Status of Genomic Sequencing of Lichen Symbionts

There are a number of genome sequencing projects being conducted with various lichen symbionts using cultured isolates. The Joint Genome Institute (JGI), of the US Department of Energy, has assembled raw data of the genome of the lichen-forming fungus, *Xanthoria parietina*, and this project is currently in the gene annotation phase. The project was initially delayed as a result of genomic DNA (which took 2 years to amass) being confiscated by US authorities as a bioterrorist threat, before processing at the JGI. However, good progress has since been made with sequencing by 454 and Illumina technologies. The assembly of a genome currently stands at 10× coverage, with 39 scaffolds for a predicted genome size of ~32 Mb. Data for average gene length (1.5 kb) and intron and exon coverage are comparable with nonlichenized Ascomycetes, and the genome is predicted to encode approximately 10,800 proteins (Kuo, Grigoriev, et al., unpublished results). There is also ongoing RNA-Seq work aiming to compare gene expression in the mycobiont alone in pure culture versus the mycobiont in the symbiotic state to identify genes that are differentially expressed and might therefore be correlated with symbiotic interactions.

Sequencing of the genome of the *Cladonia grayi* mycobiont (34 Mb) has progressed further, with some data already published and further submissions in preparation; genome sequencing of the *C. grayi* photobiont, *Asterochloris* sp. (56 Mb) has also been completed (Armaleo, Müller, et al., unpublished, http://genome.jgi.doe.gov/Clagr2/Clagr2.home.html). Initial insights from the *C. grayi* genome project have been provided in two published studies. First, it was previously known that the polyketide synthase gene *CgrPKS16* was involved in the production of the lichen depsidone grayanic acid. It was therefore of significance to discover that *CgrPKS16* clustered with a cytochrome P-450 and an O-methyltransferase gene, in agreement with a proposed pathway of grayanic acid production (Armaleo, Sun, et al., 2011). This suggested linkage of metabolic genes as has been shown elsewhere in filamentous fungi (Plumridge, Melin, et al., 2010). These findings are consistent with the proposal that a single PKS synthesizes two aromatic rings on tandem acyl carrier proteins and links them into a depside, and that the transition from depside to depsidone requires only a cytochrome P-450. Secondly, evidence for several ancient, independent horizontal gene transfers (HGTs) of the methylammonium permease family between prokaryotes and the *C. grayi* mycobiont were detected in the genomic data by McDonald, Dietrich, et al. (2012). This was consistent with previous reports by Schmitt and Lumbsch (2009), who provided evidence that PKS genes of the methylsalicylic acid synthase family (responsible for production of phenolics) are phylogenetically related to those found in soil bacteria. They suggested that the lichen fungi had gained these genes by horizontal transfer from bacteria.

Unlike the first two lichens for which the mycobionts are sequenced from haploid cultures established from ascospores, the foliose terrestrial cyanolichen *P. membranacea* has been sequenced as a metagenome including not only the mycobiont (12× coverage, ~38 Mb) and the *Nostoc* photobiont (25× coverage, ~9 Mb) but also associated bacteria. Furthermore, the metagenomic source DNA was generated from an intentional mixture of lobes from different thalli from one locality to produce a more representative whole-genome sequence. To date, combining 454 and Illumina reads with bridging mate-pair sequences has allowed assembly of the primary symbiont genomes into 3,033 and 616 scaffolds, respectively (Andrésson, Snæbjörnsson, et al., unpublished). The *P. membranacea* metagenome is complemented by metatranscriptomic data (RNA-Seq) from different tissues, as well as by methylation data obtained from bisulphite pretreated metagenomic DNA. Nearly all (>99 percent) of the expressed genes identified appear to be included on the scaffolds, and the remaining gaps appear to consist mainly of long repetitive elements (e.g., transposons) and low complexity sequences. This suggests that the scaffold collection has full use as a base for mapping RNA and for analysis of all genes in the major partners of this lichen symbiosis. In addition, a smaller genome sequencing project for another *Peltigera* species, *Peltigera malacea*, was undertaken concurrently with the expectation that this closely related taxon would provide a ready comparator to facilitate assessment of the significance of findings in *P. membranacea*.

The metagenomic approach yielded high-sequence coverage for mitochondrial genomes (mtDNA), which were readily assembled and annotated (Xavier, Miao, et al., 2012). The ~63-kb mtDNAs of the *P. membranacea* and P. *malacea* mycobionts show not only all the major elements of mtDNAs observed in most nonlichenized fungi (e.g., unidirectional transcription, conserved mt protein and tRNA encoding genes, many group I introns) but also the presence of a gene for the RNA component of RNAseP, a feature seldom found in ascomycete mtDNA. "Mining" of the partially annotated metagenome revealed the presence of unusually variable mycobiont genes encoding galectin-like proteins (Manoharan, Miao, et al., 2012); analysis of RNA-Seq data further showed that one of these genes, *lec-1*, was differentially expressed in rhizines, a purely fungal tissue, compared to the main thallus, considered a symbiotic tissue owing to the presence of both mycobiont and photobiont cells (Miao, Manoharan, et al., 2012). Although most *Peltigera* are not known for production of lichen substances, and none have been recorded for *P. membranacea*, a large number of mycobiont and photobiont genes and gene clusters associated with secondary metabolic pathways have been identified in its metagenome, and an unusual trans-AT polyketide biosynthetic pathway of a type known only from other bacterial-eukaryote symbiosis has been identified in the Nostoc photobiont (Kampa, Gagunashvili et al., 2013).

It will be interesting to compare findings from these first mycobiont genomes. All three species are members of the Lecanoromycetes, but they are quite distinct in many ways. *Xanthoria parietina* has a foliose morphology and a stratified structure typical of many highly organized lichen thalli. The species has a cosmopolitan distribution, being found in circumpolar and temperate regions worldwide, and occurs on a variety of substrata including bark, rock, and metal surfaces (Purvis, Coppins, et al., 1992). It produces a range of quinones and the depside atranorin, formed via a polyketide pathway (Huneck & Yoshimura, 1996). In comparison, *C. grayi* has a fruticose growth form, a rather more restricted growth habitat, and a distinct secondary metabolism. Because both *X. parietina* and *C. grayi* are chlorolichens, with eukaryotic green algal photobionts, comparison with the *Nostoc* sp.-carrying cyanolichen, *P. membranacea* should reveal not only differences between cyano- and chloro-lichens but also identify potentially key features in common among lichen fungi that distinguish them from other symbiotic fungi and from saprophytic fungi.

Investigation of the functional genomics of lichen mycobionts will be facilitated by complementary genome analysis of photobionts. Genome sequencing has already been undertaken for *Asterochloris* sp. from *C. grayi* and *Nostoc* sp. of *P. membranacea* as noted previously, and in addition genome sequencing projects for the cultured lichen photobionts *Trebouxia decolorans* and *Trebouxia* sp. TR9 from *Ramalina farinacea* are in progress (Casano, del Campo, et al., 2011). There is evidence that particular locally optimized strains or species are selected for thallus formation according to specific habitats (Blaha, Baloch, et al., 2006; Fernández-Mendoza, Domaschke, et al., 2011), with studies demonstrating *Trebouxia* sp. TR9 and *T. decolorans* as two coexisting but physiologically different algal partners of *R. farinacea* (Casano, del Campo, et al., 2011). Given the recent suggestions that some ecologically successful lichen fungi may "optimize" symbiotic associations across a wide range of environmental conditions, it is essential that genome analyses of lichens include complementary work on both mycobionts and photobionts.

Bacterial Communities

One of the most notable aspects that next-generation sequencing and metagenomic methods have brought to the study of lichens is a much deeper appreciation of the taxonomic distribution and potential contribution of communities of archaea and bacteria to the lichen thallus (e.g., Hodkinson, Gottell, et al., 2011; Bates, Cropsey, et al., 2011; Grube, Köberl, et al., 2012). These associated organisms colonize hydrophilic surfaces of the lichens and are to some extent also embedded in the fungal extracellular matrix. Studies using single strand conformation polymorphisms (SSCP) as community descriptors and deep sequencing have revealed specificity of the lichen-associated bacteria for their hosts (Grube, Cardinale, et al., 2009; Bates, Cropsey, et al., 2011), but differences in thallus age or the immediate environment of the host (e.g., sun or shade) may also affect the community composition (Cardinale, Berg, et al., 2011; Mushegian, Peterson, et al., 2011;Grube, Köberl, et al., 2012). The most common taxa in growing parts of lichens belong to the Alpha proteobacteria, whereas a considerably higher diversity is present in whole thalli of certain host species and habitats, with Acidobacteria (a group of mostly uncultivated bacteria) dominating in some instances (Bjelland, Grube, et al., 2011; Hodkinson, Gottell, et al., 2011; Mushegian, Peterson, et al., 2011; Grube, Köberl, et al., 2012). Using SSCP community fingerprinting on bacteria associated with *Lobaria pulmonaria*, a large, bark inhabiting foliose lichen, Cardinale, Grube, et al. (2012) found indications of isolation by distance for Alpha proteobacterial communities. Alpha proteobacteria, predominant on young parts of the lichen, are also present on sorediate isidia, the vegetative propagules of *L. pulmonaria*, but because these usually have a limited capacity to disperse, it is not surprising to find a geographical correlation within this bacterial group. Printzen, Fernández-Mendoza, et al. (2012) working with *Cetraria aculeata*, a widespread terrestrial fruticose lichen with a bipolar geographic distribution (and additional localities in mountainous regions in Europe and elsewhere such as the Andes), found that Alpha proteobacterial communities on lichens from the Arctic and Antarctica were more similar to each other than to the more diverse communities in lichens at higher altitudes from temperate regions.

Microscopic studies using DNA fluorescence in situ hybridization to detect groups of bacteria have implied an integral role of bacteria in lichen biology (Cardinale, Müller, et al., 2008), and numerous potential roles have been suggested, ranging from nutrient scavanging (Banfield, Barker, et al., 1999) to pathogen and grazing antagonism (Gonzélez, Ayuso-Sacido, et al., 2005) to substratum attachment (de los Ríos, Wierzchos, et al., 2002). Functional studies of the culturable bacterial fraction have indicated that they possess a wide range of lytic activities (including chitinolysis, glucanolysis, and proteolysis), produce hormones, contain siderophores, and can mobilize phosphates (Cardinale, Puglia, et al., 2006; Grube, Cardinale, et al., 2009; Liba, Ferrara, et al., 2006). Although in vitro assays provide valuable insights into possible functionality, their role for the whole system must not be overemphasized, owing to possible discrepancies between culturable and nonculturable fractions.

In *P. membranacea*, analysis of metagenomically derived 454 sequences remaining after subtraction of those attributable to the primary symbionts suggest that the dominant prokaryotic taxa by far belonged to the Proteobacteria (Alpha proteobacteria 59 percent, Beta proteoacteria 29 percent), followed distantly by Actinobacteria and Bacteriodetes, in general agreement with other studies (Cardinale, Puglia, et al.; 2006; Cardinale, Müller, et al., 2008; Hodkinson & Lutzoni, 2009). A small number of BLASTX hits to indoleacetimide hydrolase (most similar to those from Actinobacteria and Beta proteobacteria) suggest that some lichen-associated bacteria are capable of synthesizing indole acetic acid, a plant hormone, via the indoleacetimide pathway. Chitin is a major constituent of the *Peltigera* biomass, comprising about 13 percent of the cell wall. The few chitinase A (family 19 glycosyl hydrolase) hits were nearly exclusively actinobacterial (Fig. 9.2) in resemblance, suggesting that Actinobacteria may be the main or only group in the *Peltigera* bacterial community to metabolize this component of the mycobiont cell walls, in accordance with observations that Actinobacteria are particularly associated with senescing thalli. Several glycosyl hydrolases of families 16 (lichenanases, laminarinases, etc.) and 43 (xylanases, etc.) were found, as were some cellulases (family 5 and 6). Most of the family 43 xylanase hits were to verrucomicrobial or bacteroidetal xylanases, suggesting that most of these activities in the *Peltigera* symbiome are carried out by *Bacteroidetes* and *Verrucomicrobia* sp. Family 16 glycosyl hydrolases from the phyla Bacteroidetes, Proteobacteria, and Actinobacteria were present in the metagenome, as were a few sequences most similar to cellulases from Actinobacteria, Bacteroidetes, and Acidobacteria. Use of AppA phytase and AcpA acid phosphatase genes as query sequences yielded diverse hits, with Alpha proteobacterial *appA* and Beta proteobacterial *acpA* homologs particularly prominent (*see* Fig. 9.2), supporting the hypothesis that inorganic phosphate solubilization may be among the roles of these abundant members (Grube & Berg, 2009). Biofilm formation should be one function of interest among lichen-associated bacteria (de los

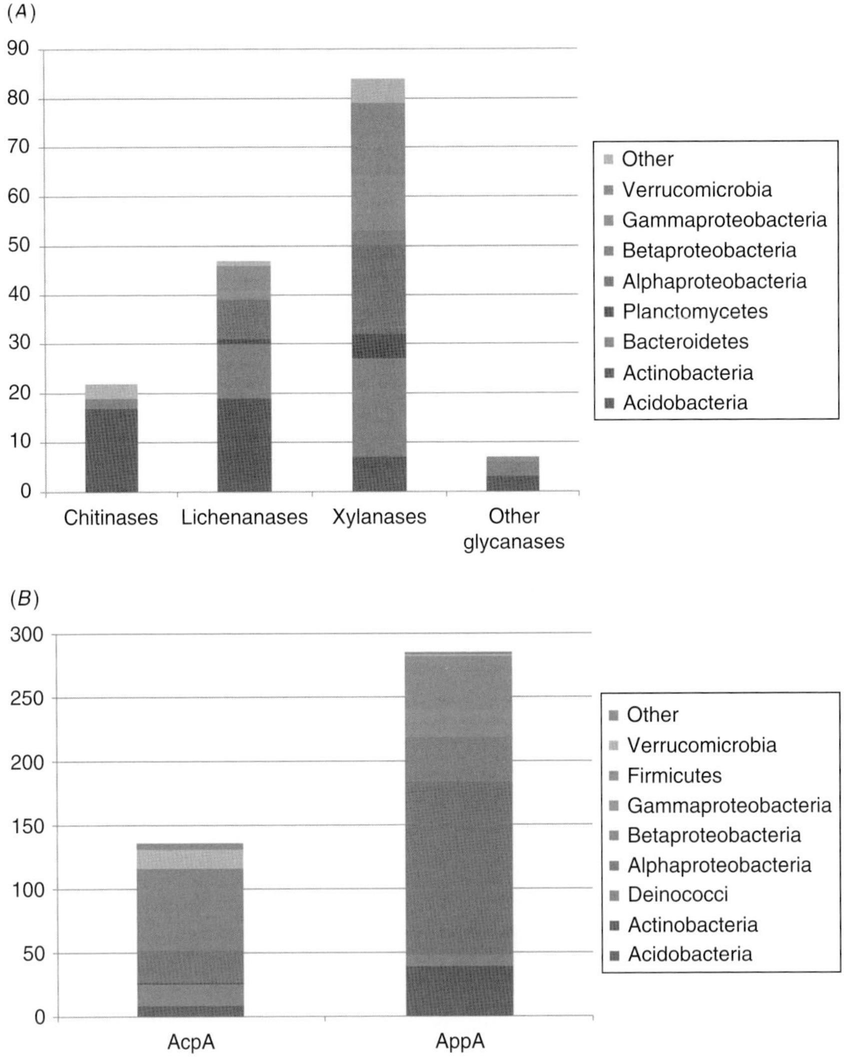

Figure 9.2 Partial functional analysis of the noncyanobacterial *Peltigera membranacea* prokaryotic metagenome. *A*, Number of 454 sequence reads extracted from *P. membranacea* metagenome based on similarity to glycolsyl hydrolase genes and taxonomic distribution of their most similar homologs in Genbank nr database. *B*, Number of *appA* phytase and *acpA* acid phosphatase homologs and taxonomic distribution. (Vilhelmsson, unpublished.) For color detail, please see color plate section.

Ríos, Wierzchos, et al., 2002). Although a search for orthologs encoding acyl homoserine lactone (AHSL) synthases, a component of quorum sensing-systems in gram-negative bacteria, has largely been negative in the lichens so far studied, it is suspected that quorum sensing does play a role in the lichen system and may be found in other species. Further insights in bacterial functionalities might also be unraveled with genome analyses of isolated lichen-associated bacteria (e.g., Lee, Shin., et al., 2012; Shin, Ahn, et al., 2012).

The cohort of lichen-associated microbes may not only interact with the primary symbionts but also with each other. Bar-coded pyrosequencing analysis of 16S rRNA genes from healthy *S. crocea* and thalli infected with the Ascomycete pathogen *Rhagadostoma lichenicola* revealed high abundances of Acidobacteria, Planctomycetes, and Proteobacteria, and analyses at the strain level by detrended correspondance analysis revealed a differentiation of communities. When data were subjected to a profile-clustering network, strain-specific abundance shifts within the Acidobacteria and hitherto unclassified bacteria were found (Grube, Köberl, et al., 2012).

Proteomics and Transcriptomics: Tools for Understanding the Process of Symbiosis

L. pulmonaria is a tri-partite lichen widely distributed in the Northern hemisphere, tropical mountains, and in South America. It contains the green alga *Dictyochloropsis reticulata* as the primary photobiont, and *Nostoc* sp. in internal cephalodia as a secondary photobiont. *L. pulmonaria* is among the ecologically and genetically best studied lichen species, being used as a flagship species for studying the conservation of primeval forests (Scheidegger & Werth, 2009) and has been featured in publications that have explored metagenomic and metaproteomic issues (Schneider, Vieira de Castro, et al., 2011; Cardinale, Grube, et al., 2012).

A metaproteomics approach can be used to analyze both taxonomic structure and function of the symbiotic consortium at the level of translated proteins. Proteins extracted from two lichen samples of *L. pulmonaria* were analyzed by one-dimensional gel electrophoresis (1-D SDS-PAGE) combined with LC-MS/MS and the resulting MS and MS/MS data were searched against a database consisting of protein sequences obtained from the public UniRef100 database (see Schneider, Riedel, et al., 2010; Schneider, Vieira de Castro, et al., 2011). Most algal proteins were assigned to energy production and conversion. Carbohydrate transport and metabolism were significant in both eukaryotic partners, but fungal functions were more diverse, with substantial read numbers suggesting biogenesis and posttranslational modification. With respect to the bacterial fraction, environmental proteomics data confirm the predominance of Alpha proteobacterial proteins in *L. pulmonaria*. Previous analyses of this lichen revealed diverse lineages of Rhizobiales (de Vieira, unpublished), which could not be resolved by metaproteomic data analyses. Bacterial proteins so far identified are primarily involved in energy conversion and carbohydrate metabolism, together with the presence of large numbers of stress-related proteins (Fig. 9.3). Also, there is first evidence for

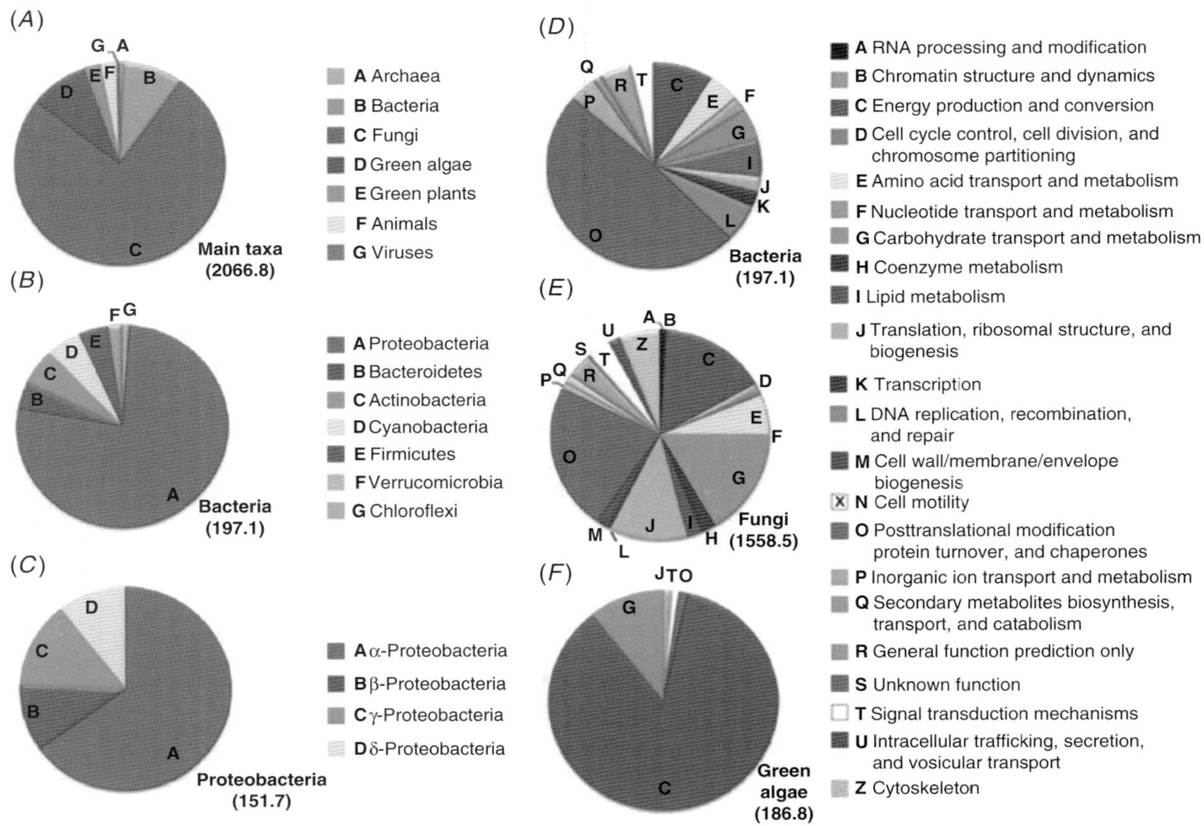

Figure 9.3 Metaproteomic profile of the *Lobaria pulmonaria* lichen symbiosis. Left side describes Taxon distribution for main taxa (*A*), bacteria (*B*), and proteobacteria (*C*). Note that the number of bacterial reads is comparable to that of the green algal partner. Right side represents gene ontology categories detected in bacteria (*D*), fungi (*E*), and green algae (*F*). (Image from Schneider, Vieira de Castro, et al., 2011.) For color detail, please see color plate section.

bacterial proteins involved in secondary metabolite synthesis. The study of Schneider, Vieira de Castro, et al. (2011) was carried out with more or less dry thalli samples, representing only one particular physiological stage of the lichen symbiosis. Because lichens are poikilohydric organisms, they must survive drastic changes in the environment. So far the influence of different physiological states on the gene expression of the participating symbionts is unknown. However, ecophysiological studies suggest that the different physiological responses to hydration and desiccation are correlated with differential enzymatic action, and certainly, transcription of genes.

An exercise in producing a full-length cDNA library of an isolated mycobiont has been provided by Wang et al. (2011), using the desert lichen *Endocarpon pusillum*. However, because a symbiotic context is missing, the significance of the detected gene expression for symbiosis is unclear. Moving a step further, Juntilla and Rudd (2012) used high-throughput next generation sequencing and EST sequence data to present a first eukaryotic transcriptome of entire thalli of the reindeer lichen *Cladonia rangiferina* (with 62.8% reads of fungal and 37.2% of algal origin). Even though a higher percentage of algal reads was found in the wetted thalli used, GO terms and identified KEGG pathways largely agreed with eukaryotic patterns found by Schneider et al. (2011).

Lichen Ecological Genomics

As new technologies are adopted by an increasing number of researchers, models that have served well to date must assimilate new findings and evolve to continue providing a conceptual framework to support and stimulate further investigations. Arguably, the traditional working description of a lichen must be expanded in many cases to encompass the concept of the lichen "symbiome" and include consideration of a larger collection of organisms and organism genotypes than the classical primary mycobiont and photobiont. This will not only generate new and more comprehensive research questions, but also guide the capture of a richer dataset by potential research collaborators (e.g., those involved with sample handling, sequencing depth, data interpretation). For example, the omnipresent and dynamic community of bacteria and archea in thalli must be considered in any whole thallus study because it may be discovered that the ecology "inside" the lichen is as critical as the more usual ecological parameters imposed by the biotic and abiotic factors of the larger environment "outside."

Because information collected for ecological genomics is ideally supported by (and supports) information from other complementary high-throughput functional analysis platforms, the experimental design stage is particularly critical. In addition to the considerations described previously ("Working with lichens and lichen mycobionts"), the fact that -omic platforms can be closely integrated necessitates careful planning for field sample processing pipelines to ensure that they can accommodate all of the multilevel downstream analyses. In addition to collecting field material for the typical herbarium voucher, an adequate amount of lichen must be collected for extraction of DNA, RNA, proteins, and possibly metabolites for chemical profiling, as well as perhaps additional material for microscopy. For example, prior knowledge of mycobiont genotypic variation would certainly affect collection of material for construction of a genome sequence for example, but consideration of thallus age, location (sun/shade), condition (e.g., infection by lichenicolous fungi), and tissue type must also be required if other levels of analysis are to be included.

A number of issues in generation of lichen metagenome assemblies have already been recognized in processing the datastream from *P. membranacea* and *P. malacea,* and they can guide new processing pipelines as more metagenomes are obtained. For example, in metagenomes obtained from field samples, the vast majority of sequence reads derive from the primary symbionts, but polymorphisms are to be expected if multiple individuals are included in the sample (if a lichen is small, for example) or if there might be multiple mycobiont genotypes within a thallus (Murtagh, Dyer, et al., 2000; Dyer, Murtagh, et al., 2001; Fahselt, 2008). For a well-demarcated species the significance may generally be low, but it could make a difference and prove much higher for any genes that are under strong positive selection (Manoharan, Miao, et al., unpublished). The same may be true for photobiont genomes; the chlorophyte symbionts may represent different populations (e.g., *R. farinacea*) and cyanobacterial photobionts may show substantial heterogeneity in many chromosomal locations (Andrésson, Gagunashvili, et al., unpublished). Even a fairly low level of polymorphism needs to be considered. The Newbler assembler (www.454.com) gives good results with 454 reads, assembling most nonrepetitive DNA from the symbionts into larger contigs, but most reads from the more heterogeneous associated organisms are poorly assembled (e.g., Proteobacteria are typically not assembled). More sequencing per se does not necessarily overcome the problem because increasing the average coverage of the primary genomes above a certain point (~50×) can impede the assembly process and result in lower average contig length. To make full use of high coverage, it is necessary to develop some kind of wet lab or bioinformatic strategy appropriate to the organism and research question, to filter and remove certain groups of reads (e.g., those from repeat elements and from genomes found at a low level or that are highly polymorphic). It is hoped that lichen mycobiont and photobiont genome sequences can be annotated to a high-quality level, but this might be an ambitious task. Whereas many model organisms have had a cadre of experienced researchers to provide a knowledge base for manual curation and annotation of genome sequence (e.g., typified by the *Neurospora* and *Aspergillus* research communities), this level of molecular-genetic expertise is generally lacking for the lichen community. Most lichen mycobiont genomes and metagenomes are thus likely to rely primarily on automatic annotation supplemented with manual annotation for specific aspects relating to the interests of particular researchers. Fortunately many non-lichenologists have expressed enthusiasm to assist with lichen genome analysis, and their insights are to be welcomed. Although it is anticipated that much will be learned from comparison with model organisms, lichen fungi and communities are also expected to have unique interactions. Therefore, some aspects must be discovered de novo, by genomic methods, by experimental manipulations on pure culture systems, or by both. To this end, mycobiont, photobionts, and associated microbes of a metagenomically sequenced lichen should be established in vitro where possible to not only assist in providing information for gap closing or structure confirmation of genome models but also to enable complementary experiments to confirm or extend ideas gained from ecological genomic studies.

Experimentation to Validate Ecological Genomic Insights

The comparative analysis of genome sequences of cultured symbionts will certainly provide a valuable tool to further identify and characterize genes that are involved in symbiotic lifestyles. Gene family expansions, notably of genes involved in transport processes, signaling, secondary metabolite synthesis, or of genes involved in as yet unknown functions might provide footprints of symbiosis. However, the functions of individual genes in a symbiosis must be assessed by subsequent experimental work that assesses their differential transcription and catalytic effect under different constraints. Such experiments may be aided by systems biology methods. They also need to consider the symbiotic context to extract the significance of gene expression for symbiosis. Symbiotic partnerships need to be resynthesized by coculture experiments or environmental thalli sampled for metatranscriptomic or metaproteomic analysis. Coculture experiments between *C. grayi* and *Asterochloris* sp. revealed fungal and algal genes that were selectively upregulated in vitro in early lichen development (Joneson, Armaleo, et al., 2011). In this study, cDNA libraries were created by suppression subtractive hybridization methods using RNA extracted from the first two stages of lichen development. Expression levels of 41 and 33 candidate fungal and algal genes, respectively, were further analyzed by real-time PCR (qPCR). Significant matches were found to fungal genes that encode proteins involved in self- and non–self-recognition, lipid metabolism, and negative regulation of glucose repressible genes, as well as to a putative D-arabitol reductase and two dioxygenases. In the algal partner other genes were upregulated, notably a chitinase-like protein, an amino acid metabolism protein, a dynein-related protein, and a protein arginine methyltransferase. Interestingly, evidence for extracellular communication without cellular contact between lichen symbionts was found, according to changes in gene expression patterns when symbionts were separated by a nitrocellulose membrane. Minor variations in expression of many other genes that could be involved in directing the development of the symbiotic phenotype were also noted.

Conclusions: Unifying Platforms and Changing Paradigms

Lichens represent a major terrestrial life form and lichen-forming fungi constitute a large component of fungal biodiversity. Despite this they remain a relatively poorly studied group of organisms. Progress in genomic studies now offers exciting prospects to gain new insights into the functional biology of lichens, with results likely to be of significance to other fungal symbioses. One of the main challenges will be to integrate data from different analytical approaches to understand the lichen symbiosis (Chaston & Douglas, 2012). Metagenomics, metatranscriptomics, metaproteomics, and such each provide insights into different pieces of the biological puzzle of symbiosis, yet, not all genes are transcribed, not all transcripts will be translated, and not all proteins need to be active under certain conditions of lichen biology. The functional contribution of genes will likely be organ-specific and modified by pertinent ecological and developmental conditions. Thus, increased knowledge of lichen ecology and ideally, the incorporation of metabolic data (e.g., using metabolic flux analysis, or the analysis of metabolites by mass spectral molecular networking; Watrous, Dorrestein, et al., 2012) are also required for systems modeling and reasonable interpretation of all the relevant data, and toward gaining a deeper understanding of the lichen symbiosis.

Acknowledgments

MG and GB are grateful to the Austrian Science Foundation FWF for financial support (I799, I882). ÓSA and OV thank the Icelandic Research Fund for support.

References

Ahmadjian V. 1995. Lichens are more important than you think. *BioScience.* 45:123–124.

Armaleo D, Clerc P. 1991. Lichen chimeras: DNA analysis suggests that one fungus forms two morphotypes. *Exp Mycol.* 15: 1–10.

Armaleo D, Miao V. 1999. Symbiosis and DNA methylation in the *Cladonia* lichen fungus. *Symbiosis.* 26: 143–163.

Armaleo D, Sun X, et al. 2011. Insights from the first putative biosynthetic gene cluster for a lichen depside and depsidone. *Mycologia.* 103: 741–754.

Banfield JF, Barker WW, et al. 1999. Biological impact on mineral dissolution: Application of the lichen model to understanding mineral weathering in the rhizosphere. *Proc Natl Acad SCI USA*. 96: 3404–3411.

Bates ST, Cropsey GWG, et al. 2011. Bacterial communities associated with the lichen symbiosis. *Appl Environ Microbiol.* 77: 1309–1314.

Bjelland T, Grube M, et al. 2011. Microbial metacommunities in the lichen–rock habitat. *Environ Microbiol Rep.* 4: 434–442.

Blaha J, Baloch E, et al. 2006. High photobiont diversity in symbioses of the euryoecious lichen *Lecanora rupicola* (Lecanoraceae, Ascomycota). *Biol J Linn Soc.* 88: 283–293.

Boddy L, Dyer PS, et al. 2010. Plant pests and perfect partners. In *From Another Kingdom* (eds. L Boddy & M Coleman), 52–65. Edinbugh: Royal Botanic Gardens.

Boustie J, Tomasi S, et al. 2011. Bioactive lichen metabolites: Alpine habitats as an untapped source. *Phytochem Rev.* 10: 287–307.

Cardinale M, Berg G, et al. 2011. Sun and age: Environmental factors triggering the bacterial communities in lichens. *Environ Microbiol Rep.* 4: 23–28.

Cardinale M, Grube M, et al. 2012. Bacterial taxa associated with the lung lichen *Lobaria pulmonaria* are differentially shaped by geography and habitat. *FEMS Microbiol Lett.* 329: 111–115.

Cardinale M, Müller H, et al. 2008. In situ analysis of the bacteria community associated with the reindeer lichen *Cladonia arbuscula* reveals predominance of *Alphaproteobacteria*. *FEMS Microbiol Ecol.* 66: 63–71.

Cardinale M, Puglia AM, et al. 2006. Molecular analysis of lichen-associated bacterial communities. *FEMS Microbiol Ecol.* 57: 484–495.

Casano LM, del Campo EM, et al. 2011. Two *Trebouxia* algae with different physiological performances are ever-present in lichen thalli of *Ramalina farinacea*. Coexistence versus competition? *Environ Microbiol.* 13: 806–818.

Chaston J & Douglas AE. 2012. Making the most of "omics" for symbiosis research. *Biol Bull.* 223: 21–29.

Crittenden PD, David JC, et al. 1995. Attempted isolation and success in the culturing of a broad spectrum of lichen-forming and lichenicolous fungi. *New Phytol.* 130: 267–297.

de los Ríos A, Wierzchos J, et al. 2002. Microhabitats and chemical microenvironments under saxicolous lichens growing on granite. *Microbial Ecol.* 43: 181–188.

DePriest PT. 2004. Early molecular investigations of lichen-forming symbionts: 1986–2001. *Annu Rev Microbiol.* 58: 273–301.

Dyer PS, Murtagh GJ, et al. 2001. Use of RAPD-PCR DNA fingerprinting and vegetative incompatibility tests to investigate genetic variation within lichen-forming fungi. *Symbiosis.* 31: 213–229.

Fahselt D. 2008. Individuals and populations of lichens. In: *Lichen Biology*, 2nd ed. (ed. TH Nash III), 252–273. Cambridge: Cambridge University Press.

Fernández-Mendoza F, Domaschke S, et al. 2011. Population structure of mycobionts and photobionts of the widespread lichen *Cetraria aculeata*. *Mol Ecol.* 20: 1208–1232.

Friedl T & Büdel B. 2008. Photobionts. In: *Lichen Biology*, 2nd ed. (ed. TH Nash III), 7–26. Cambridge: Cambridge University Press.

Gargas A, DePriest PT, et al. 1995. Multiple origins of lichen symbioses in fungi suggested by SSU rDNA phylogeny. *Science* 268: 1492–1495.

González I, Ayuso-Sacido A, et al. 2005. Actinomycetes isolated from lichens: Evaluation of their diversity and detection of biosynthetic gene sequences. *FEMS Microbiol Ecol.* 54: 401–415.

Grube M, Berg G. 2009. Microbial consortia of bacteria and fungi with focus on the lichen symbiosis. *Fung Biol Rev.* 23: 72–85.

Grube M, Cardinale M, et al. 2009. Species-specific structural and functional diversity of bacterial communities in lichen symbioses. *ISME J.* 3: 1105–1115.

Grube M, Köberl M, et al. 2012. Host parasite interaction and microbiome response: effects of fungal infections on the bacterial community of the Alpine lichen *Solorina crocea*. *FEMS Microbiol Ecol.* 82(2): 472–481.
Hawksworth DL. 2003. The lichenicolous fungi of Great Britain and Ireland: An overview and annotated checklist. *Lichenologist.* 35: 191–232.
Hodkinson BP, Gottel NR, et al. 2011. Photoautotrophic symbiont and geography are major factors affecting highly structured and diverse bacterial communities in the lichen microbiome. *Environ Microbiol.* 14: 147–161.
Hodkinson BP & Lutzoni F. 2009. A microbiotic survey of lichen-associated bacteria reveals a new lineage from the Rhizobiales. *Symbiosis.* 49: 163–180.
Honegger R & Scherrer S. 2008. Sexual reproduction in lichen-forming ascomycetes. In *Lichen Biology*, 2nd ed. (TH Nash III), 94–103. Cambridge: Cambridge University Press.
Honegger R, Zippler U, et al. 2004. Genetic diversity in *Xanthoria parietina* (L.) Th. Fr. (lichen-forming ascomycete) from worldwide locations. *Lichenologist.* 36: 381–390.
Huneck S, Yoshimura I. 1996. *Identification of Lichen Substances*. Berlin: Springer Verlag.
Huneck S. 1999. The significance of lichens and their metabolites. *Naturwissenschaften* 86: 559–570.
Itten B, Honegger R. 2010. Population genetics in the homothallic lichen-forming ascomycete *Xanthoria parietina*. *Lichenologist.* 42: 751–761.
Joneson S, Armaleo D, et al. 2011. Fungal and algal gene expression in early developmental stages of lichen-symbiosis. *Mycologia.* 103: 291–306.
Juntilla S & Rudd S. 2012. Characterization of a transcriptome from a non-model organism, *Cladonia rangiferina*, the grey reindeer lichen, using high-throughput next generation sequencing and EST sequence data. *BMC Genomics.* 13: 575.
Kampa A, Gagunashvili A, et al. 2013. Metagenomic natural product discovery in lichen provides evidence for a family of biosynthetic pathways in diverse symbioses. *Proc Natl Acad SCI USA.* doi/10.1073/pnas.1305867110.Published online before print July 29, 2013.
Kim JA, Hong SG, et al. 2012. A new reducing polyketide synthase gene from the lichen-forming fungus *Cladonia metacorallifera*. *Mycologia.* 104: 362–370.
Lawrey JD & Diederich P. 2003. Lichenicolous fungi: Interactions, evolution, and biodiversity. *The Bryologist.* 106(1): 80–120.
Lee H, Shin S-C, et al. 2012. Genome sequence of *Sphingomonas* sp. strain PAMC 26621, an arctic-lichen-associated bacterium isolated from a *Cetraria* sp. *J Bacteriol.* 194: 3030.
Liba CM, Ferrara FIS, et al. 2006. Nitrogen-fixing chemo-organotrophic bacteria isolated from cyanobacteria-deprived lichens and their ability to solubilize phosphate and to release amino acids and phytohormones. *J Appl Microbiol.* 101: 1076–1086.
Lumbsch HT, Ahti T, et al. 2011. One hundred new species of lichenized fungi: A signature of undiscovered global diversity. *Phytotaxa.* 18: 1–27.
Manoharan SS, Miao VPW, et al. 2012. LEC-2, a highly variable lectin in the lichen *Peltigera membranacea*. *Symbiosis.* 58: 91–98.
McDonald T, Dietrich F, et al. 2012. Multiple horizontal gene transfers of ammonium transporters/ammonia permeases from prokaryotes to eukaryotes: Toward a new functional and evolutionary classification. *Mol Biol Evol.* 29: 51–60.
Miao V, Coëffet-LeGal MF, et al. 2001. Genetic approaches to harvesting lichen products. *Trends Biotechnol.* 19: 349–355.
Miao, VPW, Manoharan SS, et al. 2012. Expression of *lec-1*, a mycobiont gene encoding a galectin-like protein in the lichen *Peltigera membranacea*. *Symbiosis.* 57: 23–31.
Muggia L, Schmitt I, et al. 2008. Purifying selection is a prevailing motif in the evolution of ketoacyl synthase domains of polyketide synthases from lichenized fungi. *Mycol Res.* 112: 277–288.
Murtagh GJ, Dyer PS, et al. 2000. Sex and the single lichen. *Nature* 404: 564.
Mushegian AA, Peterson CN, et al. 2011. Bacterial diversity across individual lichens. *Appl Environ Microbiol.* 77: 4249–4252.

Nash TH III. 2008. *Lichen Biology*, 2nd ed. Cambridge: Cambridge University Press.

Øvstedal DO, Lewis Smith RI. 2001. *Lichens of Antarctica and South Georgia*. Cambridge: Cambridge University Press.

Plumridge A, Melin P, et al. 2010. The decarboxylation of the weak-acid preservative, sorbic acid, is encoded by linked genes in *Aspergillus* spp. *Fungal Genet Biol.* 47: 683–692.

Printzen C, Fernández-Mendoza F, et al. 2012. Alphaproteobacterial communities in geographically distant populations of the lichen *Cetraria aculeata. FEMS Microbiol Ecol.* 2: 316–325.

Purvis OW, Coppins BJ, et al. 1992. *The Lichen Flora of Great Britain and Ireland*. London: Natural History Museum Publications.

Ruibal C, Millanes AM, et al. 2011. Molecular phylogenetic studies on the lichenicolous *Xanthoriicola physciae* reveal Antarctic rock-inhabiting fungi and *Piedraia* species among closest relatives in the Teratosphaeriaceae. *IMA Fungus.* 2: 97–103.

Scheidegger C & Werth S. 2009. Conservation strategies for lichens: Insights from population biology. *Fungal Biol Rev.* 23: 55–66.

Scherrer S, Haisch A, et al. 2002. Characterization and expression of *XPH1*, the hydrophobin gene of the lichen-forming ascomycete *Xanthoria parietina. New Phytol.* 154: 175–184.

Scherrer S, Zippler U, et al. 2005. Characterization of the mating-type locus in the genus *Xanthoria* (lichen forming ascomyctes, lecanoromyctes). *Fungal Genet Biol.* 42: 976–988.

Schmitt I & Lumbsch HT. 2009. Ancient horizontal gene transfer from bacteria enhances biosynthetic capabilities of fungi. *PLoS One.* 4: e4437.

Schneider T & Riedel K. 2010. Environmental proteomics: Analysis of structure and function of microbial communities. *Proteomics.* 10: 785–798.

Schneider T, Vieira de Castro J, et al. 2011. Structure and function of the symbiosis partners of the lung lichen (*Lobaria pulmonaria* L. Hoffm.) analyzed by metaproteomics. *Proteomics.* 11: 2752–2756.

Schoch CL, Sung GH, et al. 2009. The Ascomycota Tree of Life: A phylum wide phylogeny to address phylogenetic informativeness, ancestral character reconstruction and define novel lineages. *Syst Biol.* 58: 224–239.

Shin S-C, Ahn D-H, et al 2012. Genome sequence of *Sphingomonas* sp. strain PAMC 26605, isolated from arctic Lichen (*Ochrolechia* sp.). *J Bacteriol.* 194: 1607.

Sinnemann SJ, Andrésson ÓS, et al. 2000. Cloning and heterologous expression of *Solorina crocea* pyrG. *Curr Genet.* 37: 333–338.

Stocker-Wörgötter E & Hager A. 2008. Appendix: Culture methods for lichens and lichen symbionts. In *Lichen Biology*, 2nd ed. (ed. TH Nash III), 353–363. Cambridge: Cambridge University Press.

Stocker-Wörgötter E & Türk R 1991. Artificial resynthesis of thalli of the cyanobacterial lichen Peltigera praetextata under laboratory conditions. *Lichenologist.* 23: 127–138.

Tehler A & Wedin M. 2008. Systematics and lichenized fungi. In *Lichen Biology*, 2nd ed. (ed. TH Nash III), 336–352. Cambridge: Cambridge University Press.

Wang Y-Y, Zhang T, et al. 2011. Construction and characterization of a full-length cDNA library from mycobiont of *Endocarpon pusillum* (lichen-forming Ascomycota). *World J Microbiol Biotechnol.* 27: 2873–2884.

Waltrous JD & Dorrestein PC. 2011. Imaging mass spectrometry in microbiology. *Nat Rev Microbiol.* 9: 683–694.

Xavier BB, Miao VP, et al. 2012. Mitochondrial genomes from the lichenized fungi *Peltigera membranacea* and *Peltigera malacea*: Features and phylogeny. *Fungal Biol.* 116: 802–814.

Section 4
Animal-Interacting Fungi

10 Ecogenomics of Human and Animal Basidiomycetous Yeast Pathogens

Sheng Sun[1]*, Ferry Hagen[2]*, Jun Xu[3]*, Tom Dawson[3], Joseph Heitman[1], James Kronstad[4], Charles Saunders[3], and Teun Boekhout[5]

[1]Department of Molecular Genetics and Microbiology, Duke University Medical Center, Durham, North Carolina
[2]Department of Medical Microbiology and Infectious Diseases, Canisius Wilhelmina Hospital, Nijmegen, The Netherlands
[3]Procter & Gamble Co., Cincinnati, Ohio
[4]Michael Smith Laboratories, Department of Microbiology and Immunology, University of British Columbia, Vancouver, Canada
[5]CBS Fungal Biodiversity Centre, Utrecht, The Netherlands
**Contributed equally to the chapter*

Introduction

This chapter provides an overview on the diversity of basidiomycetous yeasts with emphasis on the human and animal pathogens. Comparative genomics studies clearly show that these yeast pathogens are well adapted to the human host and are able to circumvent the host defense systems. A discussion is provided on the diversity of mating type systems that regulate the (a)sexual development of basidiomycetes, including the human, animal, and plant pathogens. Two groups of fungi are discussed in detail as examples. The first includes *Cryptococcus neoformans*, which is causing a significant number of attributable mortalities among people infected with HIV, and its sibling species *Cryptococcus gattii* that is a primary pathogen causing outbreaks occurring in distinct locales involving a majority of individuals who have no known immunodeficiency. The second example is the adaptation of lipophilic or lipid-dependent *Malassezia* yeasts to the human and animal skin. These yeasts are phylogenetically related to the plant pathogenic smut fungi, and the adaptations of the genome that allow the species to occupy the skin habitat and that have accumulated since it divergence from the last common ancestor with the plant pathogenic smut fungi are described.

The Ecological Genomics of Fungi, First Edition. Edited by Francis Martin.

Biodiversity of Basidiomycetous Yeasts Pathogens

Basidiomycetous yeasts are unicellular or dimorphic fungi that occur in three lineages of the Basidiomycota, namely the Pucciniomycotina, Ustilaginomycotina, and Agaricomycotina (Fig. 10.1; James, Kauff, et al., 2006; Hibbett, Binder, et al., 2007; Boekhout, Fonseca, et al., 2011). At present approximately 450 species of basidiomycetous yeasts are recognized (Kurtzman, Fell, et al., 2011) and the number of newly described species is growing rapidly. Approximately equal numbers belong to Pucciniomycotina (211 spp.) and Agaricomycotina (213 spp.), whereas a much lower number belongs to the Ustilaginomycotina (29 spp.) (Kurtzman, Fell, et al., 2011). Many of the sexually characterized genera are monophyletic, which contrasts strongly with the asexual genera of which many are polyphyletic (e.g., *Cryptococcus*, *Bullera*, *Sporobolomyces*, *Bensingtonia*, and *Rhodotorula*). Three trends will strongly affect the taxonomy of this group of organisms, as well as many other groups of fungi, namely (1) the application of refined phylogenies based on multigene and phylogenomics approaches (Fitzpattrick, Logue, et al., 2006; Kuramae, Robert, et al., 2006; Robbertse, Reeves, et al., 2006; Marcet-Houben & Gabaldón, 2009), and the taxonomic inferences based on such phylogenies and the application of monophyly as a leading classification principle; (2) the ongoing species discovery (Boekhout, 2005); and (3) the application of the nomenclatural principle that each fungus will have only name (1F=1N; Hawksworth, 2011). The number of human and animal basidiomycetous yeast pathogens (BYP) is limited, and the approximately 40 pathogenic species occur mainly in the orders Tremellales and Trichosporonales (Agaricomycotina) and Malasseziales (Ustilaginomycotina) (Boekhout, Gueidan, et al., 2009).

All three main groups of Basidiomycetes show an extreme diversity in morphology and lifestyles that ranges from unicellular yeasts and yeastlike fungi to species that form complex life cycles comprising various host shifts (e.g., some rust species, see Chapter 7), and species that form highly complex multicellular fruiting bodies, such as mushrooms (see Chapter 8). Ecologically they range from saprotrophs, obligate pathogens of insects and plants, ectomycorrhiza-forming species, to facultative pathogens on humans and other vertebrate animals. Thus, the unifying characters of each of the three lineages are largely biochemical and molecular in nature. The Pucciniomycotina are characterized by the predominance of mannose and absence of xylose in the cell wall (Prillinger, Oberwinkler, et al., 1993), a type A 5S rRNA secondary structure (Gottschalk & Blanz, 1985), layered discoid spindle pole bodies (SPB), and central "simple" septal pores (Boekhout, Fonseca, et al., 2011). Unifying characteristics of the Agaricomycotina are complex dolipore septa, which are usually covered by complex septal pore caps (SPC) that, however, in the basal lineage of Cystofilobasidiales seem absent, the presence of xylose

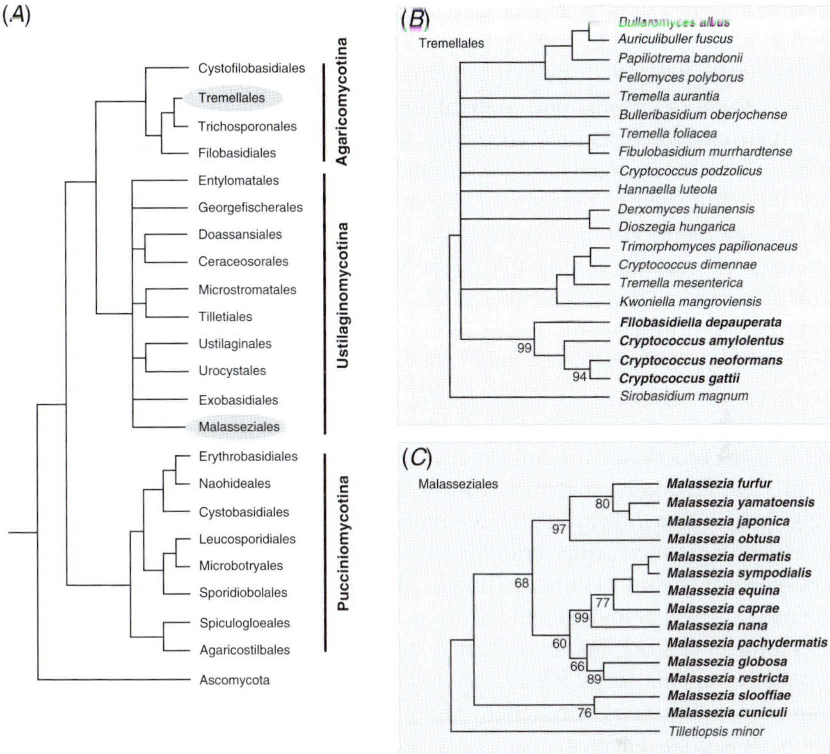

Figure 10.1 Phylogenetic placement of the *Cryptococcus neoformans/Cryptococcus gattii* complex. *A*, Simplified phylogenetic scheme of the Basidiomycetes showing the three subphyla. *B*, Phylogenetic scheme of Tremellales based on D1D2 ribosomal DNA sequences showing the unresolved position of the *C. neoformans/C. gattii* species complex. *Cryptococcus* phylogeny of *Malassezia* species in Malasseziales using D1D2 ribosomal DNA sequences.

in the cell walls with a dominant presence of glucose, a type B 5S rRNA secondary structure, and the capability of the yeast stages to assimilate D-glucuronate and usually *myo*-inositol as well, and the production of extracellular starchlike polysaccharides (Boekhout, Fonseca, et al., 2011). SPCs are complex membranous structures that cover the dolipore and that contain specific proteins (e.g., Spc33 and Spc18) that are involved in maintaining structural integrity of the SPC and probably multicellularity as well (van Peer, Wang, et al., 2009) or that are involved in pore occlusion to maintain cellular homeostasis (van Driel, van Peer, et al., 2008). The role that these structures play in the life cycle of basidiomycetous fungi, and the dimorphic BYPs in particular, needs further elucidation. Interestingly, the *SPC18* gene encoding the Spc18 protein in the *Rhizoctonia* lineage is not present in any other fungal lineage sampled so far, whereas the *SPC33* gene from *Schizophyllum commune* is not present outside the Agaricales lineage, and both are absent in the Tremellales genomes investigated, namely that of *C. neoformans* and *Tremella mesenterica* (T. Boekhout, unpublished), thus indicating a considerable divergence of the SPC-involved genes across the Basidiomycetes.

C. neoformans and *C. gattii* are among the most important animal and human pathogens. The sexual dikaryotic hyphal stage forms readily on suitable culture media but have so far not been observed in nature. Some researchers have suggested that they may represent mycoparasitic (Bandoni, 1995) or phytoparasitic stages (Xue, Tada, et al., 2007). Other important human pathogens belong to the genus *Trichosporon* that produces hyphae and arthroconidia (Chagas-Neto, Chaves, et al., 2008; Taj-Aldeen, Al-Ansari, et al., 2009).

Ustilaginomycotina are characterized by cell walls that contain glucose as the dominant sugar and that lack xylose, but galactose may be present, a type B 5S rRNA secondary structure, and a hemispherical SPB (Boekhout, Fonseca, et al., 2011). Many species of this subphylum are important plant pathogens that produce dikaryotic hyphae able to invade plant tissue (see Chapter 7). The Ustilaginomycotina comprises two classes, Ustilaginomycetes and Exobasidiomycetes (Hibbett, Binder, et al., 2007), and both contain yeast (-like) taxa. Some phylogenetically close relatives to the smuts are only known as saprobes (e.g., *Pseudozyma* spp.), and several of these have potential as biocontrol agents (e.g., *Pseudozyma* spp., *Tilletiopsis* spp., *Meira* spp., and *Acaromyces ingoldii*) (Urquehart, Menzies, et al., 1994; Belanger, Dik, et al., 1998; Boekhout, Theelen, et al., 2003; Sztejnberg, Paz, et al., 2004; Boekhout, Fonseca, et al., 2011). The human and animal pathogenic *Malassezia* species form a well-supported clade that is classified as Malasseziales (*see* Fig. 10.1) (Begerow, Bauer, et al., 2000; Begerow, Stoll, et al., 2006).

Mating Biology and Mating Types in Basidiomycetous Yeasts Pathogens

There have been several excellent recent book chapters and reviews on the topics of mating and *MAT* locus evolution in fungi in general, as well as in BYP (Heitman, Kronstad, et al., 2007; Giraud, Yockteng, et al., 2008; Butler, 2010; Lee, Ni, et al., 2010; Ni, Feretzaki, et al., 2011). Thus, here a brief summary will be provided of what is known about mating, mating types, as well as the mating type locus (MAT) evolution in BYP, focusing on novel progress with discussion of studies that have been recently published.

For BYP, as for Basidiomycetes in general, mating is normally initiated when two cells with compatible mating types encounter one another, and cell fusion ensues. The resulting zygote will then grow as a dikaryotic hyphae. Eventually, the tips of the aerial hyphae will enlarge to form the basidia, within which karyogamy (nuclear fusion) and meiosis occur and four meiotic products are generated. In the sexual stage of *C. neoformans* and *C. gattii* basidiospores are produced through repeated rounds of mitosis and emerge from the surface of the basidium to form spore chains or spore clusters. Sexual reproduction has been linked to pathogenesis. For example, basidiospores generated by sexual reproduction have long been hypothesized to be primary infectious propagules of *C. neoformans*, and both classic and recent studies document that spores are indeed infectious (Zimmer, Hempel, et al., 1984; Sukroongreung, Kitiniyom, et al., 1998; Botts, Giles, et al., 2009; Giles, Dagenais, et al., 2009; Velagapudi, Hsueh, et al., 2009). For *Ustilago maydis*, the smut fungus, it has been shown that the hyphae produced during sexual reproduction are required for infection (Böller, 2001).

In addition to opposite-sex mating, sexual reproduction can also occur between isolates that belong to the same mating type (i.e., unisexual reproduction or same-sex mating), which was first discovered in the laboratory between *MAT*α strains of *C. neoformans* (Lin, Hull, et al., 2005). Subsequent population genetics studies provided robust evidence that same-sex mating also occurs in nature (Lin, Litvintseva, et al., 2007; Bui, Lin, et al., 2008; Hiremath, Chowdhary, et al., 2008; Saul, Krockenberger, et al., 2008; Lin, Patel, et al., 2009). Furthermore, same-sex mating provides an alternative means to produce infectious spores for a species whose natural population is predominated by *MAT*α strains. On the other hand, unisexual reproduction could well have contributed to this observed skew favoring *MAT*α over *MAT***a** strains of *C. neoformans* in nature.

Within the Basidiomycota, there are two different mating systems: bipolar and tetrapolar. For species with tetrapolar mating systems, the mating type is determined by genes located within two unlinked mating type (*MAT*) loci (the homeodomain [HD] locus and the pheromone/pheromone receptor [P/R] locus, respectively). The HD locus can be multi-allelic, whereas the P/R locus can be bi-, tri-, or multi-allelic. For species with bipolar mating systems, there is only one *MAT* locus, which is bi-allelic. Within BYP, both bipolar and tetrapolar mating systems are present. For example, the species within the human pathogenic *Cryptococcus* species complex all have a bipolar mating system. Another human commensal and pathogen, *Malassezia globosa*, also appears to have two linked *MAT* loci (Xu, Saunders, et al., 2007), suggesting a bipolar mating system, although bona fide sexual reproduction has not been as yet observed for this species. Additionally, for smut fungi, *U. maydis* has a tetrapolar mating system, whereas its closely relative, *Ustilago hordei*, has two linked *MAT* loci and a bipolar mating system (Rowell & DeVay, 1954; Bakkeren, Warren, et al., 2006; Xu, Saunders, et al., 2007).

It is not known as yet whether the emergence of the tetrapolar mating system gave rise to the Basidiomycota. It could be that the origin of Basidiomycetes occurred as a result of some unknown evolutionary events that transpired first, and the appearance of the tetrapolar mating system then happened later within the Basidiomycota. Nevertheless, the universal presence of the tetrapolar mating system in all of the major groups of the Basidiomycetes suggests an ancient root in this phylum (Fig. 10.2). Interestingly, in *Sporidiobolus salmonicolor*, a red yeast species that belongs to the Pucciniomycotina, which is the earliest derived lineage of the Basidiomycota, the mating system is unlike either the bipolar or the tetrapolar mating system (Coelho, Sampaio, et al., 2010). Specifically, in *S. salmonicolor*, although the HD and P/R loci are physically linked like those seen in bipolar mating systems, occasional recombination between the two loci and the multiallelic nature of the HD locus mirror features of a tetrapolar mating system. It is not yet known if this novel arrangement represents a derived or ancestral state.

There have been multiple independent transitions from tetrapolar to bipolar mating systems in basidiomycetous species, including *U. hordei*, *M. globosa*, as well as species within the human pathogenic *Cryptococcus* species complex. The fact that the reemergence of bipolar systems occurred mostly in pathogenic species suggests that the transition from a tetrapolar to a bipolar mating system, as well as the accompanied changes within the *MAT* loci, could have contributed to the successful pathogenesis of these species. For example, it has been shown that the *MAT* locus of *U. hordei* controls the pathogenicity of this fungus (Lee, Bakkeren, et al., 1999). Interestingly, there are several recent studies providing evidence of extant transitions from tetrapolar to bipolar mating systems in some Basidiomycetous species. For example, in *Cryptococcus heveanensis* and *Cryptococcus amylolentus*, both closely related to the pathogenic *Cryptococcus* species complex, it has been shown that the P/R loci have undergone expansion, although the two *MAT* loci are not yet physically linked, consistent with transitional stages from tetrapolar to bipolar mating systems (Metin, Findley, et al., 2010; Findley, Sun, et al., 2012). This transition could be eventually achieved by fusion of the two *MAT* loci to form one contiguous *MAT* locus, either through ectopic recombination or translocation and followed by the stabilization of the newly arisen *MAT* locus configuration through backcrosses and assorted mating.

The *C. neoformans/C. gattii* Species Complex

The taxonomy of the *C. neoformans/C.gattii* complex underwent significant changes since the description of *C. neoformans* in 1896. A major step forward was the recognition of four serotypes A, B, C, and D (Evans, 1950; Wilson, Bennett, et al., 1968). During the 1970s, the sexual phase of *C. neoformans* was described and the opposite mating-types were named *MAT***a** and *MAT*α (Kwon-Chung, 1975). It was observed that the sexual state of serotype A and D isolates differed microscopically from that of serotype B and C isolates.

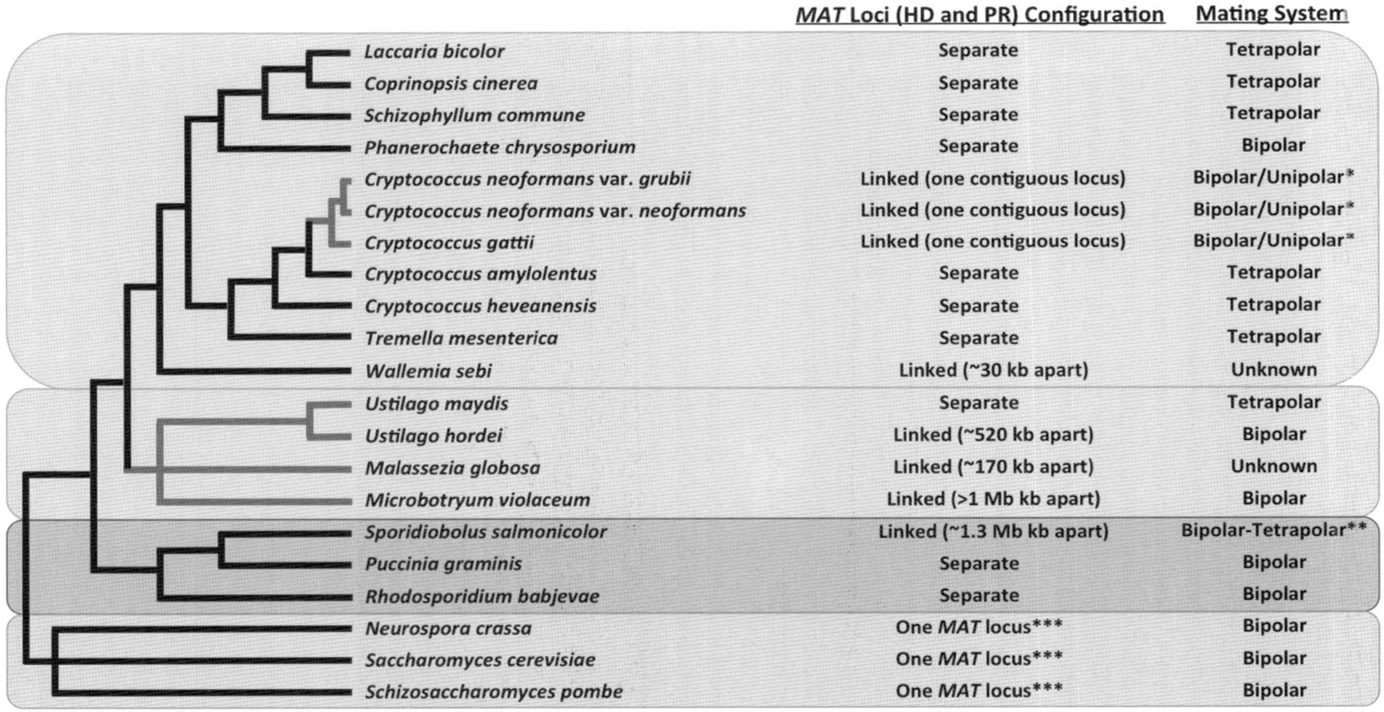

Figure 10.2 Representative Basidiomycetous species and their MAT configurations and mating systems. The phylogenetic relationship is based on Padamsee, Kumar, et al. (2012) and James, Kauff, et al. (2006). The green, pink, and blue shadings highlight species from the three supergroups of the Basidiomycota: Agaricomycotina, Ustilaginomycotina, and Pucciniomycotina, respectively. The orange shading highlights species from the Ascomycota. Branches in red color are those leading to pathogenic species. *, unipolar indicates the same-sex mating has been observed; **, the mating system in *Sporidiobolus salmonicolor* deviates from the traditional bipolar and tetrapolar mating systems, with occasional recombination between linked HD and P/R loci, as well as multiple alleles at the HD locus; ***, Ascomycete MAT loci encode transcription factors (HD, HMG, or α-domain) that control P/R gene expression, but the P/R genes are not part of the MAT locus in Ascomycete. For color detail, please see color plate section.

Therefore, both groups where accommodated in the teleomorphic genus *Filobasidiella* as *F. neoformans* for serotype A and D isolates and *F. bacillispora* for serotype B and C isolates (Kwon-Chung, 1976a, 1976b). A few years later the latter species was reduced to a variety of *F. neoformans* as *F. neoformans* variety *bacillispora*. Since then *C. neoformans* was split into *C. neoformans* variety *neoformans* (serotype A and D) with the teleomorphic phase *F. neoformans* variety *neoformans*, and *C. neoformans* variety *gattii* (serotype B and C) with the teleomorphic phase *F. neoformans* variety *bacillispora* (Kwon-Chung, Bennett, et al., 1982). The proposal to divide the two serotypic groups into two taxonomic varieties was further strengthened by biochemical characteristics and the ecological and geographical differences between serotype A and D versus serotype B and C isolates (Vanbreuseghem & Takashio, 1970; Bennett, Kwon-Chung, et al., 1977, 1978; Kwon-Chung & Bennett, 1984).

A further major change in the taxonomy of *C. neoformans* was the proposal to raise the group of serotype A isolates to variety level named *C. neoformans* variety *grubii* (Franzot, Salkin, et al., 1999). Finally, phylogenetic analyses, biochemical, ecological, and epidemiological differences were the major reasons to raise *C. neoformans* variety *gattii* to the species level as *C. gattii* (Kwon-Chung, Boekhout, et al., 2002).

The phylogeny of the *C. neoformans*/*C. gattii* species complex indicates that a taxonomic revision of the species complex is urgently required. Many molecular biological approaches have been applied to investigate the epidemiology and population structure of these pathogenic yeasts, resulting in the observation that they both contain multiple monophyletic clusters, presently recognized as genotypes that need to be interpreted as separate species. In the current complex situation, genotypic names are applied to each of these genotypically different clusters. Based on random amplification of polymorphic DNA (RAPD), polymerase chain reaction (PCR)-fingerprinting, amplified fragment length polymorphisms (AFLP) fingerprinting, and multilocus sequence typing (MLST) the *C. neoformans*/*C. gattii* complex can be differentiated into 13 monophyletic clusters/genotypes: AFLP1/VNI, AFLP1A/VNB/VNII, and AFLP1B/VNII for *C. neoformans* variety *grubii* (serotype A); genotype AFLP2/VNIV for variety *neoformans* (serotype D); genotype AFLP4/VGI (serotype B); AFLP5/VGIII (serotype C); AFLP6/VGII (serotype B); AFLP7/VGIV (serotype C); and AFLP10/VGIV (serotype B) for *C. gattii* (Boekhout, Theelen, et al., 2001; Meyer, Castañeda, et al., 2003; Bovers, Hagen, et al., 2008b; Meyer, Aanensen, et al., 2009; Hagen, Illnait-Zaragozi, et al., 2010; Hagen, Colom, et al., 2012). The understanding of the biodiversity of the complex is further complicated by the existence of hybrids between the two varieties of *C. neoformans*, the so-called serotype AD-hybrids, and between *C. neoformans* and *C. gattii*. Presently, four different types of hybrids are known, namely *C. neoformans* variety *grubii* × variety *neoformans* (serotype AD; AFLP3/VNIII); *C. neoformans* variety *neoformans* × *C. gattii* (serotype

BD; AFLP8/VNIV+VGI); *C. neoformans* variety *grubii* × *C. gattii* AFLP4/VGI (serotype AB; AFLP9/VNI+VGI); and *C. neoformans* variety *grubii* × *C.gattii* AFLP6/VGII (serotype AB; AFLP11/VNI+VGII) (Bovers, Hagen, et al., 2006, 2008a,2008b; Aminnejad, Diaz, et al., 2012). It is likely that further hybrids will be discovered because of the increasing use of refined molecular biological techniques in the field of epidemiology and taxonomy.

C. neoformans and *C. gattii* as Pathogens

Cryptococcosis, caused by the basidiomycetous yeasts *CC. gattii* and *neoformans*, is one of the most prevalent fungal diseases. Until the early 1980s *Cryptococcus* species were regarded as a rare cause of infection among individuals who were immunocompromised and immunocompetent (Mitchell & Perfect, 1995). However, the HIV/AIDS pandemic caused a dramatic increase in the number of cryptococcoal infections among individuals with an attenuated immune system. The global impact on human health as a result of cryptococcal meningitis was estimated to be approximately one million new infections annually (Park, Wannemuehler, et al., 2009). Cryptococcal infections account yearly for an estimated number of approximately 625,000 fatalities, mainly in sub-Saharan Africa where cryptococcal meningitis is the fourth most common cause of HIV-related death and responsible for approximately one-third of all deaths associated with AIDS (Park, Wannemuehler, et al., 2009; Warkentien & Crum-Cianflone, 2010).

Apparently individuals who are immunocompromised may acquire a cryptococcal infection as well. Examples are the expanding outbreaks of *C. gattii* in British Columbia (Canada) and the Pacific Northwest (United States) that caused serious disease in hundreds of people and animals (Datta, Bartlett, et al., 2009). Detailed genotyping of cryptococcal isolates involved in the outbreaks showed that they were mainly caused by the previously rare genotype AFLP6/VGII. Using different molecular approaches, three subgenotypes have been observed within this genotype, namely genotype AFLP6A/VGIIa that is referred to as the major outbreak genotype involved in the Vancouver Island outbreak and that expanded into the mainland of British Columbia (Kidd, Hagen, et al., 2004); genotype AFLP6B/VGIIb referred to as the minor outbreak genotype and occurring at a global scale (Byrnes, Li, et al., 2010; Kidd, Hagen, et al., 2004); and genotype AFLP6C/VGIIc that is involved in the Pacific Northwest outbreak in the United States (Byrnes, Li, et al., 2010). During the past years several other minor outbreaks with genotype AFLP6/VGII have been reported, namely one among psittacine birds in a bird sanctuary in southern Brazil (Raso, Werther, et al., 2004) and an outbreak among sheep in Western Australia (Carriconde, Gilgado, et al., 2011). Genotype AFLP4/VGI of *C. gattii* was found to be involved in numerous small outbreaks among goats in Spain that occurred over the past two decades

(Baró, Torres-Rodríguez, et al., 1998; Torres-Rodríguez, Baró, et al., 1999; Colom, Hagen, et al., 2012). Several clinical cases in Spain are caused by genotype AFLP4/VGI, and these isolates form a distinct clade when compared to a global set of genotype AFLP4/VGI isolates (Colom, Hagen, et al., 2012; Hagen, Colom, et al., 2012).

For a long time it was assumed that *C. neoformans* had a predilection to infect patients who are immunocompromised, whereas *C. gattii* was assumed to cause disease primarily in individuals who are immunocompetent (Springer & Chaturvedi, 2010). However, several recent studies reported that this does not hold true for the majority of cryptococcosis cases in China, Japan, and Korea. Here, the majority of cryptococcosis patients with an infection with *C. neoformans* variety *grubii* (serotype A) had no apparent underlying disease (Chen, Varma, et al., 2008; Chau, Mai, et al., 2010; Choi, Ngamskulrungroj, et al., 2010; Pan, Khayhan, et al., 2012). In contrast, the majority of *C. gattii* infections caused by genotypes AFLP5/VGIII, AFLP7/VGIV, or AFLP10/VGIV were observed to occur in patients who are immunodeficient, including those with HIV/AIDS, but infections caused by *C. gattii* genotype AFLP4/VGI and AFLP6/VGII were more frequently occurring in patients that did not have any other underlying disease (Kidd, Hagen, et al., 2004; Hagen, Illnait-Zaragozi, et al., 2010; Byrnes, Li, et al., 2011; Hagen, Colom, et al., 2012).

Clinical presentations differ between infections caused by *C. neoformans* and *C. gattii*. More often, the latter species caused cryptococcomas in the lungs or brain with high serum or cerebrospinal fluid (CSF) cryptococcal antigen titers compared to *C. neoformans* infections. Also the prognosis for individuals infected with *C. gattii* was found to be worse, despite prolonged antifungal treatment (Chen, Slavin, et al., 2012; Perfect, Dismukes, et al., 2010; see discussion below). Several studies observed significant differences in in vitro antifungal susceptibility between the different genotypes of *C. gattii*. For instance, isolates of the outbreak-related genotype AFLP6/VGII are less susceptible to azole drugs than those from genotype AFLP4/VGI, which in turn was found to be less susceptible than the other *C. gattii* genotypes (Chong, Dagg, et al., 2010; Iqbal, DeBess, et al., 2010; Hagen, Illnait-Zaragozi, et al., 2010).

Genome of the *C. neoformans/C. gattii* Complex

Currently, whole-genome sequences are available for five strains that are representative of the major serotypes/molecular groups of the *C. neoformans/C. gattii* human pathogenic *Cryptococcus* species complex—that is, strain H99 (*C. neoformans* var. *grubii*, serotype A, AFLP1/VNI, *MAT*α), JEC21 (*C. neoformans* var. *neoformans*, serotype D, AFLP2/VNIV, *MAT*α), B3501A (*C. neoformans* var. *neoformans*, serotype D, AFLP2/VNIV, *MAT*α), WM276 (*C. gattii*, serotype B, AFLP4/VGI, *MAT*α), and R265 (*C. gattii*, serotype B, AFLP6A/VGIIa, *MAT*α)

(Loftus, Fung, et al., 2005; Broad Institute, 2010; D'Souza, Kronstad, et al., 2011). In addition, genomes of several other *C. neoformans* and *C. gattii* strains have been subjected to whole-genome sequencing, and their sequencing reads have been deposited in public archives (Gillece, Schupp, et al., 2011).

Genome comparison studies have revealed that the overall levels of sequence divergence between the serotypes and species (i.e., A–D, A–B, and D–B) are 10 to 15 percent (Sun & Xu, 2009), whereas divergence between the two serotype B genomes (AFLP4/VGI-AFLP6/VGII) is 7.6 percent (D'Souza, Kronstad, et al., 2011). In addition to nucleotide polymorphisms, several studies have identified chromosomal rearrangements, such as chromosomal translocations and inversions that exist between these genomes (D'Souza, Kronstad, et al., 2011; Sun & Xu, 2009). Not surprisingly, chromosomal rearrangements are more common between genomes belonging to different serotypes (Sun & Xu, 2007, 2009), whereas the two serotype B genomes (of strains WM276 and R265) are overall syntenic (D'Souza, Kronstad, et al., 2011). Some of the chromosomal rearrangements have been shown to affect the virulence attributes of the strain (Morrow, Lee, et al., 2012). Additionally, it has been shown that many of the chromosomal rearrangements are associated with chromosomal regions experiencing the greatest reduction of recombination frequency during intervariety hybridization (Sun & Xu, 2007, 2009), consistent with the hypothesis that chromosomal rearrangements could significantly contribute to genetic isolation between diverging lineages.

Genomic analyses of *C. neoformans* and *C. gattii* revealed expansions of certain gene groups in these pathogens. For example, it has been found there is a significant increase in the number of *myo*-inositol transporters in *C. neoformans* and *C. gattii*, which has been suggested to play important roles in sexual reproduction and virulence of these human pathogenic fungi (Xue, Tada, et al., 2007; Xue, Liu, et al., 2010).

In addition, studies applying whole-genome sequence arrays and comparative genome hybridization (CGH) have identified the existence of copy number variation among clinical and environmental strains, and this variation has been shown to be associated with certain pathogenic traits, such as virulence and drug resistance (Sionov, Lee, et al., 2010; D'Souza, Kronstad, et al., 2011; Chow, Morrow, et al., 2012). This suggests the genomic contents of these pathogenic species could be more dynamic than previously thought. Furthermore, the genome sequences of *C. neoformans* and *C. gattii* also provide an opportunity to study the transcriptomes of these pathogens under a variety of virulence-related growth conditions in vitro and in vivo, such as high temperature (Steen, Lian, et al., 2002; Kraus, Boily, et al., 2004), thermal or free radical stress (Missal, Pusateri, et al., 2006; Chow, Liu, et al., 2007), hypoxia (Chang, Bien, et al., 2007; Chow, Liu, et al., 2007; Chun, Liu, et al., 2007), iron limitation (Jung, Sham, et al., 2006, 2008), as well as during phagocytosis and infection of host tissues (Steen, Zuyderduyn, et al., 2003;

Fan, Kraus, et al., 2005; Chun, Liu, et al., 2007; Hu, Steen, et al., 2007). These studies help provide a better understanding of the relationship between gene expression programs and clinical outcomes and also to identify the virulence attributes of these human pathogenic fungi that are more relevant to clinical problems.

Analysis of Virulence Traits in *C. neoformans* and *C. gattii*

The underlying mechanism of the phenotypic and clinical differences between the genotypic groups of *CC. neoformans* and *gattii* as described previously remain to be determined. It is likely that comparative and functional genomics will further show light on these differences. The annotated genome sequences of three strains of *C. neoformans* and two strains of *C. gattii* have enabled new approaches to analyze virulence including the implementation of high-throughput and whole-genome methods. The virulence traits of these species have been reviewed recently (Ma & May, 2009; Kronstad, Attarian, et al., 2011; Kronstad, Saikia, et al., 2012) and include the ability to elaborate a polysaccharide capsule, the deposition of the pigment melanin in the cell wall, and proliferation at the mammalian host temperature of 98.6° F (37° C). In addition, a number of enzymes contribute to virulence, and these include superoxide dismutase, urease, and phospholipase B. Genomic approaches that include transcriptional profiling and systematic gene deletion have allowed investigators to examine the roles and regulation of the known factors in greater detail. Additionally, these approaches are supporting and accelerating the discovery of a larger set of novel functions that contribute to virulence. Examples of the contributions of genomic approaches in selected areas of investigation are provided in the following paragraphs.

Among the virulence traits, the capsule has been the most extensively studied and it functions to modulate phagocytosis as well as the immune response. As an example of a genome-enabled approach to study capsule regulation, Haynes, Skowyra, et al. (2011) employed microarray analysis with cells that were stimulated to form capsule under a variety of conditions. These conditions included growth in low iron medium, in the presence of fetal bovine serum, in Dulbecco's Modified Eagle's Medium (DMEM) tissue culture medium with or without elevated carbon dioxide and in Littman's medium with different concentrations of thiamine. This analysis identified 316 genes whose transcript levels correlated positively with capsule size, and the encoded functions are mostly involved in the response to stress (e.g., DNA damage repair, trehalose biosynthesis, sugar transport). Some of the genes also encoded capsule-associated functions, as expected from the growth conditions; these included the transcription factors Cir1 and Hap5, as well as the kinase Ste20 and the phosphodiesterases Pde1 and Pde2. The transcript abundance for another 564 genes showed a negative correlation with capsule size; many of these genes were related to mitochondrial function and ribosome biogenesis, but no capsule-associated functions were detected.

Haynes, Skowyra, et al. (2011) went on to characterize one gene, *ADA2*, which showed a positive transcriptional correlation with capsule size and that encoded a predicted member of the Spt-Ada-Gcn5-Acetyltransferase (SAGA) complex that mediates histone acetylation. The analysis of an *ADA2* deletion mutant indicated that *ADA2* is required for establishing the wild-type capsule size, for mating, for mediating the response to stress, and for virulence in a mouse-inhalation model of cryptococcosis. Loss of *ADA2* resulted in changes in transcript abundance for 460 genes under capsule-inducing conditions, as determined by RNA-Seq analysis. The regulated genes were enriched for functions, such as ribosomal protein synthesis, sugar transport, and carbohydrate metabolism. Interestingly, subsequent ChIP-seq analysis to identify genes directly regulated by *ADA2*-dependent histone acetylation revealed regulation of *BLP1* and *GAT204*, two genes controlled by the Gat201 transcription factor (see discussion below) and capsule-related genes such as *CPL1*, *HXT1*, *STE3α*, and *UGT1*. Overall, this analysis illustrates the power of genome-wide approaches to define regulatory networks for the expression of virulence factors.

Gene expression profiling has also been used more generally to identify factors involved in virulence in the context of interactions with host cells and specific tissues. For example, Fan, Kraus, et al. (2005) used a microarray approach to identify *C. neoformans* genes expressed at 2 and 24 hours after murine macrophage infection versus growth in medium without macrophages. A total of 157 genes were found to be downregulated in the internalized cells compared with 123 upregulated genes. The upregulated genes encoded predicted membrane transporters and proteins involved in autophagy, peroxisome function, and lipid metabolism. The expression profile also indicated that phagocytosis provoked a stress response in *C. neoformans* because of the observed elevated expression of oxidative stress functions, such as the flavohemoglobin denitrosylase, Fhb1. In addition, components of the cAMP signal transduction pathway and genes clustered at the mating-type locus showed elevated expression upon phagocytosis; this result may reflect a response to nutrient limitation in the intracellular environment of the macrophage.

The transcriptional response of *C. neoformans* has also been characterized for cells collected from infected animals. These studies included cells from the cerebrospinal fluid (CSF) of infected rabbits (Steen, Zuyderduyn, et al., 2003) and cells collected from the lungs of infected mice (Hu, Cheng, et al., 2008). The method of serial analysis of gene expression was employed to examine transcript abundance. The study with cells from the CSF identified the most highly expressed transcripts during infection of the central nervous system, and these encoded functions related to translation, protein degradation, signaling, and energy production. Additional abundantly expressed genes-encoded heat shock proteins as well as proteins for carbohydrate and amino acid metabolism, and for transport (e.g., phosphate, iron, hexoses). In the case of cryptococcal cells recovered from the lungs of infected mice at 8 and 24 hours after intranasal inoculation, transcriptional profiling identified a variety of highly expressed genes. One category of particular interest encoded functions in central carbon metabolism, such as enzymes for the production and use of acetyl-CoA, for the glyoxylate cycle, and for gluconeogenesis. Genes for lipid metabolism also showed elevated expression in the cells from the pulmonary infection, and this was similar to the situation for gene expression upon phagocytosis (Fan, Kraus, et al., 2005). Overall, these results suggest that lipids and other alternative carbon sources may be important for growth in host tissue. Other categories of upregulated functions included stress proteins, known virulence factors, and transport functions (e.g., for hexose, trehalose, amino acids, copper, iron, acetate, and phosphate). Among these genes, the transcript for one candidate hexose transporter was the most abundant in cells recovered from the lung environment, although deletion of this gene did not attenuate virulence in mice. Taken together, the transcriptional profiling analysis of *C. neoformans* cells from macrophages and infected animals show similarities that highlight the importance of the stress response and specific nutritional adaptations to the host environment.

In addition to transcriptional profiling, a genome-wide genetic approach to examine virulence has been initiated using signature tagged mutagenesis (STM; Liu, Chum, et al., 2008). In this case, 1,100 deletion mutants were constructed in *C. neoformans* and screened for decreased or increased infectivity in the mouse lung environment. This study identified 33 mutants with increased infectivity and 164 mutants with decreased infectivity, and a parallel analysis of the latter strains for in vitro virulence phenotypes revealed that 85 had defects in capsule production, melanization, or growth at 98.6° F (37° C). In these strains, melanin production was influenced by 33 novel genes and 5 novel genes regulated capsule formation. An additional 40 mutants had defects in lung infectivity without altered growth, capsule production, or melanin synthesis.

Interestingly, one of the mutants identified by STM showed severely reduced infectivity and had a defect in a previously uncharacterized GATA transcription factor designated Gat201 (Liu, Chum, et al., 2008). The *gat201* mutant also showed impaired induction of capsule formation, although the defect was not striking compared with the well-studied mutants in the *CAP* genes that are required for capsule formation. However, the *gat201* mutant cells were more readily phagocytosed than the *cap* mutants. As mentioned previously, the capsule contributes to virulence in part by allowing *C. neoformans* to avoid phagocytosis and killing. In this context, *gat201 cap* double mutants showed a greater level of phagocytosis than single *cap* mutants, thus suggesting that Gat201 controls a capsule-independent antiphagocytic mechanism. Microarray analysis with the *gat201* mutant suggested that Gat201 exerts part of its influence by activating the expression of factors involved in host interactions. More recent work to examine this idea identified approximately 1,000 genes that showed statistically significant differential expression in a *gat201* strain (Chun, Brown, et al., 2011). ChIP-Chip experiments showed that Gat201 directly bound 126 of these genes and controlled the expression of 62 of them. A macrophage-uptake screen that was designed to enrich for downstream effectors involved in Gat201-dependent macrophage evasion identified mutations in a Barwin-like protein 1 (Blp1), and a transcription factor (Gat204) among the 62 genes. Loss of Blp1 or Gat204 caused a marked increase in macrophage uptake, indicating a direct role of these effectors in phagocytosis evasion. Overall, these studies revealed that Gat201 is a key regulator of virulence in *C. neoformans* and that it functions, at least in part, to regulate interactions with phagocytic cells in the host. The various studies clearly illustrate the power of large-scale genetic approaches that are enabled by the available genome sequences.

Now that annotated genome sequences are available for *C. gattii*, the types of whole-genome and genetic approaches described previously for *C. neoformans* can be employed to compare virulence functions between the species. This area of investigation is likely to be quite fruitful given recent studies that indicated substantial differences in pathogenesis during murine cryptococcosis. For example, Cheng, Sham, et al. (2009) compared three *C. gattii* strains representing the AFLP4/VGI, AFLP6A/VGIIa, and AFLP6B/VGIIb molecular subtypes with the commonly studied *C. neoformans* strain H99 and found that all strains shared the common virulence traits. However, the *C. gattii* strains all induced a less robust inflammatory response compared to the *C. neoformans* strain. This reduced response appeared to result from inhibition or failure to provoke the infiltration of neutrophils into sites of infection. The *C. gattii* strains also did not elicit production of protective cytokines (e.g., TNFα) compared to *C. neoformans*. This study suggests that *C. gattii* may proliferate in hosts who are immunocompetent by evading or suppressing the immune response. More recently, Ngamskulrungroj, Chang, et al. (2012) compared the ability of the *C. gattii* genotype AFLP6A/VGIIa strain R265 with the *C. neoformans* strain H99 in terms of pathogenesis. They found that the *C. neoformans* strain grew faster in the brain and that meningoencephalitis was the cause of death in mice infected via inhalation. In contrast, the *C. gattii* strain grew faster in the lungs and caused death without extensive brain involvement. The *C. gattii* strain also grew more slowly in the blood of mice, although the isolates of both species were able to cross the blood-brain barrier upon intravenous injection. Together, this study and the work of Cheng, Sham, et al. (2009) provide a basis for more detailed comparative studies of the immune response to the two species. Importantly, the genomic resources are now available to support the identification of cryptococcal functions that may trigger or suppress the host response.

The genome sequence of the R265 strain of *C. gattii* also enabled the discovery of an association between mitochondrial regulation/morphology and the virulence properties of isolates causing the outbreak of cryptococcosis on

Vancouver Island in British Columbia (Ma, Hagen, et al., 2009; Ma & May, 2010). Specifically, a microarray analysis of RNA from 24 outbreak and nonoutbreak strains of *C. gattii* recovered from within J774 macrophages revealed upregulation of genes encoded by the mitochondrial genome for the outbreak isolates. Interestingly, some nuclear-encoded proteins with functions in mitochondria were also upregulated in the outbreak strains. This finding of an association between mitochondrial gene expression and the outbreak isolates was extended with the discovery that the mitochondrial morphology was also different in these strains. That is, the mitochondrial morphologies of both the outbreak and nonoutbreak isolates were characterized as either diffuse or globular when the strains were grown in DMEM tissue culture medium. However, growth of the strains inside macrophages revealed that the mitochondria of only the outbreak isolates showed a distinct tubular morphology. These observations correlate with parallel experiments demonstrating that the outbreak isolates have a high intracellular proliferation rate in both the J774 macrophage-like cell line and in human primary macrophages from peripheral blood. These isolates also have higher virulence in a mouse model of cryptococcosis. Overall, the experiments of Ma, Hagen, et al. (2009) and Ma and May (2010) suggest that a recent change in mitochondrial function within a lineage of *C. gattii* isolates may have increased their capacity for intracellular proliferation in phagocytic cells and their ability to cause disease. An intriguing hypothesis is that changes in mitochondria function may enable these isolates to better withstand hypoxia and oxidative damage during proliferation in mammalian hosts.

Biodiversity and Ecology of *Malassezia* Yeasts

The genus *Malassezia* currently comprises 14 lipophilic or lipid-dependent species that are classified in the order Malasseziales in the Ustilaginomycotina *incertae sedis* (Guého-Kellermann, Boekhout, et al., 2010; Cabañes, Vega, et al., 2011). The species are well adapted to grow on human and animal skin. *Malassezia pachydermatis* is the only lipophilic species and occurs widely in the external ears of dogs. *Malassezia furfur, M. globosa*, and *Malassezia restricta* are known from healthy and diseased human skin, such as pityriasis versicolor. *Malassezia caprae, Malassezia cuniculi, Malassezia equina, Malassezia nana, Malassezia sympodialis*, and *Malassezia slooffiae* occur on animal skin. *Malassezia sympodialis, Malassezia dermatis, Malassezia japonica, Malassezia slooffiae*, and *Malassezia yamatoensis* occur mainly on healthy human skin but are also involved in atopic dermatitis and seborrheic dermatitis (Gaitanis, Mayser, et al., 2010; Guého-Kellermann, Boekhout, et al., 2010; Human Microbiome Project Consortium, 2012). *M. pachydermatis* and *M. furfur* occasionally cause nosocomial outbreaks in neonatal wards (Tragiannidis, Groll, et al., 2010). A detailed account on the biodiversity, physiology, genomics, and pathophysiology can be found in some recent treatments (Batra, Boekhout, et al., 2005; Ashbee, 2007; Boekhout, Guého, et al., 2010; Gaitanis, Magiatis, et al., 2012; Guého-Kellermann, Boekhout, et al., 2010).

Cultures of *Malassezia* species are only known from human and animal skin, but recent metagenomics studies indicated the presence of *Malassezia* DNA in a number of different terrestrial and marine ecosystems. A Finnish indoor air study revealed that 14 percent of the approximately 1,300 clones were related to *Malassezia* spp. (Pitkäranta, Meklin, et al., 2008). The presence of *Malassezia* DNA in another indoor air study made the authors conclude that humans and mammals contribute to the biodiversity present in indoor air dust (Amend, Seifert, et al., 2010). DNA of *M. globosa* was observed in Antarctic low moisture soils (Fell, Scorzetti, et al., 2006), and DNA of *M. restricta* was present in nematodes in Central European soils (Renker, Alphei, et al., 2003) and beetle gut in the southeastern region of the United States (Zhang, Suh, et al., 2003). Probably the greatest surprise was the extensive presence of *Malassezia* DNA in Hawaiian sponges (Gao, Li, et al., 2008) and corals near Samoa (Amend, Barshis, et al., 2012). All these observations may indicate alternative habitats than the human and animal skin where *Malassezia* species may occur, but caution is needed because the yeasts have not yet been observed by techniques such as fluorescence in situ hybridization (FISH) or culturing.

The *M. globosa* Genome

The genome of *M. globosa* has been sequenced with seven-fold coverage (Xu, Saunders, et al., 2007). With 9 Mb, the genome size is among the smallest of free-living fungi, similar to the size of the *Eremothecium (Ashbya) gossypii* genome (Dietrich, Voegeli, et al., 2004) and 4 Mb smaller than a particularly large bacterial genome (Schneiker, Perlova, et al., 2007). There are approximately 4,285 predicted protein-coding genes in the genome of *M. globosa*, also a small number for a free-living fungus and fewer genes than are found in some bacteria (Bentley, Chater, et al., 2002; Schneiker, Perlova, et al., 2007). Other contributions to the small genome size are (1) the paucity of introns that are present in only 27 percent of the genes, and (2) the shortage of repeat elements that comprise less than 1 percent of the genome. Small genome sizes have been associated with microbes with a highly restricted niche (Cole, Eiglmeier, et al., 2001). Possibly *M. globosa* has a more restricted niche than other free-living fungi because the species is primarily reported from the skin of warm-blooded mammals (Batra, Boekhout, et al., 2005), although there are occasional reports of detection of *M. globosa* DNA from other habitats.

Despite its small genome size, *M. globosa* encodes many of the basic metabolic processes, including glycolysis, the tricarboxylic acid cycle, the glyoxylate cycle, the pentose phosphate shunt, and synthesis of the 20 standard amino acids in proteins, and the five bases found in nucleic acids.

Unlike other free-living fungi, *M. globosa* is missing a fatty acid synthase gene (Xu, Saunders, et al., 2007), consistent with the inability to grow in the absence of exogenous lipid (Guého, Midgley, et al., 1996; Batra, Boekhout, et al., 2005; Ashbee, 2007). This suggests that *M. globosa* is dependent on host skin lipids as a source of fatty acids. Similarly, the skin bacterium *Corynebacterium jeikeium* is missing a fatty acid synthase gene and unable to grow in the absence of exogenous lipid (Tauch, Kaiser, et al., 2005). *M. globosa* is missing two other lipid metabolism genes. There is apparently no homolog (Xu, Boekhout, et al., 2010) of the *Saccharomyces cerevisiae* gene *OLE1*, whose product is a δ9 desaturase involved in the synthesis of oleic and palmitoleic acid (Stukey, McDonough, et al., 1989). There is also no homolog (Xu, Boekhout, et al., 2010) of the *S. cerevisiae* gene *ECI1* that encodes a δ^3-*cis*-δ^2-*trans*-enoyl-CoA isomerase required for complete oxidation of unsaturated fatty acids (Gurvitz, Mursula, et al., 1998). The absence of this gene in *M. globosa* raises the question of whether this yeast uses a different approach to oxidize unsaturated fatty acids or whether *M. globosa* is unable to carry out complete oxidation of unsaturated fatty acids. *M. globosa* encodes a single polyketide synthase (PKS) gene with an architecture similar to type I fatty acid synthase. The function of this gene and its product is unknown.

Interaction of *M. globosa* with the Human Host

The set of secreted proteins from *M. globosa* has been characterized to understand better how *M. globosa* might interact with host skin (Xu, Saunders, et al., 2007). Secreted proteins were predicted from the genome sequence and demonstrated by proteomics analysis. To a limited extent, the corresponding gene expression on the scalp was monitored by real-time PCR (qPCR). The presence of multiple gene copies suggests the importance of lipase, aspartyl protease, acid sphingomyelinase, and phospholipase C. The phospholipases C are similar to secreted *Pseudomonas* phospholipase C and may hydrolyze host phospholipids. Several of these proteins are among the most abundant extracellular enzymes in culture, and transcription of the corresponding genes was demonstrated on scalp. These enzymes might play nutritional roles for the yeast on the scalp and might also generate molecules that provide signals to the host, possibly causing significant changes in host tissue.

Progress is beginning to be made in understanding the role of individual genes and enzymes in *Malassezia* interactions with the host. Based on their presumed importance in obtaining host lipids, individual *Malassezia* lipases have been purified (Ran, Yoshiike, et al., 1993; Plotkin, Squiquera, et al., 1996, Brunke & Hube, 2006; Shibata, Okanuma, et al., 2006, DeAngelis, Saunders, et al., 2007). There are reports of (1) enhanced extracellular phospholipase activity from *Malassezia* isolates from diseased skin relative to

Table 10.1 Comparison of several predicted secretory proteins* in *Malassezia globosa*, *Ustilago maydis*, and *Candida albicans*.

Enzyme	Number of Predicted Secreted Proteins		
	M. globosa	*U. maydis*	*C. albicans*
Lipase	13	2	12
Phospholipase C	6	0	0
Phospholipase B	0	1	5
Acid sphingomylinase	4	1	3
Aspartyl protease	15	7	15
Total Gene Number	4,285	6,902	7,677

*Secretory proteins were predicted using SignalP3.0 (Bendtsen, Nielsen, et al., 2004). The phospholipases C were identified based on homology to *Pseudomonas aeruginosa* phospholipase C.

isolates from normal skin (Cafarchia & Otranto, 2004, Pini & Faggi, 2011) and (2) enhanced transcription of certain lipase and phospholipase genes in diseased skin (Patiño-Uzcátegui, Amado, et al., 2011). Further work will be needed to provide a molecular characterization of the interaction of *Malassezia* proteins with host skin, including in vitro characterization of *Malassezia* enzyme interaction with host substrate, the demonstration that the reaction occurs on the skin, and the indication that the alteration of the host molecule impacts the skin structure and function.

The pattern of multiple-copy genes for secreted enzymes is similar between *M. globosa* and *Candida albicans* (Table 10.1), although the two fungi are phylogenetically only distantly related. *C. albicans*, like *M. globosa*, is capable of growth on animal skin. Each of these organisms encodes multiple genes for hydrolases (phospholipase, lipase, protease, and acid sphingomyelinase) that may have a role in degrading host macromolecules. In contrast, the pattern of secreted enzymes from the phylogenetically related plant pathogen *U. maydis* is quite different from that of *M. globosa*. *U. maydis* contains many more glycosyl hydrolase genes than does *M. globosa*. *U. maydis* encodes genes for cutinases, a pectin lyase, and a pectin esterase that are all needed to degrade plant-related substrates. Each of these genes is apparently missing in *M. globosa*. This convergence of the set of *C. albicans* and *M. globosa* secreted enzymes may represent an adaptation to mammal skin, whereas the secreted enzyme pattern of *U. maydis* likely represents an adaptation to life in association with plants.

The *M. globosa* genome contains clusters of genes for secreted enzymes. Of particular interest is one cluster that contains four tandemly repeated lipase genes. These lipase genes are more similar to each other than to any other lipase gene (Xu, Boekhout, et al., 2010), suggesting that they may be products of duplication since divergence from the last common ancestor with *Ustilago*.

Hypotheses are not as easily generated for the formation of other clusters because several clusters contain genes for a diverse set of secreted enzymes.

Atopic eczema patients have immunoglobulin E (IgE) that react with *Malassezia* proteins, perhaps a consequence of a breach in the skin barrier (reviewed in Saunders, Scheynius, et al., 2012). There are at least 12 *Malassezia* proteins that can mediate allergic reactions in atopic eczema patients (Zargari, Selander, et al., 2007). In the recently sequenced genome of *M. sympodialis* 13 allergens have been identified. Four of them, Mala s1, Mala s7, Mala s8 and Mala s12, are probably secreted. Mala s1 and Mala s12 are probably exported and associated with the cell wall (Gioti, Nystedt, et al., 2013). The continued study of the interaction of allergens with the host immune system may help to provide a molecular description of atopic eczema and possibly other skin diseases.

Dandruff is a common malady, affecting about 50 percent of the world's people (Schwartz, Cardin, et al., 2004). *Malassezia* spp. have a role in the disease because they are the dominant skin fungus (Paulino, Tseng, et al., 2008) and are removed by a variety of antifungal active ingredients in antidandruff shampoos. One popular active ingredient, zinc pyrithione, was recently characterized for its mechanism of action (Reeder, Kaplan, et al., 2011) and shown to act by the pyrithione binding copper from the environment and transporting the copper across cellular membranes, inactivating the iron-sulfur (Fe-S) cluster assembly in the mitochondria.

Essentially all people contain *Malassezia* on their scalp, and yet not all people have dandruff. One explanation is that some people may have a particular sensitivity to dandruff, as indicated by the increased sensitivity to irritation by oleic acid by people that are predisposed to dandruff (DeAngelis, Gemmer, et al., 2005). Such people may be particularly susceptible to one or more unfortunate interactions with *Malassezia*. For example, *Malassezia* lipases may hydrolyze scalp triglycerides, releasing oleic acid that acts as an irritant. Another possibility is that in susceptible people the host immune system recognizes *Malassezia* allergens, perhaps because a skin barrier defect is allowing enhanced exposure of these allergens to the immune system. It will be interesting to better characterize this proposed skin barrier defect in future studies.

Concluding Remarks

Comparative genomics revealed significant differences between BYPs belonging to two major basidiomycetous lineages and that showed different adaptations to human and animal hosts (Fig. 10.3). *C. neoformans* and *C. gattii* are pathogens with a predilection for the lungs and brains, and they differ in this respect from their phylogenetic neighbors that are not readily associated with human and animal sources. Similarly, *Malassezia* spp. are adapted to human and animal skin and in this regard they differ significantly from their plant-pathogen relatives. The genome size and content of the *Cryptococcus* and *Malassezia*

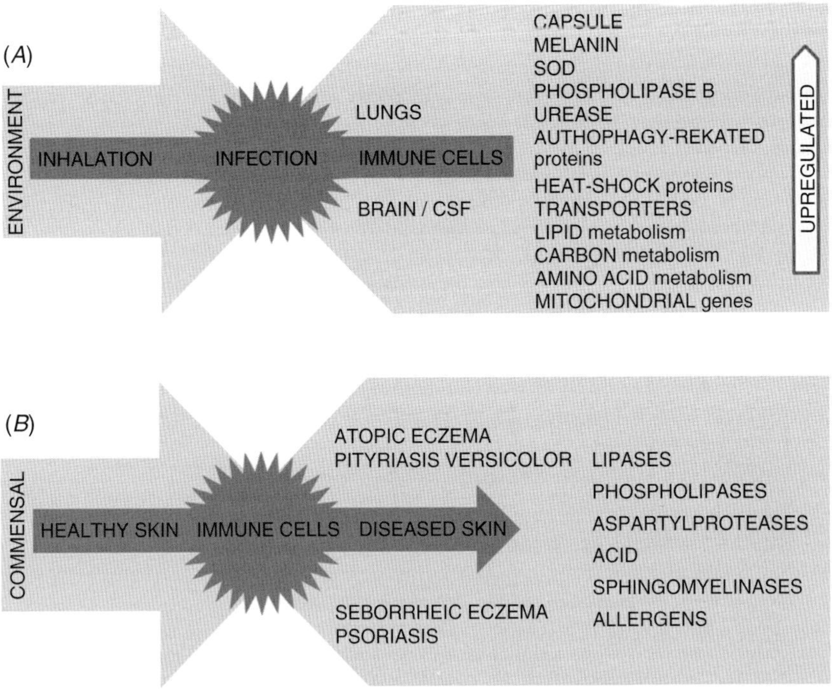

Figure 10.3 Scheme showing the major physiological and genomics differences between invasive (*A*) infections caused by *Cryptococcus* spp. and superficial skin infections (*B*) resulting from *Malassezia* spp. For further details the reader is referred to the main text. CSF, cerebrospinal fluid.

yeasts differ widely, thus indicating evolutionary adaptations of both types of pathogen to their respective hosts and body sites. Given the technological improvements in next-generation sequencing technology it is expected that soon more BYPs, which have not yet been studied by genomics approaches, will be added to the increasingly growing list of sequenced fungal species and the comparison between them will yield further surprises. Ongoing research projects aim to sequence the genomes of many isolates of *C. neoformans* and *C. gattii*, especially those of outbreak isolates, as well as several species of *Malassezia* spp. In running projects at Genoscope (France) and Joint Genome Institute (United States) several ascomycetous and basidiomycetous yeast species are being sequenced, and the results of these studies will provide a first robust phylogenomics coverage of the entire yeast domain, including many BYPs. Such projects also pave the road for an inclusive "all yeast species" comparative genomics program that is conceived by many researchers as a realistic option and that will serve both fundamental and applied sciences as well as the industry. Comparative genomics studies among BYPS and a comparison with their nonpathogen relatives will further contribute to the understanding of evolutionary adaptations of these important pathogenic fungi.

References

Amend AS, Barshis DJ et al. 2012. Coral-associated marine fungi form novel lineages and heterogeneous assemblages. *ISME J.* 6: 1291–1301.

Amend AS, Seifert KA, et al. 2010. Indoor fungal composition is geographically patterned and more diverse in temperate zones than in the tropics. *Proc Natl Acad Sci USA.* 107: 13748–13753.

Aminnejad M, Diaz M, et al. 2012. Identification of novel hybrids between *Cryptococcus neoformans* var. *grubii* VNI and *Cryptococcus gattii* VGII. *Mycopathologia.* 173: 337–346.

Ashbee HR. 2007. Update on the genus *Malassezia. Med Mycol.* 45: 287–303.

Bakkeren G, Warren R, et al. 2006. Mating factor linkage and genome evolution in basidiomycetous pathogens of cereals. *Fungal Genet Biol.* 43: 655–666.

Bandoni RJ. 1995. Dimorphic heterobasidiomycetes: Taxonomy and parasitism. *Studies Mycol.* 38: 13–27.

Baró T, Torres-Rodríguez JM. 1998. First identification of autochthonous *Cryptococcus neoformans* var. *gattii* isolated from goats with predominantly severe pulmonary disease in Spain. *J Clin Microbiol.* 36: 458–461.

Batra R, Boekhout T, et al. 2005. *Malassezia* Baillon, emerging clinical yeasts. *FEMS Yeast Res.* 5: 1101–1113.

Begerow D, Bauer R, et al. 2000. Phylogenetic placements of ustilaginomycetous anamorphs as deduced from nuclear LSU rDNA sequences. *Mycol Res.* 104: 53–60.

Begerow D, Stoll M, et al. 2006. A phylogenetic hypothesis of Ustilaginomycotina based on multiple gene analyses and morphological data. *Mycologia.* 98: 906–916.

Belangér RR, Dik AJ, et al. 1998. Powdery mildews: Recent advances towards integrated control. In Plant-Microbe Interactions and Biological Control (eds. GJ Boland & LD Kuykendall), 89–109. New York: Marcel Dekker Inc.

Bendtsen JD, Nielsen H, et al. 2004. Improved prediction of signal peptides: SignalP 3.0. *J Mol Biol.* 340: 783–795.

Bennett JE, Kwon-Chung KJ, et al. 1977. Epidemiologic differences among serotypes of *Cryptococcus neoformans. Am J Epidemiol.* 105: 582–586.

Bennett JE, Kwon-Chung KJ, et al. 1978. Biochemical differences between serotypes of *Cryptococcus neoformans. Sabouraudia.* 16: 167–174.

Bentley SD, Chater KF, et al. 2002. Complete genome sequence of the model actinomycete *Streptomyces coelicolor* AS(2). *Nature.* 417: 141–147.

Boekhout T, Theelen B, et al. 2003. New anamorphic acaropathogenic fungi belonging to the Ustilaginomycetes: *Meira geulakonigii* Boekhout, Gerson & Sztejnberg, Gen. et Spec. Nov., *Meira argovae* Boekhout, Gerson & Sztejnberg Spec. Nov. and *Acaromyces ingoldii* Boekhout, Gerson & Sztejnberg Gen. et Spec. Nov. *Int J Systemat Evol Microbiol.* 53: 1655–1664.

Boekhout T. 2005. Biodiversity: Gut feeling for yeasts. *Nature.* 434: 449–451.

Boekhout T, Theelen B, et al. 2001. Hybrid genotypes in the pathogenic yeast *Cryptococcus neoformans. Microbiol.* 147: 891–907.

Boekhout T, Guého E, et al., eds. 2010. Malassezia *and the Skin: Science and Clinical Practice.* Heidelberg, Germany: Springer-Verlag.

Boekhout T, Gueidan C, et al. 2009. Fungal taxonomy: New developments in medically important fungi. *Curr Fungal Infect Rep.* 3: 170–178.

Boekhout T, Fonseca A, et al. 2011. Discussion of teleomorphic and anamorphic basidiomycetous yeasts. In *The Yeasts: A Taxonomic Study*, 5th ed. (eds. CP Kurtzman, JW Fell, et al.), 1339–1372. Amsterdam: Elsevier.

Bölker M. 2001. *Ustilago maydis*—A valuable model system for the study of fungal dimorphism and virulence. *Microbiology.* 147: 1395–1401.

Botts MR, Giles SS, et al. 2009. Isolation and characterization of *Cryptococcus neoformans* spores reveal a critical role for capsule biosynthesis genes in spore biogenesis. *Eukaryot Cell.* 8: 595–605.

Bovers M, Hagen F, et al. 2006. Unique hybrids between the fungal pathogens *Cryptococcus neoformans* and *Cryptococcus gattii*. *FEMS Yeast Res.* 6: 599–607.
Bovers M, Hagen F, et al. 2008a. Six monophyletic lineages identified within *Cryptococcus neoformans* and *Cryptococcus gattii* by multi-locus sequence typing. *Fungal Genet Biol.* 45: 400–421.
Bovers M, Hagen F, et al. 2008b. AIDS patient death caused by novel *Cryptococcus neoformans* × *C. gattii* hybrid. *Emerg Infect Dis.* 14: 1105–1108.
Broad Institute. 2010. *Cryptococcus neoformans* var. *grubii* H99 Database. Accessed May 10, 2013, at http://www.broadinstitute.org/annotation/genome/cryptococcus_neoformans/MultiHome.html.
Brunke S & Hube B. 2006. *MfLIP1*, a gene encoding an extracellular lipase of the lipid-dependent fungus *Malassezia furfur*. *Microbiol.* 152: 547–554.
Bui T, Lin X, et al. 2008. Isolates of *Cryptococcus neoformans* from infected animals reveal genetic exchange in unisexual, alpha mating type populations. *Eukaryot Cell.* 7: 1771–1780.
Butler G. 2010. Fungal sex and pathogenesis. *Clin Microbiol Rev.* 23: 140–159.
Byrnes EJ 3rd, Li W, et al. 2010. Emergence and pathogenicity of highly virulent *Cryptococcus gattii* genotypes in the northwest United States. *PLoS Pathog.* 6: e1000850.
Byrnes EJ, Li W, et al. 2011. A diverse population of *Cryptococcus gattii* molecular type VGIII in Southern California HIV/AIDS patients. *PLOS Pathog.* 7: e1002205.
Cabañes FJ, Vega S, et al. 2011. *Malassezia* cuniculi sp. nov., a novel yeast species isolated from rabbit skin. *Med Mycol.* 49: 40–48.
Cafarchia C & Otranto D. 2004. Association between phospholipase production by *Malassezia pachydermatis* and skin lesions. *J Clin Microbiol.* 42: 4868–4869.
Carriconde F, Gilgado F, et al. 2011. Clonality and α-a recombination in the Australian *Cryptococcus gattii* VGII population—An emerging outbreak in Australia. *PLoS One.* 6: e16936.
Chagas-Neto TC, Chaves GM et al. 2008. Update on the genus *Trichosporon*. *Mycopathologia.* 166: 121–132.
Chang YC, Bien CM, et al. 2007. Sre1p, a regulator of oxygen sensing and sterol homeostasis, is required for virulence in *Cryptococcus neoformans*. *Mol Microbiol.* 64: 614–629.
Chau TT, Mai NH, et al. 2010. A prospective descriptive study of cryptococcal meningitis in HIV uninfected patients in Vietnam—High prevalence of *Cryptococcus neoformans* var *grubii* in the absence of underlying disease. *BMC Infect Dis.* 10: 199.
Chen J, Varma A, et al. 2008. *Cryptococcus neoformans* strains and infection in apparently immunocompetent patients, *China. Emerg Infect Dis.* 14: 755–762.
Chen SC, Slavin MA, et al. 2012. Clinical manifestations of *Cryptococcus gattii* infection: Determinants of neurological sequelae and death. *Clin Infect Dis.* 82: 316–325.
Cheng PY, Sham A et al. 2009. *Cryptococcus gattii* isolates from the British Columbia cryptococcosis outbreak induce less protective inflammation in a murine model of infection than *Cryptococcus neoformans*. *Infect Immun.* 77: 4284–4294.
Choi YH, Ngamskulrungroj P, et al. 2010. Prevalence of the VNIc genotype of *Cryptococcus neoformans* in non-HIV-associated cryptococcosis in the Republic of Korea. *FEMS Yeast Res.* 10: 769–778.
Chong HS, Dagg R, et al. 2010. In vitro susceptibility of the yeast pathogen *Cryptococcus* to fluconazole and other azoles varies with molecular genotype. *J Clin Microbiol.* 48: 4115–4120.
Chow ED, Liu OW, et al. 2007. Exploration of whole-genome responses of the human AIDS-associated yeast pathogen *Cryptococcus neoformans* var. *grubii*: Nitric oxide stress and body temperature. *Curr Genet.* 52: 137–148.
Chow EWL, Morrow CA, et al. 2012. Microevolution of *Cryptococcus neoformans* driven by massive tandem gene amplification. *Mol Biol Evol.* 29: 1987–2000.
Chun CD, Brown JC, et al. 2011. A major role for capsule-independent phagocytosis-inhibitory mechanisms in mammalian infection by *Cryptococcus neoformans*. *Cell Host Microbe.* 9: 243–251.
Chun CD, Liu OW, et al. 2007. A link between virulence and homeostatic responses to hypoxia during infection by the human fungal pathogen *Cryptococcus neoformans*. *PLoS Pathog.* 3: e22.
Coelho MA, Sampaio JP, et al. 2010. A deviation from the bipolar-tetrapolar mating paradigm in an early diverged basidiomycete. *PLoS Genet.* 6: e1001052.

Cole ST, Eiglmeier K, et al. 2001. Massive gene decay in the leprosy bacillus. *Nature.* 409: 1007–1011.

Colom MF, Hagen F, et al. 2012. *Ceratonia siliqua* (carob) trees as natural habitat and source of infection by *Cryptococcus gattii* in the Mediterranean environment. *Med Mycol.* 50, 67–73.

Datta K, Bartlett KH, et al. 2009. Spread of *Cryptococcus gattii* into Pacific Northwest region of the United States. *Emerg Infect Dis.* 15: 1185–1191.

DeAngelis YM, Gemmer CM, et al. 2005. Three etiologic facets of dandruff and seborrheic dermatitis: *Malassezia* fungi, sebaceous lipids, and individual sensitivity. *J Investig Dermatol Symp Proc.* 10: 295–297.

DeAngelis YM, Saunders CW, et al. 2007. Isolation and expression of a *Malassezia globosa* lipase gene, LIP1. *J Invest Dermatol.* 127: 2138–2146.

Dietrich FS, Voegeli S, et al. 2004. The *Ashbya gossypii* genome as a tool for mapping the ancient *Saccharomyces cerevisiae* genome. *Science.* 304: 304–307.

D'Souza CA, Kronstad JW, et al. 2011. Genome variation in *Cryptococcus gattii*, an emerging pathogen of immunocompetent hosts. *mBio* 2: e00342-00310.

Evans EE. 1950. The antigenic composition of *Cryptococcus neoformans*. I. A serologic classification by means of the capsular and agglutination reactions. *J Immunol.* 64: 423–430.

Fan W, Kraus PR, et al. 2005. *Cryptococcus neoformans* gene expression during murine macrophage infection. *Eukaryot Cell.* 4: 1420–1433.

Fell JW, Scorzetti G, et al. 2006. Biodiversity of micro-eukaryotes in Antarctic Dry Valley soils with <5%soil moisture. *Soil Biol Biochem.* 38: 3107–3115.

Findley K, Sun S, et al. 2012. Discovery of a modified tetrapolar sexual cycle in Cryptococcus amylolentus and the evolution of MAT in the *Cryptococcus* species complex. *PLoS Genet.* 8: e1002528.

Fitzpatrick DA, Logue ME, et al. 2006. A fungal phylogeny based on 42 complete genomes derived from supertree and combined gene analysis. *BMC Evol Biol.* 6: 99.

Franzot SP, Salkin IF, et al. 1999. *Cryptococcus neoformans* var. *grubii*: Separate varietal status for *Cryptococcus neoformans* serotype A isolates. *J Clin Microbiol.* 37: 838–840.

Gaitanis G, Magiatis P, et al. 2012. The *Malassezia* Genus in Skin and Systemic Diseases.. *Clin Microbiol Rev.* 25(1): 106–41. doi: 10.1128/CMR.00021-11.

Gaitanis G, Mayser P, et al. 2010. *Malassezia* yeasts in seborrehic and atopic dermatitis. In *Malassezia and the Skin* (eds. T Boekhout, E Guého-Kellermann, et al.), 201–228. Heidelberg, Germany: Springer Verlag.

Gao Z, Li B, et al. 2008. Molecular detection of fungal communities in the Hawaiian marine sponges *Suberites zeteki* and *Mycale armata. Appl Environ Microbiol.* 74: 6091–6101.

Giles SS, Dagenais TR, et al. 2009. Elucidating the pathogenesis of spores from the human fungal pathogen *Cryptococcus neoformans. Infect Immun.* 77: 3491–3500.

Gillece JD, Schupp JM, et al. 2011. Whole genome sequence analysis of *Cryptococcus gattii* from the Pacific Northwest reveals unexpected diversity. *PLoS ONE.* 6: e28550.

Gioti A, Nystedt B, et al. 2013. Genomic insights into the atopic eczema-associated skin commensal yeast. *Malassezia sympodialis. mBio.* 4(1): e00572–e00512. doi: 10.1128/mBio.00572–12.

Giraud T, Yockteng R, et al. 2008. Mating system of the anther smut fungus *Microbotryum violaceum*: Selfing under heterothallism. *Eukaryot Cell.* 7: 765–775.

Gottschalk M & Blanz PA 1985. Untersuchungen an 5S ribosomalen Ribonukleinsauren als Beitrag zur Klärung von Systematik und Phylogenie der Basidiomyceten. *Zeitschrift für Mykologie.* 51: 205–244.

Guého E, Midgley G, et al. 1996. The genus *Malassezia* with description of four new species. *Antonie van Leeuwenhoek.* 69: 337–355.

Guého-Kellermann E, Boekhout T, et al. 2010. Biodiversity, phylogeny, and ultrastructure. In Malassezia and the Skin (eds. T Boekhout, E Guého-Kellermann, et al.), 17–61. Heidlberg, Germany: Springer Verlag.

Guého-Kellermann E, Batra R, et al. 2011. *Malassezia* Baillon. In *The Yeasts: A Taxonomic Study*, 5th ed. (eds. CP Kurtzman, JW Fell, et al.), 1087–1832. Amsterdam: Elsevier.

Gurvitz A, Mursula AM, et al. 1998. Peroxisomal δ3-cis-δ2-trans-enoyl-CoA isomerase encoded by *ECI1* is required for growth of the yeast *Saccharomyces cerevisiae* on unsaturated fatty acids. *J Biol Chem.* 273: 31366–31374.

Hagen F, Colom MF, et al. 2012. Emerging autochthonous and dormant *Cryptococcus gattii* infections in Europe. *Emerg Infect Dis.* 18: 1618–1624.

Hagen F, Illnait-Zaragozi MT, et al. 2010. In vitro antifungal susceptibilities and amplified fragment length polymorphism genotyping of a worldwide collection of 350 clinical, veterinary, and environmental *Cryptococcus gattii* isolates. *Antimicrob Agents Chemother.* 54: 5139–5145.

Hawksworth DL. 2011. A new dawn for the naming of fungi: Impacts of decisions made in Melbourne in July 2011 on the future publication and regulation of fungal names. *MycoKeys.* 1: 7–20.

Haynes BC, Skowyra ML, et al. 2011. Toward an integrated model of capsule regulation in *Cryptococcus neoformans*. *PLoS Pathog.* 7: e1002411.

Heitman J, Kronstad JW, et al. 2007. *Sex in fungi: Molecular determination and evolutionary implications*. Washington, D.C.: ASM Press.

Hibbett DS, Binder M, et al. 2007. A higher-level phylogenetic classification of the fungi. *Mycol Res.* 111: 509–547.

Hiremath SS, Chowdhary A, et al. 2008. Long-distance dispersal and recombination in environmental populations of *Cryptococcus neoformans* var. *grubii* from India. *Microbiology.* 154: 1513–1524.

Hu G, Cheng P-Y, et al. 2008. Metabolic adaptation in *Cryptococcus neoformans* during early murine pulmonary infection. *Mol Microbiol.* 69: 1456–1475.

Hu G, Steen BR, et al. 2007. Transcriptional regulation by protein kinase A in *Cryptococcus neoformans*. *PLoS Pathog.* 3: e42.

Human Microbiome Project Consortium. 2012. Structure, function and diversity of the healthy human microbiome. *Nature.* 486(7402): 207–214.

Iqbal N, DeBess EE, et al. 2010. Correlation of genotype and in vitro susceptibilities of *Cryptococcus gattii* strains from the Pacific Northwest of the United States. *J Clin Microbiol.* 48: 539–544.

James TY, Kauff F, et al. 2006. Reconstructing the early evolution of fungi using a six-gene phylogeny. *Nature.* 443: 818–822.

Jung WH, Sham A, et al. 2008. Iron source preference and regulation of iron uptake in *Cryptococcus neoformans*. *PLoS Pathog.* 4: e45.

Jung WH, Sham A, et al. 2006. Iron regulation of the major virulence factors in the AIDS-associated pathogen *Cryptococcus neoformans*. *PLoS Biol.* 4: e410.

Kidd SE, Hagen F, et al. 2004. A rare genotype of *Cryptococcus gattii* caused the cryptococcosis outbreak on Vancouver Island (British Columbia, Canada). *Proc Natl Acad Sci USA.* 101: 17258–17263.

Kraus PR, Boily MJ, et al. 2004. Identification of *Cryptococcus* neoformans temperature-regulated genes with a genomic-DNA microarray. *Eukaryot Cell.* 3: 1249–1260.

Kronstad J, Saikia S, et al. 2012. Adaptation of *Cryptococcus neoformans* to mammalian hosts: Integrated regulation of metabolism and virulence. *Eukaryot Cell.* 11: 109–118.

Kronstad JW, Attarian R, et al. 2011. Expanding fungal pathogenesis: *Cryptococcus* breaks out of the opportunistic box. *Nat Rev Microbiol.* 9: 193–203.

Kuramae EE, Robert V, et al. 2006. Phylogenomics reveal a robust fungal tree of life. *FEMS Yeast Res.* 6: 1213–1220.

Kurtzman CP, Fell JW, et al., eds. 2011. *The Yeasts: A Taxonomic Study*, 5th ed. Amsterdam: Elsevier.

Kwon-Chung KJ & Bennett JE. 1984. Epidemiologic differences between the two varieties of *Cryptococcus neoformans*. *Am J Epid.* 120: 123–130.

Kwon-Chung KJ. 1975. A new genus, *Filobasidiella*, the perfect state of *Cryptococcus neoformans*. *Mycologia.* 67: 1197–1200.

Kwon-Chung KJ. 1976a. Morphogenesis of *Filobasidiella neoformans*, the sexual state of *Cryptococcus neoformans*. *Mycologia.* 68: 821–833.

Kwon-Chung KJ. 1976b. A new species of *Filobasidiella*, the sexual state of *Cryptococcus neoformans* B and C serotypes. *Mycologia.* 68: 943–946.

Kwon-Chung KJ, Bennett JE, et al. 1982. Taxonomic studies on *Filobasidiella* species and their anamorphs. *Antonie van Leeuwenhoek.* 48: 25–38.

Kwon-Chung KJ, Boekhout T, et al. 2002. Proposal to conserve the name *Cryptococcus gattii* against *C. hondurianus* and *C. bacillisporus* (Basidiomycota, Hymenomycetes, Tremellomycetidae). *Taxon.* 51: 804–806.

Lee N, Bakkeren G, et al. 1999. The mating-type and pathogenicity locus of the fungus *Ustilago hordei* spans a 500-kb region. *Proc Natl Acad Sci USA.* 96: 15026–15031.

Lee SC, Ni M, et al. 2010. The evolution of sex: A perspective from the fungal kingdom. *Microbiol Mol Biol Rev.* 74: 298–340.

Lin X, Hull CM, et al. 2005. Sexual reproduction between partners of the same mating type in *Cryptococcus neoformans. Nature.* 434: 1017–1021.

Lin X, Litvintseva AP, et al. 2007. αADα hybrids of *Cryptococcus neoformans*: Evidence of same-sex mating in nature and hybrid fitness. *PLoS Genet.* 3: 1975–1990.

Lin X, Patel S, et al. 2009. Diploids in the *Cryptococcus neoformans* serotype A population homozygous for the alpha mating type originate via unisexual mating. *PLoS Pathog.* 5: e1000283.

Liu OW, Chum CD, et al. 2008. Systematic genetic analysis of virulence in the human fungal pathogen *Cryptococcus neoformans. Cell.* 135: 174–188.

Loftus BJ, Fung E, et al. 2005. The genome of the basidiomycetous yeast and human pathogen *Cryptococcus neoformans. Science.* 307: 1321–1324.

Ma H & May RC. 2010. Mitochondria and the regulation of hypervirulence in the fatal fungal outbreak on Vancouver Island. *Virulence.* 1: 197–201.

Ma H, Hagen F, et al. 2009. The fatal fungal outbreak on Vancouver Island is characterized by enhanced intracellular parasitism driven by mitochondrial regulation. *Proc Natl Acad Sci USA.* 106: 12980–12985.

Ma H & May RC. 2009. Virulence in Cryptococcus species. *Adv Appl Microbiol.* 67: 131–190.

Marcet-Houben M & Gabaldón T. 2009. The tree versus the forest: The fungal tree of life and the topological diversity within the yeast phylome. *PLoS One.* 4: e4357.

Metin B, Findley K, et al. 2010. The mating type locus (MAT) and sexual reproduction of *Cryptococcus heveanensis*: Insights into the evolution of sex and sex-determining chromosomal regions in fungi. *PLoS Genet.* 6: e10000961.

Meyer W, Aanensen DM, et al. 2009. Consensus multi-locus sequence typing scheme for *Cryptococcus neoformans* and *Cryptococcus gattii. Med Mycol.* 47: 561–570.

Meyer W, Castañeda A, et al. 2003. Molecular typing of IberoAmerican *Cryptococcus neoformans* isolates. *Emerg Infect Dis.* 9: 189–195.

Missall TA, Pusateri ME, et al. 2006. Posttranslational, translational, and transcriptional responses to nitric oxide stress in *Cryptococcus neoformans*: Implications for virulence. *Eukaryotic Cell.* 5: 518–529.

Mitchell TG & Perfect JR. 1995. Cryptococcosis in the era of AIDS—100 years after the discovery of *Cryptococcus neoformans. Clin Microbiol Rev.* 8: 515–548.

Morrow CA, Lee IR, et al. 2012. A unique chromosomal rearrangement in the *Cryptococcus neoformans* var. *grubii* type strain enhances key phenotypes associated with virulence. *mBio.* 3: e00310.

Ngamskulrungroj P, Chang Y, et al. 2012. The primary target organ of *Cryptococcus gattii* is different from that of *Cryptococcus neoformans* in a murine model. *mBio.* 3: e00103-12.

Ni M, Feretzaki M, et al. 2011. Sex in fungi. *Annu Rev Genet.* 45: 405–430.

Padamsee M, Kumar TKA, et al. 2012. The genome of the xerotolerant mold *Wallemia sebi* reveals adaptations to osmotic stress and suggests cryptic sexual reproduction. *Fungal Genet Biol.* 49: 217–226.

Pan W, Khayhan K, et al. 2012. Resistance of Asian *Cryptococcus neoformans* serotype A is confined to few microsatellite genotypes. *PLoS One.* 7: e32868.

Park BJ, Wannemuehler KA, et al. 2009. Estimation of the current global burden of cryptococcal meningitis among persons living with HIV/AIDS. *AIDS.* 23: 525–530.

Patiño-Uzcátegui A, Amado Y, et al. 2011. Virulence gene expression in *Malassezia* spp from individuals with seborrheic dermatitis. *J Invest Dermatol.* 131: 2134–2136.

Paulino LC, Tseng C-H, et al. 2008. Analysis of *Malassezia* microbiota in healthy superficial human skin and in psoriatic lesions by multiplex real-time PCR. *FEMS Yeast Res.* 8: 460–471.

Perfect JF, Dismukes WE, et al. 2010. Clinical practice guidelines for the management of cryptococcal disease: 2010 update by the Infectious Diseases Society of America. *Clin Infect Dis.* 50: 291–322.

Pini G & Faggi E. 2011. Extracellular phospholipase activity of *Malassezia* strains isolated from individuals with and without dermatological disease. *Revista Iberoamericana Micologia.* 28: 179–182.

Pitkäranta M, Meklin T, et al. 2008. Analysis of fungal flora in indoor dust by ribosomal DNA sequence analysis, quantitative PCR, and culture. *Appl Environ Microbiol.* 74: 233–244.

Plotkin LI, Squiquera L et al. 1996. Characterization of the lipase activity of *Malassezia furfur*. *J Med Vet Mycol.* 34: 43–48.

Prillinger H, Oberwinkler F, et al. 1993. Analysis of cell wall carbohydrates (neutral sugars) from ascomycetous and basidiomycetous yeasts with and without derivatization. *J Gen Appl Microbiol.* 39: 1–34.

Ran Y, Yoshiike T, et al. 1993. Lipase of *Malassezia furfur*: Some properties and their relationship to growth. *J Med Vet Mycol.* 31: 77–85.

Raso TF, Werther K, et al. 2004. Cryptococcosis outbreak in psittacine birds in Brazil. *Med Mycol.* 42: 355–362.

Reeder NL, Kaplan J, et al. 2011. Zinc pyrithione inhibits yeast growth through copper influx and inactivation of iron-sulfur proteins. *Antimicrob Agents Chemother.* 55: 5753–5760.

Renker C, Alphei J, et al. 2003. Soil nematodes associated with the mammal pathogenic fungal genus *Malassezia* (Basidiomycota, Ustilaginomycetes) in Central European forests. *Biol Fertil Soils.* 37: 70–72.

Robbertse B, Reeves JB, et al. 2006. A phylogenomic analysis of the Ascomycota. *Fungal Genet Biol.* 43: 715–725.

Rowell JB & DeVay JE 1954. Genetics of *Ustilago zeae* in relation to basic problems of its pathogenicity. *Phytopathol.* 44: 356–362.

Saul N, Krockenberger M, et al. 2008. Evidence of recombination in mixed-mating-type and alpha-only populations of *Cryptococcus gattii* sourced from single eucalyptus tree hollows. *Eukaryot Cell.* 7: 727–734.

Saunders CW, Scheynius A, et al. 2012. *Malassezia* fungi are specialized to live on skin and associated with dandruff, eczema, and other skin diseases. *PLoS Pathog.* 8: e1002701.

Schneiker S, Perlova O, et al. 2007. Complete genome sequence of the myxobacterium *Sorangium cellulosum*. *Nat Biotechnol.* 25: 1281–1289.

Schwartz JR, Cardin CW, et al. 2004. Dandruff and seborrheic dermatitis. In *Textbook of Cosmetic Dermatology* (eds. R Baran & HI Maibach), pp. 259–272. London: Martin Dunitz, Ltd.

Shibata N, Okanuma N, et al. 2006. Isolation, characterization, and molecular cloning of a lipolytic enzyme secreted from *Malassezia pachydermatis*. *FEMS Microbiol Lett.* 256: 137–144.

Sionov E, Lee H, et al. 2010. *Cryptococcus neoformans* overcomes stress od azole drugs by formation of disomy in specific multiple chromosomes. *PLoS Pathog.* 6: e1000848.

Springer DJ & Chaturvedi V. 2010. Projecting global occurrence of *Cryptococcus gattii*. *Emerg Infect Dis.* 16: 14–20.

Steen BR, Lian T, et al. 2002. Temperature-regulated transcription in the pathogenic fungus *Cryptococcus neoformans*. *Genome Res.* 12: 386–1400.

Steen BR, Zuyderduyn S, et al. 2003. *Cryptococcus neoformans* gene expression during experimental cryptococcal meningitis. *Eukaryot Cell.* 2: 1336–1349.

Stukey JE, McDonough VM, et al. 1989. Isolation and characterization of *OLE1*, a gene affecting fatty acid desaturation from *Saccharomyces cerevisiae*. *J Biol Chem.* 164: 16537–16544.

Sukroongreung S, Kitiniyom K, et al. 1998. Pathogenicity of basidiospores of *Filobasidiella neoformans* var. *neoformans*. *Med Mycol.* 36: 419–424.

Sun S & Xu J. 2007. Genetic analyses of a hybrid cross between serotypes A and D strains of the human pathogenic fungus *Cryptococcus neoformans*. *Genetics*. 177: 1475–1486.

Sun S. & Xu J. 2009. Chromosomal rearrangements between serotype A and D strains in *Cryptococcus neoformans*. *PLoS ONE*. 4: e5524.

Sztejnberg A, Paz Z, et al. 2004. *Meira geulakonigii*, an unique fungus able to reduce both phytophagous mites and fungal plant pathogens. *Crop Sci.* 23: 1125–1129.

Taj-Aldeen SJ, Al-Ansari N, et al. 2009. Molecular identification and susceptibility of *Trichosporon* species isolated from clinical specimens in Qatar: Isolation of *Trichosporon dohaense* Taj-Aldeen, *Meis & Boekhout sp. nov. J Clin Microbiol.* 47: 1791–1799.

Tauch A, Kaiser O, et al. 2005. Complete genome sequence and analysis of the multiresistant nosocomial pathogen *Corynebacterium jeikeium* K411, a lipid-requiring bacterium of the human skin flora. *J Bacteriol.* 187: 4671–4682.

Torres-Rodríguez JM, Baró T, et al. 1999. Molecular characterization of *Cryptococcus neoformans* var. *gattii* causing epidemic outbreaks of cryptococcosis in goats. *Revista Iberoamericana de Micología*. 16: 164–165.

Tragiannidis A, Groll A, et al. 2010. *Malassezia* fungemia, antifungal susceptibility testing and epidemiology of nosocomila infections. In *Malassezia and the Skin* (eds. T Boekhout, E Guého-Kellermann, et al.), 229–251. Berlin: Springer Verlag.

Urquehart EJ, Menzies JG, et al. 1994. Growth and biological control activity in *Tilletiopsis* species against powdery mildew (*Sphaerotheca fuliginea*) on greenhouse cucumber. *Phytopathology.* 84: 341–351.

Vanbreuseghem R & Takashio M. 1970. An atypical strain of *Cryptococcus neoformans* (San Felice) Vuillemin 1894. *II. Cryptococcus neoformans* var. *gattii* var. nov. *Annales Societe Belges Medicinales Tropical.* 50: 695–702.

Van Driel KGA, van Peer AF, et al. 2008. The septal pore cap protein SPC18 isolated from the basidiomycetous fungus *Rhizoctonia* solani also resides in pore-plugs. *Eukaryot Cell.* 7: 1865–1873.

Van Peer A, Wang F, et al. 2009. The septal pore cap is an organelle that functions in vegetative growth and mushroom formation of a basidiomycete. *Environ Microbiol.* 12: 833–844.

Velagapudi R, Hsueh YP, et al. 2009. Spores as infectious propagules of *Cryptococcus neoformans*. *Infect Immun.* 77: 4345–4355.

Warkentien T & Crum-Cianflone NF. 2010. An update on Cryptococcus among HIV-infected patients. *Int J STD AIDS.* 21: 679–684.

Wilson DE, Bennett JE, et al. 1968. Seologic grouping of *Cryptococcus neoformans*. *Proc Soc Exper Biol Med.* 127: 820–823.

Xu J, Boekhout T, et al. 2010. Genomics and pathophysiology: Dandruff as a paradigm. In *Malassezia and the Skin* (eds. T Boekhout, E Guého-Kellermann, et al.), 253–269. Berlin: Springer.

Xu J, Saunders CW, et al. 2007. Dandruff-associated *Malassezia* genomes reveal convergent and divergent virulence traits shared with plant and human fungal pathogens. *Proc Natl Acad Sci USA*. 104: 18730–18735.

Xue C, Liu T, et al. 2010. Role of an expanded inositol transporter repertoire in *Cryptococcus neoformans* sexual reproduction and virulence. *mBio*. 1: e00084.

Xue C, Tada Y, et al. 2007. The human fungal pathogen *Cryptococcus* can complete its sexual cycle during a pathogenic association with plants. *Cell Host Microbe.* 1: 263–273.

Zargari A, Selander C, et al. 2007. Mala s12 is a major allergen in patients with atopic eczema and has sequence similarities to the GMC oxidoreductase family. *Allergy.* 62: 695–703.

Zhang N, Suh SO, et al. 2003. Microorganisms in the gut of beetles: Evidence from molecular cloning. *J Invertebrate Path.* 84: 226–233.

Zimmer BL, Hempel HO, et al. 1984. Pathogenicity of the basidiospores of *Filobasidiella neoformans*. *Mycopathologia*. 85: 149–153.

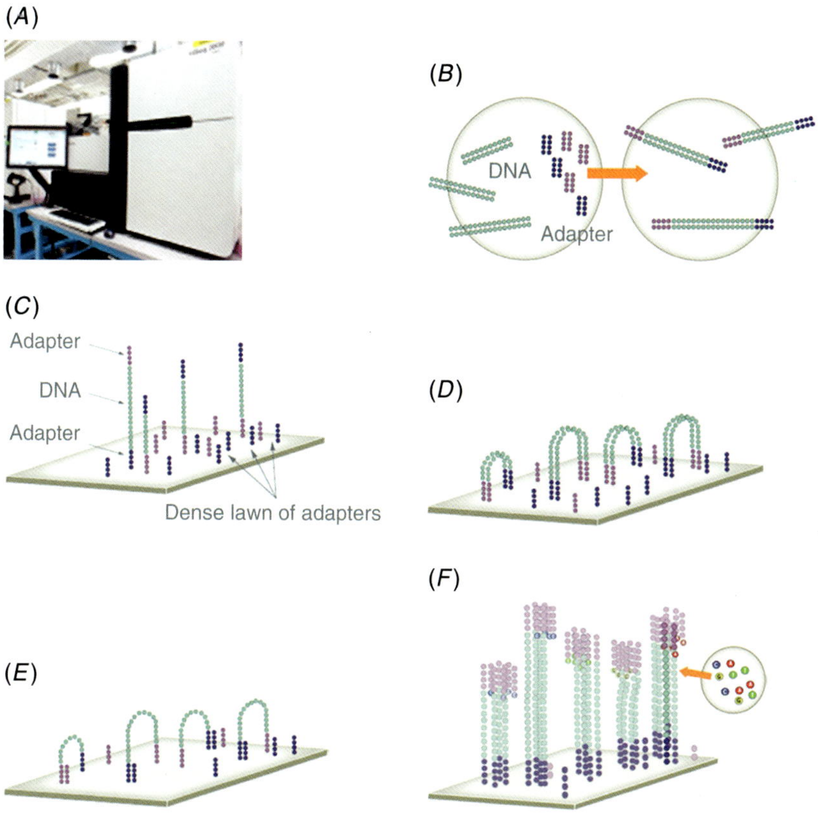

Figure 1.3 Examples of the next-generation sequencing includes (*A–F*) sequencing by synthesis using Illumina; (*G–J*) single-molecule sequencing using Pacific Biosciences; and (*K–N*) Roche's 454 pyrosequencing. (*A*) Illumina Hi-Seq process includes (*B*) adapter ligation to both ends of the random DNA fragments, (*C*) binding single-stranded DNA fragments to the inside surface of flow cell channels, (*D–E*) bridge formation and amplification, and (*F*) sequencing by detecting signals from four-labeled reversible terminators incorporated by DNA polymerase enzyme.

The Ecological Genomics of Fungi, First Edition. Edited by Francis Martin.

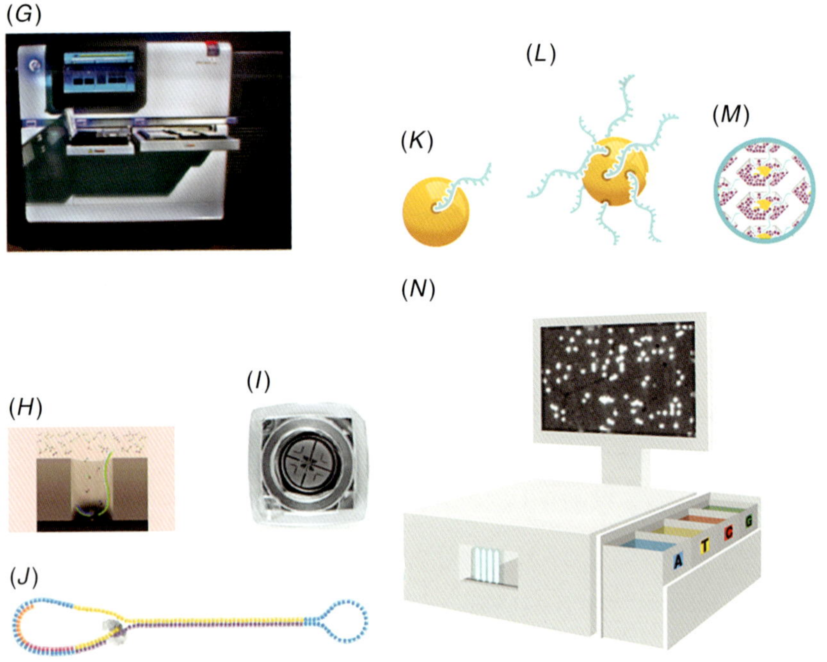

Figure 1.3 (*Continued*) (*G*) Pacific Biosciences uses (*H*) SMRT cells, each containing (*I*) a single DNA polymerase attached to the bottom, filled with diffused nucleotides with fluorescent markers on the terminal phosphate, which are illuminated during the reaction time; (*J*) double-stranded DNA linear structures (200–10,000 base pairs long) attached to the SMRT adapters produce a topologically closed circle and enable consensus sequencing of the same template. (*K*) Roche's 454 sequencing uses single-stranded DNA attached to beads, (*L*) amplified using emulsion polymerase chain reaction, (*M*) loaded into hexagonal wells of a fiber-optic slide, and (*N*) pyrosequenced by capturing light flashes from incorporation of each base (A, T, C, or G).

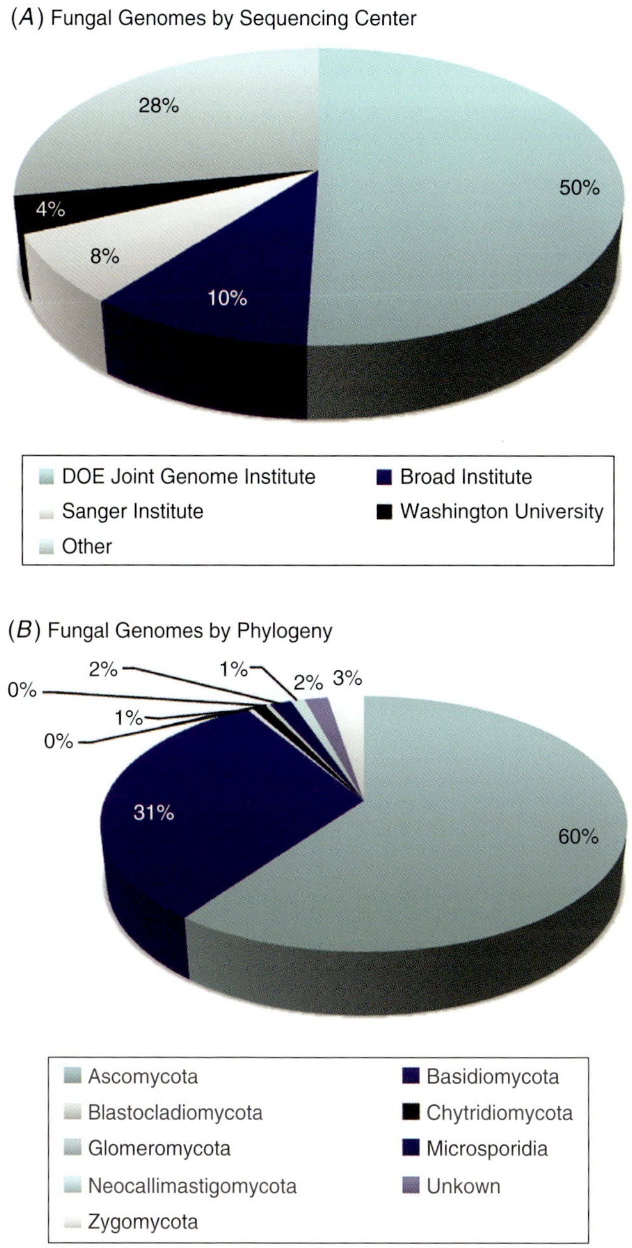

Figure 1.4 More than 1000 ongoing fungal genomic projects have been registered in GOLD database. They are grouped by (*A*) sequencing center and (*B*) represented phyla.

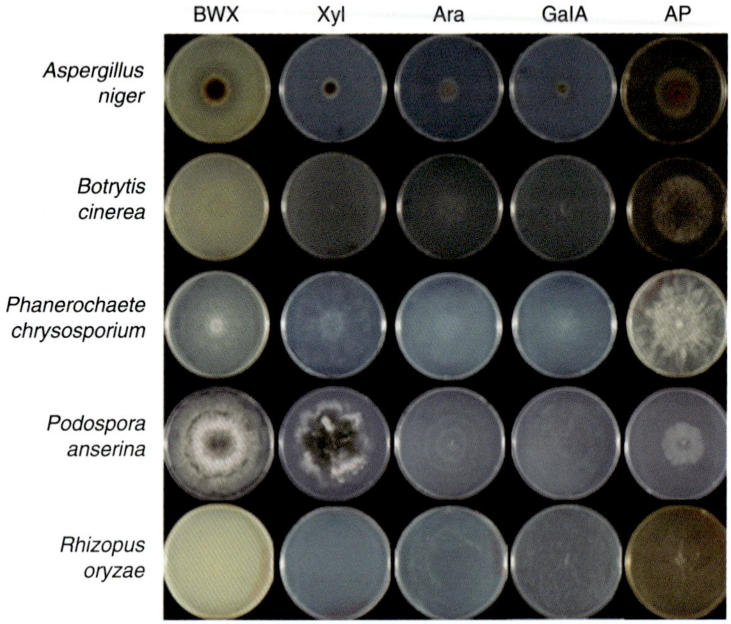

Figure 4.1 Growth profile of five fungi from general, such as *Aspergillus niger*, to more specialized life style such as *Podospora anserine* on plant cell wall polysaccharides. Ara, arabinose; AP, Apple pectin; BWX, beech wood xylan; GalA: galacturonic acid; Xyl, xylose. Corresponding CAZymes are displayed in Table 4.1.

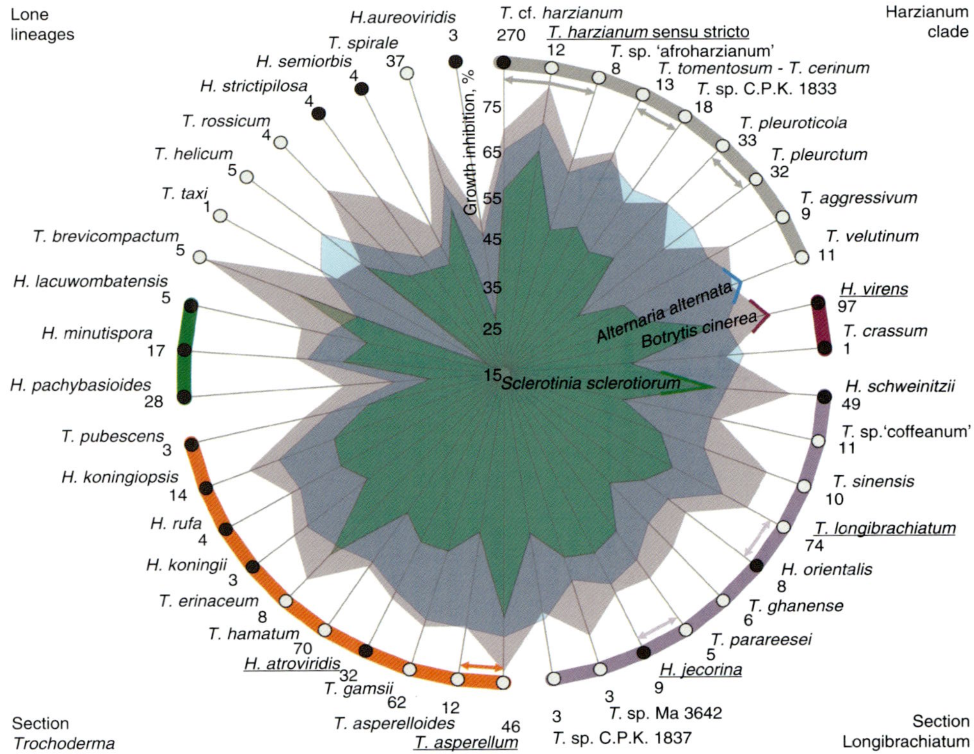

Figure 5.1 Antagonistic potential of the most common *Trichoderma* species estimated in dual confrontations after 10 days of incubation on potato dextrose agar at 77° F (25° C) in 12 hour-long illumination cycle. The scale represents the reduction of diameter of the prey colony (corrected for the growth rate in competition with itself) attributed to the mycoparasitic activity of *Trichoderma*. The pink, blue, and green areas correspond to confrontations with *Botrytis cinerea*, *Alternaria alternata*, and *Sclerotinia sclerotiorum*, respectively. All strains have been molecularly identified; species are grouped in phylogenetic clades as established on www.isth.info/biodiversity. Filled and open circles mark holomorphic and putatively clonal species, respectively. Arrows indicate sibling phylogenetic species. Digits below taxon names show number of strains analyzed per species, average and corresponding standard deviation values, respectively. Underlined species indicate completed genome sequencing initiatives (Druzhinina & Espino de Rummer, unpublished).

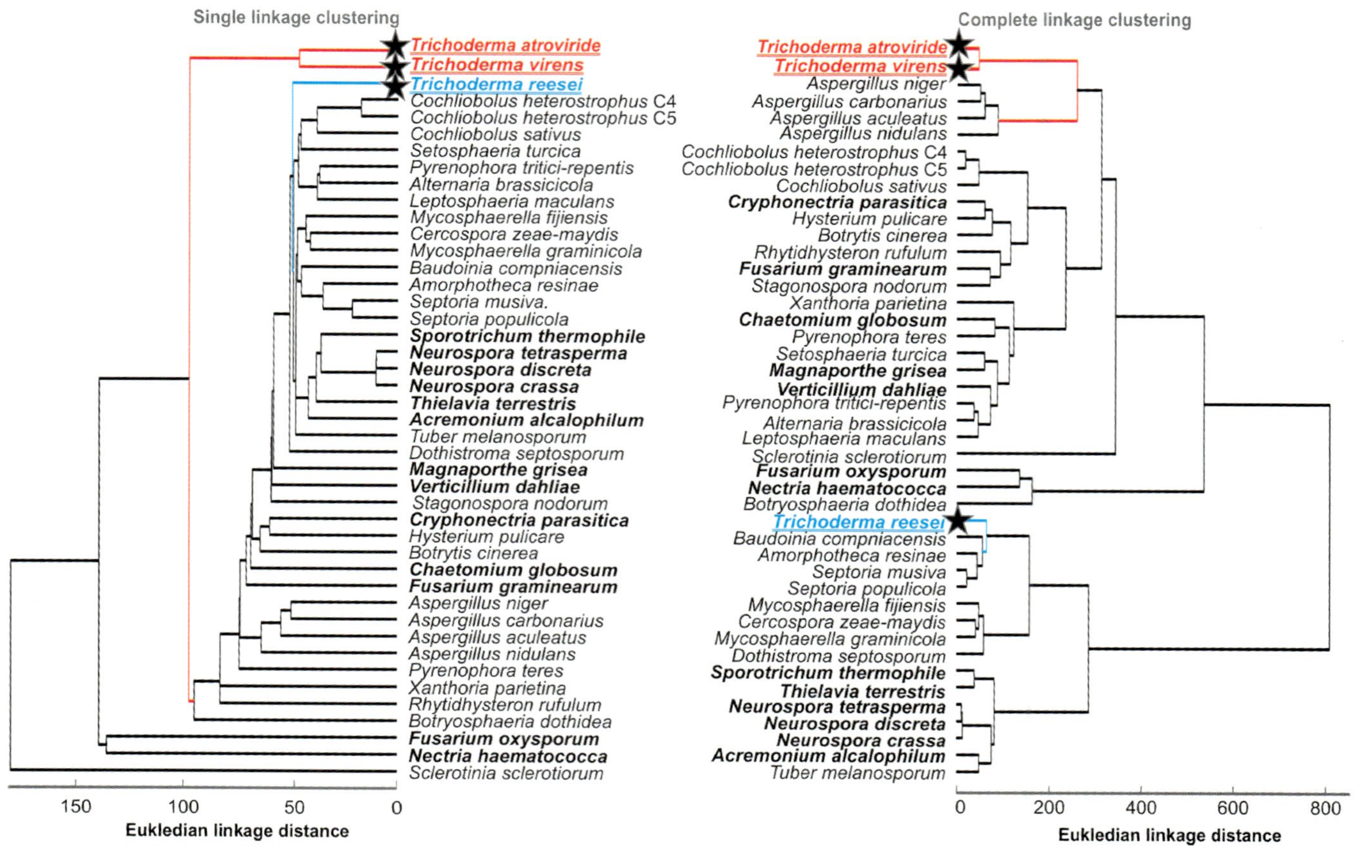

Figure 5.2 Cladograms based on the gene number per cluster matrix resulted from the MCL analysis of orthologous sequences between the three *Trichoderma* and 41 other Peziziomycotina fungi. Members of the class Sordariomycetes are in bold. Single linkage method distance between two clusters is determined by the distance of the two closest objects ("nearest neighbors") in the different clusters. Complete linkage method the distances between clusters are determined by the greatest distance between any two objects in the different clusters (i.e., by the "furthest neighbors").

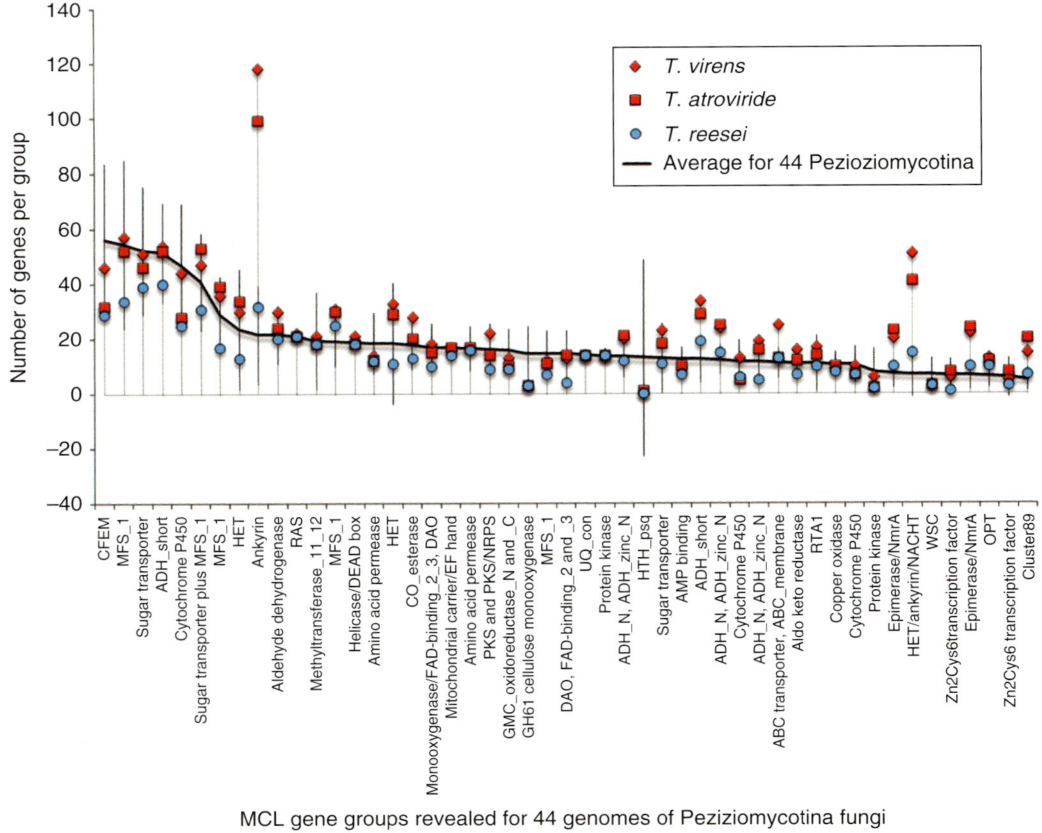

Figure 5.3 Gene numbers per cluster of orthologous genes revealed by MCL analysis for 44 Peziziomycotina fungi. Black solid horizontal bars indicate the standard deviation value for the average for Peziziomycotina, $N=44$.

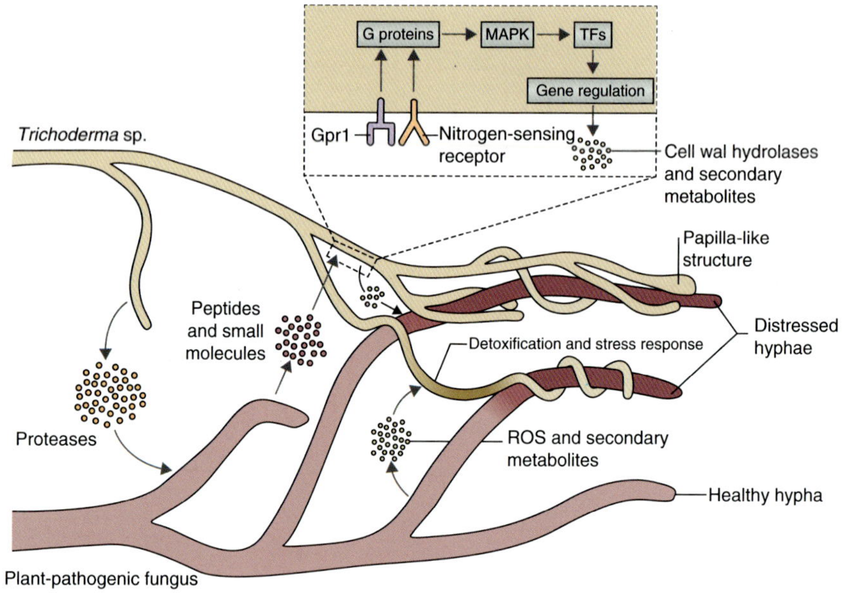

Figure 5.4 Mycoparasitism of *Trichoderma* spp. within the soil community. *Trichoderma* spp. recognize a plant-pathogenic fungus (a prey) via small molecules that are released by the pathogen; some of these molecules may be peptides that are released by the action of proteases secreted by the *Trichoderma* sp. before contact. These molecules may bind to G protein-coupled receptors (such as Gpr1) or nitrogen-sensing receptors on the surface of the *Trichoderma* sp. hyphae, thereby eliciting a signaling cascade, comprising G proteins and mitogen-activated protein kinases (MAPKs), which may ultimately modulate the activities of as-yet-unknown transcription factors (TFs). These factors then enhance the constitutive expression of genes that encode enzymes for the biosynthesis of secondary metabolites and for cell wall lysis. Lectins from the pathogenic fungus and proteins harboring cellulose-binding modules from hyphae of *Trichoderma* spp. may collaborate in the attachment of the predator to the prey. At the same time, the prey responds by forming secondary metabolites and reactive oxygen species (ROS) that elicit a stress response and detoxification in *Trichoderma* spp. *Source*: (Figure from Druzhinina IS, Seidl-Seiboth V, et al. 2011. *Trichoderma*: The genomics of opportunistic success. *Nat Rev Microbiol.* 16:749–759.)

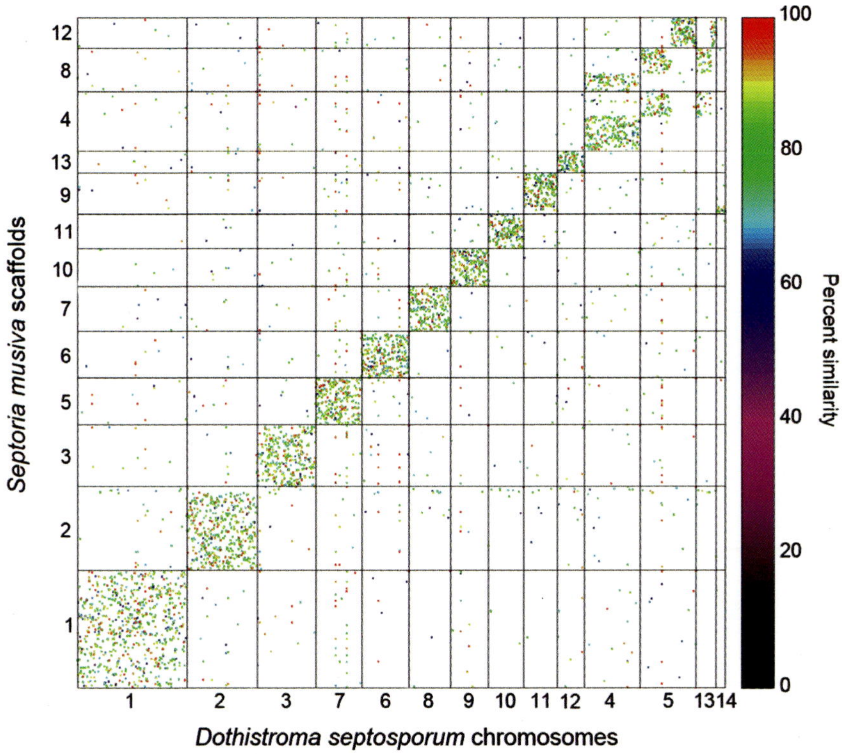

Figure 6.2 Dot plot of the 14 chromosomes of the *Dothistroma septosporum* genome assembly versus the 13 largest scaffolds of *Septoria musiva*. The content of each chromosome or scaffold is plotted from left to right or bottom to top, respectively, for the indicated chromosomes and scaffolds. Macrosynteny would be indicated as diagonal lines within each box defined by a particular chromosome-scaffold comparison. No synteny would be indicated if the contents of a particular chromosome are present over all of the scaffolds of the other genome. Instead, there is a clear pattern of mesosynteny where the contents of each chromosome of *D. septorum* are strongly conserved in a corresponding scaffold of *S. musiva*. However, there is no conservation of gene order or orientation so that the dots fill the box defined by each chromosome-scaffold combination. The results show an almost 1:1 correspondence between gene content on the chromosomes of *D. septosporum* and the largest scaffolds of *S. musiva*.

Figure 9.1 Lichens used in ecological genomics studies. Upper row: left, *Peltigera membranacea*; right, *Xanthoria parietina*. Middle row: left, *Cladonia grayi*; right, *Lobaria pulmonaria*. Lower row: left, *Solorina crocea* with lichenicolous infection by *Rhagadostoma lichenicola*; right, *Cetraria aculeata*. (Ch. Printzen.)

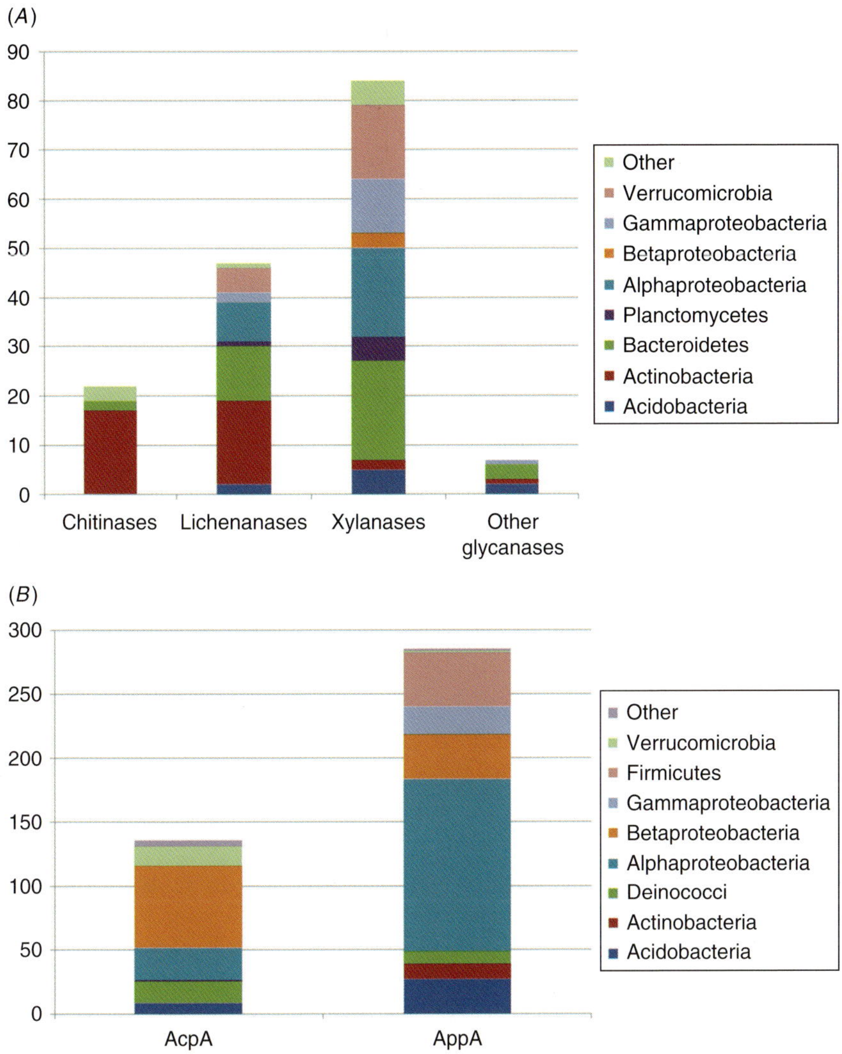

Figure 9.2 Partial functional analysis of the noncyanobacterial *Peltigera membranacea* prokaryotic metagenome. *A*, Number of 454 sequence reads extracted from *P. membranacea* metagenome based on similarity to glycolsyl hydrolase genes and taxonomic distribution of their most similar homologs in Genbank nr database. *B*, Number of *appA* phytase and *acpA* acid phosphatase homologs and taxonomic distribution. (Vilhelmsson, unpublished.)

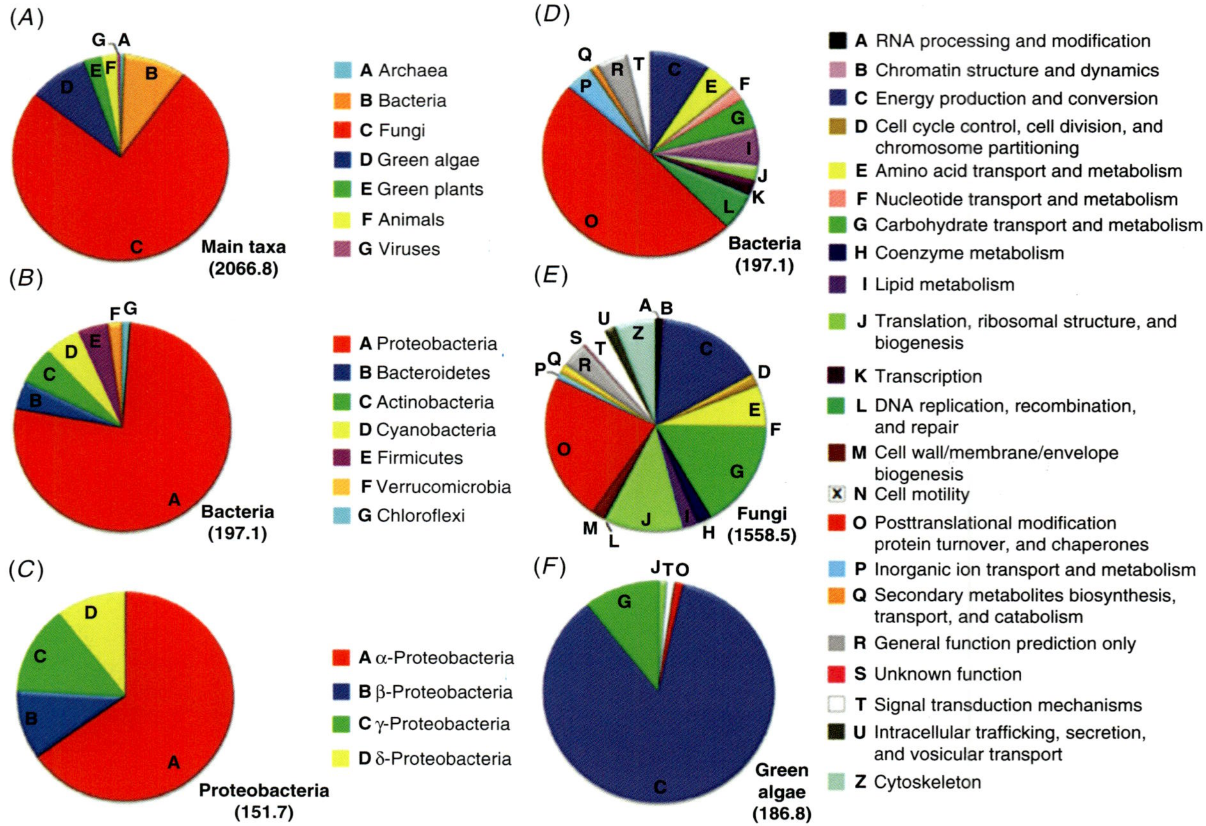

Figure 9.3 Metaproteomic profile of the *Lobaria pulmonaria* lichen symbiosis. Left side describes Taxon distribution for main taxa (*A*), bacteria (*B*), and proteobacteria (*C*). Note that the number of bacterial reads is comparable to that of the green algal partner. Right side represents gene ontology categories detected in bacteria (*D*), fungi (*E*), and algae (*G*). (Image from Schneider, Vieira de Castro, et al., 2011.)

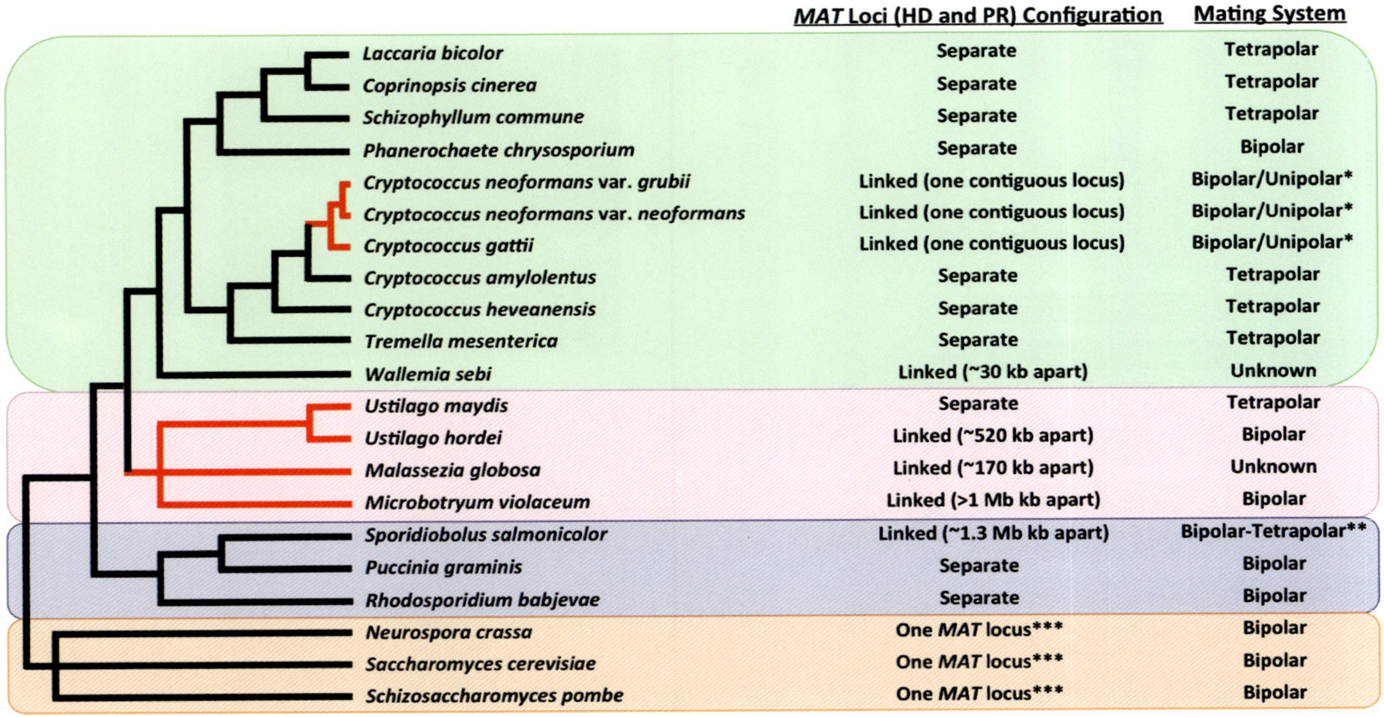

Figure 10.2 Representative Basidiomycetous species and their MAT configurations and mating systems. The phylogenetic relationship is based on Padamsee, Kumar, et al. (2012) and James, Kauff, et al. (2006). The green, pink, and blue shadings highlight species from the three supergroups of the Basidiomycota: Agaricomycotina, Ustilaginomycotina, and Pucciniomycotina, respectively. The orange shading highlights species from the Ascomycota. Branches in red color are those leading to pathogenic species. *, unipolar indicates the same-sex mating has been observed; **, the mating system in *Sporidiobolus salmonicolor* deviates from the traditional bipolar and tetrapolar mating systems, with occasional recombination between linked HD and P/R loci, as well as multiple alleles at the HD locus; ***, Ascomycete MAT loci encode transcription factors (HD, HMG, or α-domain) that control P/R gene expression, but the P/R genes are not part of the MAT locus in Ascomycete.

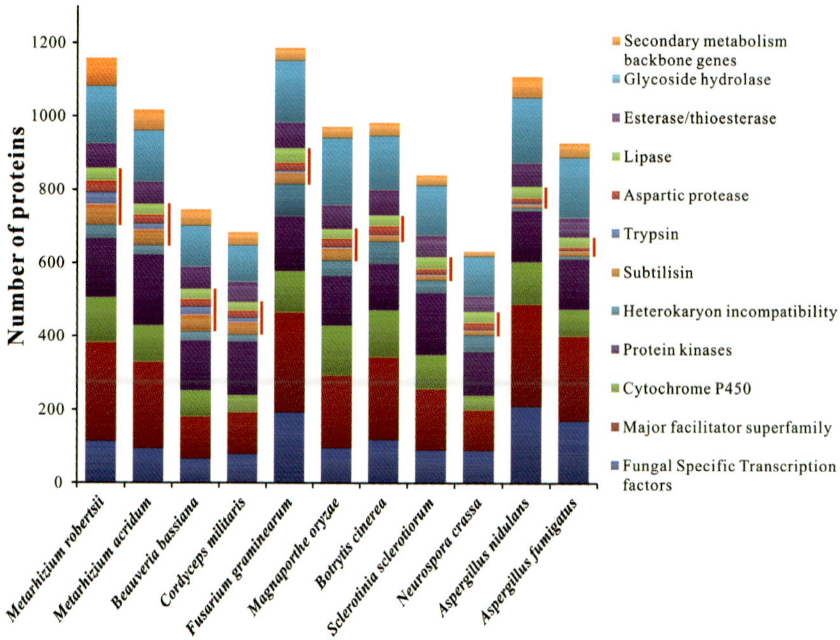

Figure 11.3 Analysis of protein family size variation between the insect pathogens and other fungi. The protein families of proteases and lipases involved in degrading insect cuticles are expanded and highlighted in red scale bar.

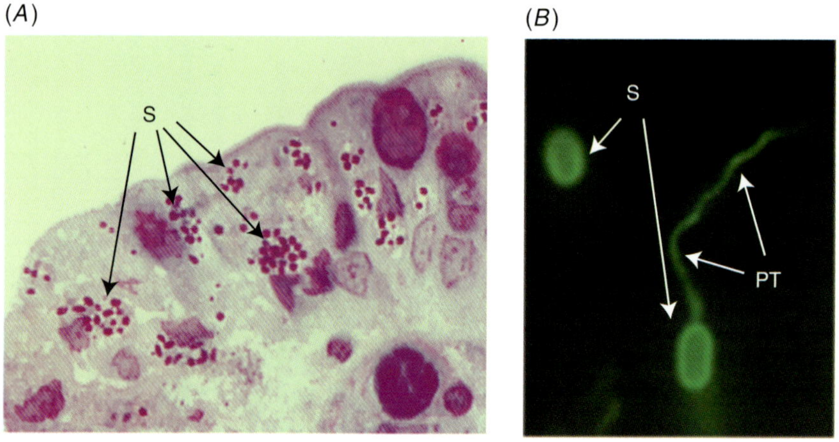

Figure 12.1 *Encephalitozoon* spp. *A*, Picture representing a toluidine blue plastic section of the intestine from a patient who is HIV positive with chronic diarrhea in the early days of antiretroviral therapy. The patient is infected with *Encephalitozoon intestinalis*, the spores of which are clearly visible(S). *B*, Resting (S) and germinated spores with polar tbe everted (PT) are shown. *Encephalitozoon* spp. spores stained with rabbit polyclonal antibody raised to *Encephalitozoon cuniculi* total spore lysate (sera raised by immunizing rabbits using Freunds Adjuvant). Secondary antibody is antiRabbit IgG-Fluorescein label. (Pictures are a courtesy of Louis Weiss.)

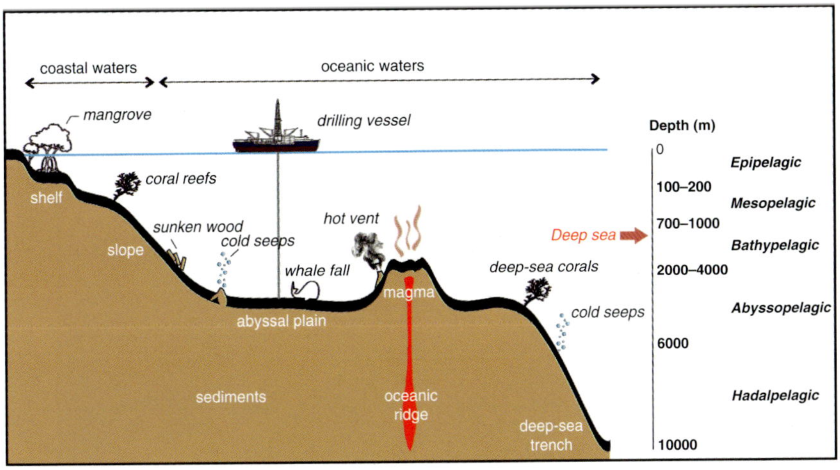

Figure 15.1 Schematic transversal section of the earth highlighting different marine ecosystems (not to scale).

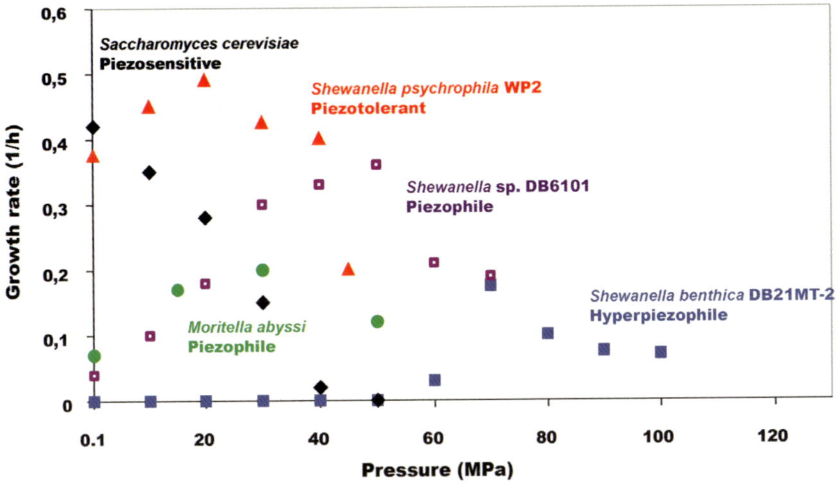

Figure 15.2 Growth rate of different microorganisms under pressure(Abe & Horikoshi, 1995; Kato, Sato, et al., 1995; Kato, Li, et al., 1998; Wang, Wang, et al., 2008; Xu, Nogi, et al., 2003).

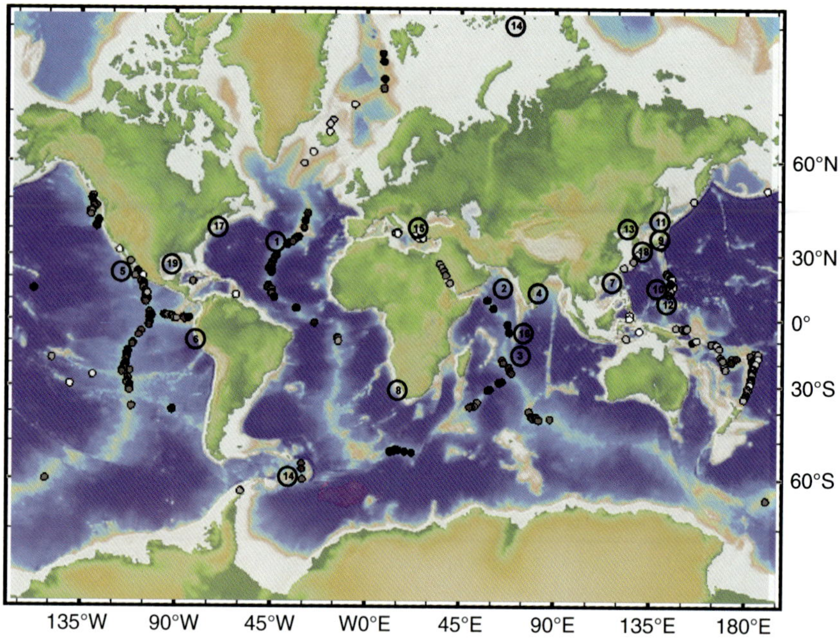

Figure 15.3 Location of known hydrothermal sites with reported depth: 0–1000 m, 1000–2000 m, 2000–3000 m, >3000 m) and location of sediment sampling sites and associated studies dealing with fungal communities. 1. Mid-Atlantic Ridge: López-García, Philip, et al. 2003; Bass, Howe, et al. 2007. 2. Arabian Sea: Raghukumar & Raghukumar, 1998; Jebaraj & Raghukumar, 2009; Jebaraj, Raghukumar, et al., 2010. 3. Central Indian Basin: Damare, Raghukumar, et al., 2006; Singh, Raghukumar, et al., 2010; Singh, Raghukumar, et al., 2011; Singh, Raghukumar, et al., 2012. 4. Bay of Bengal: Raghukumar & Raghukumar, 1998; Das, Lyla, et al., 2009. 5. Guaymas Basin: Edgcomb, Kysela, et al., 2002. 6. Peru margin & Peru trench: Edgcomb, Beaudoin, et al., 2011. 7. South China Sea: Lai, Cao, et al., 2007. 8. St Helena Bay: Mouton, Postma, et al. 2012. 9. Sagami Bay: Nagahama, Hamamoto, et al., 2001; Nagahama, Hamamoto, et al., 2003; Takishita, Yubuki, et al., 2007; Nagahama, Takahashi, et al., 2011. 10. Yap Trench: Nagahama, Hamamoto, et al., 2006. 11. Japan Trench: Nagahama, Abdel-Wahab, et al., 2008. 12. Mariana Trench: Takami, Inoue, et al., 1997; Nagano, Nagahama, et al., 2010. 13. East Sea: Park, Park, et al., 2008. 14. Arctic and Southern Ocean: Pawlowski, Christen, et al., 2011. 15. Sea of Marmara: Quaiser, Zivanovic, et al., 2011. 16. Chagos Trench: Raghukumar, Raghukumar, et al., 2004. 17. Cape Cod, Massachusetts: Stoeck & Epstein, 2003. 18. Kagoshima Bay: Takishita, Miyake, et al., 2005. 19. Gulf of Mexico: Thaler, Van Dover, et al., 2012. *This figure was generated using GeoMapApp© and the actualized InterRidge Vents Database* http://www.interridge.org/irvents/ (Beaulieu, 2010).

11 Genomics of Entomopathogenic Fungi

Chengshu Wang[1] and Raymond J. St. Leger[2]

[1] *Key Laboratory of Insect Developmental and Evolutionary Biology, Institute of Plant Physiology and Ecology, Shanghai Institutes for Biological Sciences, Chinese Academy of Sciences, Shanghai, China*
[2] *Department of Entomology, University of Maryland, College Park, Maryland*

Introduction

Fungi are the commonest insect pathogens. At least 90 genera and more than 700 species of fungi are insect pathogens, and they are distributed in virtually every major fungal taxonomic group except the higher basidiomycetes (Roberts & Humber, 1981). Fungi play a crucial role in natural ecosystems to maintain the density of insect populations, and several are being developed as environmentally friendly alternatives to chemical insecticides for the control of insect pests (de Faria & Wraight, 2007). Most of the commercially produced fungi are asexual phases in the order Hypocreales (Ascomycota): *Beauveria, Metarhizium, Nomuraea, Isaria* (formally *Paecilomyces*), *Hirsutella* and the sexual (teleomorph) phases *Cordyceps*, *Ophiocordyceps*, and *Metacordyceps* senso lato (Sung, Hywel-Jones, et al., 2007; Kepler, Sung, et al., 2012). These fungi are relatively easy to mass produce and can be used as inundative insecticides. *Metarhizium anisopliae* and *Beauveria bassiana* are both biological control agents approved by the US Environmental Protection Agency (EPA). Recent advances have identified the functions of many *Beauveria* and *Metarhizium* pathogenicity genes and technical developments have improved their virulence by using them as vehicles to carry genes-encoding toxins and antibodies (St. Leger & Wang, 2010; St. Leger, Wang, et al., 2011) or host proteins (Fan, Borovsky, et al., 2012) into insects. In addition, *Cordyceps* spp. are medicinally valued, and insect pathogens in general are prolific producers of enzymes and diverse secondary metabolites with activities against insects, fungi, bacteria, viruses, and cancer cells (Isaka, Kittakoop, et al., 2005; Kim, Song, et al., 2010). Enzymes from *Metarhizium* and *Beauveria* spp. are frequently exploited as industrial catalysts (Pereira, Noronha, et al., 2007; Silva, Santi, et al., 2009). Zygomycete entomopathogens (*Entomophthora*, *Zoophthora*, *Pandora*, *Entomophaga*, and *Erynia*) are

The Ecological Genomics of Fungi, First Edition. Edited by Francis Martin.

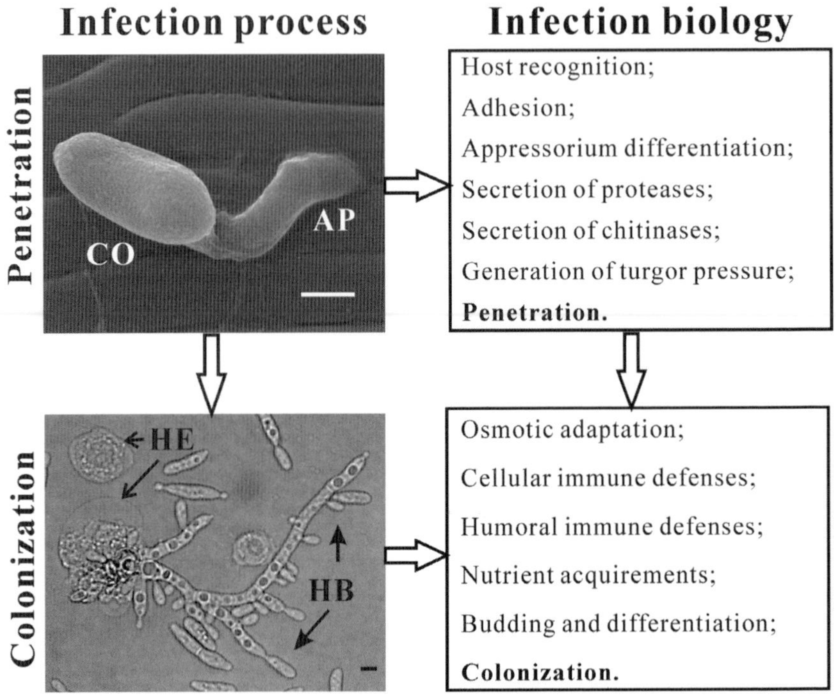

Figure 11.1 Infection processes and responses of insect pathogenic fungi against insects. The figures on the left show the representative fungus *Metarhizium robertsii*. AP, appressorium; CO, conidium; HB, hyphal body; HE, hemocyte. Bar, 2μm.

also common and usually highly virulent, but their obligate nature means they are hard to culture and so they are only employed in classical biocontrol programs (Hajek & Tobin, 2011). Oomycete entomopathogens, for example, *Lagenidium giganteum* (Kerwin, 2007) and *Aphanomyces laevis* (Patwardhan, Gandhe, et al., 2005), are mostly pathogens of mosquito larvae.

Fungi are particularly well suited for development as biopesticides because unlike bacteria and viruses that have to be ingested to cause diseases, fungi infect insects by direct penetration of the cuticle and then colonization of insect hemocoel by employing similar mechanisms (Fig. 11.1). Thus, they function as contact insecticides (Thomas & Read 2007). Biocontrol researchers have therefore made a tremendous effort to find naturally occurring fungal pathogens capable of controlling mosquitoes and other pest insects. This typically involved the selection of strains pathogenic to target insects without considering the mechanisms involved or the role of these fungi in their natural habitats. These deficiencies have hindered realization of the potential of these fungi as classical biocontrol agents that persist in the environment and recycle through pest populations (Hajek, McManus, et al., 2007). Access to genome data is paramount to advancing the knowledge of fungal infection as well as the interaction of pathogen and host. Sequence data also provide crucial information on the poorly understood ways that these organisms reproduce and persist in the environment.

Metarhizium and *Beauveria* spp. are the Best-Suited Insect Pathogenic Fungi

Most research on fungal insect pathogens has focused on *Beauveria* and *Metarhizium* spp. They have a worldwide distribution from the arctic to the tropics and colonize an impressive array of environments including forests, savannahs, swamps, coastal zones, and deserts.

Metarhizium *Species*

The genus *Metarhizium* includes the best-studied entomopathogenic fungi at the molecular and biochemical levels. Construction of *Metarhizium robertsii* deletion strains for some of the highly expressed genes has identified their roles. Some of these genes encode regulators such as the protein kinase A that controls expression of many secreted virulence factors (Fang, Pava-Ripoll, et al., 2009); an osmosensor that signals to penetrant hyphae that they have reached the hemocoel (Wang, Duan, et al., 2008); and a perilipin protein (the first characterized in a fungus) that regulates lipolysis, osmotic pressure, and formation of infection structures (Wang & St. Leger, 2007a). Some genes are highly adapted to the specific needs of *M. robertsii*, for example, MCL1 with its collagen domain is so far unique to *M. robertsii* to evade host immune responses (Wang & St. Leger, 2006). *M. robertsii* also has separate adhesins (MAD1 and MAD2) that allow it to stick to insect cuticle and plant epidermis, respectively (Wang & St. Leger, 2007b). This seems a critical point because *M. robertsii* upregulates a specific plant adhesin in the presence of plants and a specific insect adhesin in the presence of insect cuticle, demonstrating that it has specialist genes for a bifunctional lifestyle.

Infact, the principal habitat of some *Metarhizium* spp. may not be insects, but the root rhizosphere (the layer of soil influenced by root metabolism), which thus places sharp focus on the soil/root interphase as a site where plants, insects, and pathogens will interact to determine fungal efficacy, cycling, and survival (Hu & St. Leger, 2002). In retrospect, it was realized that there was evidence in the literature before the study to indicate that *Metarhizium* spp. were rhizosphere competent. Thus, general surveys have shown that although *Metarhizium* is ubiquitous, it is most abundant (~10^6 propagules/gm) in grass root soils (Milner, Lim, et al., 1992). This abundance would have been suggestive of rhizosphere competence to a soil microbiologist. Besides MAD2 (Wang & St. Leger, 2007b), other genes involved in colonizing the rhizosphere include a novel oligosaccharide transporter for root-derived nutrients, particularly raffinose, and an RNA binding protein that has important roles in both saprotrophy and pathogenicity (Fang & St. Leger, 2010a, b). Both the transporter and the RNA binding protein are the first of their kind characterized in fungi and reveal new unsuspected stratagems of adaptations to soil living, which may be relevant to all fungal biology.

The failure to appreciate the relationship between *Metarhizium* and plants seems to be an example of scientists that belong to different scientific disciplines not being familiar with each other's work. Furthermore, as shown by their antagonism to plant pathogenic fungi (Kang, Goo, et al., 1996) and pathogenicity to soil amoeba (Bidochka, Clark, et al., 2010), at least some *Metarhizium* isolates have additional unpredicted flexibility in their trophic capabilities. *Metarhizium* spp. have not yet been reported as endophytes, but the genus is closely related to the plant endophytes *Epichloe* spp. (Spatafora, Sung, et al., 2007). The genus *Metarhizium* contains biologically distinct subtypes with wide insect host ranges, for example, *M. robertsii*, formerly known as *Metarhizium anisopliae* var. *anisopliae* (Bischoff, Rehner, et al., 2009), and subtypes that show specificity for certain locusts, beetles, crickets, and hemipterans (Bidochka, Kamp, et al., 2001; Driver, Milner, et al., 2000). Different species or strains of *Metarhizium* also show differing abilities to form associations with different plant species (Bidochka, Kamp, et al., 2001; Fisher, Rehner, et al., 2011). Overall, the effects of *Metarhizium* on plants are favorable because application of conidia to corn seeds significantly increased yields (Kabaluk & Ericsson, 2007), and the fertility of soils treated with some *Metarhizium* strains can be improved beyond insect control, but there is little data as to the ecological consequences of these interactions. The fact that many genotypes of *Metarhizium* appear to be specialized to different plants (Fisher, Rehner, et al., 2011), suggests that the impact of rhizosphere competence by *Metarhizium* on plant ecology in general could be considerable with implicit co-evolutionary implications.

Beauveria *Species*

Numerous registered mycoinsecticide formulations based on *B. bassiana* and *Beauveria brongniartii* are used for control of insect pests (de Faria & Wraight, 2007). *B. bassiana* has a particularly wide host range allowing it to be used against vectors of human disease (Blanford, Chan, et al., 2005), and a wide range of agricultural pests (de Faria & Wraight, 2007). For example, in China, approximately one million hectares a year are treated with *B. bassiana* to control forest insects such as *Dendrolimus punctatus*, which are pine caterpillars (Wang, Fan, et al., 2004; Li et al. 2010). *B. bassiana* was described by Agostinio Bassi in 1835 as the cause of the devastating muscardine disease of silk worm, and it was instrumental in his development of the germ theory of disease (Steinhaus, 1956). Despite this long history, and hundreds of publications and patents, its important role as a plant endophyte and antagonist of plant pathogenic fungi has only become apparent in the last 20 years (Ownley, Griffin, et al., 2008). Studies have shown that corn, cocoa, and banana harboring *B. bassiana* endophyte are resistant to insect pests (Wagner & Lewis, 2000; Quesada-Moraga, Land, et al., 2006). These studies imply co-evolution with plants that may provide protection against insect attacks.

Beauveria is also well known for producing a large array of biologically active secondary metabolites (e.g., oosporein, bassianin, tenellin, beauvericin, bassianolides, and beauveriolides) and secreted metabolites involved in pathogenesis and virulence (e.g., oxalic acid) that have potential or realized industrial, pharmaceutical, and agricultural uses (Molnar, Gibson, et al., 2010). Silkworm larvae infected by *B. bassiana* (batryticated silkworms, also called *Bombycis corpus* or white-stiff silkworm), have been a traditional Chinese medicine for centuries, and their potential as medicines has been validated by modern technologies, for example, water extract of batryticated silkworms protect against β-amyloid-induced neurotoxicity (Koo, An, et al., 2003). The array of secondary metabolites seem to have no role in normal fungal metabolism but are highly active in animal tissues and are assumed to part of an ongoing evolutionary arms race between fungi and insects. In turn, the ability of insects to defend against *Beauveria* has illuminated many aspects of innate immunity with direct relevance to human biology (Hoffmann, Kafatos, et al., 1999). *B. bassiana* has been used to uncover immune interactions involving signaling pathways mediated by pattern recognition pathways (toll receptors or peptidoglycan recognition proteins) in *Drosophila* (Gottar, Gobert, et al., 2006).

Microbial transformation represents a series of biological reactions of xenobiotic substrates catalyzed by whole cells or enzymes obtained from microbial sources. *B. bassiana* is surpassed only by *Aspergillus niger* and brewer's yeast as a whole-cell eukaryotic catalyst in industrial and chemical applications because of its ability to catalyze a range of reactions that remain elusive to chemical approaches (Griffiths, Brown, et al., 1993). The unique reactions catalyzed by *B. bassiana* include hydroxylation, sulfoxidation, and N-acetylation reactions, epoxide and ester hydrolysis, and a series of oxidoreductase activities. Strains of *B. bassiana* are being used or developed for a number of bioremediation applications and for diverse transformations of steroids and antibiotics resulting in the production of numerous novel compounds (Orru, Archelas, et al., 1999).

Genetic Engineering

A slow kill speed is inherent for fungal biopesticides because of an evolutionary adaptive balance between pathogens and hosts. Consequently, Beauveria and Metarhizium spp. have been genetically engineered to enhance their efficacy and hence cost effectiveness. Arthropod neuropeptides are particularly attractive because they offer a high degree of biological activity and rapidly degrade in the environment providing environmental safety (Edwards & Gatehouse, 2007). Expression of a scorpion neurotoxin (AaIT) in *M. robertsii* reduced time to kill by 40 percent and lethal spore dose by up to 22-fold in caterpillars, mosquitoes, and beetles (Wang & St. Leger, 2007c; Pava-Ripoll, Posada, et al., 2008; Lu, Pava-Ripoll, et al., 2008). A recently produced strain of *Metarhizium* expresses a single-chain antibody fragment that blocks transmission of malaria (Fang, Vega-Rodríguez, et al., 2011). Recombinant antibodies also provide a vast array of potential anti-insect effectors that could target, for example, insect hormone receptors. Engineering fungi to express host proteins can also reduce time to kill (Fan, Borovsky, et al., 2012). Genetic integration of the *Bacillus thuringiensis* vegetative insecticidal protein Vip3Aa1 into *B. bassiana* generated an engineered strain with a high feeding toxicity to *Spodoptera litura* larvae in addition to the conventional virulence through cuticle infection (Qin, Ying, et al., 2010). Genetically engineering *B. bassiana* with an exogenous tyrosinase gene increased fungal production of melanins for improved conidial tolerance to ultraviolet radiation and increased virulence against diverse insects (Shang, Duan, et al., 2012).

Comparative Analysis of the Genome Sequences of the Broad-Spectrum Insect Pathogen *Metarhizium robertsii* and the Acridid-Specific *Metarhizium acridum*

The genus *Metarhizium* was recently been subdivided into 12 different species according to the sequences of several genes (Bischoff, Rehner, et al., 2009). Some of these species mostly contain strains with wide host ranges (e.g., *M. robertsii* and *M. anisopliae*), whereas others show specificity for certain locusts (*Metarhizium acridum*) or beetles (*Metarhizium majus*). *M. robertsii* and *M. acridum* in particular have emerged as excellent model organisms to explore a broad array of questions in ecology and evolution, host preference and host switching, and to investigate the mechanisms of speciation. Whole-genome analyses indicate that the genome structures of these two species are highly syntenic (Gao, Jin, et al., 2011). Comparative genomic approaches using the broad-spectrum *M. robertsii* and the locust-specific *M. acridum* confirmed that secreted proteins are markedly more numerous in *Metarhizium* spp. than in plant pathogens and non-pathogens, pointing to a greater complexity and subtlety in the interactions between *Metarhizium* spp. and their environments including insect hosts (Gao, Jin, et al., 2011). As expected, many of the secreted proteins are in families that could have roles in colonization of insect tissues, such as proteases. The trypsin family has the highest relative expansion among the proteases with 32 genes in *M. anisopliae*, almost twice as many as *M. acridum* and 6 to 10 times as many as any other fungal taxa. Overall, fewer genes were associated with plant utilization in *Metarhizium* than in plant pathogens, but almost all families of plant wall-degrading enzymes were represented in the genome. Even necrotrophs such as *Trichoderma reesei* lack many families of plant cell wall-degrading enzymes (Martinez, Berka, et al., 2008; Kubicek, Herrera-Estrella, et al., 2011), and the existence of such families in *Metarhizium* spp. implies that these species are able to use living plant tissues, which presumably could facilitate colonization of root surfaces. Consistent with their broad lifestyle options, *Metarhizium* spp. exhibits an extremely versatile metabolism, enabling growth under various environmental conditions with sparse nutrients and in the presence of compounds lethal to other fungi (Roberts & St. Leger, 2004). As expected, both *Metarhizium* genomes contain a relatively large number of genes involved in detoxification, but the broad-spectrum *M. robertsii* possesses a much greater potential for the production of secondary metabolites than *M. acridum* or most other fungi, even *Fusarium* spp. (Gao, Jin, et al., 2011; Wang, Kang, et al., 2012). Many of the additional virulence-related genes in *M. robertsii* have resulted from unique gene duplication events, but comparative genomics using microarrays also revealed divergence and loss of virulence related genes in the genomes of *Metarhizium* species specialized to beetles and crickets (Wang, Leclerque, et al., 2009). The analysis of transposase genes provided evidence of repeat-induced point (RIP) mutations and sexuality occurring in *M. acridum* but not in *M. robertsii*. It is likely that loss of RIP in *M. robertsii* facilitated gene family expansion but at the price of increased transposition.

The mechanisms of plant-fungus-insect interactions were addressed by indexing the core-set of insect and rhizosphere-induced transcripts of *M. robertsii* using EST, microarray analyses, and high-throughput transcriptomics (Freimoser, Hu, et al., 2005; Wang, Butt, et al., 2005a; Wu, Hu, et al., 2005b; Wang & St. Leger, 2005; Wang, Leclerque, et al., 2009; Gao, Jin, et al., 2011). About 20 percent of the genes most highly expressed by both *Metarhizium* species during early infection processes on their respective insect hosts show sequence similarities with experimentally verified pathogenicity, virulence, and effector genes from other fungi, particularly related plant pathogens (Gao, Jin, et al., 2011). These include many signal transduction components that provide *M. robertsii* and *M. acridum* with highly complicated finely tuned molecular mechanisms for regulating cell differentiation in response to different insect hosts. *Metarhizium* spp. also resembled *Magnaporthe oryzae* (Oh, Donofrio, et al., 2008) and the mycoparasite *Trichoderma harzianum* (Lorito, Woo, et al., 2010) in upregulating pathways associated with translation, post-translational modification, and amino acid and lipid metabolism. Formation of infection structures in all three species is associated with upregulation of genes that respond to nitrogen deprivation and related stresses initiated by different G-protein coupled receptors (Gao, Jin, et al., 2011). This is probably because of basic similarities in the fungi involved and common characteristics of the host outer surfaces (hard and wax covered in plants and insects). Microarray studies confirmed that *M. robertsii* has the ability to produce a great variety of expression patterns, which allows it to adapt to different environments and niches such as soil, water, root exudates, insects cuticles, and hemolymph (Wang, Butt, et al., 2005a; Pava-Ripoll, Angelini, et al., 2011).

Generalist and specialist *Metarhizium* species differ in the way they grow and use toxins inside hosts (Kershaw, Moorhouse, et al., 1999). For example, the generalist *M. roberstii* kills hosts quickly via toxins and grows saprophytically in the cadaver. In contrast, the specialist *M. acridum* causes a systemic infection of host tissues before the host dies. The gain and loss of the insecticidal cyclopeptide destruxin gene cluster is correlated with host specificity in *Metarhizium* spp. (Wang, Kang, et al., 2012). The genome survey also indicated that *M. roberstii* has more bacterial-like enterotoxins than *M. acridum* (Gao, Jin, et al., 2011).

Comparative Genomic Analysis of *Metarhizium* Genomes with the Caterpillar-Specific Medicinal Fungus *Cordyceps militaris* and the Broad-Spectrum Insect Pathogen *Beauveria bassiana*

A genomic analysis of *B. bassiana* and *Cordyceps militaris* showed them to be closely related and that they evolved into insect pathogens independently of the *Metarhizium* lineage (Fig. 11.2). The split between the *Cordyceps* (including *B. bassiana*) and *Metarhizium* lineages occurred before *Metarhizium* diverged from the plant endophytic *Epichloe* lineage (Zheng, Xia, et al., 2011; Xiao, Ying, et al., 2012). Nevertheless, each lineage demonstrated similar expansion of certain gene families, such as proteases and chitinases (Fig. 11.3). The ability to degrade protein- and chitin-rich insect cuticles is likely to be crucial for any pathogen that infects via this route, suggesting convergent evolution of functions necessary for pathogenesis. These expansions through gene duplication, horizontal gene transfer from bacteria, and even insect hosts may therefore identify prerequisites for entomopathogenic fungi (Xiao, Ying, et al., 2012).

Expansion or contraction of protein family size has occurred in different fungal species in association with evolutionary adaptations for different hosts and lifestyles. Thus, plant pathogens have expanded families of glycoside hydrolases, carbohydrate esterases, cutinases, and pectin lyases to degrade

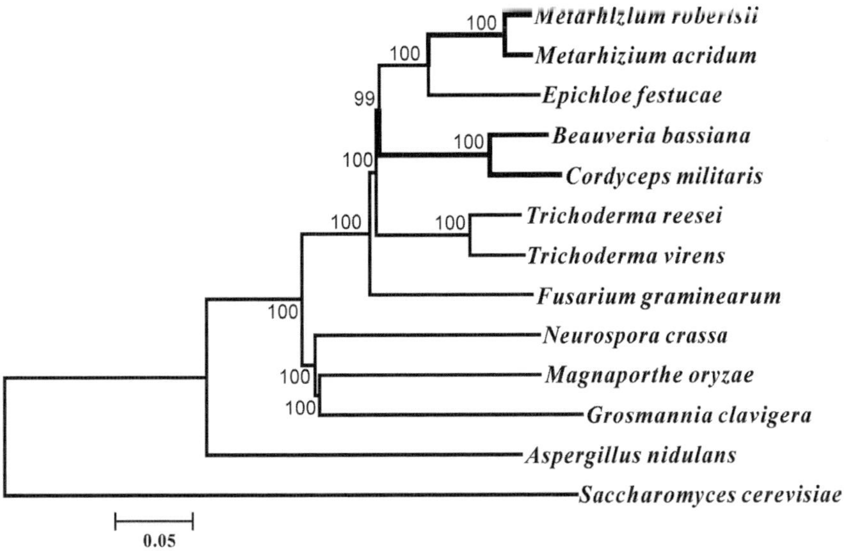

Figure 11.2 Phylogenomic relationships of insect pathogenic fungi (branches highlighted in thicker lines) with other fungi.

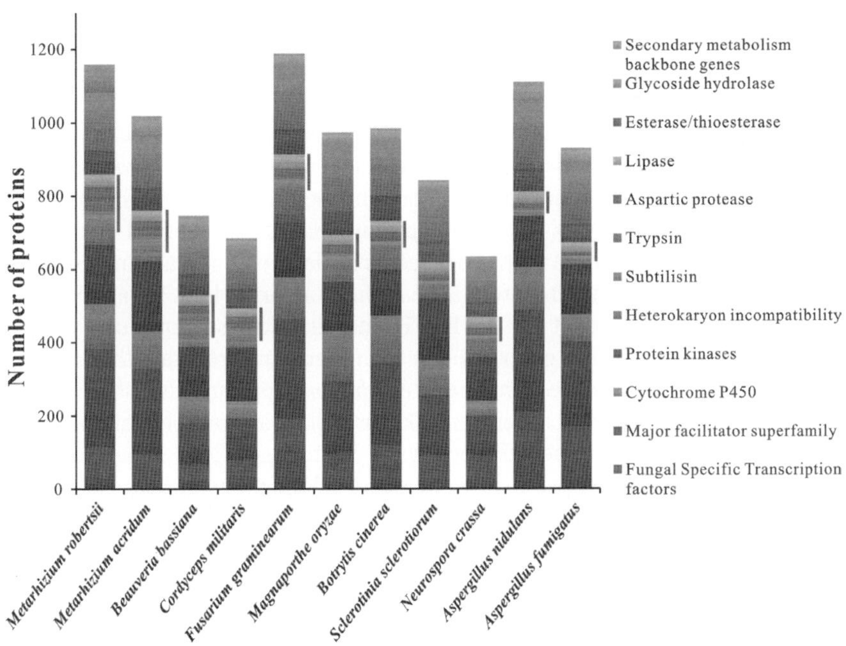

Figure 11.3 Analysis of protein family size variation between the insect pathogens and other fungi. The protein families of proteases and lipases involved in degrading insect cuticles are expanded and highlighted in red scale bar. For color detail, please see color plate section.

plant materials (Xu, Peng, et al., 2006). Mammalian pathogens are enriched for aspartyl proteases and phospholipases (van Asbeck, Clemons, et al., 2009). Mycoparasitic fungi have expanded numbers of chitinases to degrade fungal cell walls (Kubicek, Herrera-Estrella, et al., 2011). *B. bassiana* genome resembles *Metarhizium* spp. and *C. militaris* (Gao, Jin, et al., 2011; Zheng, Xia, et al., 2011), in the expansion of gene families of proteases, chitinases, lipases, fatty acid hydroxylases, and acyl-CoA dehydrogenases (for β-oxidation of fatty acids), which all have potential targets in insect hosts (Xiao, Ying, et al., 2012). Relative to plant pathogens, the expansions of amidohydrolases, glyoxalases, and monooxygenases in insect pathogens implies that the latter are better able to detoxify corresponding compounds. The genomes of the broad host range *M. robertsii* and *B. bassiana* code for even more of these enzymes than do the narrow host range species. Thus, as with plant pathogens (Ma, van der Does, et al., 2010; Stukenbrock, Bataillon, et al., 2011), differences between the insect pathogens in protein family size appear related to their insect-killing strategies and host range (*see* Fig. 11.3).

Virulence-related genes already characterized in *B. bassiana* include MAP kinases controlling cell growth, appressorium formation, abiotic stress responses, and virulence (Zhang, Zhao, et al., 2009; Zhang, Zhang, et al., 2010; Luo, Keyhani, et al., 2012). A neuronal calcium sensor was found involved in pre-penetration or early penetration events to contribute to virulence by regulating extracellular acidification (Fan, Borovsky, et al., 2012). A new cytochrome P450 subfamily enzyme CYP52X1 displays the highest activity against insect cuticular midrange fatty acids and thus contributes to the penetration and virulence (Pedrini, Zhang, et al., 2010; Zhang, Widemann, et al., 2012). A GH73 family of β-1,3-glucanosyltransferase of *B. bassiana* maintains cell well integrity and contributes to conidial thermotolerance and virulence (Zhang, Xia, et al., 2011). Two dehydrogenases (i.e., mannitol-1-phosphate dehydrogenase and manitol dehydrogenase) regulate mannitol accumulation in *B. bassiana*, and thus the stress tolerance abilities against H_2O_2, ultraviolet, and heat stresses (Wang, Lu, et al., 2011b). Homologs of these genes can also be found in the *Metarhizium* genomes. On the other hand, several other experimentally verified virulence genes in *Metarhizium* spp. are also shared with *B. bassiana*, for example, a perilipin-like protein that controls cellular lipid storage and appressorium penetration (Wang & St. Leger, 2007a), an osmosensor to mediate adaptation to the insect hemocoel (Wang, Duan, et al., 2008) and an esterase gene that is involved in mobilizing nutrients (Wang, Fang, et al., 2011a). The presence of these genes in both *B. bassiana* and *Metarhizium* spp. suggests that some strategies for interacting with plants and insects are shared. The identification of highly conserved secondary metabolite biosynthetic gene clusters in the four insect pathogens that are absent in other fungi implies that the evolution of fungal entomopathogenicity may be associated with the production of some similar secondary metabolites (Xiao, Ying, et al., 2012).

The presence of many insect pathogen-specific small secreted cysteine-rich protein (SSCP) clusters suggests that some of these shared strategies are currently unknown (Xiao, Ying, et al., 2012). However, *Beauveria* and *Cordyceps* lack a homolog to the collagen-like protein used by *Metarhizium* to evade the insect immune system (Wang & St. Leger, 2006). *Beauveria* blastospores are known to evade insect cellular immune responses (Pendland, Hung, et al., 1993) suggesting that it has evolved alternative species-specific strategies. The *Metarhizium dtxS1* gene cluster involved in biosynthesis of the insecticidal destruxins (Wang, Kang, et al., 2012), is absent from the *Beauveria* and *Cordyceps* genome. In addition, both *B. bassiana* and *C. militaris* lack a GPR1-like G-protein coupled receptor (GPCR) for sensing nitrogenous nutrients (Xue, Batlle, et al., 1998), whereas GPR1-like GPCR homologs in *Metarhizium* respond to nutrient levels on the insect surface (Gao, Jin, et al., 2011). In *Metarhizium* and the rice blast fungus *Metarhizium oryzae*, orphan genes are responsible for important species-specific processes during development or pathogenicity (Wang & St. Leger, 2006; Jeon, Park, et al., 2007). It is likely that some of the genes unique to *B. bassiana*, will likewise play important and novel roles as *B. bassiana* overcomes challenges in the dynamic microenvironments it will encounter in insect or plant hosts. As entomopathogenicity is polyphyletic and each pathogen will have evolved its own multifaceted and robust mechanisms to overcome these challenges this raises the interesting possibility of switching genes between strains of *Metarhizium* and *Beauveria* to determine if that increases their ability to colonize insects or plants.

As an endophyte, *B. bassiana* presumably possesses additional mechanisms to avoid stimulating plant defenses. Fungal endoxylanases (GH11) are known to trigger plant immune responses (Dean & Anderson, 1991), and these are absent in *B. bassiana*, which could facilitate immune evasion in plants. Like the basidiomycete plant symbiont *Laccaria bicolor* (Martin, Aerts, et al., 2008), *B. bassiana* and *Ephichloë festucae* each have a large battery of SSCPs. Unlike *B. bassiana*, more than half of *E. festucae* SSCPs are species-specific, implying many specific functions are required for specialization to endophytism. Of particular interest, *E. festucae* has sequences similar to *Metarhizium* adhesin MAD2 that mediates spore adhesion to plant surfaces (Wang, & St. Leger, 2007b). In addition, *B. bassiana* and *E. festucae* have homologs to a *Metarhizium* oligosaccharide transporter that facilitates rhizosphere competency by taking up sucrose and raffinose, the two most abundant soluble sugars in plants and root exudates (Fang & St. Leger, 2010a). Sucrose is the primary metabolite used by most plants to translocate carbon throughout their tissues. To acquire the host sucrose, it is crucial for plant interacting fungi to possess the necessary enzymes, such as extracellular invertase(s), to split sucrose into its constituent monosaccharides, glucose, and fructose (Parrent, James, et al., 2009). Some *Trichoderma* spp. such as *Trichoderma reesei* lack invertase and cannot grow on sucrose, whereas rhizosphere competent

Trichoderma spp. unusually produce intracellular (but not extracellular) invertases so sucrose must be taken up by a sucrose transporter (Vargas, Crutcher, et al., 2011). GH32 invertase genes are typically lacking in animal pathogens (Parrent, James, et al., 2009), but the insect pathogens and *E. festucae* each have a single GH32 invertase (β-fructosidase) with a signal peptide indicative of secretion. They also possess an intracellular invertase, an enzyme that degrades sucrose, so they are adapted for extracellular and intracellular conversion of sucrose to fructose and glucose.

Asexual *Aspergillus* species usually arise from sexual lineages (Geiser, Timberlake, et al., 1996). If this finding is broadly applicable, then *Beauveria* spp. are probably asexual derivations from a *Cordyceps* lineage. Host switching is particularly common in *Cordyceps* spp. accounting for their wide variety of associations with animals, plants, and fungi (Suh, Noda, et al., 2001). Some *Trichoderma* spp., such as *Trichoderma strigosum*, have a *Cordyceps* teleomorph. The phylogenomic data suggests that the insect pathogenic *C. militaris* and *B. bassiana* diverged from mycoparasitic *Trichoderma* 74-97 MYA (Xiao, Ying, et al., 2012). A degree of genome structure divergence was observed between *B. bassiana* and *C. militaris*, which is unexpected given their close phylogenetic relationship. Transposable elements (TEs) are a major force driving genetic variation and genome evolution (Daboussi & Capy, 2003; Cordaux & Batzer, 2009). *B. bassiana* has many more TEs than *C. militaris*, that is, 88 versus 4 (Xiao, Ying, et al., 2012), apparently because *B. bassiana* lacks the RIP genome defense mechanism. However, the genomes of *Metarhizium* species are highly syntenic despite a similar difference in the number of TEs, that is, 148 TEs in *M. roberstii* versus 20 TEs in *M. acridum* (Gao, Jin, et al., 2011). Most field populations of *Beauveria* and *Metarhizium* species reproduce clonally (Meyling, Lübeck, et al., 2009; Wang, Fang, et al., 2011a). In contrast, *C. militaris* readily reproduces sexually (Zheng, Xia, et al., 2011), thereby facilitating genome structure reorganization because of frequent genetic or chromosomal recombination. Thus, differences in life cycle might have led to the genome structure disparities between *B. bassiana* and *C. militaris* (Xiao, Ying, et al., 2012). Consistent with the previous microarray analysis of *Metarhizium* (Wang, Butt, et al., 2005a), high-throughput transcriptomics indicated that *Metarhizium*, *Beauveria*, and *Cordyceps* could finely tune gene transcription to adapt to different environmental niches or stage-specific developments (Gao, Jin, et al., 2011; Zheng, Xia, et al., 2011; Xiao, Ying, et al., 2012).

Future Sequencing Needs and Major Questions

DNA sequence data from any individual genome is only a snapshot in evolutionary time and space. To really understand the dynamics of genomes, we need:

(a) to understand the balance as well as the processes whereby new genes are acquired by duplication and old genes are being removed and (b) to determine the extent to which shared genes are regulated in new ways in different strains. A great deal of biodiversity among insect pathogens is currently being explored at deep taxonomic levels with the sequencing of eight additional entomopathogens: *Beauveria brogniarti*, *Metarhizium album*, *Isaria fumosoroseus*, *Nomuraea rileyi*, *Lecanicillium lecanii*, *Sporothrix insectorum*, *Aschersonia aleyrodis*, and *Ascosphaera apis* among others. These taxa could be too divergent to be useful in evaluating many of these important evolutionary processes that occur on a much shorter time scale. This has already been demonstrated by multispecies exploration of genome evolution of *Aspergillus* spp. (Galagan, Calvo, et al., 2005). In addition, the more than one million different fungal species display extraordinary diversity, especially given the number of different pathogens and their products that can be studied (Isaka, Kittakoop, et al., 2005). To increase the accuracy of comparative analysis, much more extensive sampling of related fungi is needed. Thus, we also intend to examine pathogen genome evolution and host range usage by confining the comparisons within a single genus, while exploring its evolutionary range as far as possible. *Metarhizium* is a particularly good model system for studying evolutionary processes because it consists of lineages that in terms of developmental processes are almost indistinguishable from each other but differ dramatically in one key factor, host range. Given that specialization has occurred many times in *Metarhizium* it provides an unusual and innovative opportunity to study a genus with species containing a large number of independently evolved models of adaptation and response. These should provide a novel perspective on the evolution and strategies of host selectivity and host switching. Although host selectivity and host switching are widely documented phenomena in diverse pathogens, in most cases the underlying mechanisms are poorly understood. As a radiating lineage, the natural molecular variation of *Metarhizium* spp. offers the chance of finding processes of both adaptive change and phylogenetic differentiation still in operation, even in intermediate states. We should thus be able to: (a) correlate genetic differences with adaptations to specific hosts and identify the underlying regulatory, metabolic, and biosynthetic differences that define host preferences; (b) determine what roles do changes in gene complement or expression profiles play in generating differences in virulence and host range; (c) identify mechanisms by which novel pathogens emerge with either wide or narrow host ranges, and (d) identify genes that are involved in interactions with plants and soil biota.

Conclusions

In conclusion, the genomes of several well-known insect pathogenic fungi have been sequenced. Sequencing related species that have evolved specialist or generalist lifestyles has increased their use as models and provided insights into the evolution of pathogenicity. Such sequences are also allowing for more rapid identification of genes-encoding biologically active molecules and genes responsible for interactions between fungi, plants, and insects. The resulting information will benefit future molecular studies of insect-fungus interactions and will facilitate the development of insect pathogens as cost-effective mycoinsecticides. The new information on the abundant enzymes of these fungi will also facilitate more extensive work to determine mechanisms of the biotransformation reactions that make these fungi such useful industrial catalysts. Overall, therefore, the entomopathogen genome sequences will help realize the still-undeveloped potential possessed by these fungi both as insect pathogens and as microbial biocatalysts, as well as illuminate their poorly understood role as endophytes and plant symbionts.

References

Blanford S, Chan BHK, et al. 2005. Fungal pathogen reduces potential for malaria transmission. *Science*. 308:1638–1641.

Bidochka MJ, Clark DC, et al. 2010. Could insect phagocytic avoidance by entomogenous fungi have evolved via selection against soil amoeboid predators? *Microbiol.* 156: 2164–2171.

Bidochka MJ, Kamp AM, et al. 2001. Habitat association in two genetic groups of the insect-pathogenic fungus *Metarhizium anisopliae*: Uncovering cryptic species? *Appl Environ Microbiol.* 67: 1335–1342.

Bischoff JF, Rehner SA, et al. 2009. A multilocus phylogeny of the *Metarhizium anisopliae* lineage. *Mycologia.* 101: 512–530.

Cordaux R & Batzer MA. 2009. The impact of retrotransposons on human genome evolution. *Nat Rev Genet.* 10: 691–703.

Daboussi MJ & Capy P. 2003. Transposable elements in filamentous fungi. *Annu Rev Microbiol.* 57: 275–299.

Dean JFD & Anderson JD. 1991. Ethylene biosynthesis-inducing xylanase. 2. Purification and physical characterization of the enzyme produced by Trichoderma viride. *Plant Physiol.* 95: 316–323.

de Faria MR & Wraight SP. 2007. Mycoinsecticides and Mycoacaricides: A comprehensive list with worldwide coverage and international classification of formulation types. *Biol Control.* 43: 237–256.

Driver F, Milner RJ, et al. 2000. A taxonomic revision of *Metarhizium* based on a phylogenetic analysis of rDNA sequence data. *Mycol Res.* 104: 134–150.

Edwards MG & Gatehouse AMR. 2007. Biotechnology in crop protection: Towards sustainable insect control. In: *Novel Biotechnologies for Biocontrol Agent Enhancement and Management* (eds. M Vurro & J Gressel), 1–24. New York: Springer.

Fan YH, Borovsky D, et al. 2012. Exploiting host molecules to augment mycoinsecticide virulence. *Nat Biotechnol.* 30: 35–37.

Fang WG, Pava-Ripoll M, et al. 2009. Protein kinase A regulates production of virulence determinants by the entomopathogenic fungus, *Metarhizium anisopliae*. *Fungal Genet Biol.* 46: 277–285.

Fang WG & St Leger RJ. 2010a. Mrt, a gene unique to fungi, encodes an oligosaccharide transporter and facilitates rhizosphere competency in *Metarhizium robertsii*. *Plant Physiol.* 154: 1549–1557.

Fang WG & St Leger RJ. 2010b. RNA binding proteins mediate the ability of a fungus to adapt to the cold. *Environ Microbiol.* 12: 810–820.

Fang WG, Vega-Rodríguez J, et al. 2011. Development of transgenic fungi that kill human malaria parasites in mosquitoes. *Science.* 331: 1074–1077.

Fisher JJ, Rehner SA, et al. 2011. Diversity of rhizosphere associated entomopathogenic fungi of perennial herbs, shrubs and coniferous trees. *J Invertebr Pathol.* 106: 289–295.

Freimoser FM, Hu G, et al. 2005. Variation in gene expression patterns as the insect pathogen *Metarhizium anisopliae* adapts to different host cuticles or nutrient deprivation in vitro. *Microbiol.* 151: 361–371.

Galagan JE, Calvo SE, et al. 2005. Sequencing of *Aspergillus nidulans* and comparative analysis with *A. fumigatus* and *A. oryzae*. *Nature.* 438: 1105–1115.

Gao Q, Jin K, et al. 2011. Genome sequencing and comparative transcriptomics of the model entomopathogenic fungi *Metarhizium anisopliae* and *M. acridum*. *PLoS Genet.* 7: e1001264.

Geiser DM, Timberlake WE, et al. 1996. Loss of meiosis in *Aspergillus*. *Mol Biol Evol.* 13: 809–817.

Gottar M, Gobert V, et al. 2006. Dual detection of fungal infections in *Drosophila* via recognition of glucans and sensing of virulence factors. *Cell.* 127: 1425–1437.

Griffiths DA, Brown DE, et al. 1993. Metabolism of xenobiotics by *Beauveria bassiana*. *Xenobiotica.* 23: 1085–1100.

Hajek AE & Tobin PC. 2011. Introduced pathogens follow the invasion front of a spreading alien host. *J Anim Ecol.* 80: 1217–1226.

Hajek AE, McManus ML, et al. 2007. A review of introductions of pathogens and nematodes for classical biological control of insects and mites. *Biol Control.* 41: 1–13.

Hoffmann JA, Kafatos FC, et al. 1999. Phylogenetic perspectives in innate immunity. *Science.* 284: 1313–1318.

Hu G & St Leger RJ. 2002. Field studies using a recombinant mycoinsecticide (*Metarhizium anisopliae*) reveal that it is rhizosphere competent. *Appl Environ Microbiol.* 68: 6383–6387.

Isaka M, Kittakoop P, et al. 2005. Bioactive substances from insect pathogenic fungi. *Acc Chem Res.* 38: 813–823.

Jeon J, Park SY, et al. 2007. Genome-wide functional analysis of pathogenicity genes in the rice blast fungus. *Nat Genet.* 39: 561–565.

Kabaluk JT & Ericsson JD. 2007. Environmental and behavioral constraints on the infection of wireworms by *Metarhizium anisopliae*. *Environ Entomol.* 36: 1415–1420.

Kang CS, Goo BY, et al. 1996. Antifungal activities of *Metarhizium anisopliae* against *Fusarium oxysporum*, *Botrytis cinerea* and *Alternaria solani*. *Korean J Mycol.* 24: 49–55.

Kepler RM, Sung GH, et al. 2012. New teleomorph combinations in the entomopathogenic genus Metacordyceps. *Mycologia.* 104: 182–197.

Kershaw MJ, Moorhouse ER, et al. 1999. The role of destruxins in the pathogenicity of *Metarhizium anisopliae* for three species of insects. *J Invertebr Pathol.* 74: 213–223.

Kerwin JL. 2007. Oomycetes: *Lagenidium giganteum*. *J Am Mosq Control Assoc.* 23: 50–57.

Kim HG, Song H, et al. 2010. *Cordyceps pruinosa* extracts induce apoptosis of HeLa cells by a caspase dependent pathway. *J Ethnopharm.* 128: 342–351.

Koo BS, An HG, et al. 2003. Bombycis corpus extract (BCE) protects hippocampal neurons against excitatory amino acid-induced neurotoxicity. *Immunopharm Immunotoxicol.* 25: 191–201.

Kubicek CP, Herrera-Estrella A, et al. 2011. Comparative genome sequence analysis underscores mycoparasitism as the ancestral life style of *Trichoderma*. *Genome Biol.* 12: R40.

Lorito M, Woo SL, et al. 2010. Translational research on *Trichoderma*: from 'omics to the field. *Annu Rev Phytopathol.* 48: 395–417.

Lu D, Pava-Ripoll M, et al. 2008. Insecticidal evaluation of *Beauveria bassiana* engineered to express a scorpion neurotoxin and a cuticle degrading protease. *Appl Microbiol Biotechnol.* 81: 515–522.

Luo X, Keyhani NO, et al. 2012. The MAP kinase Bbslt2 controls growth, conidiation, cell wall integrity, and virulence in the insect pathogenic fungus *Beauveria bassiana*. *Fungal Genet Biol.* 49: 544–555.

Ma LJ, van der Does HC, et al. 2010. Comparative genomics reveals mobile pathogenicity chromosomes in *Fusarium*. *Nature*. 464: 367–373.

Martin F, Aerts A, et al. 2008. The genome of *Laccaria bicolor* provides insights into mycorrhizal symbiosis. *Nature*. 452: 88–92.

Martinez D, Berka RM, et al. 2008. Genome sequencing and analysis of the biomass-degrading fungus *Trichoderma reesei* (syn. *Hypocrea jecorina*). *Nature Biotechnol*. 26: 553–560.

Meyling NV, Lübeck M, et al. 2009. Community composition, host range and genetic structure of the fungal entomopathogen *Beauveria* in adjoining agricultural and seminatural habitats. *Mol Ecol*. 18: 1282–1293.

Milner RJ, Lim RP, et al. 2002. Risks to the aquatic ecosystem from the application of *Metarhizium anisopliae* for locust control in Australia. *Pest Manag Sci*. 58: 718–723.

Molnár I, Gibson DM, et al. 2010. Secondary metabolites from entomopathogenic Hypocrealean fungi. *Nat Prod Rep*. 27: 1241–1275.

Oh Y, Donofrio N, et al. 2008. Transcriptome analysis reveals new insight into appressorium formation and function in the rice blast fungus *Magnaporthe oryzae*. *Genome Biol*. 9: R85.

Orru RV, Archelas A, et al. 1999. Epoxide hydrolases and their synthetic applications. *Adv Biochem Eng Biotechnol*. 63: 145–167.

Ownley BH, Griffin MR, et al. 2008. *Beauveria bassiana*: endophytic colonization and plant disease control. *J Invertebr Pathol*. 98: 267–270.

Parrent JL, James TY, et al. 2009. Friend or foe? Evolutionary history of glycoside hydrolase family 32 genes encoding for sucrolytic activity in fungi and its implications for plant-fungal symbioses. *BMC Evol Biol*. 9: 148.

Patwardhan A, Gandhe R, et al. 2005. Larvicidal activity of the fungus Aphanomyces (oomycetes: Saprolegniales) against *Culex quinquefasciatus*. *J Commun Dis*. 37: 269–274.

Pava-Ripoll M, Posada FJ, et al. 2008. Increased pathogenicity against coffee berry borer, *Hypothenemus hampei* (Coleoptera: Curculionidae) by *Metarhizium anisopliae* expressing the scorpion toxin (AaIT) gene. *J Invertebr Pathol*. 99: 220–226.

Pava-Ripoll M, Angelini C, et al. 2011. The rhizosphere-competent entomopathogen *Metarhizium anisopliae* expresses a specific subset of genes in plant root exudate. *Microbiol*. 157: 47–55.

Pedrini N, Zhang S, et al. 2010. Molecular characterization and expression analysis of a suite of cytochrome P450 enzymes implicated in insect hydrocarbon degradation in the entomopathogenic fungus *Beauveria bassiana*. *Microbiol*. 156: 2549–2557.

Pendland JC, Hung SY, et al. 1993. Evasion of host defense by in vivo-produced protoplast-like cells of the insect mycopathogen *Beauveria bassiana*. *J Bacteriol*. 175: 5962–5969.

Pereira JL, Noronha EF, et al. 2007. Novel insights in the use of hydrolytic enzymes secreted by fungi with biotechnological potential. *Lett Appl Microbiol*. 44: 573–581.

Qin Y, Ying SH, et al. 2010. Integration of insecticidal protein Vip3Aa1 into *Beauveria bassiana* enhances fungal virulence to *Spodoptera litura* larvae by cuticle and per Os infection. *Appl Environ Microbiol*. 76: 4611–4618.

Quesada-Moraga E, Landa BB, et al. 2006. Endophytic colonisation of opium poppy, Papaver somniferum, by an entomopathogenic *Beauveria bassiana* strain. *Mycopathologia*. 161: 323–329.

Roberts DW & Humber RA. 1981. Entomogenous Fungi. In: Biology of Conidial Fungi. (eds. GT Cole & B Kendrick), 201–236. New York: Academic Press.

Roberts DW & St. Leger RJ. 2004. *Metarhizium* spp., cosmopolitan insect-pathogenic fungi: mycological aspects. *Adv Appl Microbiol*. 54: 1–70.

Shang YF, Duan ZB, et al. 2012. Improving UV resistance and virulence of *Beauveria bassiana* by genetic engineering with an exogenous tyrosinase gene. *J Invertebr Pathol*. 109: 105–109.

Silva WOB, Santi L, et al. 2009. Characterization of a spore surface lipase from the biocontrol agent *Metarhizium anisopliae*. *Proc Biochem*. 44: 829–834.

Spatafora JW, Sung GH, et al. 2007. Phylogenetic evidence for an animal pathogen origin of ergot and the grass endophytes. *Mol Ecol*. 16: 1701–1711.

St. Leger RJ & Wang CS. 2010. Genetic engineering of fungal biocontrol agents to achieve greater efficacy against insect pests. *Appl Microbiol Biotech*. 85: 901–907.

St. Leger RJ, Wang CS, et al. 2011.New perspectives on insect pathogens. *Fungal Biol Rev*. 25: 84–88.

Steinhaus EA. 1956. Microbial control—The emergence of an idea: A brief history of insect pathology through the nineteenth century. *Hilgardia*. 26: 107–160.

Stukenbrock EH, Bataillon T, et al. 2011. The making of a new pathogen: Insights from comparative population genomics of the domesticated wheat pathogen *Mycosphaerella graminicola* and its wild sister species. *Genome Res*. 21: 2157–2166.

Suh SO, Noda H, et al. 2001. Insect symbiosis: Derivation of yeast-like endosymbionts within an entomopathogenic filamentous lineage. *Mol Biol Evol*. 18: 995–1000.

Sung GH, Hywel-Jones NL, et al. 2007. Phylogenetic classification of *Cordyceps* and the clavicipitaceous fungi. *Stud Mycol*. 57: 5–59.

Thomas MB & Read AF. 2007. Can fungal biopesticides control malaria? *Nat Rev Microbiol*. 5: 377–383.

van Asbeck EC, Clemons KV, et al. 2009. *Candida* parapsilosis: a review of its epidemiology, pathogenesis, clinical aspects, typing and antimicrobial susceptibility. *Crit Rev Microbiol*. 35: 283–309.

Vargas WA, Crutcher FK, et al. 2011. Functional characterization of a plant-like sucrose transporter from the beneficial fungus *Trichoderma virens*. Regulation of the symbiotic association with plants by sucrose metabolism inside the fungal cells. *New Phytol*. 189: 777–789.

Wagner BL & Lewis LC. 2000. Colonization of corn, *Zea mays*, by the entomopathogenic fungus *Beauveria bassiana*. *Appl Environ Microbiol*. 66: 3468–3473.

Wang B, Kang QJ, et al. 2012. Unveiling the biosynthetic puzzle of destruxins in *Metarhizium* species. *Proc Natl Acad Sci USA*. 109: 1287–1292.

Wang CS, Fan MZ, et al. 2004. Molecular monitoring and evaluation of the application of the insect-pathogenic fungus *Beauveria bassiana* in southeast China. *J Appl Microbiol*. 96: 861–870.

Wang CS, Butt TM, et al. 2005a. Colony sectorization of *Metarhizium anisopliae* is a sign of ageing. *Microbiol*. 151: 3223–3236.

Wang CS, Hu G, et al. 2005b. Differential gene expression by *Metarhizium anisopliae* growing in root exudate and host (*Manduca sexta*) cuticle or hemolymph reveals mechanisms of physiological adaptation. *Fungal Genet Biol*. 42: 704–718.

Wang CS & St. Leger RJ. 2005. Developmental and transcriptional responses to host and non-host cuticles by the specific locust pathogen *Metarhizium anisopliae* sf. *acridum*. *Eukaryot Cell*. 4: 937–947.

Wang CS & St. Leger RJ. 2006. A collagenous protective coat enables *Metarhizium anisopliae* to evade insect immune responses. *Proc Natl Acad Sci USA*. 103: 6647–6652.

Wang CS & St. Leger RJ. 2007a. The *Metarhizium anisopliae* perilipin homolog MPL1 regulates lipid metabolism, appressorial turgor pressure, and virulence. *J Biol Chem*. 282: 21110–21115.

Wang CS & St. Leger RJ. 2007. The MAD1 adhesin of *Metarhizium anisopliae* links adhesion with blastospore production and virulence to insects, and the MAD2 adhesin enables attachment to plants. *Eukaryotic Cell*. 6: 808–816.

Wang CS & St. Leger RJ. 2007b. A scorpion neurotoxin increases the potency of a fungal insecticide. *Nat Biotechnol*. 25: 1455–1456.

Wang CS, Duan ZB, et al. 2008. MOS1 osmosensor of *Metarhizium anisopliae* is required for adaptation to insect host hemolymph. *Eukaryot Cell*. 7: 302–309.

Wang SB, Leclerque A, et al. 2009. Comparative genomics using microarrays reveals divergence and loss of virulence-associated genes in host-specific strains of the insect pathogen *Metarhizium anisopliae*. *Eukaryot Cell*. 8: 888–898.

Wang SB, Fang WG, et al. 2011a. Insertion of an esterase gene into a specific locust pathogen (*Metarhizium acridum*) enables it to infect caterpillars. *PLoS Pathol*. 7: e1002097.

Wang ZL, Lu JD, et al. 2011b. Primary roles of two dehydrogenases in the mannitol metabolism and multi-stress tolerance of entomopathogenic fungus *Beauveria bassiana*. *Environ Microbiol*. 14: 2139–2150.

Xiao GH, Ying S-H, et al. 2012. Genomic perspectives on the evolution of fungal entomopathogenicity in *Beauveria bassiana*. *Sci Rep*. 2: 483.

Xu JR, Peng YL, et al. 2006. The dawn of fungal pathogen genomics. *Annu Rev Phytopathol*. 44: 337–366.

Xue Y, Batlle M, et al. 1998. GPR1 encodes a putative G protein-coupled receptor that associates with the Gpa2p G-alpha subunit and functions in a Ras-independent pathway. *EMBO J*. 17: 1996–2007.

Zhang Y, Zhao J, et al. 2009. Mitogen-activated protein kinase hog1 in the entomopathogenic fungus *Beauveria bassiana* regulates environmental stress responses and virulence to insects. *Appl Environ Microbiol*. 75: 3787–3795.

Zhang Y, Zhang J, et al. 2010. Requirement of a mitogen-activated protein kinase for appressorium formation and penetration of insect cuticle by the entomopathogenic fungus *Beauveria bassiana*. *Appl Environ Microbiol*. 76: 2262–2270.

Zhang SZ, Xia YX, et al. 2011. Contribution of the gas1 gene of the entomopathogenic fungus *Beauveria bassiana*, encoding a putative glycosylphosphatidylinositol-anchored beta-1, 3-glucanosyltransferase, to conidial thermotolerance and virulence. *Appl Environ Microbiol*. 77: 2676–2684.

Zhang S, Widemann E, et al. 2012. CYP52X1, representing new cytochrome P450 subfamily, displays fatty acid hydroxylase activity and contributes to virulence and growth on insect cuticular substrates in entomopathogenic fungus *Beauveria bassiana*. *J Biol Chem*. 287: 13477–13486.

Zheng P, Xia YL, et al. 2011. Genome sequence of the insect pathogenic fungus *Cordyceps militaris*, a valued traditional Chinese medicine. *Genome Biol*. 12: R116.

12 Ecological Genomics of the Microsporidia

Nicolas Corradi[1] and Patrick J. Keeling[2]

[1] *Canadian Institute for Advanced Research, Department of Biology, University of Ottawa, Ontario, Canada*
[2] *Canadian Institute for Advanced Research, Department of Botany, University of British Columbia, Vancouver, Canada*

Introduction

Microsporidia represent a group of obligate intracellular parasites that can infect a large spectrum of animal lineages; including humans (Larsson, 1999; Weiss, 2001; Richards, Hirt, et al., 2003). In the environment, microsporidia are present as spores, the cytoplasm of which is dominated by their most recognizable characteristic, the remarkable apparatus for host-invasion called the polar filament (or polar tube). When triggered by environmental cues, the polar filament is everted from the spore following osmotic pressure, which results in the discharge of the microsporidian cytoplasmic contents through the erupted tube and into the recipient host; if the tube has penetrated a nearby cell (Fig. 12.1; Vavra, 1965; Vávra & Larsson, 1999). Once inside a host cell, the parasite life cycle enters the merogony stage, which is characterized by the rapid multiplication of microsporidian cells (called *meronts*), generally within a parasitophorous vacuole. Ultimately, the parasites develop into new spores and the host cell lyses, releasing the spores into the environment (Kudo, 1947; Kudo & Daniels, 1963; Larsson, 1999).

Microsporidia are ubiquitous and diverse, with a total of more than 1,200 species in 160 genera described in the literature. These were isolated from a large variety of hosts, including vertebrates, invertebrates, and even protists, and different species are known to vary in host specificity. Indeed, some microsporidia seem to exclusively parasitize a single host, (e.g., *Nosema ceranae* infecting the honey bee *Apis mellifera*, or *Hamiltosporidium tvaerminnensis* infecting the crustacean *Daphnia magna*) (Vizoso, Lass et al., 2005; Corradi, Akiyoshi, et al. 2007; Cornman, Chen, et al. 2009), whereas others are commonly found to infect many members of one lineage (e.g., different species in the genus *Encephalitozoon* have been found in several vertebrate lineages) (Didier & Bertucci, 1996; Didier, Bertucci, et al., 2001; Didier, 2005).

The Ecological Genomics of Fungi, First Edition. Edited by Francis Martin.

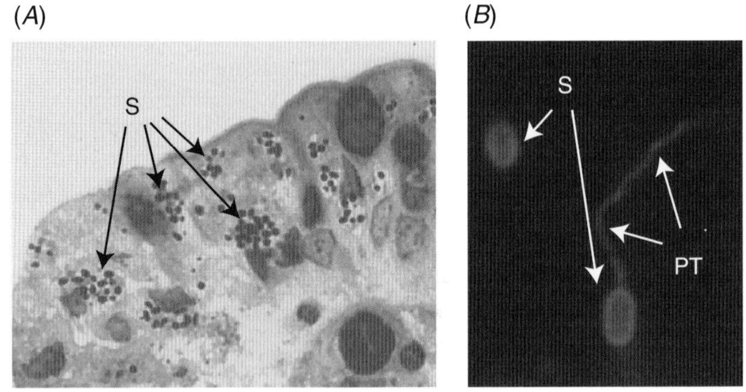

Figure 12.1 *Encephalitozoon* spp. *A*, Picture representing a toluidine blue plastic section of the intestine from a patient who is HIV positive with chronic diarrhea in the early days of antiretroviral therapy. The patient is infected with *Encephalitozoon intestinalis*, the spores of which are clearly visible (S). *B*, Resting (S) and germinated spores with polar tbe everted (PT) are shown. *Encephalitozoon* spp. spores stained with rabbit polyclonal antibody raised to *Encephalitozoon cuniculi* total spore lysate (sera raised by immunizing rabbits using Freunds Adjuvant). Secondary antibody is antiRabbit IgG-Fluorescein label. (Pictures are a courtesy of Louis Weiss.) For color detail, please see color plate section.

The wide range of hosts characteristic of the group has had obvious impacts on the heath and wealth of many humans, both directly and indirectly. For instance, species in the genus *Nosema* have been linked with population decline in bees and silkworms, resulting in important economic losses for apiculture and sericulture industries (Higes, Martin, et al., 2006; Cox-Foster, Conlan, et al., 2007; Anderson & East, 2008). Microsporidia are also considered important emerging human pathogens and represent a potential relevant threat to human health. *Enterocytozoon bieneusi* and *Encephalitozoon intestinalis* are the most prevalent human-infecting species, especially in sub-Saharan countries. Most cases of microsporidiosis (the disease associated with microsporidian infection) are found in patients who are immunocompromised, although infections of healthy patients are also frequently reported (Weiss & Vossbrinck, 1998; Weiss, 2001). Microsporidiosis is rarely fatal, however, and its symptoms are usually similar to those resulting from other opportunistic pathogens, ranging from encephalitis to severe chronic diarrhea.

A Seemingly Simplistic Eukaryotic Cell

The obligate intracellular parasitic lifestyle of microsporidia has shaped their evolution, sometimes in drastic ways, most obviously resulting in the simplification of many cellular components. As a result, a typical microsporidian cell is missing a number of cellular features common to other eukaryotes or has altered features so that they are radically different from those of the canonical eukaryote. These include the presence of unconventional mitochondria in microsporidia, which have been reduced in physical size, complexity, and metabolic function and are now called *mitosomes* (Vavra, 1965; Vávra & Larsson, 1999). Mitosomes have lost their genome altogether and have been found to only function in the production of iron-sulfur clusters (Williams, Hirt, et al., 2002; Williams & Keeling, 2005; Goldberg, Molik, et al., 2008). Other unconventional features of microsporidian cells include the absence of Golgi bodies, peroxysomes, and 9+2 microtubular structures (Vavra, 1965; Vávra & Larsson, 1999), as well as the presence of prokaryote-sized 70S ribosomes (as opposed to the 80S ribosomes that are usually found in eukaryotes) (Ishihara & Hayashi, 1968; Curgy, Vávra, et al., 1980).

For some time, the seemingly "simple" nature of microsporidian cells was interpreted as "primitive," supporting the idea that they represented ancient eukaryotes. This notion received compelling support from the first eukaryotic phylogenies reconstructed using ribosomal RNA genes (Vossbrinck, Maddox, et al., 1987), and was further supported by the first analyses of protein-coding genes from the lineage (Kamaishi, Hashimoto, et al., 1996a; Kamaishi, Hashimoto, et al., 1996b). As a result, microsporidia were considered to be members of a group called Archezoa, which consisted of a number of other supposedly amitochondriate lineages that were all thought to have diverged before the origin of mitochondria through primary endosymbiosis (Cavalier-Smith, 1983, 1987). The idea that microsporidia represented an ancient lineage was, however, rather short lived. Specifically, a new hypothesis emerged from two sources. First, as the phylogeny of more genes was analyzed, the deep position was challenged by an alternative relationship to fungi (Keeling & Doolittle, 1996; Hirt, Logsdon, et al., 1999; Keeling, Luker, et al., 2000; Van de Peer, Ben Ali, et al., 2000). At the same time, nucleus-encoded genes of mitochondrial origin were identified (Peyretaillade, Broussolle, et al., 1998), and eventually the relict mitosome itself was found (Williams, Hirt, et al. 2002), ruling out an ancestral amitochondriate nature. Currently, most studies based on comparative genomics and phylogenomics support the idea that these parasites represent an early offshoot of the fungal kingdom, though exactly how the two groups are related remains somewhat contentious (James, Kauff, et al., 2006; Lee, Corradi, et al., 2008; Lee, Corradi, et al., 2010; Capella-Gutierrez, Marcet-Houben, et al., 2012; Cuomo, Desjardins, et al., 2012).

The atypical cells of microsporidia, coupled with their enigmatic evolutionary origin, and their medical importance of microsporidia, have resulted in the acquisition of extensive sequence data from many lineages; with different hosts and genome sizes. This sequencing effort has provided essential information about the origin and evolution of the content and structure of microsporidian genomes and also how these intracellular parasites interact with, and benefit from, their host cells (Table 12.1; Richards, Hirt, et al., 2003).

Table 12.1 General characteristics of microsporidian genomes.

	E.i	*E. h*	*E. r*	*E. c*	*E.b.*	*N.p.*	*H.t*	*N. c*
Chromosomes (#)	11	11	11	11	N.A.	N.A.	N.A.	N.A.
Genome size (Mbp)	2.3	2.5	2.5	2.9	6	4	24	7.86
G+C content (%)	41.4	43.4	40.3	47	25	34.4	26	N.A.
Gene density (gene/kbp)	0.86	0.86	0.84	0.83	0.87	0.66	0.23	0.6
Mean gene length	1041 bp	1080 bp	1061 bp	1041 bp	1002	N.A.	1056	N.A.
Mean intergenic length	120 bp	124 bp	130 bp	166 bp	127	N.A.	429	N.A.
Presence overlap. genes	yes	Yes	yes	yes	Yes	N.A.	no	yes
SSU-LSU rRNA genes	22	22	22	22	N.A.	N.A.	>2	N.A.
tRNAs	46	46	46	46	46	52	37	N.A.
tRNA synthetases	21	21	22	21	21	N.A.	21.	20
Splic. introns	36	36	36	36	0	0	> 6	6
Predicted ORFs	1848	1848	1835	2010	3,804	2660	2174	2614

The characteristics of the genomes listed here are based the respective published data. E.i., *Encephalitozoon intestinalis* (Corradi, Pombert, et al., 2010); E.h, *Encephalitozoon hellem* (Pombert, Selman, et al., 2012); E.r., *Encephalitozoon romaleae* (Pombert, Selman, et al. 2012); E.c., *Encephalitozoon cuniculi* (Katinka, Duprat, et al., 2001); E.b., *Enterocytozoon bieneusi* (Akiyoshi, Morrison, et al., 2009); N. p., *Nematocida parisii* (Aurrecoechea, Barreto, et al., 2011); H.t., *Hamiltosporidium tvaerminnensis* (Corradi, Haag, et al., 2009); N.c.; *Nosema ceranae* (Cornman, Chen, et al., 2009).

Comparative Genomics of Microsporidia

The obligate intracellular lifestyle of microsporidia has left an impact on most aspects of their biology, including at the molecular level. Microsporidian genomes have attracted a great deal of attention since they were first investigated because the reduction in their content and size is unprecedented (Katinka, Duprat, et al. 2001; Fig. 12.2). Indeed, all microsporidian genomes published to date have been found to encode a tiny set of proteins, 2,500 at most, which highlights their ability to steal energy and most essential metabolites from their hosts (Peyretaillade, El Alaoui, et al., 2010; Corradi & Slamovits, 2011).

This reduction in gene content is generally mirrored by a massive reduction in the overall size of the genome itself. Some microsporidia harbor the smallest nuclear genomes ever characterized in any eukaryote: at the low end of the range of microsporidian genome sizes, the genome of *E. intestinalis* is only 2.3 Mb (Corradi, Pombert, et al. 2010), which is considerably smaller than the genomes of many free-living prokaryotes. Sibling species in the genus *Encephalitozoon* are only slightly larger (the largest being 2.9 Mb) and are all remarkably similar in content and structure; they all harbor about 2,000 genes that encode for an extremely reduced set of cellular pathways and are strikingly conserved in order along their homologous chromosomes.

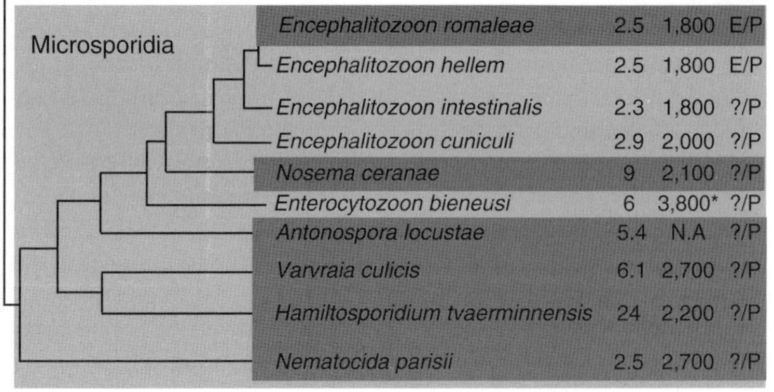

Figure 12.2 Microsporidian species with sequenced genomes. Schematic representation based on (Capella-Gutierrez, Marcet-Houben, et al., 2012) of the microsporidian phylogeny based on species with available genome data. Species highlighted in dark grey infect insects, whereas all other are notorious pathogens of vertebrates; including humans. The genome size (G), number of open reading frames (N), and known events of lateral gene transfers from eukaryotic (E), or prokaryotic donors (P) are shown. HGT, horizontal gene transfer.

Genome Reduction

Genome reduction in microsporidia involved a number of different changes, probably beginning with a massive loss of genes early in their evolution as they adapted to a new, intracellular parasitic mode of life (Williams, Lee, et al. 2008; Corradi, Haag, et al. 2009). As the intracellular parasites became more and more capable of stealing energy and nutrients from their hosts, independent biochemical functions became unnecessary and genes encoding those pathways were lost. Accordingly, reduction in gene content resulted in the simplification of regulation networks and in further gene loss. Loss of sequence seems to have equally affected coding and noncoding regions, however, so introns and transposons are both rare or absent from the most reduced members of the group, but the reduction of or complete loss of otherwise highly conserved biochemical pathways remains the most striking cause of their host dependence. Examples of reduced pathways include those encoding for the tricarboxylic acids cycle (TCA), de novo biosynthesis of amino acids and nucleotides, and the oxidation of fatty acids, all of which are absent or incomplete (Katinka, Duprat, et al. 2001; Akiyoshi, Morrison, et al. 2009; Cornman, Chen, et al. 2009; Corradi, Haag, et al. 2009; Corradi, Pombert, et al. 2010; Cuomo, Desjardins, et al. 2012; Pombert, Selman, et al. 2012). The link between biochemical shrinkage and a growing reliance on the organisms they infect to obtain much needed cellular supplies has not been shown directly because the range of tools to study the microsporidia during their intracellular stage are woefully limited. However, there is indirect evidence in the range of transporters encoded by their genomes and how this relates to genome size and metabolic complexity. Specifically, investigations along the large genome sequence of the *Daphnia magna* microsporidian parasite *H. tvaerminnensis* (formerly known as *Octosporea bayeri*) resulted in the identification of one single ATP transporter, whereas species with much smaller genomes harbor many paralogs of such genes (Corradi, Haag, et al. 2009).

At the opposite extreme, is the case of host dependence represented by the human pathogen *E. bieneusi*, whose genome has been found to lack genes related to several pathways that are otherwise universally conserved among eukaryotes, including other microsporidia. In addition to completely lacking introns and splicing machinery, the most outstanding of these missing pathways are glycolysis, trehalose metabolism, and the pentose phophate pathway (Akiyoshi, Morrison, et al. 2009; Keeling, Corradi, et al. 2010; Keeling & Corradi, 2011). These are significant because they represent the only known pathways in microsporidia that produce the ATP and reducing equivalents (NADH) necessary for the survival of the cell. The loss of these pathways in *E. bieneusi* suggests this species is incapable of producing ATP; the only such case known in eukaryotes (Akiyoshi, Morrison, et al. 2009; Keeling, Corradi, et al. 2010; Keeling & Corradi, 2011).

The Smallest Eukaryotic Genomes

The most reduced microsporidian genomes are found in species of the genus *Encephalitozoon*. These genomes contain a mere 2,000 open reading frames (ORFs), encoding proteins that are significantly shorter compared to orthologs from other organisms, and that are typically separated by shrunken intergenic regions that average less than 200 base pairs. Interestingly, even microsporidians with large genomes tend to encode for proteins that are shorter than their eukaryotic homologues, so the shortening of proteins in microsporidia does not appear to be correlated with the size of their genomes (Corradi, Haag, et al., 2009). Instead, the shortening is likely related to the reduced complexity of the proteome and the simpler regulatory networks that this requires (Zhang, 2000; Katinka, Duprat, et al., 2001; Metenier & Vivares, 2001).

The smallest, nonendosymbiotic eukaryotic genome ever sequenced from *E. intestinalis* further supports the notion that the gene repertoire of these hyper-adapted intracellular parasites has probably reached the lowest level required for their survival. Indeed, at 2.3 Mb, this genome has been found to harbor a gene set that is strikingly similar to that of its sibling species, *Encephalitozoon cuniculi*, despite a genome size that is 20 percent smaller. Genome inspections identified most losses as affecting ORFs without known function (i.e., hypothetical proteins) and possibly spurious. Similarly, the internal portion of the 11 homologous chromosomes of *E. cuniculi* and *E. intestinalis* were found to be almost identical in order between both species, with the most drastic losses of genome sequence occurring within the subtelomeric regions in *E. intestinalis* where biochemically important genes are lacking (Corradi, Pombert, et al., 2010).

The shortening of both coding and noncoding regions in the smallest microsporidian genomes have resulted in an extreme compression of the genome, and this elevated gene density has impacted important cellular and evolutionary mechanisms, namely the transcription of messenger RNAs (mRNAs) and the rate of gene rearrangement. In the vertebrate pathogen *E. cuniculi* and in the insect parasite *Antonospora locustae*, the most gene-dense regions have been repeatedly found to present increased levels of "overlapping transcription." This phenomenon represents an atypical transcriptional process that produces several mRNAs overlapping up to four adjacent genes, on different strands (Williams, Slamovits, et al., 2005; Corradi, Burri, et al., 2008a; Corradi, Gangaeva, et al., 2008a; Peyretaillade, Goncalves, et al. 2009; Gill, Lee, et al. 2010). This unique cellular phenomenon is, therefore, different from that found in eukaryotic operons (Cutter & Agrawal, 2010), and it has been proposed to have evolved following the displacement or removal of transcription initiators and terminators during the process of genome reduction (Williams, Slamovits, et al., 2005). Importantly, however, overlapping transcription does not seem to be universally present within the group because recent mRNA-seq analyses of one basal fungal lineage (i.e., *Nematocida* spp.) failed to identify overlapping mRNAs (Cuomo, Desjardins, et al., 2012).

In parallel, the close proximity of genes in the most dense microsporidian genomes seems to have also impacted the frequency at which the genome is rearranged. Indeed, high gene order conservation across different lineages is an important hallmark of microsporidian genomes, and this structural preservation of the genome is thought to have resulted from the elevated probability of genetic disruption following gene shuffling (Slamovits, Fast, et al., 2004; Corradi, Akiyoshi, et al., 2007). Interestingly, this conservation in gene order has been proposed to extend right back to the most recent common fungal ancestor of microsporidia (Lee, Corradi, et al., 2008; Lee, Corradi, et al., 2010), although this claim has been recently challenged (Koestler & Ebersberger, 2011). Nevertheless, the conservation within microsporidia is clear and is also of practical use because it may be a tool to identify orthologous genes by position, a key piece of evidence in the identification of two genes that play important roles in the interaction between microsporidia and their hosts: one gene encoding part of the polar tube infection apparatus (Polonais, Prensier, et al., 2005) and another gene involved in secretion (Slamovits, Burri, et al., 2006).

Microsporidia with Large Genomes

Reduction in genome size is so extreme among a subset of microsporidian lineages, that the more "normal-sized" genomes found in many other members of the group have been often overlooked. This is unfortunate because without additional data on the nature of these genomes, we are unable to conclude which type is the ancestral to the other, whether genome reduction has occurred more than once or not, or even whether there is a strong correlation between genome size, proteomic complexity, and host dependence. At the high end of the size spectrum, the genomes of the fish parasite *Glugea*, the mosquito parasite *Edhazardia aedis*, and the *Daphnia* pathogen *H. tvaerminnensis* have been estimated to be 12.5, 51, and 24 Mbps, respectively, or up to 20 times larger than the smallest microsporidian genome known (*E. intestinalis*) (Biderre, Pages, et al., 1994; Biderre, Mathis, et al., 1999; Didier, Stovall, et al., 2004; Gill, Becnel, et al., 2008; Williams, Lee, et al., 2008; Corradi, Haag, et al., 2009).

Large genomes are intuitively thought to harbor more genes compared to those that are smaller. However, recent studies on the coding capacity of some of the largest microsporidian genomes show this to be only partly true (Gill, Becnel, et al., 2008; Williams, Lee, et al., 2008; Corradi, Haag, et al., 2009). The most thoroughly sequenced of these is *H. tvaerminnensis*, and compared to *E. cuniculi* and *E. bieneusi* (whose genomes are 21 Mbps and 18 Mbp smaller than that of *H. tvaerminnensis*, respectively) this genome underscores the conclusion that variation in genome size is better explained by variation in the length of intergenic regions rather than gene content (Corradi, Haag, et al., 2009). Indeed, conservative annotation of the *H. tvaerminnensis* genome draft resulted in the identification of only 2,200 ORFs accounting for a range of cellular processes that were remarkably similar to those previously identified in *E. cuniculi* and *E. bieneusi*. The *H. tvaerminnensis* genome was found to harbor many putative genes with no known function (i.e., hypothetical proteins), and their presence is also likely to have contributed to increase the genome size of this species (Corradi, Haag, et al., 2009). The same overall tendencies were also found in the genome of *E. aedis*, which was estimated to be 51 Mbp by extrapolating the results of a genome sequence survey, which is an error-prone estimate but no physical evidence of genome size is available (Williams, Lee, et al., 2008). The results of this low-coverage survey together with the results of an expressed sequence tag survey are consistent with the conclusion that the great majority of genome size variation in the larger microsporidian genomes derives from variation in the size of intergenic regions, and not in the presence of large numbers of novel genes (Gill, Becnel, et al., 2008; Williams, Lee, et al., 2008).

Genome Content and Metabolic Independence

Extreme gene loss appears to have impacted all microsporidia, regardless of their genome size and to have resulted in a reduced biochemical repertoire that is fairly well conserved among the different microsporidian genomes sequenced to date. Altogether, these findings suggest that a massive loss of genes is likely to have happened before the diversification of the group. This, however, does not mean to say that all microsporidian genomes encode exactly the same range of biochemical and cellular pathways. Indeed, variation in metabolic versatility has been observed and sometimes with interesting implications.

In general, microsporidian species with larger genomes seem to have slightly more genes involved in additional metabolic pathways, which in turn suggests that there is some variation in host dependence within the group. This is best illustrated in the case of *H. tvaerminnensis*, in which additional genes with known functions include those encoding for proteins involved in fatty acid metabolism and glycolysis (Corradi, Haag, et al., 2009). These proteins are evidence of a more elaborated core metabolism, which is also consistent with a reduced number of transporters in this species. Acquiring deep-sequence coverage from many more species with larger genomes will be essential to confirm whether a strong correlation between the genome size and metabolic power in microsporidian species truly exists. Recently, representative sequences from many additional species with large genomes have been released (i.e., *Anncallia algerae*, *E. aedes*, *Vavraia culicis*, *Vittaforma corneae*) (Aurrecoechea, Barreto, et al., 2011). These new genomes are currently un-annotated, but there is reason to believe that they will soon provide sufficient data to reveal the true extent of metabolic diversity that is present across this parasitic lineage.

Horizontal Gene Transfers and Microsporidian Genomes

Obligate intracellular parasitism has been often linked with gene loss, especially in microsporidia. In contrast, gene gains are relatively rare. This makes sense in the context of reduction, but at the same time an intracellular lifestyle allows parasites to remain in close contact with the cellular content of other organisms, most notably those of their hosts or co-infecting bacteria, for long periods of time. This "genetic" proximity could potentially enable nonsexual genetic exchanges, called horizontal gene transfers (HGTs), between the microsporidia and coexisting organisms; allowing the parasites to pick-up genes from different sources and use them for their own benefit.

Some of the genes acquired by HGT have played a dramatic role in the ecology and evolution of microsporidia. Probably the most significant examples are the genes encoding the ATP transporters (or translocators) in microsporidia. ATP transporters are located at the interface between the cellular membranes of microsporidia and their hosts and are key elements in the parasites' system for scavenging energy (in the form of ATP) from their hosts. Surprisingly, microsporidia are the only group of eukaryotes that is currently known to harbor those genes; they are absent in all other known intracellular parasites, such as *Plasmodium* and are otherwise only found in intracellular prokaryotes. The narrow distribution of these transporters in eukaryotic organisms suggests that these are unlikely to have been acquired vertically from a common ancestor, so an evolutionary scenario involving their acquisition by means of HGT, possibly from coexisting prokaryotes, has been proposed (Richards, Hirt, et al., 2003). This hypothesis is supported by phylogenies reconstruction, which suggests the microsporidian ATP transproters are related to homologues in bacterial pathogens in the genus *Chamydia* (Richards, Hirt, et al., 2003). Interestingly, co-infection of mammalian cells by both *Chlamydia* and microsporidia has also been recently reported, so coexistence and exchange between both organisms in one host appears possible (Lee, Weiss, et al., 2009). These genes play a number of important roles in the intracellular stages of microsporidia. In addition to laying in the membrane separating the parasite from the parasitophorous vesicle and host cytoplasm (Tsaousis, Kunji, et al., 2008), some copies have also been shown to be located in the membrane of the mitosome, where they import ATP to provide the energy necessary to produce iron-sulfur clusters (Williams & Keeling, 2005; Tsaousis, Kunji, et al., 2008). These transporters are therefore essential for the survival and propagation of any microsporidian parasite and their origin by HGT was probably a key event in the origin of microsporidian parasitism and their interactions with the host cells.

Prokaryotes have also provided other genes to subsets of microsporidian lineages that are now used to protect the parasites against a number of environmental insults. These genes include important cell detoxifiers, such as *catalase* and *superoxide dismutase* (Fast, Law, et al., 2003; Corradi, Haag, et al., 2009; Xiang, Pan, et al., 2010). To date, these have only been found in the genera *Antonospora*, *Nosema*, and *Hamiltosporidium* and are absent from the genomes of more derived species in the genus *Encephalitozoon*. Both genes are functionally related because they both play an important role in the detoxification of reactive oxygen species. Phylogenies of both genes strongly indicate a prokaryotic origin, suggesting that their presence in microsporidian is a consequence of HGT. Furthermore, both genes are located within genomic regions that are rich in genes of obvious eukaryotic descent, suggesting their identification in the genome of microsporidia is not a result of contamination from bacterial sources. All three genomes also encode for a photolyase gene that also appears to have been acquired by HGT. The photolyase is a photon-driven protein that repairs ultraviolet-induced thymine dimers (Slamovits & Keeling, 2004; Corradi, Haag, et al., 2009). It is therefore thought that this gene provides an essential protection to the cell against DNA damage in the environmental spore stage. The distribution of the three aforementioned HGTs could either indicate a recent origin or may indicate an ancient origin with subsequent loss in some of the more reduced lineages, but in either case it suggests that some lineages of microsporidia may be better protected against environmental damage than others.

The Impact of Horizontal Gene Transfer on Metabolism

As more eukaryotic genomes are sequenced, one of the emerging features is that many HGTs have had obvious ecological implications for other microsporidian lineages (Keeling & Palmer, 2008). Interestingly, there is often a good case for movement of genes between the genomes of organisms that coexist in the same environment. Movement of genes between host and parasite is comparatively rare, but there are a few notable exceptions (Anderson & Seifert, 2011), one of which has been recently reported from two highly derived microsporidia (Selman, Pombert, et al., 2011). Indeed, the genomes of the sister species *Encephalitoozoon romaleae* and *Encephalitozoon hellem* were shown to encode one gene that is absent from any other sequenced genome from the genus. The gene encodes for a purine nucleotide phosphorylase, an enzyme that plays a key role in the salvage of purines, so its presence in the genome of *E. romaleae* and *E. hellem* is likely beneficial by creating the capacity to salvage purines from different precursors.

Inspections of both microsporidian PNP genes revealed that they were likely acquired by HGT, and possibly from an ancestral host, as both sequences were highly similar to orthologues from arthropods. The inclusion of these genes in the genomes of the two microsporidia was confirmed by polymerase chain reaction, and their animal origin strongly supported by a wide variety of phylogenetic analyses (Selman & Corradi, 2011; Selman, Pombert, et al., 2011).

The genome of *E. hellem* has also revealed the presence of three other genes (GTP cyclohydrolase I, GTPCH; folic acid synthase, FASP; dihydrofolate synthase, DHFS) that were also previously unknown in microsporidia. Their products reconstruct a pathway that is absent from other species of the group and which is involved in the de novo synthesis of folate. This cellular compound feeds into the one carbon core (C1) metabolism and can be synthesized in plants, fungi, and many protists. However, the hosts that are usually infected by *E. hellem* (vertebrates) cannot produce folate, so the ability to synthesize it de novo certainly represents an important advantage for the parasite. Other known *Encephalitozoon* species acquire folate from the host using folate transporters, but none of them is capable of producing it de novo. Interestingly this includes the sister-species of *E. hellem*, *E. romaleae*, whose genome contains pseudo-genized versions of those three genes (GTPCH, FASP, DHFS).

The increased metabolic capacity of *E. hellem* compared to any other microsporidian is remarkable, but more surprising is the origin of these genes. Phylogenetic analyses suggest many or all of these genes were acquired by HGT, but not as a functional unit; instead, the genes appear to have been acquired from several different lineages. Specifically, orthologues of GTPCH and FASP are only found in bacterial genomes, whereas the DHFS appears to be of either metazoan origin or fungal origin.

The discovery of these new genes in two highly derived species opens up the exciting perspective that microsporidian genomes could be much more malleable than previously anticipated. Certainly, these recent findings of HGTs in microsporidia warrant further genome inspections, especially across the natural populations of these critical parasites.

Potential Horizons: Molecular Ecology of Microsporidia

Sequencing the genomes of many microsporidia have helped us better comprehend how these organisms have adapted to living within other cells. Specifically, their genome sequences have revealed how their gene arsenal originated and instances in which their biochemical machinery has adapted to use their hosts more efficiently. In all known cases, microsporidian genomes have lost many genes and related biochemical pathways, and at the extreme, these losses have been paralleled with massive reduction in genome size. However, many microsporidia have also gained genes by means of HGT, and these acquisitions have also played an important role in the ecology and evolution of these parasites; either by helping them acquire metabolic compounds from their hosts or, in some cases, by improving their overall metabolism. Indeed, one could argue that the most significant change in the ecology of microsporidia was the acquisition of their ATP transporters via HGT because importing energy was potentially the key step in their evolving ability to survive within other cells.

Despite this growing wealth of sequence data, however, most fundamental questions about the ecology and evolution of these pathogens remained unanswered. In particular, the evolutionary processes that occur at the level of populations in the field are still virtually unknown, and so are the selective forces that have shaped the content and structure of their hyper-adapted genomes. In addition, the diversity of microsporidia has been scarcely investigated, leaving us not only in the dark about much of the ecology and evolution of the group, but also potentially unprepared to recognize newly emerging opportunistic pathogens. These open questions could be addressed by sequencing and comparing a number of genomes from different strains of many species, a process that is currently underway. At the same time, however, the level of natural variation found across populations of these parasites in specific samples collected during field work would address different but equally important questions.

The lack of studies regarding the population genomics of microsporidian parasites is not surprising in light of the difficulty sometimes associated with working on this group, but in other ways is somewhat surprising because they would make an interesting model for population genomic questions. Indeed, the compact nature of microsporidian genomes, combined with the availability of several cultured strains and reference genomes, offers a unique opportunity to study the adaptive processes that occur across the genomes of eukaryotic parasites. In particular, their miniaturized genomes allow explorations of genome diversity that are otherwise difficult in other fields of parasitology where species of interest have genomes that are predominantly large (e.g., in *Plasmodium*, the agent of malaria or *Trypanosoma*, the agent of the sleeping sickness).

Furthermore, reference genomes are currently available from microsporidian species with many different genome sizes and that infect independent hosts, including representatives from many human infecting species (i.e., *Encephalitozoon* spp.). These reference sequences represent a tremendous asset to study the population genomics of parasites in a broad sense because sequencing reads generated from different strains of one species using next-generation technologies can be rapidly aligned against the reference and searched for polymorphisms. This type of investigation typically leads to the identification of single nucleotide polymorphism (SNPs), small-scale deletions and insertions (indels), chromosomal inversions, and ideally, gene gains and losses, which can be readily linked with adaption to different hosts, response to their hosts' immune systems, strain biogeography, or neutral evolutionary processes.

Finally, even though new microsporidian species are described on a regular basis, studies targeted at describing the overall diversity of these organisms in the field are still lacking at many levels. To date, most studies of microsporidian diversity have focused on species of known medical or zoonotic interest (i.e., *Encephalitozoon* spp., *Enterocytozoon* spp.), and on those ecological areas that are in close proximity to human activities (i.e., sewage, recreational parks). As a consequence, most microsporidian diversity is at high risk of being left undetected, which is significant given the potential threat that such lineages represent for the future health of many humans worldwide.

Moreover, studies of microsporidian diversity have typically used targeted polymerase chain reaction and DNA sequencing on environmental samples to identify the species of interest, and for this reason, the results that are usually reported can only be restricted to a few lineages for which specific primers are readily available (Izquierdo, Castro Hermida, et al., 2011; Fournier, Liguory, et al., 2000; Slifko, Smith, et al., 2000; Coupe, Delabre, et al., 2006). Overall, these highly selective approaches could underestimate the real diversity of microsporidia, so finding alternative approaches to study their natural diversity seem necessary.

Future studies of microsporidia diversity should take stock on the most recent advances in DNA sequencing technologies (next-generation sequencing). Indeed, these are frequently used to study the biodiversity of eukaryotes across many terrestrial and aquatic ecosystems, and in many cases, have resulted in the discovery of many cryptic lineages that were previously unknown to exist (Lara, Moreira, et al., 2010; Jones, Forn, et al., 2011; Kim, Harrison, et al., 2011). Similar approaches could therefore result in the detection of several unknown microsporidian lineages that have long been left unidentified.

Intriguingly, environmental sequences related to microsporidia are rare among environmental eukaryotic DNA sequences present in public databases, a feature that may be related to current procedures used for environmental DNA extractions. On one hand, microsporidia are present in the environment in the form of highly-resistant spores that are hard to crush (and thus release DNA), and this may contribute to the current lack of environmental sequence data from these parasites. On the other hand, microsporidian small subunit ribosomal genes (SSU rRNA) are also considerably shorter than the canonical eukaryotic SSU rRNA (i.e., as small as 1200 bp rather than 1800 bp). Consequently, clone-based methodologies are likely to overlook such sequences and may result in a near-total systematic exclusion of microsporidia from surveys even from environments where they are common. For all these reasons, future studies focusing on the natural environmental diversity of these parasites should take particular attention on the protocols used to acquire sequences.

Conclusions

Research on microsporidian parasites has long centered on identification and diagnosis, taxonomy, and at the cellular level their atypical and seemingly primitive features, but the last decade has also seen special scientific interest arising to study their content and structure of their genomes. These studies have revealed the reduced nature of many microsporidian biochemical pathways and have shown how these organisms have evolved to offset these reductive processes. However, it must be remembered that current knowledge about their genomes is still based on few lineages, most of which are characterized by particularly small genomes, and future work could potentially reveal microspordian genomes with contents that are far greater than are currently anticipated. Certainly, recent efforts to sequence the genomes of many new species with larger genomes represent a great step forward to understand the genome complexity of these parasites (Aurrecoechea, Barreto, et al., 2011), and these should provide essential insights into their origin, ecology and evolution.

References

Akiyoshi DE, Morrison HG, et al. 2009. Genomic survey of the non-cultivatable opportunistic human pathogen, *Enterocytozoon bieneusi*. *PLoS Pathog*. 5(1): e1000261.

Anderson D & East IJ. 2008. The latest buzz about colony collapse disorder. *Science*. 319(5864): 724–725; author reply 724–725.

Anderson MT & Seifert HS. 2011. Opportunity and means: horizontal gene transfer from the human host to a bacterial pathogen. *mBio*. 2(1): e00005–00011.

Aurrecoechea C, Barreto A, et al. 2011. AmoebaDB and MicrosporidiaDB: Functional genomic resources for Amoebozoa and Microsporidia species. *Nucl Acids Res*. 39(Database issue): D612–D619.

Biderre C, Mathis A, et al. 1999. Molecular karyotype diversity in the microsporidian *Encephalitozoon cuniculi*. *Parasitol*. 118 (Pt 5): 439–445.

Biderre C, Pages M, et al. 1994. On small genomes in eukaryotic organisms: molecular karyotypes of two microsporidian species (Protozoa) parasites of vertebrates. *C R Acad Sci III* 317(5): 399–404.

Capella-Gutierrez S, Marcet-Houben M, et al. 2012. Phylogenomics supports microsporidia as the earliest diverging clade of sequenced fungi. *BMC Biol*. 10(1): 47.

Cavalier-Smith T. 1983. A 6-kingdom classification and a unified phylogeny. In: *Endocytobiology. II. Intracellular Space as Oligogenetic* (eds. HEA Shenck & WS Schwemmler), 1027–1034. Berlin: Walter de Gruyter.

Cavalier-Smith T. 1987. Eukaryotes with no mitochondria. *Nature*. 326(6111): 332–333.

Cornman RS, Chen YP, et al. 2009. Genomic analyses of the microsporidian *Nosema ceranae*, an emergent pathogen of honey bees. *PLoS Pathog*. 5(6): e1000466.

Corradi N, Akiyoshi DE, et al. 2007. Patterns of genome evolution among the microsporidian parasites Encephalitozoon cuniculi, *Antonospora locustae* and *Enterocytozoon bieneusi*. *PLoS ONE*. 2(12): e1277.

Corradi N, Burri L, et al. 2008a. mRNA processing in *Antonospora locustae* spores. *Mol Genet Genom*. 280(6): 565–574.

Corradi N, Gangaeva A, et al. 2008b. Comparative profiling of overlapping transcription in the compacted genomes of microsporidia *Antonospora locustae* and *Encephalitozoon cuniculi*. *Genomics*. 91(4): 388–393.

Corradi N, Haag KL, et al. 2009. Draft genome sequence of the *Daphnia* pathogen *Octosporea bayeri*: Insights into the gene content of a large microsporidian genome and a model for host-parasite interactions. *Genome Biol*. 10(10): R106.

Corradi N, Pombert JF, et al. 2010. The complete sequence of the smallest known nuclear genome from the microsporidian *Encephalitozoon intestinalis*. *Nat Commun*. 1: 77.

Corradi N & Slamovits CH. 2011. The intriguing nature of microsporidian genomes. *Brief Funct Genom*. 10(3): 115–124.

Coupe S, Delabre K, et al. 2006. Detection of *Cryptosporidium*, *Giardia* and *Enterocytozoon bieneusi* in surface water, including recreational areas: a one-year prospective study. *FEMS Immunol Med Microbiol*. 47(3): 351–359.

Cox-Foster DL, Conlan S, et al. 2007. A metagenomic survey of microbes in honey bee colony collapse disorder. *Science*. 318(5848): 283–287.

Cuomo CA, Desjardins CA, et al. 2012. Microsporidian genome analysis reveals evolutionary strategies for obligate intracellular growth. *Genome Res*. 22(12): 2478–2488.

Curgy JJ, Vávra J, et al. 1980. Presence of ribosomal RNAs with prokaryotic properties in Microsporidia, eukaryotic organisms. *Biol Cell*. 38: 49–51.

Cutter AD & Agrawal AF. 2010. The evolutionary dynamics of operon distributions in eukaryote genomes. *Genetics*. 185(2): 685–693.

Didier ES. 2005. Microsporidiosis: an emerging and opportunistic infection in humans and animals. *Acta Trop* 94(1): 61–76.

Didier ES & Bertucci DC. 1996. Identification of *Encephalitozoon intestinalis* proteins that induce proliferation of sensitized murine spleen cells. *J Eukaryot Microbiol*. 43(5): 92S.

Didier ES, Bertucci DC, et al. 2001. *Encephalitozoon cuniculi* infection in mice with the chronic granulomatous disease (CGD) disorder. *J Eukaryot Microbiol Suppl*. 79S-80S.

Didier ES, Stovall ME, et al. 2004. Epidemiology of microsporidiosis: sources and modes of transmission. *Vet Parasitol*. 126(1–2): 145–166.

Fast NM, Law JS, et al. 2003. Bacterial catalase in the microsporidian *Nosema locustae*: Implications for microsporidian metabolism and genome evolution. *Eukaryot Cell*. 2(5): 1069–1075.

Fournier S, Liguory O, et al. 2000. Detection of microsporidia in surface water: A one-year follow-up study. *FEMS Immunol Med Microbiol*. 29(2): 95–100.

Gill EE, Becnel JJ, et al. 2008. ESTs from the microsporidian Edhazardia aedis. *BMC Genomics*. 9: 296.

Gill EE, Lee RC, et al. 2010. Splicing and transcription differ between spore and intracellular life stages in the parasitic microsporidia. *Mol Biol Evol*. 27(7): 1579–1584.

Goldberg AV, Molik S, et al. 2008. Localization and functionality of microsporidian iron-sulphur cluster assembly proteins. *Nature*. 452(7187): 624–628.

Higes M, Martin R, et al. 2006. *Nosema ceranae*, a new microsporidian parasite in honeybees in Europe. *J Invertebr Pathol*. 92(2): 93–95.

Hirt RP, Logsdon JM, Jr, et al. 1999. Microsporidia are related to Fungi: Evidence from the largest subunit of RNA polymerase II and other proteins. *Proc Natl Acad Sci USA*. 96(2): 580–585.

Ishihara R & Hayashi YJ. 1968. Some properties of ribosomes from the sporoplasm of *Nosema bombycis*. *J Invertebr Pathol*. 11: 377–385.

Izquierdo F, Castro Hermida JA, et al. 2011. Detection of microsporidia in drinking water, wastewater and recreational rivers. *Water Res*. 45(16): 4837–4843.

James TY, Kauff F, et al. 2006. Reconstructing the early evolution of Fungi using a six-gene phylogeny. *Nature*. 443(7113): 818–822.

Jones MD, Forn I, et al. 2011. Discovery of novel intermediate forms redefines the fungal tree of life. *Nature*. 474(7350): 200–203.

Kamaishi T, Hashimoto T, et al. 1996a. Complete nucleotide sequences of the genes encoding translation elongation factors 1 alpha and 2 from a microsporidian parasite, *Glugea plecoglossi*: implications for the deepest branching of eukaryotes. *J Biochem*. 120(6): 1095–1103.

Kamaishi T, Hashimoto T, et al. 1996b. Protein phylogeny of translation elongation factor EF-1 alpha suggests microsporidians are extremely ancient eukaryotes. *J Mol Evol*. 42(2): 257–263.

Katinka MD, Duprat S, et al. 2001. Genome sequence and gene compaction of the eukaryote parasite *Encephalitozoon cuniculi*. *Nature*. 414(6862): 450–453.

Keeling PJ & Corradi N. 2011. Shrink it or lose it: Balancing loss of function with shrinking genomes in the microsporidia. *Virulence*. 2(1): 67–70.

Keeling PJ, Corradi N, et al. 2010. The reduced genome of the parasitic microsporidian *Enterocytozoon bieneusi* lacks genes for core carbon metabolism. *Genome Biol Evol*. 2: 304–309.

Keeling PJ & Doolittle WF. 1996. Alpha-tubulin from early-diverging eukaryotic lineages and the evolution of the tubulin family. *Mol Biol Evol*. 13(10): 1297–1305.

Keeling PH, Luker MA, et al. 2000. Evidence from beta-tubulin phylogeny that microsporidia evolved from within the fungi. *Mol Biol Evol*. 17(1): 23–31.

Keeling PJ & Palmer JD. 2008. Horizontal gene transfer in eukaryotic evolution. *Nat Rev Genet*. 9(8): 605–618.

Kim E, Harrison JW, et al. 2011. Newly identified and diverse plastid-bearing branch on the eukaryotic tree of life. *Proc Natl Acad Sci USA*. 108(4): 1496–1500.

Koestler T & Ebersberger I. 2011. Zygomycetes, microsporidia, and the evolutionary ancestry of sex determination. *Genome Biol Evol*. 3: 186–194.

Kudo RR. 1947. Protozoology, 3th ed. Springfield, Illinois.

Kudo RR & Daniels EW. 1963. An electron microscope study of the spore of a Microsporidian, Thelohania californica. *J Protozool*. 10: 112–120.

Lara E, Moreira D, et al. 2010. The environmental clade LKM11 and Rozella form the deepest branching clade of fungi. *Protist*. 161(1): 116–121.

Larsson JIR. 1999. Identification of microsporidia. *Acta Protozool*. 38(3): 161–197.

Lee SC, Corradi N, et al. 2008. Microsporidia evolved from ancestral sexual fungi. *Curr Biol*. 18(21): 1675–1679.

Lee SC, Corradi N, et al. 2010. Evolution of the sex-related locus and genomic features shared in microsporidia and fungi. *PLoS One*. 5(5): e10539.

Lee, SC, Weiss LM, et al. 2009. Generation of genetic diversity in microsporidia via sexual reproduction and horizontal gene transfer. *Commun Integr Biol*. 2(5): 414–417.

Metenier G. & Vivares CP. 2001. Molecular characteristics and physiology of microsporidia. *Microbes Infect*. 3(5): 407–415.

Peyretaillade E, Broussolle V, et al. 1998. Microsporidia, amitochondrial protists, possess a 70-kDa heat shock protein gene of mitochondrial evolutionary origin. *Mol Biol Evol*. 15(6): 683–689.

Peyretaillade E, El Alaoui H, et al. 2010. Extreme reduction and compaction of microsporidian genomes. *Res Microbiol*. 162(6): 598–606.

Peyretaillade E, Goncalves O, et al. 2009. Identification of transcriptional signals in *Encephalitozoon cuniculi* widespread among Microsporidia phylum: Support for accurate structural genome annotation. *BMC Genomics*. 10: 607.

Polonais V, Prensier G, et al. 2005. Microsporidian polar tube proteins: Highly divergent but closely linked genes encode PTP1 and PTP2 in members of the evolutionarily distant *Antonospora* and *Encephalitozoon* groups. *Fungal Genet Biol.* 42(9): 791–803.

Pombert JF, Selman M, et al. 2012. Gain and loss of multiple functionally related, horizontally transferred genes in the reduced genomes of two microsporidian parasites. *Proc Natl Acad Sci USA.* 109(31): 12638–12643.

Richards TA, Hirt RP, et al. 2003. Horizontal gene transfer and the evolution of parasitic protozoa. *Protist.* 154(1): 17–32.

Selman M & Corradi N. 2011. Microsporidia: Horizontal gene transfers in vicious parasites. *Mob Genet Elements.* 1(4): 251–255.

Selman M, Pombert JF, et al. 2011. Acquisition of an animal gene by microsporidian intracellular parasites. *Curr Biol.* 21(15): R576–577.

Slamovits CH, Burri L, et al. 2006. Characterization of a divergent Sec61beta gene in microsporidia. *J Mol Biol.* 359(5): 1196–1202.

Slamovits CH, Fast NM, et al. 2004. Genome compaction and stability in microsporidian intracellular parasites. *Curr Biol.* 14(10): 891–896.

Slamovits CH & Keeling PJ. 2004. Class II photolyase in a microsporidian intracellular parasite. *J Mol Biol.* 341(3): 713–721.

Slifko TR, Smith HV, et al. 2000. Emerging parasite zoonoses associated with water and food. *Int J Parasitol.* 30(12–13): 1379–1393.

Tsaousis AD, Kunji ER, et al. 2008. A novel route for ATP acquisition by the remnant mitochondria of *Encephalitozoon cuniculi. Nature.* 453(7194): 553–556.

Van de Peer Y, Ben Ali A, et al. 2000. Microsporidia: Accumulating molecular evidence that a group of amitochondriate and suspectedly primitive eukaryotes are just curious fungi. *Gene.* 246(1–2): 1–8.

Vavra J. 1965. [Study by electron microscope of the morphology and development of some Microsporidia]. *C R Acad Sci Hebd Seances Acad Sci D.* 261(17): 3467–3470.

Vávra J & Larsson JIR. 1999. Structure of the microsporidia. In: The Microsporidia and Microsporidiosis (eds. M Wittner & LM Weiss), 7–84. Washington, DC: ASM Press.

Vizoso DB, Lass S, et al. 2005. Different mechanisms of transmission of the microsporidium *Octosporea bayeri*: A cocktail of solutions for the problem of parasite permanence. *Parasitology.* 130(Pt 5): 501–509.

Vossbrinck CR, Maddox JV, et al. 1987. Ribosomal RNA sequence suggests microsporidia are extremely ancient eukaryotes. *Nature.* 326(6111): 411–414.

Weiss LM, 2001. Microsporidia: Emerging pathogenic protists. *Acta Trop.* 78(2): 89–102.

Weiss LM & Vossbrinck CR. 1998. Microsporidiosis: Molecular and diagnostic aspects. *Adv Parasitol.* 40: 351–395.

Williams BA, Hirt RP, et al. 2002. A mitochondrial remnant in the microsporidian Trachipleistophora hominis. *Nature.* 418(6900): 865–869.

Williams BA & Keeling PJ. 2005. Microsporidian mitochondrial proteins: Expression in *Antonospora locustae* spores and identification of genes coding for two further proteins. *J Eukaryot Microbiol.* 52(3): 271–276.

Williams BA, Lee RC, et al. 2008. Genome sequence surveys of *Brachiola algerae* and *Edhazardia aedis* reveal microsporidia with low gene densities. *BMC Genomics.* 9: 200.

Williams BA, Slamovits CH, et al. 2005. A high frequency of overlapping gene expression in compacted eukaryotic genomes. *Proc Natl Acad Sci USA.* 102(31): 10936–10941.

Xiang H, Pan G, et al. 2010. A tandem duplication of manganese superoxide dismutase in *Nosema bombycis* and its evolutionary origins. *J Mol Evol.* 71(5–6): 401–414.

Zhang J. 2000. Protein-length distributions for the three domains of life. *Trends Genet.* 16(3): 107–109.

Section 5
Metagenomics and Biogeography of Fungi

13 Metagenomics for Study of Fungal Ecology

Björn D. Lindahl[1] and Cheryl R. Kuske[2]

[1] *Swedish University of Agricultural Sciences, Department of Forest Mycology and Plant Pathology, Uppsala, Sweden*
[2] *Environmental Microbiology Team, Bioscience Division, Los Alamos National Laboratory, Los Alamos, New Mexico*

From Single Genomes to Fungal Communities

Exploration of the gene content and regulation of fungal genomes representing the wide diversity of fungi is providing information on fungal metabolic capabilities at an unprecedented level of resolution. In addition to enabling phylogenetic, metabolic, and physiological comparisons of particular fungi, the increasingly cost-effective high-throughput DNA sequencing approaches enable investigation of fungal communities in their natural environments. Investigation of pools of nucleic acids that represent complex biotic communities, including members from all three domains of life—Bacteria, Archaea, and a wide variety of Eukarya, including the fungi—has been termed *metagenomics*. Metagenomics is defined herein as sequence-based approaches applied *across* genomes in an environment, essentially providing a comparative assessment of a community in situ. This type of investigation greatly expands the breadth of genomic studies by encompassing thousands or more organisms in a survey, but with a consequent reduction in the depth of genomic information typically obtained in a single genome sequencing effort.

Although single genomes and transcriptomes provide information on potential metabolic and functional contributions of a single organism, the goal of metagenomic assessment is to provide information about the functional capabilities and responses of organism assemblages in situ. This is important because an organism's performance in axenic culture does not encapsulate its role(s) in an environmental context in which competition or other interactions, resource utilization, and environmental heterogeneity become critical factors. Furthermore, to obtain information on identities and functional properties of organisms that are not easily isolated and cultivated, which are often major components of natural fungal communities, metagenomic approaches are often the only option.

The Ecological Genomics of Fungi, First Edition. Edited by Francis Martin.

Two metagenomic approaches have been applied to the study of fungal communities. *Targeted metagenomics*, in which sequencing of a single genetic marker, such as a gene with known function, provides a comprehensive survey for that marker across the community. *Shotgun metagenomics*, in which total nucleic acids from an environmental sample are extracted and randomly sequenced without regard for phylogeny or function, provides a snapshot of the total community (Tringe, Von Mering, et al., 2005). Recent applications of both approaches to the study of fungi as members of complex communities are discussed in this chapter, along with their benefits and limitations.

Importance of Fungi in Ecosystems

Fungi represent a major source of global biodiversity and are an important source of highly valued pharmaceuticals, foods, and industrial enzymes. In terrestrial environments, fungi comprise a diverse, abundant biomass that have major impacts on ecosystem (natural or agricultural) performance (e.g., via mutualistic or pathogenic interactions with plant hosts, via organic matter degradation, soil stabilization, mineral weathering, and simply by contributing their biomass to the organic matter pool). By facilitating plant nutrient uptake or causing disease, fungi are important regulators of plant performance. In many terrestrial environments, fungi are the principal degraders of dead plant tissues (litter) and soil organic matter and are as such pivotal for regulation of terrestrial and global biogeochemical cycles. Yet the establishment of reliable models of carbon and nutrient cycling in terrestrial ecosystems has been hampered by the inability to develop a holistic understanding of fungal community structure and functioning (Lindahl, Taylor, et al., 2002).

The structure and composition of fungal communities depend largely on properties of the environment. Plant interactions, abiotic factors (e.g., water availability and temperature), chemical environment (e.g., pH), and the availability and quality of organic substrates collectively determine which fungi can inhabit an environment and their relative competitive success. By using genetic markers in combination with high-throughput sequencing techniques, fungal communities may now be analyzed in depth and in their environmental context. This enables identification of factors that maintain or threaten fungal diversity as well as exploration of relationships between community composition and abiotic and biotic interactions in terrestrial ecosystems, which will lead to a better understanding of fungal roles in terrestrial ecosystems. We are now poised to determine how different functional groups of fungi interact in complex communities to influence the productivity of agricultural and forest crops, knowledge that may be used to guide development of sustainable management policies (Johansson, Paul, et al., 2004).

The ability to conduct metagenomic studies focused on fungi in the environment provides opportunities to explore major ecological questions with higher depth of coverage and precision. Broad topics where improved understanding of fungal community structure in soils is critical include species and functional diversity, redundancy, and evolution of new strains; the influence of soil physical and geochemical factors across many spatial scales, including variability across centimeter distances to regional characteristics; and the influence of plants on fungal community structure and plant-fungal metabolic interactions that span from beneficial to pathogenic outcomes. Addressing questions of fungal community responses and resiliency to environmental perturbations is also within reach by including molecular and metagenomic approaches to assess community structure.

Genetic Markers in Fungal Ecology

To explore the identity, dynamics, and collective processes of fungi in the environment, one must be able to reliably document the community structure (diversity, richness, evenness, composition) and to compare that structure over a series of experimental or survey criteria to identify patterns of change or response. The fact that protocols to amplify fungal genetic markers (White, Bruns, et al., 1990) were developed only a few years after the advent of polymerase chain reaction (PCR) testing (Mullis & Fallona, 1987) reflects the urgency by which fungal biologists searched for new approaches to establish phylogenetic relationships and facilitate identification of unknown fungi from soil and other environments. Morphological characteristics have largely proven unreliable as phylogenetic markers, and the increasing dependence on genetic markers has enabled a complete revision of fungal taxonomy (James, Kauff, et al., 2006a; James, Letcher PM, et al., 2006b). New major branches, recently added to the fungal kingdom (Schadt, Martin, et al., 2003; Rosling, Cox, et al., 2011), were discovered and characterized using genetic markers, and discovery of new fungal species is now primarily derived by screening environmental samples for novel genetic signatures (Hibbett, Ohman, et al., 2009). Thus, genetic markers have rapidly become the principal tools to analyze the structure and diversity of fungal communities, as well as to identify community members. Molecular biology is now a central component of almost all fungal ecology and phylotaxonomy. Mycologists are thus making a concerted effort to deconstruct the barrier between field ecology and genetics. Having suffered the concealed nature of fungal mycelia and the scarcity of characters for morphological recognition, clever employment of genetic markers in field surveys and manipulative experiments now enable fungal ecologists to access the full biodiversity and ecological functioning of fungal communities.

Ambiguities associated with classical methods for studying fungal communities have driven the need for more accurate molecular tools in fungal taxonomy and ecology. For examples, from morphological examination of ectomycorrhizal root tips it became clear that inventories of above ground sporocarps provided a highly distorted reflection of the belowground community (Dahlberg, Jonsson, et al., 1997), and although Basidiomycetes were frequently observed in the microscope as clamped hyphae colonizing plant litter, they rarely turned up among cultured strains obtained from the same litter (Frankland, 1998). Plant diseases caused by fungi, monitored using extent of disease symptoms or fungal spore counts, made early preventive and control measures difficult to implement. The need for early diagnosis and control measures has driven the use of molecular tools in plant pathology to improve accuracy and enable earlier detection of fungal diseases of plants.

From Restriction Patterns to Next-Generation Sequencing

Early molecular-based surveys were limited to generation of DNA restriction patterns (i.e., fingerprints), in cases in which mono-specific DNA could be extracted from discrete substrates, such as ectomycorrhizal root tips. Restriction fragment length polymorphism (RFLP) profiling of PCR amplified markers enabled putative identification of fungi that colonized root tips (e.g., Gardes, Fortin, et al., 1990), and sequencing of PCR products provided increased resolution (reviewed by Horton & Bruns, 2001). Early applications of PCR in detection of plant pathogens were reviewed by Henson and French (1993). The first molecular glimpse into complex fungal communities was offered by denaturing gradient gel electrophoresis (DGGE) analysis, which provides a community fingerprint as bands on electrophoresis gels, in which individual bands may be excised, reamplified, and sequenced (e.g., Kowalchuk, Gerards, et al., 1997). Community fingerprinting may also be conducted by terminal fragment length polymorphism (T-RFLP) analysis, which is based on capillary electrophoresis of fluorescently labeled PCR products after cutting with restriction enzymes (e.g., Buchan, Newell, et al., 2002; Dickie, Xu, et al., 2002). Fungal community analysis based on metagenomic sequencing was first conducted by cloning of single-molecule PCR products into bacteria and subsequent reamplification of cloned fragments, followed by Sanger sequencing (e.g., Smit, Leeflang, et al., 1999; Chen & Cairney, 2002; O'Brien, Parrent, et al., 2005). These initial *targeted* sequencing efforts provided high-resolution, high-quality, long sequences but at high cost relative to today's standards and were necessarily limited in scope. However, by combining the detailed community information offered by cloning and sequencing with the larger sample numbers enabled by T-RFLP, large-scale field studies of fungal communities could be carried out (e.g., Lindahl, Ihrmark, et al., 2007).

The cloning-based approach to community analysis has recently been replaced by an explosion of high-throughput sequencing techniques, such as 454-pyrosequencing (Margulies, Egholm, et al., 2005), Illumina, PacBio, and Ion Torrent, all with continued increased sequence output, but with widely varying read lengths and quality (Shokralla, Spall, et al., 2012; Scholz, Lo, et al., 2012). The newer sequence-based approaches generate large sequence data sets that overcome early issues with small sample numbers and small clone library representation. The increased sequence output allows simultaneous analysis of large numbers of samples, where sequences are assigned to the samples from which they originated by specific sequence tags, which are usually added as an extension of the primers (Binladen, Gilbert, et al., 2007). The increased sequencing depth enables detection and identification of rare community members, aiming for estimation of the true species richness of fungal communities (e.g., Buée, Reich, et al., 2009; Jumpponen, Jones, et al., 2009). With thousands of sequences from each sample, even moderately abundant community members are represented by many sequences, providing quantitative information on relative abundances in the amplicon pool (Ihrmark, Bödeker, et al., 2012).

Ribosome-Encoding Genes as Taxonomic Markers

Although amplification-free shotgun sequencing of metagenomic DNA is likely to partly or fully replace sequencing of PCR-amplified markers in the near future, current community analysis depends on primers that are able to amplify markers from a wide range of different fungal groups. Because of the degeneracy in the genetic code, particularly in the third position of codon triplets, genes coding for proteins are rarely fully conserved within or among phylogenetic groups. In contrast, the genes coding for the ribosomes are highly conserved, and it is possible to identify primer binding sites shared by a wide range of eukaryotes. Therefore, development of markers to be amplified by universal primers has focused on the ribosome-encoding genes (e.g., White, Bruns, et al., 1990; Gardes & Bruns, 1993; van Tuinen, Jacquot, et al., 1998; Ihrmark, Bödeker, et al., 2012).

In bacterial ecology, the small subunit ribosomal RNA gene (SSU) is regularly used as a universal phylogenetic marker, and it may seem tempting to use the same approach for fungi. However, the evolutionary history of many fungal groups is short compared to bacteria, and the fungal SSU contains too little variation to provide adequate resolution in the fungi (Schoch, Seifert, et al., 2012). Despite this, the SSU has been used as a marker to characterize microbial communities in the environment. Kunin, Engelbrektson, et al. (2010) designed a SSU primer set that detects representatives from all three domains (bacteria, archaea, eukarya) for use in calibrating coverage in shotgun metagenomes. Although potentially useful as a shotgun metagenomics screening tool, use of the SSU gene for fungal community analysis can give misleading results that invites misinterpretations. The conservation of identical SSU sequences throughout large groups of related fungi leads to that the "best blast match" is often selected at random among a large number of equally good alternatives. For some fungal clades, the SSU does provide adequate and accurate phylogenetic resolution, and Öpik, Vanatoa, et al. (2010) have used the SSU gene successfully for the Glomeromycota.

The large subunit ribosomal RNA gene (LSU) contains discrete variable regions flanked by sequences conserved among all fungi (and some with broader conservation among nonfungal eukarya). This arrangement allows sequences to be aligned, facilitating establishment of a phylogenetic backbone for the fungi (AFTOL project, see James, Lecher, et al. 2006b; Arnold, Miadlikowska, et al. 2009; Aime, Ball, et al., 2011), and phylogenetic placement of newly described species. The LSU gene, especially the 5' region, has been used to provide phylogenetic context to ITS sequences (described herein). Liu, Porras-Alfaro, et al. (2012) recently curated a LSU database (approx. 8,500 sequences) and described the influence of different sequence lengths (mimicking 454-titanium and Illumina read lengths) and PCR priming sites on accuracy of taxonomic calls from the class to genus level. Sequence length and PCR priming site significantly affected accuracy. Anchored to the LR3 primer (http://www.biology.duke.edu/fungi/mycolab/primers.htm), sequence lengths of 150 bp or longer were more than 99 percent accurate at the order level, 90 percent accurate at the family level and 70 to 80 percent accurate at the genus level when a naíve Bayesian classifier was used. More conserved markers, such as the LSU, may be useful in environmental surveys, where a large proportion of obtained sequences do not match well with any representatives in databases, and alignment of sequences from distantly related taxa is required to infer phylogenetic placement.

The internal transcribed spacer regions of the rRNA operon (ITS1 and ITS2) have proven to be high resolution taxonomic markers for the fungi because they vary greatly in length and in sequence composition. Referred to as a "barcode," these sequences provide species and strain level identification (but can also be hypervariable within a given species). These noncoding regions separate the ribosome-encoding genes and are spliced off shortly after transcription. They are little constrained by conserving selection and therefore evolve rapidly. Because of high sequence variation, amplification of the ITS regions from fungal communities depends on primer sites located in the adjoining, more conserved coding genes. The ITS region has gained increasing popularity in finer branch taxonomy, for identification of fungal species and for analysis of fungal communities. More than 10,000 species are represented by their ITS sequence at the NCBI database (Nilsson, Ryberg, et al. 2009), and other high-quality reference databases are also available (e.g., UNITE, see Abarenkov, Nilsson, et al., 2010). The ITS region was recently proposed as the universal genetic barcode for Basidiomycota and Ascomycota (Begerow, Nilsson, et al., 2010; Schoch, Seifert, et al., 2012). However, for some other groups of fungi, for example, Glomeromycota, the ITS regions are too variable to provide phylogenetic information, and the LSU and SSU are used more frequently (Öpik, Vanatoa, et al., 2010, Stockinger, Kruger, et al., 2010). In addition to its utility as a phylogenetic marker, the pool of ITS RNA, that is, the parts that are spliced of after transcription, may be used to as a marker for the momentarily active community because of its relatively quick turnover rate (Rajala, Peltoniemi, et al., 2011).

Read lengths in 454-pyrosequencing now suffice to cover the entire ITS regions for many fungi, but there are nevertheless reasons to aim for short (<300 bp) amplicons, because long amplicons increase the risk that community structure is distorted during PCR, particularly if the length varies between species. By amplifying and sequencing an artificial community of ITS templates, Ihrmark, Bödeker, et al. (2012) found severe discrimination against species with long ITS regions. This problem was overcome by using priming sites in the 5.8S rRNA gene—the short, coding, and conserved region that is situated in between the ITS1 and ITS2 regions. When only the ITS2 region was amplified, there was improved concordance between abundance of 454-sequences and composition of the template community. This supports the use of ITS-PCR and high-throughput sequencing for quantitative assessments of relative abundances (of rDNA operons but not necessarily of biomass; see below and Amend, Seiferet, et al., 2010). These variabilities and constraints highlight the need to tailor a sequencing approach to the research question, the target population(s), and the taxonomic resolution predicted to provide an ecological answer.

Enzyme-Encoding Genes as Markers of Ecological Functions

With increasing interest in the functional diversity of fungi, focus turns to the evolution of functional traits within the fungal phylogeny, and community analysis conjoins with evolutionary genomics. Most ecological life strategies in the fungi (e.g., ectomycorrhizal symbiosis or pathogenicity) have evolved repeatedly within relatively small phylogenetic clades (Hibbett, Gilbert, et al., 2000, Wolfe, Tulloss, et al., 2012). This requires analysis of fungal communities at the level of species and genera to understand their integrated functional properties. Potentially, enzyme-encoding genes coding for ecologically relevant functions may provide simultaneous information on phylogenetic community composition and functional capabilities. Extracellular enzymes are particularly interesting as functional markers because these enzymes enable fungi to interact with their environment, and the connection to coding genes is relatively straightforward. Targeting of expressed genes through molecular analysis of expressed RNA pools also offers the opportunity to analyze gene transcription and active populations at the community level. Courty, Poletto, et al. (2008) investigated transcription of four target genes in oak/*Lactarius quietus* ectomycorrhizas and surrounding rhizosphere soil, and Damon, Barroso, et al. (2010) used a central metabolic gene, cytochrome c oxidase 1 (cox1), as a general marker for active fungi in soil. With a focus on plant biomass decomposition, Kellner, Zak, et al. (2010) used a screening approach and an impressive number of new primers to demonstrate transcription of 26 different groups of fungal extracellular enzymes potentially involved in degradation of plant and fungal cell walls in the organic horizon of a forest soil.

Before genome sequencing became a feasible alternative, degenerate primers were employed to amplify specific functional genes and investigate their distribution among cultured isolates representing different fungal groups. For example, the distribution of genes coding for phenol oxidizing laccase enzymes was investigated in cultured wood-rotting and ectomycorrhizal fungi (Dsousa, Boominathan, et al., 1996; Chen, Bastias, et al., 2003). With primers designed to amplify the range of laccases found among cultured fungi in hand, a wide range of studies were subsequently initiated to investigate the diversity and distribution of laccase genes in environmental samples (Lyons, Newell, et al., 2003; Luis, Walther, et al., 2004; Luis, Kellner et al., 2005; reviewed by Theuerl & Buscot, 2010). Although proposed to play a role in degradation of recalcitrant organic matter, the functional roles of laccases in fungi and ecosystems remain highly uncertain (Baldrian, 2006; Kües & Rühl, 2011). Nevertheless, Edwards, Zak, et al. (2011) were able to demonstrate reduced abundance of laccase transcripts in litter samples exposed to simulated N-deposition, in accordance with retarded litter decomposition rates.

The involvement of fungal cellobiohydrolases (*cbh*I) in cellulose degradation seems more straightforward. Edwards, Upchurch, et al. (2008) designed primers to amplify the *cbh*I gene from fungi, and this marker has been used to demonstrate changes in the cellulolytic fungal community in response to carbon dioxide (CO_2) enrichment and nitrogen (N) fertilization (Edwards, Zak, et al., 2011; Weber, Zak, et al., 2011; Weber, Balasch, et al., 2012). Comparisons of the fungal community harboring the *cbh*I gene (DNA analysis), and the expressed pool of *cbh*I genes (RNA analysis) demonstrate the utility of a functional gene approach. Baldrian, Kolarik, et al. (2012) showed that the *cbh*I gene pool in forest litter was more diverse and differed in composition from that of the humus horizon and that a higher proportion of the detected *cbh*I genes were transcribed in the litter layer, in accordance with a higher cellulose content of the litter. Weber, Balasch, et al. (2012) showed that only a subset of the resident *cbh*I genes represented in the DNA pool was transcribed (at the time sampled) and that most of the soil *cbh*I genes represented potentially novel fungal lineages.

An important research gap noted in these studies is the difficulty of linking the functional profiles to the organisms responsible for those processes. Because gene regulation and enzyme activity are highly dependent on the lifestyle and ecophysiology of the host organism, this is a consideration that must be addressed to obtain ecologically meaningful information from metagenomic surveys. Currently, there is a lack of well-populated databases of reference sequences for functionally relevant genes, and it remains difficult to assign functional gene sequences to fungal species or genera (Weber, Zak, et al., 2011; Baldrian, Kolarik, et al., 2012). Even the rRNA sequences databases are fraught with misidentified sequences, requiring careful curation (Vilgalys, 2003, Liu, Porras-Alfaro, et al. 2012; Tedersoo, Abarenkov, et al., 2011), and the importance of accurately curated databases cannot be underestimated. For example, after initial fungal community analysis by 454-sequencing of ITS-amplicons, Bödeker (2012) collected herbarium material from the dominant ectomycorrhizal taxa. With relevant reference material in hand it was then possible to confidently associate soil transcripts of Mn-peroxidase (MnP) genes with ectomycorrhizal *Cortinarius* species. This study is an example of how gene expression may be studied directly in the field to provide functional information for organisms that are difficult to isolate and cultivate. The rapid sequencing of entire fungal genomes (*see* Chapter 1), the continued investigation of genes involved in specific physiological processes in key fungi, and the continued establishment of publicly available database resources, such as the Joint Genome Institute MycoCosm (http://genome.jgi-psf.org/programs/fungi/index.jsf), will undoubtedly facilitate better identification of genetic marker sequences in the near future.

Another way to identify fungal populations with specific functional capacities is to use stable isotope probing (Radajewski, Inseon, et al., 2000). Specific substrates labeled with heavy isotopes (^{13}C or ^{15}N) are supplied to fungal communities in the field or under controlled laboratory conditions. Fungi that are able to utilize these substrates will incorporate the heavier isotopes in nucleic acids, which may be separated from those of nonutilizing fungi by density gradient centrifugation and analyzed for composition by *targeted* metagenomic techniques. Some recent examples of how the technique may be used include the identification of fungal subcommunities involved in degradation of cellulose (Eichorst & Kuske, 2012; Stursova, Zifcakova, et al., 2012), fungal mycelium (Drigo, Anderson, et al., 2012), plant litter (Espana, Rasche, et al., 2011; Murase, Shibata, et al., 2012) as well as in rhizosphere interactions (Hannula, Boschker, et al., 2012).

Targeted amplification of functional genes is complicated by the decisive influence of the primers. Most functional genes occur as gene families with many representative paralogs in each genome (e.g., the ectomycorrhizal symbiont *Laccaria bicolor*, Martin, Aerts, et al., 2008). Such paralogs within species often diverge in sequence within the same range as homologous genes from different fungal species (Lindahl & Taylor, 2004; Bödeker, Nygren, et al., 2009). To target gene-families throughout the entire fungal kingdom, primers usually have to be degenerated to the extent that they also amplify unrelated nontarget genes (e.g., Bödeker, Nygren, et al., 2009). Such primers are not practical to use with complex community templates, and one has to assume that all usable gene-specific primers discriminate both against subgroups within gene families, as well as against certain groups of fungi. PCR-based approaches targeting functional genes may, thus, be excellent to provide positive evidence for the occurrence and expression of genes, whereas failed detection and quantitative estimates have to be treated with caution. In addition, most functions cannot be directly assigned to single gene families, as strikingly illustrated by extracellular degradation of proteins. Although fungal secreted endopeptidases (proteases) belong to at least six major families (Martin, Aerts, et al., 2008) with different evolutionary origin and biochemical mechanisms, they all contribute to the same ecological process. Such convergent evolution and complementarity imply that gene-to-function relationships are often complex. Another dimension of complexity is added for processes that are not mediated by enzymes, but rather by secondary metabolites or other nonprotein exudates. As sequence databases become larger, design of appropriate primer sets can become complex and computationally intensive. Gans, Dunbar, et al. (2012) recently developed an algorithm for accurate design of rRNA primer sets for a complex bacterial target group that would be equally useful for fungal rRNA or functional gene sets.

Shotgun Metagenomics to Survey Environmental Fungal Communities

Targeted genetic marker approaches can be used to address specific hypotheses about fungal community structure or functions. One limitation of the functional marker is that one only sees what one knows to look for, and admittedly, knowledge of genes-encoding fungal functions of interest is rudimentary. To overcome this constraint, shotgun metagenome approaches have been explored for assessment of fungal communities in soils and other environments. A shotgun metagenome approach provides the opportunity to holistically assess the environmental gene pool.

Environmental shotgun metagenomes, particularly from soils, are a potentially rich sources for discovery of novel genes-encoding potentially useful enzymes or enzymes with altered specificity or activity (Daniel, 2004; Ferrer, Martinez-Abarca, et al., 2005; Shendure & Ji, 2008). Shotgun metagenome data provides a source of total community DNA from which information on target organisms can be mined. For example, Steven, Gallegos-Graves, et al. (2012b) recently used replicate shotgun metagenomes from soils to compare the relative abundance of carbon (C)- and N-fixing cyanobacteria in a large field experiment, by comparing pools of shotgun sequence fragments that matched (by BLAST homology) total genome sequences from reference cyanobacteria known to be dominant in that soil. By comparing relative abundance of cyanobacteria sequences in the shotgun metagenome surveys, they were able to confirm observations made using PCR-based approaches.

The promise of shotgun metagenomic approaches is to capture the functional potential of an environmental microbiome. Recent progress in mining shotgun metagenomes from soils has been made with bacterial genes and genomes (Mackelprang, Waldrop, et al., 2011; Steven, Gallegos-Graves, et al., 2012a). Adequately surveying fungal genes or the fungal community using environmental shotgun metagenomes poses greater challenges than the bacterial community. First, fungal genomes are typically 6 to 10 times the size of the average bacterial genome (about 5 mb) and exist as multiple nuclear states (haploid, diploid, dikaryotic). Second, fungal genes contain introns, which may have differential splicing patterns under particular environmental conditions, and genomes contain a high percentage of intergenic noncoding sequences that are equally (or over) represented in shotgun metagenomes. Ability to parse shotgun sequences into broad functional distinctions using metabolic and biochemical information (e.g., SEED annotation environment, Overbeen, Begley, et al., 2005; Kyoto Encyclopedia of Genes and Genomes or KEGG database, Kanehisa, Goto, et al., 2012), or sequence-based similarities (Clusters of Orthologous Groups of proteins or COG/KOG database, Tatusov, Fedorova, et al., 2003) has shown that the fraction of total sequences classifiable by any metric of the current understanding of metabolism or cell components is low (typically ranging from 30 to 60 percent of reads from full 454 titanium metagenomes), and representation of recognizable fungal sequences in shotgun metagenomes is low (a few percent; Fig. 13.1*A*).

To date, the complexity of the soil community has precluded detection or assembly of most of the soil community, even in studies focused solely on the smaller-genome bacterial component (Tringe, Von Mering, et al., 2005; Mackelprang, Waldrop, et al., 2011; Delmont, Prestat, et al., 2012). Even terabase-level sequencing has resulted in limited coverage, which poses a severe constraint on use in ecological studies in which multiple samples and replication are essential. The use of shotgun metagenomes has been explored as a complex fingerprint of the soil microbiome in comparative assessment of different soils and experimental field treatments (Steven, Gallegos-Graves, et al. 2012a; Fig. 13.1*B*). Analyzed using feature-based (kmer) or annotation-based approaches, even low-coverage metagenomes were able to distinguish among geochemically and physically different soils (not shown), but were less able to differentiate more subtle differences, such as between soils associated with plant root zones or nonplant interspaces, than parallel targeted approaches (*see* Fig. 13.1*B*). Two factors that strongly limit the ability to functionally

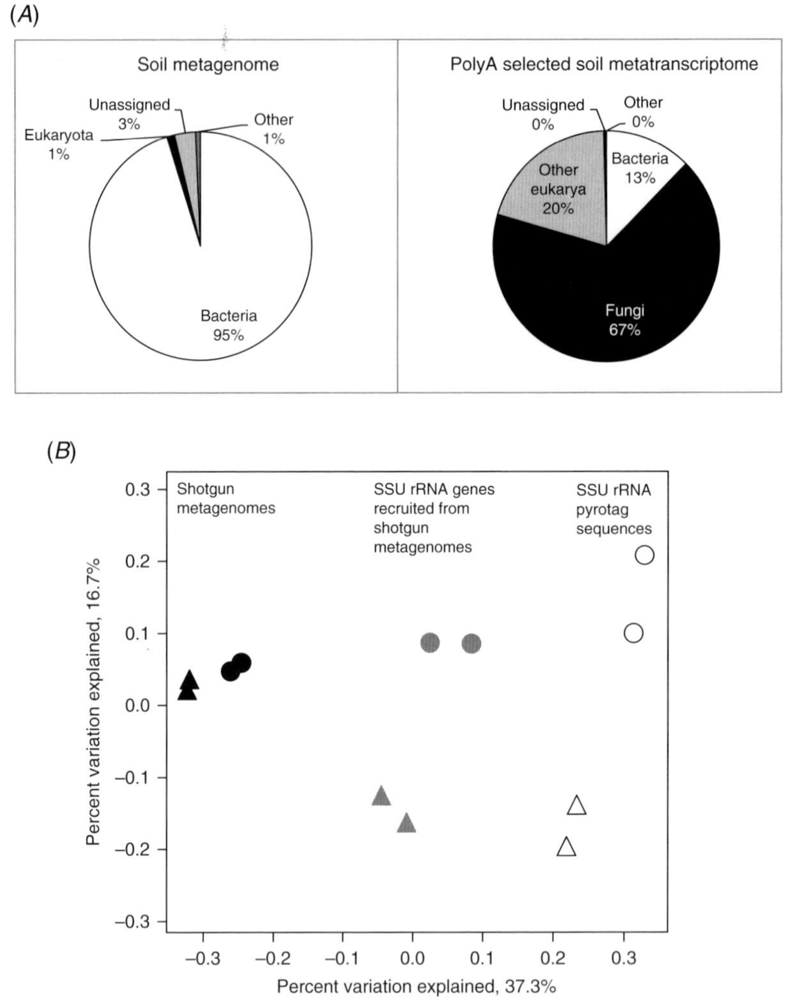

Figure 13.1 *A*, Distribution of shotgun metagenome (*left panel*) and polyA selected shotgun metatranscriptome (*right panel*) sequences generated from a pine forest soil. Sequences were parsed using gene homology searches through the Integrated Microbial Genomes/Microbiomes website (http://img.jgi.doe.gov/cgi-bin/m/main.cgi). The DNA-based metagenome is dominated by bacterial sequences whereas the metatransriptome is dominated by eukaryotic sequences. About two thirds of the eukaryotic sequences mapped to fungi (Kuske, unpublished data). *B*, Nonmetric MDS plot comparing the ability of rRNA pyrotag sequencing (tri-domain primers that include eukarya, bacteria, and archaea) and two shotgun metagenome approaches to discern differences soil communities between two environments. Duplicate samples were collected from *Larrea tridentata* shrub root zones (*triangles*) or from biological soil crusts (*circles*) in the interspaces of an arid shrub land. The three approaches provided different snapshots of the total soil community and differed in their ability to detect differences between the two soil environments. Comparisons based on shotgun metagenome fragments that could be identified and binned into SEED functional categories provided the least differentiation (*black circles and triangles*), followed by comparisons of SSU sequences recruited from shotgun metagenomes (*grey circles and triangles*). The rRNA gene pyrotag datasets provided the most discriminatory comparison (*open circles and triangles*). The three approaches differ in sequencing depth and availability of underpinning databases. These databases are rapidly improving for fungi. PCR, polymerase chain reaction; RT-PCR, real-time polymerase chain reaction. (From Steven, Gallegos-Graves, et al. 2012a.)

analyze and compare shotgun metagenomes are low-sequencing coverage and limited ability to accurately annotate gene fragments into functionally relevant categories. Many genes in bacteria and fungi remain functionally uncategorized (i.e., hypothetical proteins) and those that *can* be assigned to functional groups or domain families often span a broad functional repertoire. This makes shotgun metagenome data sets from different environments look similar because the genes identified tend to be in conserved, central metabolic pathways.

Sequence analysis of expressed sequences (ESTs) has been a highly successful approach through which to identify gene-coding regions and new genes and to compare genetic regulation of metabolic processes in eukaryotes. Applied to plants (Rudd, et al., 2003), yeast, and more recently to filamentous fungi (Chapter 3) and the ectomycorrhizosphere (Courty, Poletto, et al., 2008), this RNA-based sequence approach focuses on the expressed genes after intron splicing has occurred and thus avoids much of the genome noncoding DNA. A parallel approach applied to complex communities in environmental samples, termed *metatranscriptomic assessment*, is a promising alternate shotgun approach to enrich for fungal genes to enable more in-depth assessment of potential functions (Bailly, Fraissinet-Tachet, et al., 2007; Damon, Lehembre, et al., 2012) and to focus comparisons on the active gene pool (*see* Chapter 14). Figure 13.1*A* illustrates the distribution of cDNA sequences obtained from a soil shotgun metagenome and a corresponding metatranscriptome, in which a polyA mRNA selection was used to enrich for fungal cDNAs. Mining the transcribed gene pool from environmental samples should provide better representation of fungal genes and community activities.

Interpreting Metagenomic Surveys in an Ecological Context

With the availability of low-cost sequencing platforms, it is not difficult to sequence nucleic acids from environmental samples. However, designing a metagenomic survey that is informative in an ecological context can be challenging and must be guided by the specific objectives of the scientific inquiry and knowledge of the inherent biases associated with molecular surveys. Some factors to consider at each step in the process are included in Figure 13.2. The reduced cost of high-throughput sequencing now allows one to use a targeted metagenomic approach to compare many environmental samples within a field-scale experiment and combining highly replicated surveys (hundreds to thousands of samples) with more in-depth sequencing of fewer (tens of samples) can be a useful approach.

Initial shotgun metagenomic attempts were largely descriptive of single samples, and attempts to use this approach in replicated comparisons of soil communities are few (Steven, Gallegos-Graves, et al. 2012a). Proper replication of sampling is an issue that has often been neglected in microbial ecology

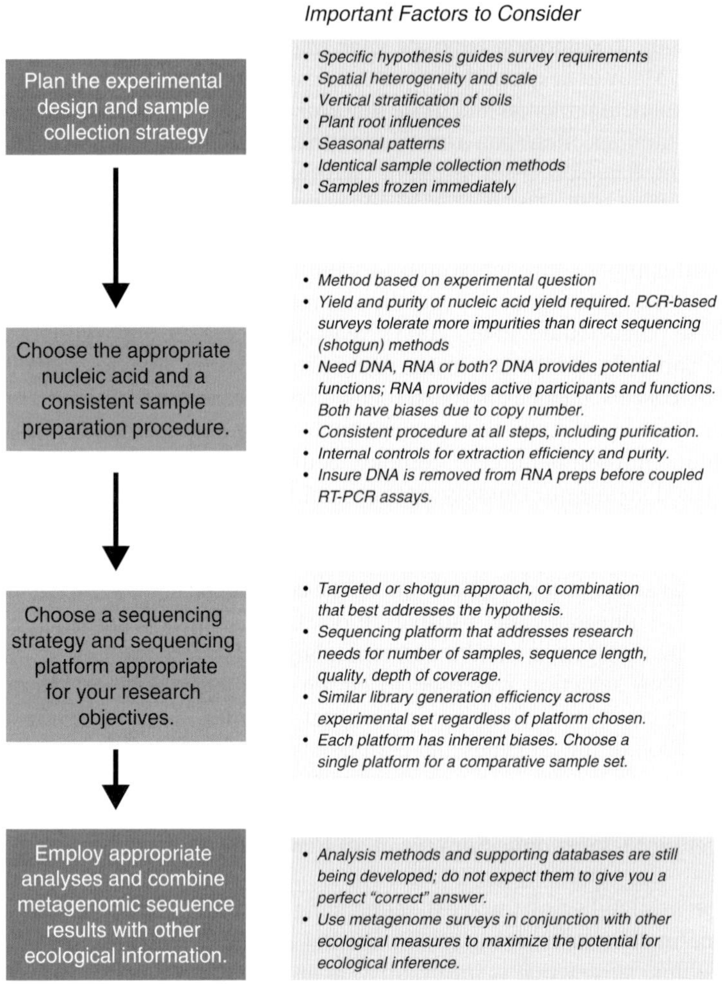

Figure 13.2 Design and execution of an informative metagenomic survey.

(Prosser, 2010). Fungal communities are usually spatially heterogeneous with practical sample sizes often much smaller than individual mycelia, requiring large numbers of samples to assess ecological features over and above the confounding heterogeneity effects. In soils, strong vertical stratification in community composition and function may add to the complexity (e.g., Lindahl, Ihrmark, et al., 2007). Temporal variation may also hamper assessments of the full functional potential of fungal communities, particularly when analyzed on the level of transcriptomes because only a subset of the gene pool is likely to be expressed at a specific point in time (Weber, Balasch, et al. 2012).

Technical issues associated with high-throughput technologies include lower sequence quality, homopolymer artifacts, and short sequence lengths (Scholz, Lo, et al., 2012), features that confound accuracy of sequence identification and can enormously inflate diversity estimates (e.g., Tedersoo, Nilsson, et al., 2010; Dickie, 2010). Accuracy and algorithms for removal of artifacts are constantly improving (Glenn, 2011; Gomez-Alvarez, Teal, et al., 2009), but these factors need to be taken into consideration during experimental design and interpretation of results. Because all of the metagenomic and sequencing approaches contain inherent biases, it is important to control for artifacts by using identical nucleic acid preparation methods (*see* Fig. 13.2; *see* for example, Zhou, Burns, et al., 1996; Kuske, Banton, et al., 1998; Dunbar, Eichorst, et al. 2012; Delmont, Prestat, et al., 2012).

To be ecologically informative, metagenomic surveys must provide relevant indications of fungal biomass, activity, and diversity, but there are several issues that hamper direct translation of metagenomic data. Some examples: (a) fungi encompass a broad range of growth habits, cellular forms and nuclear states (filamentous versus unicellular, large hyphal mats versus localized spore-forming colonies, nondifferentiated versus cords or rhizomorphs, haploid/diploid/dikaryotic) and congruence of metagenomic surveys with actual biomass of the fungal community remains unclear (Amend, Seiferet, et al., 2010). (b) Even after accounting for PCR and sequencing artefacts, estimation of species diversity from ITS amplicons is not a straight forward undertaking because of significant ITS sequence variation even within single fungal individuals (Pawlowska & Taylor, 2004; Lindner & Banik, 2011). (c) The problems associated with incongruences between transcript abundances, protein production, and enzyme activities are well known from single-organisms studies (i.e., post transcriptional regulation, enzyme turnover). These issues are even more valid in metatrascriptional studies, in which transcription-process relationships may diverge for different organisms. Even so, metagenomic approaches are clearly the optimal choice for comparison and identification of ecological trends or shifts, but subsequent in-depth assessments and validations may be required to determine the underlying causes and effects.

Metagenomic surveys enable comparison of the relative abundance of gene copies that can represent changes in community structure and potential functional properties. However, outside a phylogenetic context, data on presence, abundance, and expression of genes provide little ecological understanding above the descriptive level, and direct analysis of ecosystem processes (i.e., as biochemical transformations or potential enzyme activities) can strengthen ecological interpretations considerably. Additionally, as proteomic approaches are developed for applications to complex environmental systems (metaproteomics), parallel analysis of system genomes and proteomes have the potential to bridge community structure with in situ functions. The major benefit of metagenomic approaches is that processes may be connected to the responsible organisms, and mechanistic understanding may be obtained from how genes and functions interact within genomes and organisms. To make optimal use of metagenomic methods, data must be analyzed within a theoretical framework in which knowledge from physiological studies of model organisms is integrated with phylogenetic taxonomy and evolutionary genomics. Here, fungi may be advantageous objects of study, compared to bacteria, because horizontal gene transfer is less frequent and relationships between function and phylogeny consequently more stable. The concept of functional guilds, that is, groups of organisms that share a similar set of ecological properties, is essential to be able to discern structures and raise hypothesis about functional properties of complex fungal communities.

Conclusions and Future Prospectus

Community-based analyses are critical for ecological assessments of fungal roles in the environment, but issues of spatial scale and consequent sampling ability, temporal dynamics, the definition of taxonomic or operational "units," the assignment of genes to taxa and the assembly of functional guilds remain elusive topics. High-throughput sequencing, used with targeted phylogenetic or functional markers (Table 13.1) is rapidly improving the understanding of soil fungal community structure and compositional responses to altered environmental conditions. The ability to generate sequence data sets currently outpaces the ability to computationally analyze them or identify them against accurate comprehensive reference databases. Increased sequence data alone will not advance the science, but it must be coupled with curated databases that follow a phylogenetically sound nomenclature, both with respect to organism taxonomy and functional annotation of genes. Although the availability of curated, validated sequence databases for fungal genes has lagged behind bacterial databases, recent efforts are resulting in more comprehensive and publicly available resources (for target genes, see Table 13.1; for total genomes see Chapter 3 and the Integrated Microbial Genomes [IMG] website: http://img.jgi.doe.gov/cgi-bin/w/main.cgi). Expanded genome coverage of cultured fungi, together with environmental surveys of enzyme-encoding genes and soil metagenomes and metatransriptomes will provide the prerequisite sequence datasets from which to develop hypotheses about in situ fungal metabolic capability and ecosystem functions. Furthermore, efficient approaches for computational analysis are continually being developed. Coupled with appropriate physiological and molecular studies of gene and enzyme function, targeted and shotgun metagenome surveys have increasing potential

Table 13.1 Target genes and resources for their analysis of soil fungal communities.

Target Genes	Application	Primer Set Design	Supporting Databases and Resources for Sequence Classification
Taxonomic and Phylogenetic Markers (rRNA operon)	Compare Community Diversity and Composition in DNA or RNA Pool		
Small subunit rRNA (SSU)	All fungi (but too low specificity for Dikarya)	(Tri-domain specificity includes fungi) Kunin, Engelbrektson, et al., 2010.	SILVA SSU classifier, www.arb-silva.de, Pruesse, Quast, et al., 2007.
		http://www.biology.duke.edu/fungi/mycolab/primers.htm	Werner, Koren, et al., 2012. MaarjAM, for Glomeromycota öpik, Vanatoa, et al., 2010.
Large subunit rRNA (LSU)	All fungi	http://www.biology.duke.edu/fungi/mycolab/primers.htm	SILVA LSU classifier, www.arb-silva.de, Pruesse, Quast, et al., 2007.
			Liu, Porras-Alfaro, et al., 2012. Ribosomal Database Project http://rdp.cme.msu.edu/classifier/classifier.jsp
Intergenic transcribed spacers (ITS1, ITS2)	All fungi (too variable for Glomeromycota)	http://www.biology.duke.edu/fungi/mycolab/primers.htm	UNITE, for ectomycorrhizal fungi, Abarenkov, Nilsson, et al., 2010.
		Gardes & Bruns, 1993.	Schoch, Seifert, et al., 2012 www.fungalbarcoding.org

(Continued)

Table 13.1 (*Continued*)

Target Genes	Application	Primer Set Design	Supporting Databases and Resources for Sequence Classification
Enzyme-encoding (functional markers)	Compare active communities and functional guilds in DNA or RNA pool		
Mitochondrial cytochrome c oxidase (cox1)	Active populations of Agaricomycetes and Pezizomycotina	Damon, Barroso, et al., 2010	
Cellobiohydrolase I (cbhI)		Edwards, Upchurch, et al., 2008	Ribosomal Database Project http://fungene.cme.msu.edu//index.spr
Ascomycota laccase (lcc)	For Ascomycota	Lyons, Newell, et al., 2003	
Basidiomycota laccase (lcc)	For Basidiomycota	Luis, Walther, et al., 2004 Luis, Kellner, et al., 2005	Ribosomal Database Project http://fungene.cme.msu.edu//index.spr
Ligninolytic enzymes(multiple)		Kellner, Zak, et al., 2010.	
Cellulolytic/hemicellulolytic enzymes (multiple)		Kellner, Zak, et al., 2010.	
Chitinolytic enzymes (multiple)		Kellner, Zak, et al., 2010.	
Type I Polyketide synthase (multiple)		Kellner & Zak, 2009.	

Analysis tools for SSU and LSU rRNA genes are available for use through consolidated software. For example, MG-RAST, see Meyer, Aerts, et al., 2008; Mothur, see Schloss, Westcott, et al., 2009.

to provide interpretable and accurate assessments of the fungal community, its ecological functions, and relevance.

References

Aime C, Ball B, et al. 2011. Assembling the Fungal Tree of Life. Accessed May 16, 2013, at http://aftol.org.

Abarenkov K, Nilsson RH, et al. 2010. The UNITE database for molecular identification of fungi—recent updates and future perspectives. *New Phytol.* 186: 281–285.

Amend AS, Seiferet KA, et al. 2010. Quantifying microbial communities with 454 pyrosequencing: does read abundance count? *Mol Ecol.* doi: 10.1111/j.1365-294X.2010.04898.x.

Arnold AE, Miadlikowska J, et al. 2009. A phylogenetic estimation of trophic transition networks for ascomycetous fungi: Are lichens cradles of symbiotrophic fungal diversification? *Syst Biol.* 58: 283–297.

Bailly J, Fraissinet-Tachet L, et al. 2007. Soil eukaryotic functional diversity, a metatranscriptomic approach. *ISME J.* 1: 632–642.

Baldrian P. 2006. Fungal laccases—occurrence and properties. *FEMS Microbiol Rev.* 30: 215–242.

Baldrian P, Kolarik M, et al. 2012. Active and total microbial communities in forest soil are largely different and highly stratified during decomposition. *ISME J.* 6: 248–258.

Begerow D, Nilsson H, et al. 2010. Current state and perspectives of fungal DNA barcoding and rapid identification procedures. *Appl Microbiol Biotechnol.* 87: 99–108.

Binladen J, Gilbert MTP, et al. 2007. The use of coded PCR primers enables high-throughput sequencing of multiple homolog amplification products by 454 parallel sequencing. *PLoS ONE.* 2: e197

Bödeker I. 2012. Functional ecology of ectomycorrhizal fungi—Peroxidases, decomposition, spatial community datterns. Ph.D. thesis. Acta Universitatis Agriculturae Sueciae, Uppsala, Sweden.

Bödeker ITM, Nygren CMR, et al. 2009. Class II peroxidase-encoding genes are present in a phylogenetically wide range of ectomycorrhizal fungi. *ISME J.* 3: 1387–1395.

Buchan A, Newell SY, et al. 2002. Analysis of internal transcribed spacer (ITS) regions of rRNA genes in fungal communities in a southeastern US salt marsh. *Microbial Ecol.* 43: 329–340.

Buée M, Reich M, et al. 2009. 454 pyrosequencing analyses of forest soils reveal an unexpectedly high fungal diversity. *New Phytol.* 184: 449–456.

Chen DM, Bastias BA, et al. 2003. Identification of laccase-like genes in ectomycorrhizal basidiomycetes and transcriptional regulation by nitrogen in *Piloderma byssinum. New Phytol.* 157: 547–554.

Chen DM & Cairney JWG. 2002. Investigation of the influence of prescribed burning on ITS profiles of ectomycorrhizal and other soil fungi at three Australian sclerophyll forest sites. *Mycol Res.* 106: 532–540.

Courty PE, Poletto M, et al. 2008. Gene transcription in Lactarius quietus-Quercus petraea ectomycorrhizas from a forest soil. *Appl Environ Microbiol.*74: 6598–6605.

Dahlberg A, Jonsson L, et al. 1997. Species diversity and distribution of biomass above and below ground among ectomycorrhizal fungi in an old-growth Norway spruce forest in south Sweden. *Can J Bot.* 75: 1323–1335.

Damon C, Barroso G, et al. 2010. Performance of the COX1 gene as a marker for the study of metabolically active Pezizomycotina and Agaricomycetes fungal communities from the analysis of soil RNA. *FEMS Microbiol Ecol.* 74: 693–705.

Damon C, Lehembre F, et al. 2012. Metatranscriptomics reveals the diversity of genes expressed by eukaryotes in forest soils. *PLoS ONE.* 7(1): e28967.

Daniel R. 2004. The soil metagenome—a rich resource for the discovery of novel natural products. *Curr Opin Biotechnol.* 15: 199–204.

Delmont T, Prestat E, et al. 2012. Structure, fluctuation and magnitude of a natural grassland soil metagenome. *ISME J*. 6(9): 1677–1687.

Dickie IA. 2010. Insidious effects of sequencing errors on perceived diversity in molecular surveys. *New Phytol*. 188: 916–918.

Dickie IA, Xu B, et al. 2002. Vertical niche differentiation of ectomycorrhizal hyphae in soil as shown by T-RFLP analysis. *New Phytol*. 156: 527–535.

Drigo B, Anderson IC, et al. 2012. Rapid incorporation of carbon from ectomycorrhizal mycelial necromass into soil fungal communities. *Soil Biol Biochem*. 49: 4–10.

Dsouza TM, Boominathan K, et al. 1996 Isolation of laccase gene-specific sequences from white rot and brown rot fungi by PCR. *Appl Environ Microbiol*. 62: 3739–3744.

Dunbar J, Eichorst SA, et al. 2012. Common bacterial responses in six ecosystems exposed to ten years of elevated atmospheric carbon dioxide. *Environ Microbiol*. 14: 1145–1158.

Edwards IP, Upchurch RA, et al.. 2008. Isolation of fungal cellobiohydrolase I genes from sporocarps and forest soils by PCR. *Appl Environ Microbiol*. 74: 3481–3489.

Edwards IP, Zak DR, et al. 2011. Simulated atmospheric N deposition alters fungal community composition and suppresses ligninolytic gene expression in a northern hardwood forest. *PLoS ONE*. 6: e20421.

Eichorst SA & Kuske CR. 2012. Identification of cellulose-responsive bacterial and fungal communities in geographically and edaphically different soils by using stable isotope probing. *Appl Environ Microbiol*. 78: 2316–2327.

Espana M, Rasche F, et al. 2011. Assessing the effect of organic residue quality on active decomposing fungi in a tropical Vertisol using N-15-DNA stable isotope probing. *Fungal Ecol*. 4: 115–119.

Ferrer M, Martinez-Abarca F, et al. 2005. Mining genomes and 'metagenomes' for novel catalysts. *Curr Opin Biotechnol*. 16(6): 588–593.

Frankland JC. 1998. Fungal succession—unravelling the unpredictable. *Mycol Res*. 102: 1–15.

Gans JD, Dunbar J, et al. 2012. A robust PCR primer design platform applied to the detection of Acidobacteria Group 1 in soil. *Nucl Acids Res*. 40(12): e96.

Gardes M & Bruns TD. 1993. ITS primers with enhanced specificity for basidiomycetes—Application to the identification of mycorrhizae and rusts. *Mol Ecol*. 2: 113–118.

Gardes M, Fortin JA, et al. 1990 Restriction fragment length polymorphisms in the nuclear ribosomal DNA of four *Laccaria* spp.: *L. bicolor*, *L. laccata*, *L. proxima*, and *L. amethystina*. *Phytopathology*. 80: 1312–1317.

Glenn TC. 2011. Field guide to next-generation DNA sequencers. *Mol Ecol Res*. 11: 759–769.

Gomez-Alvarez V, Teal TK, et al. 2009. Systematic artifacts in metagenomes from complex microbial communities. *ISME J*. 3: 1314–1317.

Hannula SE, Boschker HTS, et al. 2012. ^{13}C pulse-labelling assessment of the community structure of active fungi in the rhizosphere of a genetically starch-modified potato (*Solanum tuberosum*) cultivar and its parental isoline. *New Phytol*. 194: 784–799.

Henson JM & French R. 1993. The polymerase chain reaction and plant disease diagnosis. *Annu Rev Phytopathol*. 31: 81–109

Hibbett DS, Gilbert LB, et al. 2000. Evolutionary instability of ectomycorrhizal symbioses in basidiomycetes. *Nature*. 407: 506–508.

Hibbett DS, Ohman A, et al. 2009. Fungal ecology catches fire. *New Phytol*. 184: 279–282.

Horton TR & Bruns TD. 2001. The molecular revolution in ectomycorrhizal ecology: Peeking into the black-box. *Mol Ecol*. 10: 1855–1871.

Ihrmark K, Bödeker ITM, et al. 2012. New primers to amplify the fungal ITS2 region—evaluation by 454-sequencing of artificial and natural communities. *FEMS Microbiol Ecol*. 82(3): 666–677.

James TY, Kauff F, et al. 2006a. Reconstructing the early evolution of Fungi using a six-gene phylogeny. *Nature*. 443: 818–822.

James TY, Letcher PM, et al. 2006b. A molecular phylogeny of the flagellated fungi (*Chytridiomycota*) and description of a new phylum (*Blastocladiomycota*). *Mycologia*. 98: 860–871.

Johansson JF, Paul LR, et al. 2004. Microbial interactions in the mycorrhizosphere and their significance for sustainable agriculture. *FEMS Microbiol Ecol*. 48: 1–13.

Jumpponen A & Jones KL. 2009. Massively parallel 454 sequencing indicates hyperdiverse fungal communities in temperate quercus macrocarpa phyllosphere. *New Phytol*. 184: 438–448.

Kellner H & Zak DR. 2009. Detection of expressed fungal Type I polyketide synthase genes in a forest soil. *Soil Biol Biochem*. 41: 1344–1347.

Kellner H, Zak DR, et al. 2010. Fungi unearthed: Transcripts encoding lignocellulolytic and chitinolytic enzymes in forest soil. *PLoS One*. 5(6): e10971.

Kanehisa M, Goto S, et al. 2012. KEGG for integration and interpretation of large-scale molecular data sets. *Nucl Acids Res*. 40 Database issue D109–D114. doi:10.1093/nar/gkr988.

Kowalchuk GA, Gerards S, et al. 1997. Detection and characterization of fungal infections of *Ammophila arenaria* (marram grass) roots by denaturing gradient gel electrophoresis of specifically amplified 18S rDNA. *Appl Environ Microbiol*. 63: 3858–3865.

Kües U & Rühl M. 2011. Multiple multi-copper oxidase gene families in Basidiomycetes—What for? *Curr Genomics*. 12: 72–94.

Kuske CR, Banton KL, et al. 1998. Small-scale DNA sample preparation method for field PCR detection of microbial cells and spores in soil. *Appl Environ Microbiol*. 64: 2463–2474.

Kunin V, Engelbrektson A, et al. 2010. Wrinkles in the rare biosphere: Pyrosequencing errors can lead to artificial inflation of diversity estimates. *Environ Microbiol*. 12: 118–123.

Lindahl B, Taylor AFS, et al. 2002. Defining nutritional constraints on carbon cycling in boreal forests—towards a less "phytocentric" perspective. *Plant Soil*. 242: 123–135.

Lindahl BD & Taylor AFS. 2004. Occurrence of N-acetylhexosaminidase-encoding genes in ectomycorrhizal basidiomycetes. *New Phytol*. 164: 193–199.

Lindahl BD, Ihrmark K, et al. 2007. Spatial separation of litter decomposition and mycorrhizal nitrogen uptake in a boreal forest. *New Phytol*. 173: 611–620.

Lindner DL & Banik MT. 2011. Intragenomic variation in the ITS rDNA region obscures phylogenetic relationships and inflates estimates of operational taxonomic units in genus *Laetiporus*. *Mycologia*. 103: 731–740.

Liu K-L, Porras-Alfaro A, et al. 2012. Accurate, rapid taxonomic classification of fungal large subunit rRNA genes. *Appl Environ Microbiol*. 78: 1523–1533.

Luis P, Walther G, et al. 2004. Diversity of laccase genes from basidiomycetes in a forest soil. *Soil Biol Biochem*. 36: 1025–1036.

Luis P, Kellner H, et al. 2005. A molecular method to evaluate basidiomycete laccase gene expression in forest soils. *Geoderma*. 128: 18–27.

Lyons JI, Newell SY, et al. 2003. Diversity of ascomycete laccase gene sequences in a southeastern US salt marsh. *Microbial Ecol*. 45: 270–281.

Mackelprang R, Waldrop MP, et al. 2011. Metagenomic analysis of a permafrost microbial community reveals a rapid response to thaw. *Nature*. 480: 368–U120.

Margulies M, Egholm M, et al. 2005. Genome sequencing in microfabricated high-density picolitre reactors. *Nature*. 437: 376–380.

Martin F, Aerts A, et al. 2008. The genome of *Laccaria* bicolor provides insights into mycorrhizal symbiosis. *Nature*. 452: 88–93.

Mullis KB & Faloona FA. 1987. Specific synthesis of dna invitro via a polymerase-catalyzed chain-reaction. *Methods Enzymol*. 155: 335–350.

Murase J, Shibata M, et al. 2012. Incorporation of plant residue-derived carbon into the microeukaryotic community in a rice field soil revealed by DNA stable-isotope probing. *FEMS Microbiol Ecol*. 79: 371–379.

Nilsson RH, Ryberg M, et al. 2009. The ITS region as target for characterization of fungal communities using emerging sequencing technologies. *FEMS Microbiol Lett*. 296: 97–101

O'Brien HE, Parrent JL, et al. 2005. Fungal community analysis by large-scale sequencing of environmental samples. *Appl Environ Microbiol*. 71: 5544–5550.

Öpik M, Vanatoa A, et al. 2010. The online database MaarjAM reveals global and ecosystemic distribution patterns in arbuscular mycorrhizal fungi (*Glomeromycota*). *New Phytol.* 188: 223–241.

Overbeen R, Begley T, et al. 2005. The subsystems approach to genome annotation and its use in the project to annotate 1000 genomes. *Nucl Acids Res.* 33: 5691–5702.

Pawlowska TE & Taylor JW. 2004. Organization of genetic variation in individuals of arbuscular mycorrhizal fungi. *Nature*. 427: 733–737.

Prosser JI. 2010. Replicate or lie. *Environ Microbiol.* 12: 1806–1810.

Pruesse E, Quast C, et al. 2007. SILVA: A comprehensive online resource for quality checked and aligned ribosomal RNA sequence data compatible with ARB. *Nucl Acids Res.* 35: 7188–7196.

Radajewski S, Ineson P, et al. 2000. Stable-isotope probing as a tool in microbial ecology. *Nature*. 403: 646–649.

Rajala T, Peltoniemi M, et al. 2011. RNA reveals a succession of active fungi during the decay of Norway spruce logs. *Fungal Ecol.* 4: 437–444.

Rosling A, Cox F, et al. 2011. Archaeorhizomycetes: Unearthing an ancient class of ubiquitous soil fungi. *Science*. 333: 876–879.

Rudd S. 2003. Expressed sequence tags: Alternative or complement to whole genome sequences? *Trends Plant Sci.* 8(7): 321–329. doi: 10.1016/S1360-1385(03)00131-6.

Schadt CW, Martin AP, et al. 2003. Seasonal dynamics of previously unknown fungal lineages in tundra soils. *Science*. 301: 1359–1361.

Schloss PD, Westcott SL, et al. 2009. Introducing mothur: Open-source, platform-independent, community-supported software for describing and comparing microbial communities. *Appl Environ Microbiol.* 75(23): 7537–7541.

Shendure J & Ji HL. 2008. Next-generation DNA sequencing. *Nat Biotechnol.* 26: 1135–1145.

Schoch CL, Seifert KA, et al. 2012. Nuclear ribosomal internal transcribed spacer (ITS) region as a universal DNA barcode marker for fungi. *Proc Natl Acad Sci USA*. 109: 6241–6246.

Scholtz MB, Lo C-C, et al. 2012. Next generation sequencing and bioinformatic bottlenecks: The current state of metagenomic analysis. *Curr Opin Biotechnol.* 23: 9–15.

Shokralla S, Spall JL, et al. 2012. Next-generation sequencing technologies for environmental DNA research. *Mol Ecol.* 21: 1794–1805.

Smit E, Leeflang P, et al. 1999. Analysis of fungal diversity in the wheat rhizosphere by sequencing of cloned PCR-amplified genes encoding 18S rRNA and temperature gradient gel electrophoresis. *Appl Environ Microbiol.* 65: 2614–2621.

Steven B, Gallegos-Graves L, et al. 2012a. Targeted and shotgun metagenomic approaches provide different descriptions of dryland soil microbial communities in a manipulated field study. *Environ Microbiol Rep.* 4: 248–256.

Steven B, Gallegos-Graves LV, et al. 2012b. Dryland biological soil crust cyanobacteria show unexpected decreases in abundance under long term elevated CO_2. *Environ Microbiol.* 14(12): 3247–3258.

Stockinger H, Kruger M, et al. 2010. DNA barcoding of arbuscular mycorrhizal fungi. *New Phytol.* 187: 461–474.

Stursova M, Zifcakova L, et al. 2012. Cellulose utilization in forest litter and soil: identification of bacterial and fungal decomposers. *FEMS Microbiol Ecol.* 80: 735–746.

Tatusov RL, Fedorova ND, et al. 2003. The COG database: an updated version includes eukaryotes. *BMC Bioinformatics*. 4: 41.

Tedersoo L, Abarenkov K, et al. 2011. Tidying up international nucleotide sequence databases: Ecological, geographical and sequence quality annotation of ITS sequences of mycorrhizal fungi. *PLoS ONE*. 6: e24940

Tedersoo L, Nilsson RH, et al. 2010. 454 Pyrosequencing and Sanger sequencing of tropical mycorrhizal fungi provide similar results but reveal substantial methodological biases *New Phytol.* 188: 291–301.

Theuerl S & Buscot F. 2010. Laccases: toward disentangling their diversity and functions in relation to soil organic matter cycling. *Biol Fertil Soils*. 46: 215–225.

Tringe SG, Von Mering C, et al. 2005. Comparative metagenomics of microbial communities. *Science*. 308: 554–557.

Van Tuinen D, Jacquot E, et al. 1998. Charecterization of root colonization profiles by a microcosm community of arbuscular mycorrhizal fungi using 25S rDNA-targeted nested PCR. *Mol Ecol*. 7: 879–887.

Vilgalys R. 2003. Taxonomic misidentification in public DNA databases. *New Phytol*. 160: 4–5.

White TJ, Bruns T, et al. 1990. Amplification and direct sequencing of fungal ribosomal RNA genes for phylogenetics. In: *PCR Protocols: A Guide to Methods and Applications* (eds. MA Innis, DH Gelfland, et al.), 315–322. San Diego: Academic Press, San Diego.

Weber CF, Balasch MM, et al. 2012. Soil fungal cellobiohydrolase I gene (cbhI) composition and expression in a loblolly pine plantation under conditions of elevated atmospheric CO_2 and nitrogen fertilization. *Appl Environ Microbiol*. 78: 3950–3957.

Weber CF, Zak DR, et al. 2011. Responses of soil cellulolytic fungal communities to elevated atmospheric CO_2 are complex and variable across five ecosystems. *Environ Microbiol*. 13: 2778–2793.

Werner JJ, Koren O, et al. 2012. Impact of training sets on classification of high-throughput bacterial 16s rRNA gene surveys. *ISME J*. 6: 94–103.

Wolfe BE, Tulloss RE, et al. 2012. The irreversible loss of a decomposition pathway marks the single origin of an ectomycorrhizal symbiosis. *PLoS ONE*. 7: e39597. doi: 10.1371/journal.pone.0039597.

Zhou J, Bruns MA, et al. 1996. DNA recovery form soils of diverse composition. *Appl Environ Microbiol*. 62: 316–322.

14 Metatranscriptomics of Soil Eukaryotic Communities

Laurence Fraissinet-Tachet[1], Roland Marmeisse[1], Lucie Zinger[2], and Patricia Luis[1]

[1]*Ecologie Microbienne, UMR CNRS 5557 – USC INRA 1364, Université de Lyon, Université Lyon 1, Villeurbanne, France*
[2]*Laboratoire d'Ecologie Alpine, UMR CNRS 5553, Université Joseph Fourrier, Grenoble, France*

Environmental Genomics: its Contribution to the Understanding of Ecosystem Functioning

Individual genomes are frequently explored to infer the range of biological activities accomplished by a given fungal strain or species. Similarly, comparing two or more genomes allows evaluating the diversity and evolution of biochemical pathways adopted by individual species to perform a common function. Genomics and comparative genomics are, however, poor predictors when it comes to characterizing biogeochemical processes. Indeed, these complex processes are mainly carried out by a diverse set of microorganisms—including fungi—that interact with each other. Currently, gaining insights into the molecular machinery that controls such processes consists in deciphering the genetic make-up of whole microbial communities instead of focusing on selected species. This is one of the objectives of "environmental genomics" popularized by the metagenome concept. Such approaches became recently of particular interest with the advent of next-generation sequencing (NGS) techniques that can generate billions of DNA sequences from a given environmental sample.

A metagenome (Handelsman, Rondon, et al., 1998) can be defined as the sum of the genomes of all different microorganisms that co-occur and interact with each other in a specific environment. It is accessed by extracting DNA from an environmental sample without any prior attempt to isolate microorganisms in pure culture, which, for a majority of them, remain uncultivable or undescribed (Rappé & Giovannoni, 2003). Analyzing the gene content of metagenomes gives insights not only into the functions likely to be performed by microbial communities but also into the microbial taxonomic groups that

The Ecological Genomics of Fungi, First Edition. Edited by Francis Martin.

preferentially participate to specific biological activities. Striking differences were indeed observed between metagenomes from distinct environments. One of the first and most illustrative studies was performed by Tringe, von Mering, et al. (2005) comparing microbial metagenomes from a farmland soil, surface seawater, and whale carcasses. As expected, gene categories related to light-driven processes were more abundant in the seawater metagenome. In contrast, the soil metagenome was enriched in genes involved in potassium homeostasis and antibiotic biosynthesis, in agreement with the higher levels of potassium and interspecific interactions reported in soils. These observations, which have now been extended to more ecologically relevant comparisons (e.g., Shi, Tyson, et al., 2011; Toulza, Tagliabue, et al., 2012), demonstrate that the environment exerts a strong selection pressure on the resident microbial communities, affecting their global gene content.

Metagenomic studies are limited for two reasons. First, they are only indicative of the potential functions that communities are susceptible to perform, not of those that are actually being performed. Second, eukaryotes, and more specifically fungi, are poorly represented in the existing metagenomic DNA sequence data, including those from soils (e.g., <1 percent of sequences in organic or mineral horizons of a temperate forest soil, Stéphane Uroz, personal communication and Chapter 13 of this book). This latter observation may not necessarily reflect lower eukaryotic biomass levels compared to bacterial ones, but instead a far lower DNA-to-biomass ratio for eukaryotes or the dominance of "anonymous" intronic or intergenic regions in eukaryotic genomes. A promising alternative to metagenomics consists in accessing the pools of protein-coding genes expressed by the different members of the microbial communities—i.e., their metatranscriptomes, a term that first appeared in 2007 (Bailly, Fraissinet-Tachet, et al., 2007; Maron, Ranjard, et al., 2007). Environmental RNA-based metatranscriptomics is assumed to better reflect ongoing biological activities and has been developed for environments as diverse as freshwater and seawater (Gilbert, Field, et al., 2010; Shi, Tyson, et al., 2011; Stewart, Ulloa, et al., 2012; Rinta-Kanto, Sun, et al., 2012) or soils (Urich, Lanzen, et al., 2008; Shrestha, Kube, et al., 2009; Damon, Lehembre, et al., 2012). Furthermore, the eukaryotic 3' polyadenylated messenger RNAs can be isolated from ribosomal, noncoding RNAs and bacterial mRNAs that dominate environmental metatranscriptomes (Fig. 14.1, see discussion this chapter). Proofs of concept for this protocol have now been established for various ecosystems such as soils (Grant, Grant, et al., 2006; Bailly, Fraissinet-Tachet, et al., 2007; Kellner, Luis, et al., 2011; Damon, Vallon, et al., 2011; Damon, Lehembre, et al., 2012) or bovine (Qi, Wang, et al., 2011) and termite digesters (Todaka, Moriya, et al., 2007; Tartar, Wheeler, et al., 2009). However, eukaryotic metratranscriptomics is still in its infancy. Researchers who engage in this field have therefore to face methodological and conceptual issues that will be addressed in this chapter.

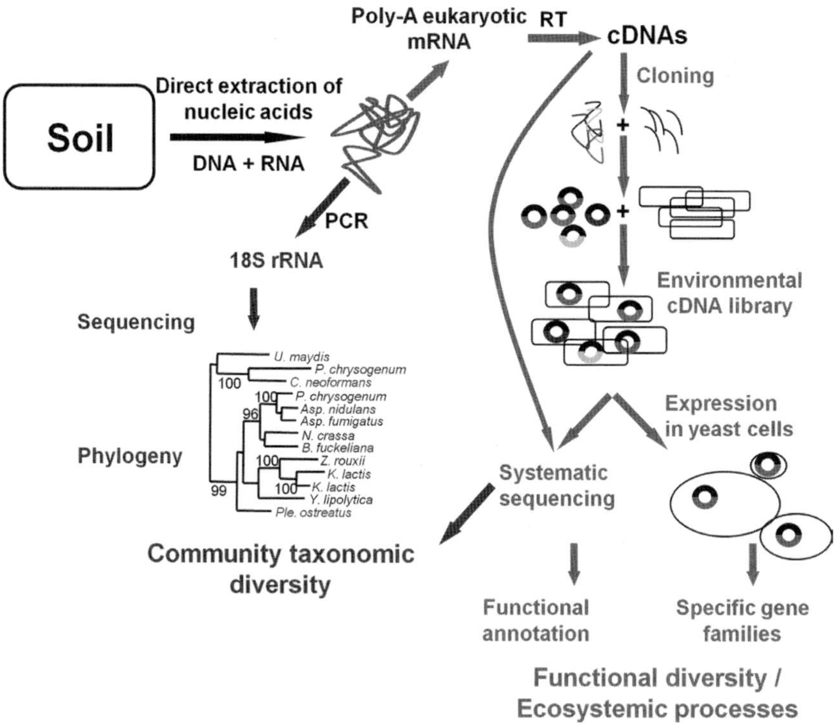

Figure 14.1 The metatranscriptomic work flow (soil sample–RNA extraction–systematic sequencing–data analysis–cDNA cloning–expression in heterologous hosts).

Diversity and Presumed Roles of Soil Eukaryotes and Fungi

No established experimental protocol exists to specifically isolate intact fungal hyphae and cells from complex environmental matrices such as soils. The only alternative consists in extracting total RNA from environmental samples, which inevitably leads to the isolation of both prokaryotic and eukaryotic RNA molecules. Furthermore, molecular diversity surveys based on the nuclear 18S rRNA gene showed that all major eukaryotic clades are represented in soil samples (Richards & Bass, 2005; Bailly, Fraissinet-Tachet, et al., 2007; Lesaulnier, Papamichail, et al., 2008; Damon, Lehembre, et al., 2012).

According to several independent global phylogenetic and phylogenomic analyses, it is generally admitted that the domain Eukarya is divided into 7 to 10 major clades whose positions relative to one another are still debated (Burki, Shalchian-Tabrizi, et al., 2007; Parfrey, Grant, et al., 2010). A majority of these clades, namely the Amoebozoa, the Rhizaria (e.g., Cercozoa and Foraminifera), the Stramenopiles (with the Oomycetes), the Alveolata (comprising ciliates and Apicomplexa), and the Excavata (e.g., the Euglenozoa), are essentially represented by single-cell organisms and constitute an artificial polyphyletic assemblage referred to as the "protists". Two other clades contain numerous multicellular organisms; the Plantae (or Archaeplastidia), that include green plants, and the Opistokonts, which encompass the kingdoms Metazoa, Fungi, and other minor groups such as the Choanoflagellates. In soils, these different clades are not equally diversified and many of their members remain undescribed. For instance, a recent report provided molecular evidences of the widespread occurrence of Foraminiferas in soils although this group was classically considered as strictly aquatic (Lejzerowicz, Pawlowski, et al., 2010).

Although soil hosts an incredible diversity of eukaryotes, studies using either 18S rRNA sequences amplified from soil DNA (O'Brien, Parrent, et al., 2005; Lesaulnier, Papamichail, et al., 2008; Damon, Lehembre, et al., 2012) or soil cDNAs (Urich, Lanzen, et al., 2008; Damon, Lehembre, et al., 2012) all consistently show that the phyla Metazoa and Fungi dominate the soil eukaryotic communities. In one study, the fungi contributed up to 40 percent of 18S soil sequences and 70 percent of the cDNA sequences (Damon, Lehembre, et al. 2012).

From a functional point of view, Bacteria, Archaea, and eukaryotes can all participate in major soil processes. For example, fungi are directly involved in plant nutrition through the widespread mycorrhizal association that improves plant nutrient uptake (Smith & Read, 2008 and Chapter 8 of this book). A major soil process attributed primarily to fungal activity is the decomposition of soil organic matter (SOM). Together, these processes make available considerable amounts of essential nutrients (such as organic nitrogen [N], phosphorus [P], and sulfur [S]) trapped in the several tons per hectares of above- and belowground plant litter that return to the soil each year (e.g., 3.5 t ha-1 per year in forests; Lützow, Kögel-Knabner, et al., 2006). SOM mineralization is usually incomplete, leading to accumulation of humus, a factor correlated to soil fertility and making the soil environment as one of the main carbon sinks on Earth (~1600 Pg C for the top 1 m; Spalding, Kendirli, et al., 2012).

Plant litter is comprised of polysaccharides protected by lignin and is assumed to be degraded primarily by fungi, which possess the enzymatic machinery to efficiently degrade all these different components, including lignin (Valášková, Šnajdr, et al., 2007 and Chapter 3, this book). This was recently confirmed by Schneider, Keiblinger, et al. (2012) who identified proteins directly extracted from beech litters by mass spectrometry. All proteins identified in this study and annotated as extracellular hydrolytic enzymes (including cellulases, xylanases, pectinases) were of fungal origin although bacterial proteins were present in other functional classes. Thus, one of the main current motivations for studying eukaryotic soil metatranscriptomes is to peek into the black box of the biochemistry of SOM degradation and to understand how this major ecosystemic process is affected by various environmental factors, especially in a context of global changes.

Experimental Strategies

In soil, fungi and other eukaryotes are not spatially uniformly distributed at a local scale such as in a single forest stand (Horton & Bruns, 2001). It is therefore necessary to collect multiple soil samples, or cores, distributed over the whole study site to picture the complexity of the corresponding eukaryotic community. Samples are then classically sieved to remove plant roots and large soil debris, pooled to constitute a single composite sample, and stored frozen. All these steps must be carried out as quickly as possible to leave unchanged the gene expression pattern in the soil cores. Although not evaluated for soil samples yet, it has been shown for water samples that the different collection steps (pumping and filtering) can have a significant effect on bacterial gene expression, for example, by increasing the stress-responsive *recA* gene transcription level (Feike, Jürgens, et al., 2012).

One of the most challenging aspects in environmental transcriptomics remains RNA extraction from environmental samples. The total soil RNA extraction yields vary significantly depending on soil types and composition. Usually a large amount of soil (e.g., >100 g) may be required to obtain sufficient quantities of total RNA (e.g., 100 μg). Several methods for soil RNA extraction have been reported in literature (Griffiths, Whiteley, et al., 2000; Hurt, Qiu, et al., 2001; Luis, Kellner, et al., 2005a; Bailly, Fraissinet-Tachet, et al., 2007; Mettel, Kim, et al., 2010), and commercial extraction kits are available. However, the extraction of high-quality RNA from soils remains challenging because of the coextraction of humic acids and others organic compounds that strongly inhibit downstream reactions such as mRNA amplification or reverse transcription. Solubility of humic acids in aqueous solutions and organic solvents is pH-dependent, and the use of low-pH lysis buffers (pH 5) and organic solvents (pH 4.5) can minimize the coextraction of humic acids (Mettel, Kim, et al., 2010). The presence of polyvinylpolypyrrolidone, a complexing agent of phenolic compounds, or guanidine isothiocyanate in extraction buffers can also limit the amount of humic acids extracted. Thus, by changing the final concentrations of guanidine isothiocyanate and sodium dodecyl sulphate in extraction buffers, Damon, Lehembre, et al. (2012) were able to extract high-quality RNAs from several forest soils. Different other methods were also reported and can be used for purifying soil RNA, including sample pretreatment with aluminum sulfate (Dong, Yan, et al., 2006), the use of active charcoal to adsorb contaminating molecules (Luis, Kellner, et al., 2005a), Sephadex G-50 (Bailly, Fraissinet-Tachet, et al., 2007), or Q-Sepharose chromatography (Mettel, Kim, et al., 2010).

The next steps in metatranscriptomics are, at first sight, not different from those followed in single culture transcriptomics, consisting into the conversion of RNA into cDNAs that can be subsequently cloned or sequenced directly (*see* Fig. 14.1). Urich, Lanzen, et al. (2008) reported the direct pyrosequencing of total, reversed-transcribed RNA, including rRNAs and mRNAs, extracted from a paddy soil. This approach, which presents the advantage of avoiding polymerase chain reaction (PCR) or cloning biases, largely failed to uncover eukaryotic mRNA sequence diversity because of the predominance of rRNAs and bacterial mRNAs. mRNAs from both prokaryotic and eukaryotic origins can be specifically isolated by rRNA subtractive hybridization (Mettel, Kim, et al., 2010; Stewart, Sharma, et al., 2011), but this approach appears insufficient to obtain reliable amounts of eukaryotic mRNA. This limitation can be overcome by specifically isolating eukaryotic poly-A mRNAs from total soil RNAs by affinity capture using poly-dT oligonucleotides grafted on a solid matrix (Grant, Grant, et al., 2006; Bailly, Fraissinet-Tachet, et al., 2007).

Soil eukaryotic mRNAs are then converted into cDNAs, usually by using the SMART™ cDNA Library Construction Kit (Clontech), which can generate micrograms of cDNA from nanograms of input total or purified poly-A mRNA (Grant, Grant, et al., 2006; Bailly, Fraissinet-Tachet, et al., 2007; Kellner, Luis, et al., 2011; Damon, Vallon, et al., 2011). cDNA amplification using this kit can however modify the initial abundance of individual sequences (Weber & Kuske, 2011). cDNA can then be sequenced using Sanger, or NGS, which can generate huge amount of sequence data and are therefore more appropriate to apprehend the high complexity of soil metatranscriptomes. Alternatively, cDNA libraries can be cloned in an oriented manner in plasmid vectors for expressing the environmental cDNAs in the yeast *Saccharomyces cerevisiae*, and hence potentially identify new enzymes (Bailly, Fraissinet-Tachet, et al., 2007; Kellner, Luis, et al., 2011; Damon, Vallon, et al., 2011).

Systematic Sequencing: Whole Metatranscriptomes

The Bioinformatics Challenge

The analysis of meta-"omic" and single genome or transcriptome sequence data obtained with NGS share many aspects because they are both constituted of short sequences randomly distributed across the metagenomes or transcriptomes. Each of the different successive steps of sequence trimming, clustering, and assembly and annotation and prediction are complicated in the case of meta-omic data by the taxonomic and functional diversity of the biotic assemblages considered, for which genome's composition, size and diversity are a priori unknown (see previous discussion). Although sequencing technologies are moving forward at an incredible pace (MacLean, Jones, et al., 2009), the coverage of the most representative transcriptomes from such complex meta-populations remains incomplete (but see Tamames, de la Peña, et al., 2012 for an estimation tool). Detailing the principles, pro and cons of all existing bioinformatic resources is out of the scope of this section (interested readers will refer to Kunin, Copeland, et al., 2008; Scholz, Lo, et al., 2012), which rather aims at providing an overview of the main steps for processing metatranscriptomic data.

Metatranscriptomic data sets can contain a considerable amount of undesirable or noninformative sequences as a result of experimental artifacts (e.g., sequencing/PCR errors) or the dominance of noncoding nucleic acids (e.g., rRNA, tRNAs, introns). The proportion of these undesirable sequences largely depends on the sequencing method and the mRNA enrichment procedure (see preceding discussion; He et al., 2010). Apart from the obvious initial screening and removal of sequence regions corresponding to the sequencing adaptors, amplification primers, cloning vector or poly-A/T tails (Table 14.1; Kunin, Copeland, et al., 2008), one has also to lower the amount of artifactual sequences in the data set. This can be done first by removing sequences of insufficient length or with low phred quality scores (estimates of per-base error probabilities; *see* Table 14.1). Concerning rRNA sequences, their presence during the functional annotation process (see discussion following) can artificially inflate the amount of functional genes annotated. This, together with the deposition of numerous false-positive coding-sequences in international databases can lead to cascading mistakes (Tripp, Hewson, et al., 2011). Efforts are currently invested in improving rRNA sequences detection, which remains challenging because it relies on ribosomal reference databases that are far from being exhaustive, especially for micro-eukaryotes (*see* Table 14.1 for a few examples).

The complexity of metatranscriptomic data sets has also to be lowered, primarily for computational reasons. This consists first in grouping homologous sequences (clustering step; *see* Table 14.1), a poorly efficient step because metatranscriptomic data sets usually contain mostly unique sequences (~ 80 to 90 percent singletons; Damon, Lehembre, et al., 2012; Gilbert, Field, et al., 2008). In any cases, strictly replicated sequences have to be clustered because they are suspected to be generated during the emulsion PCR steps carried out before 454 pyrosequencing, and hence artificially inflate a given gene abundance (Gomez-Alvarez, Teal, et al., 2009). In complement, partially overlapping sequences can be combined to form contiguous DNA stretches, so-called contigs (i.e., assembly step; *see* Table 14.1). This step should improve the functional annotation quality because sequence identification is strongly improved for long sequences (Götz, García-Gómez, et al., 2008). However, sequence assembly also usually fails in significantly simplifying the dataset because of low-coverage sequencing. Furthermore, the obtained contigs need to be inspected before making any biological inference as a result of the versatile nature of metatranscriptomes. Sequence assemblers are likely to generate numerous chimeric contigs (Kunin, Copeland, et al., 2008; Scholz, Lo, et al., 2012) either strictly artifactual or corresponding to proteins sequence fragments present in distinct but related taxa. These latter contigs nevertheless

Table 14.1 Examples of bioinformatics resources for analyzing fungal metatranscriptomes.*

Resource type	Resource	Brief description	Website
Sequence trimming	cutadapt	Removal of sequence adaptors and low-quality sequence	http://code.google.com/p/cutadapt/
	EMBOSS's trimest	Removal of poly-A/T tails	http://emboss.bioinformatics.nl/cgi-bin/emboss/trimest
	SeqClean	Sequence trimming and validation by screening for contaminants, low quality and low-complexity sequences	http://compbio.dfci.harvard.edu/tgi/software/
	MeTaxa	Detection and identification of ribosomal sequences; based on BLASTn and various Hidden Markov models	http://microbiology.se/software/metaxa/
	riboPicker	Detection of ribosomal sequences; based on BWA-SW	http://ribopicker.sourceforge.net/
Clustering/Assembly	CD-Hit	Clustering of homologous sequences by word counting	http://www.bioinformatics.org/cd-hit/
	Phrap	Sequence assembler	http://www.phrap.org
	Trinity	Sequence assembler	http://trinityrnaseq.sourceforge.net/
Databases	ARB-SILVA	High quality repository of large and small subunit RNA sequences	http://www.arb-silva.de/download/arb-files/
	RefSeq†	NCBI's Sequence Database	http://www.ncbi.nlm.nih.gov/RefSeq/
	Swiss-prot†	Repository of manually annotated and reviewed peptide sequences	http://www.uniprot.org/
Pipelines/Packages	Galaxy	Web platform for large scale genome analysis	https://main.g2.bx.psu.edu/
	EMBOSS	Analysis package for molecular biology.	http://emboss.sourceforge.net/
	MEGAN	Metagenomic analysis and visualization tool	http://ab.inf.uni-tuebingen.de/software/megan/
	MG-RAST	Metagenomic analysis server; uses the SEED environment	http://metagenomics.anl.gov/

*See also Kunin, Copeland, et al., 2008, and Scholz, Lo, et al., 2012, for additional methods and references.
†These databases are grouped with others in the nr database, available at ftp://ftp.ncbi.nlm.nih.gov/blast/db/.

provide information about ecosystem functioning but would require further post-analyses to reattribute these shared functions to their respective taxa.

Finally, the trimmed data set can be used for functional annotation or gene prediction. Whereas the first process consists in searching sequences that are homologous to proteins of known functions, the second aims at predicting coding regions and hence, contributes at identifying potentially new protein families. More concretely, sequence functional annotation is done by comparing sequences against reference databases (*see* Table 14.1), usually by using BLASTx (Altschul, Gish, et al., 1990), or PLASTx, a faster alternative (Nguyen & Lavenier, 2009) and defined as robust for a bitscore less than 40 (Frias-Lopez, Shi, et al., 2008). BLAST-hits are most of the time annotated using different functional classifications (e.g., KEGG orthology [Kanehisa, Goto, et al., 2004] or GO terms [Ashburner, Ball, et al., 2000]), rendering possible the comparison of different metatranscriptomes at multiple functional levels (e.g., proteins, pathways). Functional annotation can be complemented by a gene prediction step, for identifying new putative proteins. This approach is hampered by the fragmented nature of metatransciptome sequences and the existence of chimeric contigs. Several tools have been proposed for prokaryote metatranscriptomes (e.g., MetaGene; Noguchi, Park, et al., 2006), but, to present knowledge, none have yet been developed for eukaryote metatranscriptomes. This, together with the aforementioned current limitations in the processing of metatranscritpomes emphasizes the need for further bioinformatics developments in that area.

Examples of Systematically Sequenced Eukaryotic Metatranscriptomes

Systematic eukaryotic metatranscriptome sequencing has been reported for the termite gut (Tartar, Wheeler, et al., 2009; Xie, Zhang, et al., 2012), bovine rumen (Qi, Wang, et al., 2011), and forest soil (Damon, Lehembre, et al., 2012) by using different sequencing techniques. All four studies specifically focused on plant biomass degradation, but the corresponding data are not directly comparable because they concern communities of varying complexity, either dominated by a few protists (termite gut) or anaerobic ciliate and fungal taxa (bovine rumen) or encompassing several hundreds of organisms from the entire eukaryotic domain (soil). Accordingly, a significant proportion of the bovine (Muskoxen) rumen metatranscriptomic sequences could be assembled compared to the soil metatranscriptome. This observation suggests that only deep coverage sequencing can picture the full complexity of eukaryotic metatranscriptomes. In the case of the bovine rumen metatranscriptome, collector's curves plotting the number of contigs *versus* the number of sequences reached an asymptotic value for 18,000 to 20,000 genes, suggesting that the sequencing effort was sufficient to grasp the diversity of genes expressed in this environment.

Another salient feature of these studies is that a significant proportion of the sequences were homologous to NCBI nr entries in BLASTX searches; from 37 percent for the bovine rumen up to 52 percent in the termite gut (Tartar, Wheeler, et al., 2009) and 40 to 48 percent for the soil metatranscriptome. The large fraction of unannotated sequences could correspond to genes of unknown function with a restricted phylogenetic distribution. In this respect, this large proportion of unknown sequences should not prevent the ability to infer ecological processes from metatranscriptomic gene sequences because this inference is purely based on current knowledge of protein function. As for most transcriptomes, metatranscriptomes are dominated by house-keeping genes sequences that are of little value to infer ecological processes (e.g., gene involved in ribosome biogenesis and protein or cytoskeleton synthesis). Nevertheless, all studies also identified genes whose protein products are essential to the ecosystem functioning. This is the case of several genes-encoding biomass-degrading enzymes but also of membrane transporters participating to the assimilation of essential nutrients as well as of downstream metabolic enzymes (Damon, Lehembre, et al., 2012). As pointed out in soil and bovine rumen metratranscriptomes (Qi, Wang, et al., 2011; Damon, Lehembre, et al., 2012), some gene categories such as those coding for the carbohydrate active enzymes (CAZymes) are quite different from their relatives in databases. This observation, illustrated in Figure 14.2 for GH61 proteins from soil metatranscriptomes illustrates the fact that many well-studied and fully sequenced fungal species are not necessarily representative of most active organisms in the environment.

As with metagenomic surveys, metatranscriptomic surveys offer the possibility to attribute biological activities to unsuspected taxonomic groups. Thus, careful phylogenetic examination of 12 full-length CAZymes suggested that two of them, a GH45 (putative cellulase) and a GH7 (putative cellobiohydrolase) could originate from the soil microfauna and not from fungi; thus suggesting that the microfauna could play a significant, yet unsuspected, direct role in SOM hydrolysis (Damon, Lehembre, et al., 2012).

Data Interpretation: The Obvious and Potential Pitfalls

Transcriptome sequence data from single organisms can be interpreted both quantitatively and qualitatively. Similarly, for metatranscriptomes, transcript abundance may reflect gene expression level and can be used as a proxy to infer the level of activity of a specific process and the prevailing environmental conditions influencing that process. However, sequence clusters from metatranscriptomes can encompass homologous transcripts expressed by several related or unrelated organisms. Therefore, the assumption that the "size" of a cluster of transcripts reflects globally the level of activity of a specific

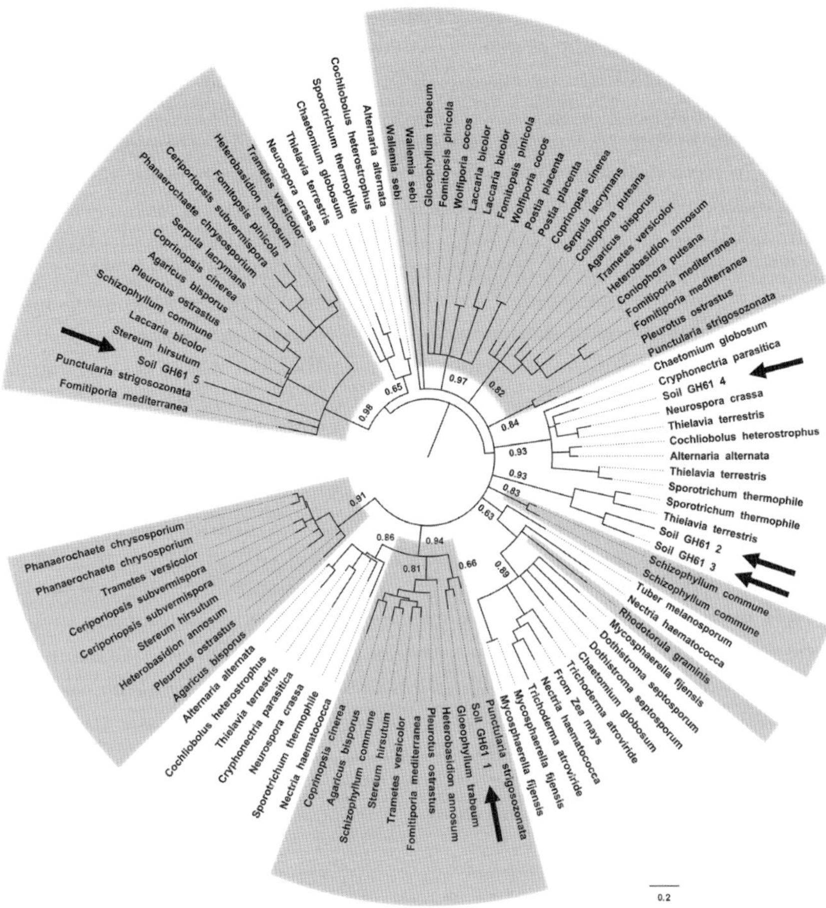

Figure 14.2 Phylogenetic assignation of forest soil sequences (*arrows*) encoding GH61 enzymes known to participate to cellulose degradation (Quinlan, Sweeney, et al., 2011). Five full-length GH61 cDNA sequences were identified among systematically sequenced cDNAs obtained from spruce and beech soil-extracted RNAs. Soil sequences seemed to originate from either Basidiomycota (shaded in grey) or Ascomycota (in white) fungal species. Branch supports (posterior probability) values above 0.6 are indicated for the most internal branches. Adapted from Damon, Lehembre, et al. 2012.

process is valid only if transcription of homologous genes from different organisms responds similarly to the same environmental or developmental clues. This is evidently not the case as can be illustrated by, for example, the regulation of fungal transporter genes that are essential to the use of soil nutrients. In the case of nitrate uptake, it has been demonstrated in many Ascomycota that nitrate ion (NO_3) transporter genes are both repressed by reduced N sources (e.g., ammonium, glutamine) and induced by NO_3 (Marzluf,

1997). In contrast, in other species such as the Basidiomycota *Hebeloma cylindrosporum*, only the presence of reduced N repressed transporter gene expression that does not require NO_3 for induction (Jargeat, Rekangalt, et al., 2003). From this example, it is obvious that absence or on the contrary a high prevalence of NO_3 transporter gene transcripts within a soil metatranscriptome can have different environmental origins depending on the taxonomic composition of the corresponding fungal communities.

In addition to comparing gene expression levels, metatranscriptomic sequence data sets can be interpreted in a more qualitative way by reconstructing putative biochemical or developmental pathways. In this case, taxonomic annotation of sequences attributed to a common pathway must be carefully evaluated to avoid reconstructing pathways where distinct steps are performed by proteins originating from distinct unrelated taxonomic groups. This is particularly important for developmental or biochemical pathways comprising strictly intracellular reactions, which are intrinsic of one single species. Regarding pathways that concern the biochemical modification or conversion of extracellular compounds, realization of specific steps by enzymes from specific taxonomic groups is on the contrary ecologically relevant information indicative of taxonomic specialization and of complementarities between divergent taxa for the realization of a common process. This would be relevant for processes such as SOM degradation and has been suggested for heavy metal inactivation by bacteria (Bertin, Heinrich-Salmeron, et al., 2011).

Systematic Sequencing: Target Genes

Whole metatranscriptome sequencing reports show that gene categories of particular interest (e.g., involved in SOM degradation) are detected, but often represent a minor fraction of the whole data set. For example, only 7 cellobiohydrolase (Glycoside Hydrolase GH7 family) and one endoxylanase (GH10 family) sequences were obtained from a metatranscriptome of 40,000 sequences from forest soils (Damon, Lehembre, et al., 2012). To get a deeper view into a function such as SOM degradation, one can focus on specific gene families whose cDNAs are PCR-amplified and sequenced using gene-specific degenerate primers targeting the gene of interest. Advantages of this approach compared to the whole metatranscriptome shotgun sequencing include (a) its capacity to produce thousands of sequences for a single gene family and (b) the possibility of inferring precisely gene diversity using phylogenetic approaches because a single unique DNA region is sequenced for all gene variants. Disadvantages include a) potential taxonomic biases resulting from primer design, (b) the focus on a limited number of gene categories, which may not be fully representative of the studied biological process, and (c) the overestimation of gene diversity as a result of PCR or sequencing errors (see also Chapter 13 of this book).

The recent development of degenerate primer sets for amplification of many different fungal ligninase, cellulase, and hemicellulase gene families and their use on nucleic acids extracted from soil samples provide cultivation-independent tools for assessing the genetic diversity and activity of lignocellulolytic degrading guild within fungal communities (Luis, Walther, et al., 2004; Edwards, Upchurch, et al., 2008; Bödeker, Nygren, et al., 2009; Kellner, Zak, et al., 2010; but see also Chapter 13 of this book). The development of such primers requires the availability of protein sequences in public databases. These degenerate primers target highly conserved regions bordering variable ones. For example, concerning fungal laccase proteins, sequences localized around the four domains involved in the copper-atom binding are highly conserved and allowed the design of efficient primers (Luis, Walther, et al., 2004).

Gene-specific amplification and sequencing of cDNAs has been used to follow SOM degradation at a fine spatial scale across neighboring cm-thick soil horizons. By using high-throughput 454 pyrosequencing of amplified cellobiohydrolase I transcripts from a spruce forest soil, Baldrian, Kolařík, et al. (2012) showed that few of the sequences were common between the litter and organic horizons, suggesting that different fungal species produced these proteins in these horizons. Similar horizon specificity was also independently observed by Luis, Kellner, et al. (2005b) for fungal laccase genes. Interestingly, some of the most abundant cellobiohydrolase I sequences were from saprotrophic fungal species such as *Mycena* spp. that were not among the most abundant species in the corresponding forest soil. This suggests that "minor" species may play a prominent role in cellulose hydrolysis (Baldrian, Kolařík, et al., 2012). Cellobiohydrolase I transcripts were nevertheless mostly from Ascomycota, thus underlining the importance of this fungal group in SOM decomposition in forest soils (Baldrian, Kolařík, et al., 2012).

Laccase and cellobiohydrolase I primers were also used to quantify the corresponding gene expression levels in forest soils by real-time PCR (qPCR). Higher N availability is known to affect the transcription of fungal ligninolytic genes, which can lead to a decline of enzyme activities and an accumulation of organic matter in soils, as evidenced in a sugar maple forest exposed to N deposition (Waldrop, Zak, et al., 2004). To determine if lower decomposition rates result from down-transcription of key lignocellulolytic genes or from changes in fungal community composition, Edwards, Zak, et al. (2011) simultaneously studied fungal community composition by sequencing 28S rRNA genes and laccase and cellobiohydrolase I gene expression levels by qRT-PCR using degenerate primers on soil-extracted RNA. Simulated N deposition increased the proportion of basidiomycota 28S rRNA sequences and significantly lowered the proportion of Ascomycota ones. The laccase gene expression was reduced under N deposition whereas the expression level of the cellobiohydrolase I gene did not significantly differ between N treatments. These results suggest that N deposition could lower decomposition rates through a combination of reduced expression of ligninolytic genes such as laccase and compositional changes in the fungal community.

Expression in Heterologous Hosts

Metagenomes or metatranscriptomes systematic sequencing approaches are limited when it comes to unknown or undescribed proteins. Gene discovery from metagenomes or metatranscriptomes can be achieved with functional metagenomics and metatranscriptomics, two powerful approaches that allow the identification of environmental clones that express a specific function. Functional metagenomics is based on the direct extraction of whole environmental DNA and on its direct cloning in appropriate vectors propagated in domesticated bacterial cells (e.g., *Escherichia coli*; Dinsdale, Edwards, et al., 2008). Screening of these libraries by different approaches has led to the discovery of novel genes originating from unknown, uncultured bacterial species (Rondon, August, et al., 2000; Daniel, 2005; Ferrer, Beloqui, et al., 2009; Chistoserdova, 2010). Functional metatranscriptomics is based on the conversion of eukaryotic poly-A mRNAs into cDNAs that are cloned in a vector, which allow their expression in a cultured heterologous host such as the yeast *Saccharomyces cerevisiae* (*see* Fig. 14.1).

Two main metagenomic or metatranscriptomic screening approaches have been followed. The first approach corresponds to the phenotypical detection of a desired biochemical activity (Liaw, Cheng, et al., 2010; Kellner, Luis, et al., 2011; Kang, Oh et al., 2011). It usually consists in incorporating in the growth medium a specific substrate conjugated to a chromophore. Enzymatic activities are then detected by means of release of a dye. The second approach is the phenotypic complementation of host strains or mutants unable to grow under specific culture conditions (Riesenfeld, Goodman, et al., 2004; Bailly, Fraissinet-Tachet, et al., 2007; Donato, Moe, et al., 2010; Damon, Vallon, et al., 2011). A significant limitation of both approaches is that many heterologous proteins may not be either expressed or functional in heterologous hosts. The scarcity of active clones is another limitation and necessitates the development of high-throughput screening strategies. For example, 1,2 10^6 clones containing soil metagenomic DNA were necessary to identify only 10 unique clones that conferred antibiotic resistance (Riesenfeld, Goodman, et al., 2004). Functional metagenomics allowed the isolation of several novel genes encoding for degradative enzymes (Rondon, August, et al., 2000; Gupta, Berg, et al., 2002; Kang, Oh, et al., 2011), or involved in antibiotic resistance (Riesenfeld, Goodman, et al., 2004; Donato, Moe, et al., 2010) and antibiotic synthesis pathways (MacNeil, Tiong, et al., 2001; Courtois, Cappellano, et al., 2003). Many of these genes have biotechnological applications or exhibit original activities. For example, Kanaya, Sakabe, et al. (2010) identified 12 type 1 RNases H encoding genes from a compost metagenomic library. Among them, 11 genes were novel including one lacking the typical type 1 RNase H active-site motif.

Proof of concept for functional eukaryotic metatranscriptomics was established by the cloning of two histidine biosynthetic genes from a forest soil cDNA library by complementation of a histidine auxotrophic yeast mutant (Bailly, Fraissinet-Tachet, et al., 2007). Subsequently, Kellner, Luis, et al. (2011) isolated a Basidiomycota-secreted acid phosphatase from a soil metatranscriptomic library by expression in a *Saccharomyces cerevisiae* phosphatase minus mutant. Similar screening was reported for eukaryotic cDNAs from the bovine rumen (Findley, Mormile, et al., 2011). These reports open the way to screening eukaryotic soil cDNA libraries for genes of biotech potential such as genes-encoding enzymes active on plant biomass. One of the scientifically most promising aspects of functional metatranscriptomic is its use to discover novel gene categories. This has been reported by Damon, Vallon, et al. (2011) who identified a novel fungal family of broad-specificity di/tripeptide membrane transporters by complementation of a *S. cerevisiae* mutant defective in di/tripeptide transport. In databases, homologous sequences to these transporter genes had all been automatically annotated as putative amino-acid transporters because of their weak similarities to known fungal amino acid transporters.

Future Prospects

Metatranscriptomics is an emerging research field that could rapidly represent a standard approach to study ecosystemic processes. Besides keeping up with constant progresses in high-throughput sequencing technologies and their associated bioinformatics tools, researchers who engage in this field must also make sure to comply with standards in ecology and environmental sciences. Among these basic standards is the necessity to adopt a rigorous sampling design and to replicate analyses on independent environmental samples collected in different locations in both space and time (Prosser, 2010). It is also necessary to adopt guidelines between laboratories to facilitate future meta-analyses, as already proposed for metagenomes (Field, Amaral-Zettler, et al., 2011). It is therefore, not only necessary to make sequence data available in a standard format, but also environmental meta-data (Field, Amaral-Zettler, et al., 2011).

In the field of ecology, metatranscriptomics should also not be regarded as a standalone approach; it must be combined with additional field measurements (e.g., for estimating global fluxes). In that sense, developing a metatranscriptomic program to study degradation of SOM must be performed in parallel and not as a substitute to other more traditional field measurements such soil respiration or litter mass loss.

Metatranscriptomics represents also one facet of environmental transcriptomics. Although not yet explored, transcriptomes of individual microorganisms collected in the field could be obtained thanks to recent advances in single-cell genomics, for instance by using laser capture microdissection to study single fungal hyphae transcriptomes (de Bekker, Brunig, et al., 2011). Such approaches open new perspectives to understand the relation between genotypic diversity and adaptation to environmental change, as already done for macroorganisms.

References

Altschul SF, Gish W, et al. 1990. Basic local alignment search tool. *J Mol Biol*. 215: 403–410.

Ashburner M, Ball CA, et al. 2000. Gene ontology: Tool for the unification of biology. *Nature Genet*. 25: 25–29.

Bailly J, Fraissinet-Tachet L, et al. 2007. Soil eukaryotic functional diversity, a metatranscriptomic approach. *ISME J*. 1: 632–642.

Baldrian P, Kolařík M, et al. 2012. Active and total microbial communities in forest soil are largely different and highly stratified during decomposition. *ISME J*. 6: 248–258.

Bertin PN, Heinrich-Salmeron A, et al. 2011. Metabolic diversity among main microorganisms inside an arsenic-rich ecosystem revealed by meta- and proteo-genomics. *ISME J*. 5: 1735–1747.

Bödeker IT, Nygren CM, et al. 2009. Class II peroxidase-encoding genes are present in a phylogenetically wide range of ectomycorrhizal fungi. *ISME J*. 3: 1387–1395.

Burki F, Shalchian-Tabrizi K, et al. 2007. Phylogenomics reshuffles the eukaryotic supergroups. *PLoS One*. 2: e790.

Chistoserdova L. 2010. Recent progress and new challenges in metagenomics for biotechnology. *Biotechnol Lett*. 32: 1351–1359.

Courtois S, Cappellano CM, et al. 2003. Recombinant environmental libraries provide access to microbial diversity for drug discovery from natural products. *Appl Environ Microbiol*. 69: 49–55.

Damon C, Vallon L, et al. 2011. A novel fungal family of oligopeptide transporters identified by functional metatranscriptomics of soil eukaryotes. *ISME J*. 5: 1871–1880.

Damon C, Lehembre F, et al. 2012. Metatranscriptomics reveals the diversity of genes expressed by eukaryotes in forest soils. *PLoS One*. 7: e28967.

Daniel R. 2005. The metagenomics of soil. *Nat Rev Microbiol*. 3: 470–478.

de Bekker C, Bruning O, et al. 2011. Single cell transcriptomics of neighboring hyphae of *Aspergillus niger*. *Genome Biol*. 12: R71.

Dinsdale EA, Edwards RA, al. 2008. Functional metagenomic profiling of nine biomes. *Nature*. 452: 629–632.

Donato JJ, Moe LA, et al. 2010. Metagenomic analysis of apple orchad soil reveals antibiotic resistance genes encoding predicted bifunctional proteins. *Appl Environ Microbiol*. 76: 4396–4401.

Dong D, Yan A, et al. 2006. Removal of humic substances from soil DNA using aluminium sulfate. *J Microbiol Methods*. 66: 217–222.

Edwards IP, Upchurch RA, et al. 2008. Isolation of fungal cellobiohydrolase I genes from sporocarps and forest soils by PCR. *Appl Environ Microbiol*. 74: 3481–3489.

Edwards IP, Zak DR, et al. 2011. Simulated atmospheric N deposition alters fungal community composition and suppresses ligninolytic gene expression in a northern hardwood forest. *PLoS One*. 6: e20421.

Feike J, Jürgens K, et al. 2012. Measuring unbiased metatranscriptomics in suboxic waters of the central Baltic sea using a new in situ fixation system. *ISME J*. 6: 461–470.

Ferrer M, Beloqui A, et al. 2009. Metagenomics for mining new genetic resources of microbial communities. *J Mol Microbiol Biotechnol*. 16: 109–123.

Field D, Amaral-Zettler L, et al. 2011. The genomic standards consortium. *PLoS Biol*. 9: e1001088.

Findley SD, Mormile MR, et al. 2011. Activity-based metagenomic screening and biochemical characterization of bovine ruminal protozoan glycoside hydrolases. *Appl Environ Microbiol*. 77: 8106–8113.

Frias-Lopez J, Shi Y, et al. 2008. Microbial community gene expression in ocean surface waters. *Proc Natl Acad Sci USA*. 105: 3805–3810.

Gilbert JA, Field D, et al. 2008. Detection of large numbers of novel sequences in the metatranscriptomes of complex marine microbial communities. *PLoS One*. 3: e3042.

Gilbert JA, Field D, et al. 2010. The taxonomic and functional diversity of microbes at a temperate coastal site: a 'multi-omic' study of seasonal and diel temporal variation. *PLoS One*. 5: e15545.

Gomez-Alvarez V, Teal TK, et al. 2009. Systematic artifacts in metagenomes from complex microbial communities. *ISME J*. 3: 1314–1317.

Götz S, García-Gómez JM, et al. 2008. High-throughput functional annotation and data mining with the Blast2GO suite. *Nucl Acids Res*. 36: 3420–3435.

Grant S, Grant WD, et al. 2006. Identification of eukaryotic open reading frames in metagenomic cDNA libraries made from environmental samples. *Appl Environ Microbiol*. 72: 135–143.

Griffiths RI, Whiteley AS, et al. 2000. Rapid method for coextraction of DNA and RNA from natural environments for analysis of ribosomal DNA- and rRNA-based microbial community composition. *Appl Environ Microbiol*. 66: 5488–5491.

Gupta R, Berg QK et al. 2002. Bacterial alkaline proteases: Molecular approaches and industrial applications. *Appl Microbiol Biotechnol*. 59: 15–32.

Handelsman J, Rondon MR, et al. 1998. Molecular biological access to the chemistry of unknown soil microbes: a new frontier for natural products. *Chem Biol*. 5: R245–R249.

He S, Wurtzel O, et al. 2010. Validation of two ribosomal RNA removal methods for microbial metatranscriptomics. *Nat Methods*. 7: 807–812.

Horton TR & Bruns TD. 2001. The molecular revolution in ectomycorrhizal ecology: Peeking into the black-box. *Mol Ecol*. 10: 1855–1871.

Hurt RA, Qiu X, et al. 2001. Simultaneous recovery of RNA and DNA from soils and sediments. *Appl Environ Microbiol*. 67: 4495–4503.

Jargeat P, Rekangalt D, et al. 2003. Characterisation and expression analysis of a nitrate transporter and nitrite reductase genes, two members of a gene cluster for nitrate assimilation from the symbiotic basidiomycete Hebeloma cylindrosporum. *Curr Genet*. 43: 199–205.

Kanaya E, Sakabe T, et al. 2010. Cloning of the RNase H genes from a metagenomic DNA library: Identification of a new type 1 RNase H without a typical active-site motif. *J Appl Microbiol*. 109: 974–983.

Kanehisa M, Goto S, et al. 2004. The KEGG resource for deciphering the genome. *Nucl Acids Res*. 32: D277–D280.

Kang CH, Oh KH, et al. 2011. A novel family VII esterase with industrial potential from compost metagenomic library. *Microbial Cell Fact*. 10: 41.

Kellner H, Luis P, et al. 2011. Screening of a soil metatranscriptomic library by functional complementation of Saccharomyces cerevisiae mutants. *Microbiol Res*. 166: 360–368.

Kellner H, Zak RD, et al. 2010. Fungi unearth: transcripts encoding lignocellulosic and chitinolytic enzymes in forest soil. *PLoS One*. 5: e10971

Kunin V, Copeland A, et al. 2008. A bioinformatician's guide to metagenomics. *Microbiol Mol Biol Rev*. 72: 557–578.

Lejzerowicz F, Pawlowski J, et al. 2010. Molecular evidence for widespread occurrence of Foraminifera in soils. *Environ Microbiol*. 12: 2518–2526.

Lesaulnier C, Papamichail D, et al. 2008. Elevated atmospheric CO_2 affects soil microbial diversity associated with trembling aspen. *Environ Microbiol.* 10: 926–941.
Liaw RB, Cheng MP, et al. 2010. Use of metagenomic approaches to isolate lipolytic genes from activated sludge. *Bioresources Technol.* 101: 8323–9329.
Luis P, Walther G, et al. 2004. Diversity of laccase genes from basidiomycetes in a forest soil. *Soil Biol Biochem.* 36: 1025–1036.
Luis P, Kellner H, et al. 2005a. A molecular method to evaluate basidiomycete laccase geneexpression in forest soils. *Geoderma.* 128: 18–27.
Luis P, Kellner H, et al. 2005b. Patchiness and spatial distribution of laccase genes of ectomycorrhizal, saprotrophic, and unknown basidiomycetes in the upper horizons of a mixed forest cambisol. *Microbial Ecol.* 50: 570–579.
Lützow MV, Kögel-Knabner I, et al. 2006. Stabilization of organic matter in temperate soils: Mechanisms and their relevance under different soil conditions—a review. *Eur J Soil Sci.* 57: 426–445.
MacLean D, Jones JD et al. 2009. Application of 'next-generation' sequencing technologies to microbial genetics. *Nat Rev Microbiol.* 7: 287–296.
MacNeil IA, Tiong CL, et al. 2001. Expression and isolation of antimicrobial small molecules from soil DNA libraries. *J Mol Microbiol Biotechnol.* 3: 301–308.
Maron PA, Ranjard L, et al. 2007. Metaproteomics: A new approach for studying functional microbial ecology. *Microbial Ecol.* 53: 486–493.
Marzluf GA. 1997. Genetic regulation of nitrogen metabolism in the fungi. *Microbiol Mol Biol Rev.* 61: 17–32.
Mettel C, Kim Y, et al. 2010. Extraction of mRNA from soil. *Appl Environ Microbiol.* 76: 5995–6000.
Nguyen VH & Lavenier D. 2009. PLAST: Parallel local alignment search tool for database comparison. *BMC Bioinformatics.* 10: 329.
Noguchi H, Park J, et al. 2006. MetaGene: Prokaryotic gene finding from environmental genome shotgun sequences. *Nucl Acids Res.* 34: 5623–5630.
O'Brien HE, Parrent JL, et al. 2005. Fungal community analysis by large-scale sequencing of environmental samples. *Appl Environ Microbiol.* 71: 5544–5550.
Parfrey LW, Grant J, et al. 2010. Broadly sampled multigene analyses yield a well-resolved eukaryotic tree of life. *Syst Biol.* 59: 518–533.
Prosser JI. 2010. Replicate or lie. *Environ Microbiol.* 12: 1806–1810.
Qi M, Wang P, et al. 2011. Snapshot of the eukaryotic gene expression in muskoxen rumen—a metatranscriptomic approach. *PLoS One.* 6: e20521.
Quinlan RJ, Sweeney MD, et al. 2011. Insight into oxidative degradation of cellulose by a copper metalloenzyme that exploits biomass components. *Proc Natl Acad Sci USA.* 108: 15079–15084.
Rappé MS & Giovannoni SJ. 2003. The uncultured microbial majority. *Annu Rev Microbiol.* 57: 369–394.
Richards TA & Bass D. 2005. Molecular screening of free-living microbial eukaryotes: Diversity and distribution using a meta-analysis. *Curr Opin Microbiol.* 8: 240–252.
Riesenfeld CS, Goodman RM, et al. 2004. Uncultured soil bacteria are a reservoir of new antibiotic resistance genes. *Environ Microbiol.* 6: 981–989.
Rinta-Kanto JM, Sun S, et al. 2012. Bacterial community transcription patterns during a marine phytoplankton bloom. *Environ Microbiol.* 14: 228–239.
Rondon MR, August PR, et al. 2000. Cloning the soil metagenome: A strategy for accessing the genetic and functional diversity of uncultured microorganisms. *Appl Environ Microbiol.* 66: 2541–2547.
Schneider T, Keiblinger KM, et al. 2012. Who is who in litter decomposition? Metaproteomics reveals major microbial players and their biogeochemical functions. *ISME J.* doi: 10.1038/ismej.2012.11.

Scholz MB, Lo C-C, et al. 2012. Next generation sequencing and bioinformatic bottlenecks: the current state of metagenomic data analysis. *Curr Opin Biotechnol*. 23: 9–15.

Shi Y, Tyson GW, et al. 2011. Integrated metatranscriptomic and metagenomic analyses of stratified microbial assemblages in the open ocean. *ISME J*. 5: 999–1013.

Shrestha PM, Kube M, et al. 2009. Transcriptional activity of paddy soil bacterial communities. *Environ Microbiol*. 11: 960–970.

Smith SE & Read DJ. 2008. Mycorrhizal Symbiosis, 3rd ed. London: Academic Press.

Spalding D, Kendirli E, et al. 2012. The role of forests in global carbon budgeting. In: Managing Forest Carbon in a Changing Climate (eds. MS Ashton, ML Tyrrell, et al.), 165–179. Berlin: Springer.

Stewart FJ, Ulloa O, et al. 2012. Microbial metatranscriptomics in a permanent marine oxygen minimum zone. *Environ Microbiol*. 14: 23–40.

Stewart FJ, Sharma AK, et al. 2011. Community transcriptomics reveals universal patterns of protein sequence conservation in natural microbial communities. *Genome Biol*. 12: R26.

Tamames J, de la Peña S, et al. 2012. COVER: A priori estimation of coverage for metagenomic sequencing. *Environ Microbiol Rep*. doi: 10.1111/j.1758-2229.2012.00338.x.

Tartar A, Wheeler MM, et al. 2009. Parallel metatranscriptome analyses of host and symbiont gene expression in the gut of the termite *Reticulitermes flavipes*. *Biotechnol Biofuels*. 2: 25.

Todaka N, Moriya S, et al. 2007. Environmental cDNA analysis of the genes involved in lignocellulose digestion in the symbiotic protist community of *Reticulitermes speratus*. *FEMS Microbiol Ecol*. 59: 592–599.

Toulza E, Tagliabue A, et al. 2012. Analysis of the global ocean sampling (GOS) project for trends in iron uptake by surface ocean microbes. *PLoS One*. 7: e30931.

Tringe SG, von Mering C, et al. 2005. Comparative metagenomics of microbial communities. *Science*. 308: 554–557.

Tripp HJ, Hewson I, et al. 2011. Misannotations of rRNA can now generate 90% false positive protein matches in metatranscriptomic studies. *Nucl Acids Res*. 39: 8792–8802.

Urich T, Lanzen A, et al. 2008. Simultaneous assessment of soil microbial community structure and function through analysis of the metatranscriptome. *PLoS One*. 3: e2527.

Valášková V, Šnajdr J, et al. 2007. Production of lignocellulose-degrading enzymes and degradation of leaf litter by saprotrophic basidiomycetes isolated from a Quercus petraea forest. *Soil Biol Biochem*. 39: 2651–2660.

Waldrop MP, Zak DR, et al. 2004. Nitrogen deposition modifies soil carbon storage through changes in microbial enzymatic activity. *Ecol Appl*. 14: 1172–1177.

Weber CF & Kuske CR. 2011. Reverse transcription-PCR methods significantly impact richness and composition measures of expressed fungal cellobiohydrolase I genes in soil and litter. *J Microbiol Methods*. 86: 344–350.

Xie L, Zhang L, et al. 2012. Profiling the metatranscriptome of the protistan community in Coptotermes formosanus with emphasis on the lignocellulolytic system. *Genomics*. 99: 2246–2255.

15 Fungi in Deep-Sea Environments and Metagenomics

Stéphane Mahé[1], Vanessa Rédou[2], Thomas Le Calvez[1], Philippe Vandenkoornhuyse[1], and Gaëtan Burgaud[2]

[1] Université de Rennes 1, CNRS, UMR6553 EcoBio, Observatoire Des Sciences de l'Univers de Rennes (OSUR), Campus de Beaulieu, Rennes, France

[2] Laboratoire Universitaire de Biodiversité et Ecologie Microbienne, Université Européenne de Bretagne, Université de Brest, ESIAB Technopôle Brest-Iroise, Plouzané, France

Fungi in Oceans: An Overview

Oceans harbor a broad diversity of photosynthesis- and chemosynthesis-based ecosystems, from coastal waters to the deep biosphere that is, oceanic waters below 1000 meters (Jannasch & Taylor, 1984), and a wide diversity of microorganisms involved in all biogeochemical cycles (Fig. 15.1). Fungi in oceans form an ecologically defined group of filamentous ascomycetes, their anamorphs, and yeasts (Kohlmeyer & Kohlmeyer, 1979). Although ecologically important relationships with other organisms (e.g., pathogens and symbionts of algae, higher plants and animals, as well as an important role as decomposers) have been clearly demonstrated, few species of marine fungi have been listed to date. Recent reports of fungi in the deep oceans have provided insights into their global diversity and ecological role. The constant increase of molecular data using clone libraries or high-throughput sequencing is permitting a revision of the definition of marine fungi.

History and Definitions: The Concept of Marine Fungi

From the first comprehensive study on marine fungi by Barghoorn and Linder (1944) to the first exhaustive and comprehensive book dealing with marine mycology (Kohlmeyer & Kohlmeyer, 1979), and from the initial recovery of fungal communities at deep-sea extreme environments (Nagahama, Hamamoto, et al., 2001) to one recent exhaustive review dealing with deep fungal communities (Nagano & Nagahama, 2012), marine fungi have

The Ecological Genomics of Fungi, First Edition. Edited by Francis Martin.

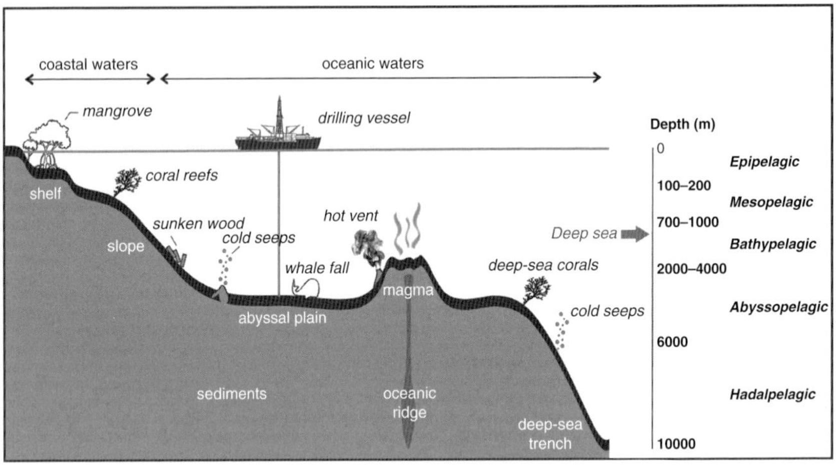

Figure 15.1 Schematic transversal section of the earth highlighting different marine ecosystems (not to scale). For color detail, please see color plate section.

usually been considered as exotic microorganisms that only fascinate a limited panel of scientists. The first definition of a marine fungus was produced by Johnson and Sparrow (1961) and based on an ability to grow at seawater concentrations. This physiologically based postulate was later criticized by Kohlmeyer and Kohlmeyer (1979) who suggested a broad ecological definition dividing marine fungi into obligate and facultative groups. This strict dichotomy between obligate and facultative marine fungi led to a distinction between fungi "that grow and sporulate exclusively in a marine or estuarine habitat from those from freshwater or terrestrial milieus able to grow and possibly to sporulate in the marine environment" (Kohlmeyer & Kohlmeyer, 1979: 42). Currently, the interest in marine fungi has clearly exploded, and this is reflected in several recent reviews (Shearer, Descals, et al., 2007; Jones, Sakayaroj, et al., 2009; Jones, 2011; Nagahama & Nagano, 2012; Nagano & Nagahama, 2012; Richards, Jones, et al., 2012). The idea here is not to provide yet another updated review dealing with marine mycology but rather to discuss the actual concept of marine fungi and to attempt to update the preceding definition.

The definition proposed by Kohlmeyer and Kohlmeyer (1979) has been widely accepted by the scientific community but, as is usual, has not escaped debate. The dilemma is always whether or not to consider facultative marine fungi because the split generated by this definition has led to a kind of segregation against facultative marine fungi. This can be easily explained from the analytical strategy used by scientists: (a) historically, direct observations of in situ fungal structures have allowed the identification and characterization of numerous obligate marine fungi and (b) untargeted culture-based approaches have mostly revealed strains commonly isolated in terrestrial habitats and thus defined as facultative marine fungi. The fact that facultative marine fungi were receiving little attention led some scientists to propose an initial exegesis, stating that "terrestrial species turn out far too regularly from macrophytic detritus to be dismissed lightly" (Raghukumar & Raghukumar, 1999: 26). In a form of response, Kohlmeyer and Volkmann-Kohlmeyer (2003) commented on the plate method, which often reveals common dust and wind-born forms and concluded that advocates of the untargeted culture-based method must be responsible for clarifying whether such facultative marine fungi occur as dormant spores or not.

Some studies have proven facultative marine fungi with terrestrial representatives to have an ecological role in marine environments, for example, *Aspergillus sydowii* pathogen of corals (Geiser, Taylor, et al., 1998) or *Fusarium oxysporum* pathogen of crustaceans (Khoa & Hatai, 2005). Metabolic profile analyses of facultative marine fungi have also helped to clarify their ecological role because marine-derived fungi synthesize a broad spectrum of secondary metabolites. Some have been retrieved from both terrestrial and marine strains (e.g., wentilactone from *Aspergillus wentii*). But in general, many compounds harvested from marine-derived fungi differ from those of their terrestrial representatives (Bhakuni & Rawat, 2005), for example, nitrogenous compounds such as heptapeptides from *Acremonium persicinum* showing cytotoxicity against brine shrimps and toward tumor cell lines (Chen, Song, et al., 2012), polyketides from *Phoma herbarum* showing antiviral activity (Zhang, Han, et al., 2012), and terpenoids from *Penicillium chrysogenum* showing antibacterial activity (Gao, Li, et al., 2011). This hidden wealth of resources has not only opened up a new era in drug research but also suggests that secondary metabolites from marine-derived fungi may act as a chemical defense mechanism in oceans and that marine-derived fungi are not only dormant spores in this ecosystem (Damare, Singh, et al., 2012). Although the synthesis of exotic secondary bioactive molecules may indicate a stress response, it definitely indicates metabolic activity. Going round in circles, Capon, Ratnayake, et al. (2005) revealed that the terrestrial fungal strain *Aspergillus unilateralis* was able to synthesize Trichodermamide B dipeptide, a typical marine-derived metabolite, when cultured in media enriched with sodium chloride (NaCl). This result suggests that metabolic expression profiles are influenced by environmental rather than genetic factors.

What if this old debate regarding obligate and facultative marine fungi can be laid to rest just by accepting that it is a non-issue? Scaling up the methods used to define marine fungi might provide a kind of consensus. Most marine mycologists have discussed Kohlmeyer and Kohlmeyer's definition summarizing marine fungi as an ecological rather than taxonomic group (Hyde, Jones, et al., 1998; Shearer, Descals, et al., 2007; Gao, Li, et al., 2008;

Burgaud, Le Calvez, et al., 2009; Jones, Sakayaroj, et al., 2009; Das, Lyla, et al., 2009; Nagahama & Nagano, 2012). However, based on this definition, marine fungi are truly initially defined by culture efficiency or direct observations rather than by ecological traits. Culture-based and in situ observation approaches are biased by unculturable fungi, endophytes, cryptic species, strict parasites, fast growing strains, dormant spores germination and indeed are strongly selective because they only allow detection of a tiny fraction of fungi in environmental samples. The huge advances in marine molecular ecology have led to a reversal of these common approaches and thus constrain to update Kohlmeyer and Kohlmeyer's definition through the use of informative keywords. The keywords mostly used to describe fungal communities in marine environments are *isolation*, *direct detection*, *diversity*, *adaptation*, and *ecological role*. *Activity* appears as another keyword that fits well to an updated version of the Kohlmeyer and Kohlmeyer's definition. Fungal communities in oceans can thus be divided into three levels of occurrence: (a) strict endemic active marine fungi, (b) ubiquist metabolically active marine fungi, and (c) ubiquist passive fungi. By combining this three-level classification with Kohlmeyer and Kohlmeyer's definition, a pattern of fungal diversity, based on a functional scale, can be drawn up. The current new era of metagenomics and metatranscriptomics is providing enough power to fit this pattern.

Diversity and Ecological Roles of Fungi in Oceans

Marine as compared with terrestrial ecosystems have been little studied by mycologists. Currently, 549 obligate marine species of higher fungi (i.e., *Ascomycota* and *Basidiomycota*) have been described based on the Kohlmeyer and Kohlmeyer's definition, with only 54 species since the year 2000 (Jones, 2011). Only a few basal lineage species have been identified from marine environments (Gleason, Küpper, et al., 2011). In littoral and sublittoral regions, fungal species are mainly lignicolous (Jones, 2011) and algicolous (Bugni & Ireland, 2004), but have also been isolated from substrates such as marine plants, for example, partly submerged plants such as *Spartina* or permanently submerged plants such as *Posidonia*, sand, corals, calcareous algae, mollusk shells, hydrozoan exoskeletons, or annelid tubes (Jones, 2011). Microscopic fruiting bodies were directly examined on marine substrates or isolated on culture media using ascospores from ascoma as inoculum. Cultivation and molecular approaches were proved more appropriate in open-sea ecosystems. Indeed, the first deep-sea studies were based on the observation of fruiting bodies on wood- or polyurethane-covered panels (Jones, 2011), which made it difficult to differentiate between indigenous fungi and contaminants (Kohlmeyer & Kohlmeyer, 1979). Although descriptions based on an examination of fruiting bodies on marine substrates have been widely used, a culture-based approach has led to the description of many marine fungi in various marine niches at different depths (Table 15.1).

When molecular methods were used, with or without a fungal-specific approach (Table 15.2), the sequences of higher fungi retrieved were often closely affiliated to terrestrial fungi (Richards, Jones, et al., 2012). Yeasts appeared dominant because many genera were retrieved, for example, *Cryptoccocus*, *Rhodotorula*, *Candida*, and *Debaryomyces* (Bass, Howe, et al., 2007; Le Calvez, Burgaud, et al., 2009). Environmental sequences close to the yeast *Malassezia* were frequently found in eukaryotic- or fungal-specific studies in different marine extreme environments and have been clustered in a group formally known as "hydrothermal and/or anoxic marine yeasts" (Dawson & Pace, 2002; López-García, Vereshchaka, et al., 2007; Bass, Howe, et al., 2007; Le Calvez, Burgaud, et al., 2009; Jebaraj, Raghukumar, et al., 2010; Edgcomb, Beaudoin, et al., 2011). Significant numbers of *Malassezia* sequences were retrieved in hypersaline anoxic basins in the Mediterranean Sea (Alexander, Stock, et al., 2009; Edgcomb, Orsi, et al., 2009). Filamentous forms, such as *Aspergillus*, *Exophiala*, and *Tilletiopsis* are fairly well represented in cultures (Damare, Raghukumar, et al., 2006; Burgaud, Le Calvez, et al., 2009) and their presence has been confirmed using molecular methods (López-García, Rodríguez-Valera, et al., 2001; Stoeck, Hayward, et al., 2006; Bass, Howe, et al., 2007). Sequences close to terrestrial *Agaricomycetes* were also reported in deep marine ecosystems, for example, *Exidia*, *Coprinopsis*, or *Antrodia* (Bass, Howe, et al., 2007; Alexander, Stock, et al., 2009; Le Calvez, Burgaud, et al., 2009; Jebaraj, Raghukumar, et al., 2010) leading to debate about the activity of terrestrial fungi in such extreme environments and the need to investigate functional diversity rather than genetic diversity. Basal fungal lineages were detected in various marine systems by using molecular methods (i.e., *Blastocladiomycota*, *Chytridiomycota*, and *Cryptomycota*) but most of them have not yet been affiliated to known isolated representatives (Richards, Jones, et al., 2012).

Another level of fungal occurrence was revealed by pyrosequencing eukaryotic V4 and V9 tags from marine anoxic waters. Potential sources of error that might inflate the apparent level of diversity were discussed (nucleotide misincorporation, read errors, chimaera formation) but fungal diversity was much higher than expected with 1.5 to 4 percent of unique fungal tags depending on whether the markers, V4 or V9 marker was used (Stoeck, Bass, et al., 2010). Sequencing of the eukaryotic V9 diversity tags from sediment samples (686–6,326 m) using Roche 454-pyrosequencing revealed 101 operational taxonomic units (OTUs) in six samples, as compared to the total eukaryotic richness of 8309 OTUs representing 1.2 percent of the total eukaryotic diversity (Pawlowski, Christen, et al., 2011). A vertical pattern became apparent because most of the OTUs were obtained at lower depths. Fungi do not constitute a major component of the total eukaryotic OTUs recovered

Table 15.1 Fungal phyla retrieved by cultivation methods from different marine samples at various depths.

Reference	Location and Depth	Kind of Sample Processed	Phylogenetic Affinities
Nagahama, Hamamoto, et al., 2001	Sagami and Suruga Bay, Japan (1,000 to 11,000 m)	Superficial sediments Animals	*Ascomycota* *Basidiomycota*
Biddle, House, et al., 2005	Peru Margin (150 m)	Sediments (30 to 157 m below subfloor)	*Ascomycota*
Raghukumar, Raghukumar, et al., 2004	Chagos Trench Indian Ocean (5,904 m)	Sediments (0 to 370 cm below subfloor)	*Ascomycota*
Damare, Raghukumar, et al., 2006	Central Indian Basin (4,900 to 5,390 m)	Superficial sediments	*Ascomycota*
Singh, Raghukumar, et al., 2010	Central Indian Basin (4,000 to 5,700 m)	Superficial sediments	*Ascomycota* *Basidiomycota*
Gadanho & Sampaio, 2005	Hydrothermal sites, MAR (800 to 3,150 m)	Water: 3 to 5 m above the sea floor	*Ascomycota* *Basidiomycota*
Burgaud, LeCalvez, et al., 2009	Hydrothermal sites, MAR and EPR (700 to 3,650 m)	Water, sediments, mineral samples, animals	*Ascomycota* *Basidiomycota*
Le Calvez, Burgaud, et al., 2009	Hydrothermal sites, MAR (860 & 1,700 m) and EPR (2,630 m)	Animals, mineral samples	*Ascomycota*
Burgaud, Arzur, et al., 2010	Hydrothermal sites, MAR and Lau Basin: Pacifique (900 to 3,630 m)	Water, animals, experimental microcolonizer, shrimp sloughs	*Ascomycota* *Basidiomycota*
Jebaraj & Raghukumar, 2009	Arabian Sea (14 & 26 m)	Anoxic superficial sediments	*Ascomycota*
Jebaraj, Raghukumar, et al., 2010	Arabian Sea (3, 25 & 200 m)	Anoxic superficial sediments, water	*Ascomycota* *Basidiomycota*
Mouton, Potsma, et al., 2012	St Helena Bay (South Africa) (8, 15, 28, 32 61 m)	Superficial sediments	*Ascomycota* *Basidiomycota*

EPR, East Pacific Rise, MAR, Mid-Atlantic Ridge.

Table 15.2 Fungal phyla recovered by fungal-specific molecular methods from different marine samples at various depths.

Reference	Location and Depth	Kind of Sample Processed	Phylogenetic Affinities
Bass, Howe, et al., 2007	Wreck of *Bismarck* (3,000 & 4,000 m)	Water	*Ascomycota*
	Wreck of *Titanic* (3,000 & 3,700 m)	Water	*Basidiomycota*
	Hydrothermal site, MAR (2,264 m)	Sediments, experimental microcolonizers	*Chytridiomycota*
	Drake Passage (250, 500, 2,000 & 3,000 m)	Water	
	Gulf of California (1,575 m)	Anoxic bacterial mat	
Lai, Cao, et al., 2007	South China Sea (350, 884, 1,123, 2,965 & 3,011 m)	Methane hydrate-bearing sediments	*Ascomycota* *Basidiomycota*
Le Calvez, Burgaud, et al., 2009	Hydrothermal sites, MAR and EPR (860 to 2,630 m)	Animals, mineral samples	*Basidiomycota* *Chytridiomycota*
Nagano, Nagahama, et al., 2010	Izu-Ogasawara Trench (7,111 & 9,760 m)	Sediments (0 to 110 cm below seafloor)	*Ascomycota* *Basidiomycota*
	Mariana Trench (10,131 m)	Sediments	*Cryptomycota*
	Sagami Bay, methane cold-seep (1,174 m)	Bacterial mats	
Nagahama, Takahashi, et al., 2011	Sagami Bay, methane cold-seep (830 to 1,200 m)	Sediments	*Ascomycota* *Basidiomycota* *Blastocladiomycota* *Cryptomycota*
Thaler, Van Dover, et al., 2012	Gulf of Mexico, methane cold-seep (2,400 m)	Sediments (0 to 30 cm below seafloor)	*Ascomycota* *Basidiomycota* *Chytridiomycota*

EPR East Pacific Rise, MAR Mid-Atlantic Ridge.

from marine environments with only 1 percent of fungal OTUs obtained from coastal seawaters (Monchy, Grattepanche, et al., 2012). However, new high-throughput technologies are definitely enhancing the assessment of the marine fungal diversity.

The functions of fungi in oceans and their impact have been poorly investigated because the attention has only been focused on describing the diversity. However, major terrestrial fungal functions have already been observed in ocean systems. Fungi play pivotal roles as lignocellulolytic decomposers of floating or sunken woody substrates in marine ecosystems such as mangroves or other coastal areas (Hyde, Jones, et al., 1998). Fungal decomposers split complex polymers into particulate organic matter that then becomes available to other organisms. Planktonic marine fungi (i.e., ascomycetous and basidiomycetous mycoplankton) were found associated with the decomposition of organic matter, nutrient, and carbon cycling (Gao, Johnson, et al., 2010) and displayed distinct lateral and vertical patterns. Mycoplankton diversity and composition were relatively well correlated to phytoplankton biomass and primary production in Hawaii coastal waters from 5 to 200 m depth. Gutiérrez, Pantoja, et al. (2011) indicated a third dynamic temporal pattern of mycoplankton distribution represented by the Humboldt Current System, a cold, low-salinity and nutrient-rich ocean current rising to the surface. A large fungal biomass was retrieved during productive periods and was associated with a bloom of extracellular enzymes hydrolyzing up to 90 percent of the organic molecules from marine photosynthetic producers during the seasonal peak of production. Mycoplankton, such as prokaryotes, appeared as important actors in marine processes because of their abundance, diversity, and active participation to the carbon cycle through the release of dissolved organic molecules. Fungi also intervene in other biogeochemical patterns, such as the N cycle occurring in marine sediments. Fungal strains isolated from anoxic or suboxic sediments were able to participate in anaerobic denitrification processes, reducing nitrate or nitrite under anaerobic conditions (Jebaraj & Raghukumar, 2009; Mouton, Postma, et al., 2012).

As in terrestrial ecosystems, marine fungi can be found associated with macro and microorganisms. Some marine fungal species (e.g., *Leiophloea*, *Pharcidia*, or *Mycosphaerella* genera) contract symbioses with microscopic algae or cyanobacteria to form lichens. Some are associated with macroalgae to form mycophycobioses, such as members of the *Blodgettia* and *Turgidosculum* genera (Kohlmeyer & Kohlmeyer, 1979; Hyde, Jones, et al., 1998). Sponges host an astonishing diversity of fungal species but studies have tended to concentrate on the identification of secondary metabolites rather than their ecological role (Bugni & Ireland, 2004). Some fungal communities associated with sponges have been described (Gao, Li, et al., 2008), but the type of relationship between the two organisms has not been determined. Fungal pathogens of marine animals or algae were reported and were able to cause severe infections (Hyde, Jones, et al., 1998). In deep-sea ecosystems, the yeast-like fungus *Exophiala* was identified as the causative agent of mass mortalities in deep-sea endemic mussels (Van Dover, Ward, et al., 2007). Some other marine fungi are described, as skeletal-components of healthy, partially dead, and diseased corals (Ravindran, Raghukumar, et al., 2001). *A. sydowii* was found to cause an epizooty among sea fan corals (Alker, Smith, et al., 2001). *Scolecobasidium* sp. was the causative agent of necrotic patches on five different corals, *Porites lutea, Porites lichen, Montipora tuberculosa, Goniopora* sp., and *Goniastra* sp. (Raghukumar & Raghukumar, 1991). A similar fungus was retrieved in deep-sea corals but no infections were reported (Burgaud, Le Calvez, et al., 2009). Golubic, Radtke, et al. (2005) observed that fungi were able to penetrate the live tissues of different coral species and feed on organic matter using digestive enzymes.

Knowledge of marine fungi based on their function is still in its infancy. At present, only the well-known functions are observed although some evidence (See "Fungal metagenomics" in this chapter) suggests more diversified roles.

Culture and Isolation Techniques Used to Study Marine Fungi in Past Decades and at Present

Deep-sea conditions are different from those typically applied in laboratories. One of the major challenges in marine mycology is to isolate the fungal communities that are present and active in this extreme ecosystem characterized by a set of environmental restrictions such as high and low temperatures and high hydrostatic pressures. The culture media typically used in marine mycology are the same ones used to isolate terrestrial fungi and are composed of organic matter such as malt extract, potato-carrot, yeast extract, peptone, or glucose. The addition of antibiotics (chloramphenicol, streptomycin, penicillin, or vancomycin) is conventionally used to suppress bacterial growth to a generation time generally shorter. Zuccaro, Schulz, et al. (2003) compared culture-based and culture-independent methods and demonstrated a major dichotomy between strain collection and the molecular database generated. Most of the fungal isolates on the aforementioned media were thus presumed to be inactive or present in small amounts. Some exceptions occurred because *Corollospora angusta, Emericellopsis minima* (Zuccaro, Schulz, et al., 2003) and *Acremonium fuci* (Zuccaro, Schoch, et al., 2008) were active marine fungi detected using both methods on seaweed samples.

Different strategies can be imagined to search for fungi in deep-sea habitats. As proposed by Kohlmeyer and Kohlmeyer (1979), a low throughput method based on the observation of fungal structures can be used directly on marine substrates to characterize active marine fungal communities growing in situ. Direct detections using the specific fluorescent stain Calcofluor or immunofluorescence using antibodies have also been applied to reveal *Aspergillus* strains in deep sediments from the Indian Ocean (Raghukumar, Raghukumar, et al., 2004; Damare, Raghukumar, et al., 2006). An untargeted approach allows the characterization of culturable fungal communities present on the surface or inside a marine substrate. The low availability of nutrients in deep-sea sediments requires the use of one-fifth-strength media to simulate the relative oligotrophic in situ conditions (Damare, Raghukumar, et al., 2006). Enrichment cultures were used to highlight fungi in marine sediments (Biddle, House, et al., 2005), deep-sea hydrothermal vents (Burgaud, Le Calvez, et al., 2009), and methane cold-seep sediments (Takishita, Yubuki, et al., 2007). Particle plating methods described by Bills and Polishook (1994) and also dilution plating along with pressure incubation were used to obtain fungal strains (Damare, Raghukumar, et al., 2006; Singh, Raghukumar, et al., 2010). Such methods have been used to isolate filamentous fungi and yeasts from marine sediments. To date, no endemic active marine fungi have been detected in deep marine sediments using a culture-based approach and only a minor fraction of the marine fungal community has been recovered by conventional selective media. Van Dover, Ward, et al. (2007) reported an epizootic event occurring on deep-sea mussels and hypothesized that the detected fungi could be facultative parasites or opportunistic pathogens. Chytrids were revealed using molecular methods in deep-sea environments (Bass, Howe, et al., 2007; Le Calvez, Burgaud, et al., 2009) with close representatives defined as parasites. The putative prevalence of parasitic fungal forms in deep-sea environments may explain the gap between cultural and molecular methods but also drastically hinders the experimental strategy to be used for cultures.

Depending on the nature of the samples, the estimated cultivation efficiency with standard techniques is between 0.001 and 1 percent (Amman, Ludwig, et al., 1995). Thus, the microorganisms cultured do not reflect the diversity present in the ecosystem studied. The evolution of innovative culture techniques mimicking deep-sea conditions will probably enhance the number of deep marine fungal strains that have currently been identified only by applying molecular methods. In the future, to improve cultivation efficiency one will need to consider (a) the chemistry of natural habitats, (b) the natural biotic and abiotic interactions, and (c) cell-to-cell communication (Alain & Querellou, 2009). One way to preserve endogenous cell-to-cell communication mechanisms is to process cultures in gel micro-droplets using micro-encapsulation. Gel micro-droplets were incubated under a constant stream of depleted medium nutrients and then sorted by flow cytometry. This approach preserves an exchange of metabolites and signals between cells and can provide more than 10,000 bacterial and fungal isolates from environmental samples (Zengler, Walcher, et al., 2005). The extinction-dilution technique in liquid medium described by Stingl, Tripp, et al. (2007) appears as another effective method to feed culture collections.

Uncultured Fungi and Molecular Approaches

Molecular approaches enable the description of an extended diversity because they allow uncovering sequences from species not yet cultured or the distinction between cryptic species. The description of fungal diversity has been enhanced by molecular techniques (e.g., Richards & Bass [2005]). Recently, the *Cryptomycota* (Jones, Forn, et al., 2011a) was described mainly from molecular data and characterized as a new phylum composed of one known genus (*Rozella*) and environmental sequences from different habitats including marine systems and more precisely deep-sea sediments (Nagano, Nagahama, et al., 2010; Nagahama, Takahashi, et al., 2011) and microaerobic seawater (Stoeck & Epstein, 2003).

Regarding marine systems, the presence of fungi was initially serendipitously detected in microeukaryote inventories and the fungal diversity described was low, that is, one to six OTUs (López-García, Rodríguez-Valera, et al., 2001; Stoeck & Epstein, 2003; Stoeck, Hayward, et al., 2006) with some exceptions of dominance in deep-sea sediments (Takishita, Tsuchiya, et al., 2006; Edgcomb, Beaudoin, et al., 2011). However, such studies were targeting a wide spectrum of microorganisms, mainly protists. The description of fungi and the weight of fungal communities may have been biased because of their poor representation in the Eukaryota biomass at the sites examined (Massana & Pedrós-Alió, 2008). This bias is also enhanced by the low sequencing effort in clone libraries, which rarely attains saturation (Edgcomb, Kysela, et al., 2002; Richards & Bass, 2005; Stoeck, Hayward, et al., 2006). Finally it has been shown that the described diversity mainly depends on the primers used. Indeed, an overlap of only 4 percent was observed between species lists obtained with three different primer pairs used (Stoeck, Hayward, et al., 2006). Another issue is contamination by organisms from other habitats such as autotrophs from the upper seawater columns. Indeed, comparison of the SSU rRNA gene using DNA and RNA-based approaches has revealed an absence of active autotrophs in deep-sea sediments even though their DNA sequences were retrieved from the same samples (Edgcomb, Beaudoin, et al., 2011). On the contrary, fungal sequences were abundant with *Basidiomycota* representing most of the microeukaryotic diversity described using DNA and cDNA. The repeated occurrence of fungal sequences in microeukaryote inventories motivated the use of specific fungal primers rather than universal eukaryotic primers to describe fungi in oceans.

The molecular approaches specifically designed to assess marine fungal diversity were focused on the 18S rRNA gene (Bass, Howe, et al., 2007; Le Calvez, Burgaud, et al., 2009; Nagahama, Takahashi, et al., 2011) but also internal transcribed spacers (Lai, Cao, et al., 2007; Nagano, Nagahama, et al., 2010), 5.8S (Nagano, Nagahama, et al., 2010), and 28S (Nagahama, Takahashi, et al., 2011). Most of the retrieved sequences show affinities with upper fungi.

To date, 35 OTUs belonging to undescribed fungal species have been retrieved (Richards, Jones, et al., 2012). Novel marine fungal OTUs from basal lineages (*Chytridiomycota, Blastocladiomycota,* and *Cryptomycota*) represent one third of the 35 OTUs retrieved and reveal a greater divergence from terrestrial fungal sequences than other OTUs (Richards, Jones, et al., 2012). One explanation could be the bias resulting from the primer pairs used as shown previously for eukaryote marine inventories. Another could be the global knowledge of the diversity of basal lineages, whatever the ecosystem. This has a direct impact on the sequences of basal lineages present in the molecular databases such as GenBank.

Molecular approaches are cogent methods to describe fungal diversity in deep-sea ecosystems. The continual recovery of novel OTUs using different kind of samples and different primer pairs reinforce the hypothesis that marine fungal diversity is higher than previously thought (Edgcomb, Kysela, et al., 2002; Stoeck, Hayward, et al., 2006; Jebaraj, Raghukumar, et al., 2010). Among marine fungi, basal lineages appear more abundant than expected. Use of specific primers in both marine and terrestrial ecosystems will help clarifying their diversity and ecological role.

Required Adaptation to Deep-Sea Conditions: The Hydrostatic Pressure

Many factors govern biodiversity in the oceans (e.g., temperature, salinity, pH, and nutrient availability). The majority of the biosphere is occupied by oceans with an average depth of 3800 m (Abe, 2007) indicating, as a Gaussian distribution, that most of the biosphere is subjected to a pressure of 38 megapascals (MPa), that is, 380-fold higher than the atmospheric pressure (0.1 MPa). Hence, the hydrostatic pressure appears as a key physical parameter, even defined as the most unique physical parameter, in the dark cold abyss (Lauro & Bartlett, 2008). Deep-sea microorganisms can be defined as piezophilic, piezotolerant or piezosensitive: (a) piezophiles display optimal growth rates at pressures higher than 0.1 MPa and below 60 MPa whereas hyperpiezophiles are defined as showing optimal growth at a pressure above 60 MPa, (b) piezotolerant display an optimal growth rate at pressures between 0.1 and 50 MPa but are able to grow at a pressure above 50 MPa, and (c) piezosensitives are sensitive to elevated hydrostatic pressures (Abe & Horikoshi, 2001; Bartlett, 2002).

Although many piezotolerant prokaryotes (Wang, Wang, et al., 2008), piezophilic prokaryotes (Kato, Sato, et al., 1995; Bernhardt, Jaenicke, et al., 1988), and even piezophilic hyperthermophilic archaeon (Zeng, Birrien, et al., 2009) have been isolated and characterized, only a few studies are focusing on the adaptation of fungi to hydrostatic pressure exception made for the model yeast *Saccharomyces cerevisiae* that has been intensively analyzed and characterized as piezosensitive (Fig.15.2).

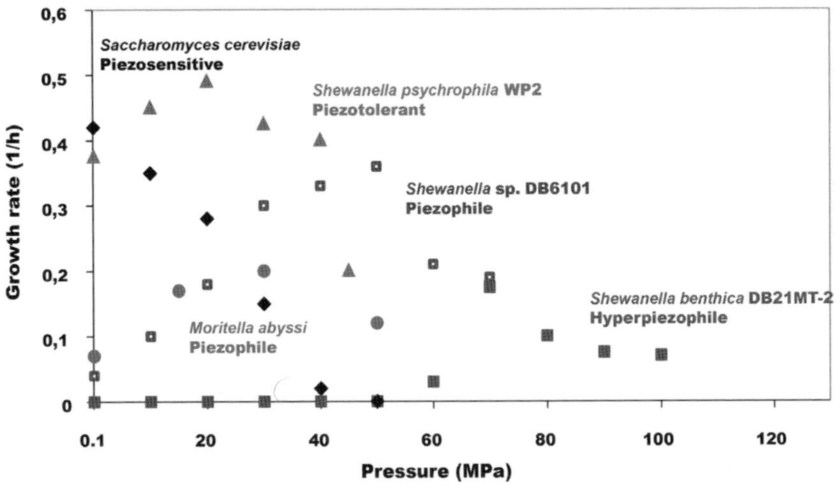

Figure 15.2 Growth rate of different microorganisms under pressure (Abe & Horikoshi, 1995; Kato, Sato, et al., 1995; Kato, Li, et al., 1998; Wang, Wang, et al., 2008; Xu, Nogi, et al., 2003). For color detail, please see color plate section.

Kohlmeyer and Kohlmeyer (1979) indicated that a simple method to obtain indigenous deep-sea fungi is to search for them directly on substrates that had been submerged in the deep sea at known depths. Wood infested by marine borers, chitin of hydrozoa, and calcareous shells appeared as ecological niches for fungal decomposers (Kohlmeyer, 1977; Raghukumar & Raghukumar, 1998). Five indigenous deep-sea fungi, *Abyssomyces hydrozoicus*, *Bathyascus vermisporus*, *Oceanitis scuticella*, *Allescheriella bathygena*, and *Periconia abyssa*, were respectively visualized directly on hydrozoa (1) at 600-m depth or wood (4) at depths below 1600 m but were not cultured. Ascospores of *A. hydrozoicus* and *O. scuticella* have mucilaginous appendages appearing as floating and attachment devices indicating an adaptation to marine habitats but not to deep high-pressurized environments. Dupont, Magnin, et al. (2009) collected wood fragments in the Vanuatu archipelago at depths between 100 and 1200 meters. Excised sporocarps from small twigs and sugar cane debris were analyzed using molecular methods and microscopy (i.e., scanning electron microscopy [SEM] and transmission electron microscopy [TEM]). The occurrence and wide distribution of *O. scuticella* in deep oceans was confirmed and a novel fungal pyrenomycete named *Alisea longicolla* was described and placed in the *Halosphaeriaceae,* the largest and most diverse lineage of marine ascomycetes to date (Sakayaroj, Pang, et al., 2011). Although fungal spores were plated on 2% malt extract agar or 2% water agar at 50° F (10° C), no growth was ever visualized (Dupont, Magnin, et al., 2009), which could be the result of the absence of elevated hydrostatic pressure during culture or the short incubation time. As noted by Kohlmeyer (1969), after 13 months immersion of wood panels at 2000-m depth, only sterile hyphae were observed. Ascocarps were only visualized after 2 to 3 years. Other species of *Bathyascus* and *Periconia* were respectively isolated from shallow water marine habitats (Nambiar, Raveendran, et al., 2008) or found in soil or marine habitats on wood or sea hares (Nambiar, Raveendran, et al., 2008; Usami, Ichikawa, et al., 2008). Such species are good candidates to assess growth rate and morphological adaptations between shallow-waters and deep-sea representatives. Calcareous shells of animals recovered at 860- and 965-m depth were examined for endolithic fungi (Raghukumar & Raghukumar, 1998), and some stained fungal borers were visualized after dissolution of carbonate shells, suggesting an ecological role of decomposers. Epifluorescence microscopy can also be applied to sediment samples using Calcofluor staining or an immunofluorescence technique, for example, antibodies raised for *Aspergillus terreus* commonly retrieved in Central Indian Basin sediments (Damare, Raghukumar, et al., 2006). Fungi were detected in low abundance when such techniques were used. Recently, Damare and Raghukumar (2007) hypothesized that this low abundance was the result of macroaggregation of fungi in deep-sea sediments because ethylenediaminetetraacetic acid (EDTA) treatment of sediment particles released highest fungal particles. The authors suggested that fungal biomass was much higher than expected in sediments and might be involved in humic aggregate formation in deep-sea sediments. Such techniques may be enhanced with other methods classically used to reveal and count prokaryotes in deep sediments using EDTA, Tween 80, sodium-pyrophosphate, methanol, and ultrasonic treatment with or without a carbonate dissolution step with an acidic acetate buffer (Kallmeyer, Smith, et al., 2008) to reveal all the fungal elements in sediments.

A culture-based approach on calcareous shells led to the isolation of filamentous fungi commonly retrieved from terrestrial environments, for example, *Aspergillus* sp., *Cladosporium* sp., and *Penicillium* sp. (Raghukumar, Raghukumar, et al., 1992; Raghukumar & Raghukumar, 1998). Although *Aspergillus restrictus* was able to penetrate shells and release calcium, these studies were unable to allow precise strain definition because only two pressures were tested (0.1 and 10 MPa). However, based on the current definition of Abe and Horikoshi (2001), such terrestrial strains are definitely piezosensitive. The same culture-based approach led to the isolation of numerous *Aspergillus* sp. strains (Damare, Raghukumar, et al., 2006; Damare, Nagarajan, et al., 2008) or a wider diversity with *Aspergillus* sp., *Cladosporium* sp., *Exophiala* sp., and *Acremonium* sp. (Singh, Raghukumar, et al., 2010) from deep-sea sediments but only piezosensitive fungi were retrieved. By testing several pressures, Lorenz and Molitoris (1997) defined precisely strains such as the piezosensitive facultative marine yeasts *Rhodotorula rubra*, *Debaryomyces hansenii*, and *Rhodosporidium sphaerocarpum* and thus respected the postulate of Kohlmeyer and Kohlmeyer (1979:42) that "tests for tolerance of high pressures ... can indicate whether the isolated fungal species are indigenous deep-sea forms or aliens from other habitats." The only issue remains the incubation time because hyphae may need several months to grow and reproductive forms several years. An in situ hybridization technique may be applied on fungal strains growing under different hydrostatic pressures to assess and quantify targeted-rRNA that can be correlated to the cellular ribosome content which, in turn, reflects the relative activity of cells (Poulsen, Ballard, et al., 1993; Binder & Liu, 1998; Daims, Lücker, et al., 2006).

High-pressure is mainly used by microbial ecologists to understand the ecological role of marine microorganisms but also by food technologists to inactivate microorganisms and by biotechnologists to enhance the productivity of bioprocesses (Follonier & Zinn, 2012). The effects of high-hydrostatic pressure on cells and cellular components are diverse: (a) pressure-sensitive lipids modifying the fluidity, permeability, and functioning of cell membranes; (b) pressure-sensitive proteins affecting multimer association and stability, (c) pressure-stabilized DNA hydrogen bonds affecting the replication and transcription steps requiring formation of single-strand DNA, and (d) loss of flagellar motility (Bartlett, 2002; MacGregor, 2002; Winter & Jeworrek, 2009; Oger & Jebbar, 2010; Follonier & Zinn, 2012). Piezophiles adapted to high pressures display specific adaptations, as compared to piezosensitive strains such as *Escherichia coli*, showing pressure-sensitive processes such as motility (Meganathan & Marquis, 1973), cell division leading to filamentation under pressure (Zobell & Cobet, 1964), growth, DNA replication, and translation (Yayanos & Pollard, 1969).

The model yeast *S. cerevisiae* has been widely analyzed to describe cellular responses to high pressure, named piezophysiology (Abe, 2004). *S. cerevisiae* is piezosensitive, meaning that no growth occurs at pressures higher than 50 MPa because of disruption of the microtubules, actin filaments, and nuclear membranes (Kobori, Sato, et al., 1995). Stress treatment of yeast cells with temperature (heat-shock or cold shock), ethanol (6%), or H_2O_2 (0.4 M) increases piezotolerance (Palhano, Gomes, et al., 2004). This cross-protection is mainly explained by heat shock protein (HSP) synthesis and enhancement of trehalose metabolism (Singer & Lindquist, 1998) acting as a whole stress response. The frontiers of basic metabolism are apparently not defined since Abe and Horikoshi (1998) and Abe (2004) argue that low cytoplasmic pH slows down glycolysis and to some extent stops ethanol fermentation around 50 MPa, whereas Picard, Daniel, et al. (2007) reported an ethanolic fermentation of 30 percent at 65 MPa that was definitely inhibited at 87 MPa. Studies of this model yeast may provide clues about the adaptation of fungi to high-pressure conditions. Even if *S. cerevisiae* is piezosensitive, this yeast is able to modify its membrane composition to tolerate high-hydrostatic pressure but only under short-term treatment. After

30 minutes at 200 MPa, *S. cerevisiae* up-regulates *ole1* gene expression to increase the proportion of unsaturated fatty acids (Fernandes, Domitrovic, et al., 2004) and prevents the hydrostatic pressure effect by increasing membrane fluidity. *S. cerevisiae* also upregulates *erg25* gene expression involved in ergosterol biosynthesis, and it is hypothesized that ergosterol may be an important protector of the cell membrane (Fernandes, 2004). Such adaptations may explain the presence of *S. cerevisiae* at 250- to 500-m depth in the water column (Bass, Howe, et al., 2007).

Filamentous fungi cultured under elevated hydrostatic pressure display abnormal structures; for example, *Aspergillus* from deep-sea sediments has altered reproduction forms with long hyphae instead of conidial heads or even hyphal swellings (Raghukumar, Raghukumar, et al., 2004; Damare & Raghukumar, 2007). Morphological anomalies occurred on piezosensitive microorganisms such as *Escherichia coli* that forms filament under increased hydrostatic pressure (Zobell & Cobet, 1964). Hydrostatic pressure was found to dissociate FtsZ protein polymers that normally play the most important role in the cell-division process (Ishii, Sato, et al., 2004). More data must be gathered to understand the abnormal morphologies of *Aspergillus* and to find out whether some yeasts become filamentous under high-pressure conditions.

Some molecular trends are directly correlated to piezophily. The occurrence of elongated helices in the 16S rRNA genes was mostly detected in piezophiles (Lauro, Chastain, et al., 2007) indicating a piezospecificity. A 454-pyrosequencing metagenomic data set from the Puerto Rico Trench at 6,000-m depth demonstrates that this deep microbial community possesses large genomes enriched in signal transduction, transcriptional regulators, and transporter mechanisms (Eloe, Fadrosh, et al., 2011). But, to the current knowledge, no universal genetic marker of piezophily is yet known and metagenomics are still unable to clearly segregate piezosensitive, piezotolerant, and piezophiles.

Fungi in Deep-Sea Sediment and Hydrothermal Ecosystems: Toward a New Evolutionary Paradigm

The deep subsurface biosphere and hydrothermal ecosystems represent large biomes on Earth characterized by a set of extreme conditions (darkness, low or high temperatures, anoxia, elevated hydrostatic pressure, relative oligotrophy or organic-richness). Whereas hydrothermal vent ecosystems have been exhaustively characterized by geologists, chemists, biologists, and microbiologists since 1979, only recent studies have estimated that the deep subseafloor biosphere comprises a significant fraction of the Earth's microbial biomass. However, the fungal communities occurring in those extreme ecosystems remain mostly undescribed.

Hydrothermal Ecosystems: Ecological Niches for Fungi?

Life on the seafloor of deep oceans relies on organic matter mainly sinking from the upper layers. Productivity in these environments is generally low and abyssal ecosystems are considered as deserts with a biomass of about 1 g.m^{-2} (Jahnke & Jackson, 1992). In contrast, deep-sea hydrothermal ecosystems are considered as hot spots of productivity on Earth with a biomass exceeding 50 kg/m^{-2} (Desbruyeres, Almeida, et al., 2000). The discovery of hydrothermal sites ranging from shallow waters—Vulcano active site in Italy (Gugliandolo & Maugeri, 1993)—to 4960-meter depth (Connelly, Copley, et al., 2012) have totally refined the biogeosciences. Hydrothermalism is directly correlated to the volcanic activities occurring in different parts of the oceanic ridges (Fig. 15.3). The hydrothermal fluids emitted by smokers result from the heating of seawater, in contact with the magma mantle, and are enriched with reduced compounds, such as hydrogen sulfide (H_2S), methane (CH_4), or carbon dioxide (CO_2) (Jannasch, 1989; Karl, 1995). The primary production on which the biological communities in these ecosystems depend is finely tuned by chemoautotrophic prokaryotes, which constitute the basis of the food web (Jannasch, 1989; Fisher, 1995). These microorganisms harvest their energy by oxidizing the reduced compounds emitted by the fluids (Van Dover & Fry, 1994; Childress, Fisher, et al., 1991). The energy released can be used to convert mineral carbon to organic carbon that may subsequently be used by a large panel of heterotrophic macroorganisms (mussels, shrimps, clams, tubeworms).

It is well established that the hydrothermal vent biosphere harbors a high microbial biomass mainly composed of prokaryotes. The occurrence of fungal sequences in hydrothermal vent environments of the Guaymas Basin (Gulf of California) was assessed about a decade ago (Edgcomb, Kysela, et al., 2002). Fungi were exclusively retrieved from the sediment-seawater interface layer and not from deeper cores at higher temperature (149° F [65° C]). Using similar approaches, the incidence of fungi in hydrothermal vents was confirmed and a wider diversity was revealed at the Mid-Atlantic Ridge and East Pacific Rise hydrothermal sites (López-García, Philip, et al., 2003; López-García, Vereshchaka, et al., 2007; Bass, Howe, et al., 2007; Le Calvez, Burgaud, et al., 2009; Sauvadet, Gobet, et al., 2010). The group "hydrothermal and/or anoxic marine yeasts" appeared as the most consistently detected fungi in clone libraries and the initial occurrence of chytrids in deep-sea ecosystems was reported. The first culture-based approaches on samples collected from deep-sea environments in the Mid-Atlantic Ridge and in the North-West Pacific Ocean led to the isolation of yeasts composed of known and previously undescribed species (Gadanho & Sampaio, 2005; Nagahama, Hamamoto, et al., 2006; Burgaud, Arzur, et al., 2011). Deep-sea hydrothermal yeasts were frequently found associated with endemic fauna, which raised questions regarding their ecological role (Burgaud, Arzur, et al., 2010). Such yeasts may be facultative parasites or

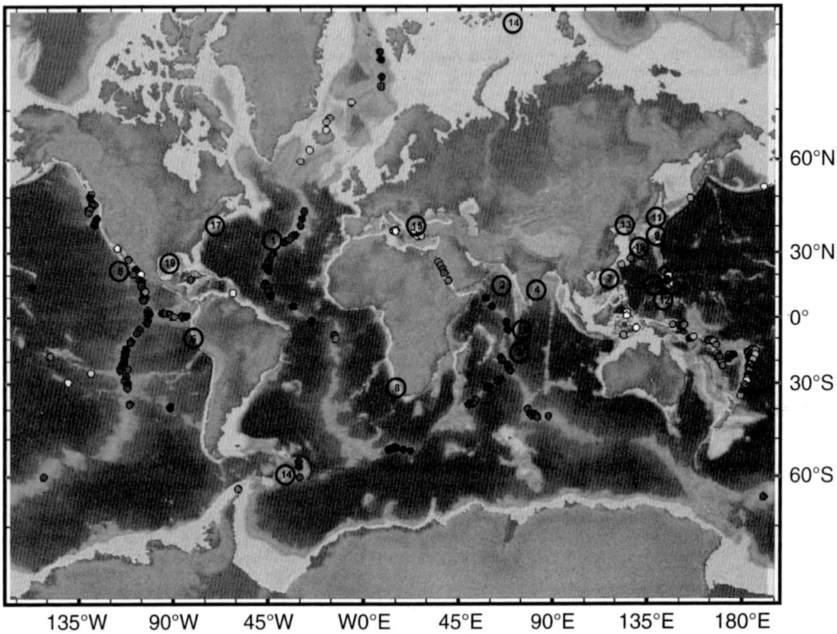

Figure 15.3 Location of known hydrothermal sites with reported depth: 0–1000 m, 1000–2000 m, 2000–3000 m, >3000 m) and location of sediment sampling sites and associated studies dealing with fungal communities. 1. Mid-Atlantic Ridge: López-García, Philip, et al. 2003; Bass, Howe, et al. 2007. 2. Arabian Sea: Raghukumar & Raghukumar, 1998; Jebaraj & Raghukumar, 2009; Jebaraj, Raghukumar, et al., 2010. 3. Central Indian Basin: Damare, Raghukumar, et al., 2006; Singh, Raghukumar, et al., 2010; Singh, Raghukumar, et al., 2011; Singh, Raghukumar, et al., 2012. 4. Bay of Bengal: Raghukumar & Raghukumar, 1998; Das, Lyla, et al., 2009. 5. Guaymas Basin: Edgcomb, Kysela, et al., 2002. 6. Peru margin & Peru trench: Edgcomb, Beaudoin, et al., 2011. 7. South China Sea: Lai, Cao, et al., 2007. 8. St Helena Bay: Mouton, Postma, et al. 2012. 9. Sagami Bay: Nagahama, Hamamoto, et al., 2001; Nagahama, Hamamoto, et al., 2003; Takishita, Yubuki, et al., 2007; Nagahama, Takahashi, et al., 2011. 10. Yap Trench: Nagahama, Hamamoto, et al., 2006. 11. Japan Trench: Nagahama, Abdel-Wahab, et al., 2008. 12. Mariana Trench: Takami, Inoue, et al., 1997; Nagano, Nagahama, et al., 2010. 13. East Sea: Park, Park, et al., 2008. 14. Arctic and Southern Ocean: Pawlowski, Christen, et al., 2011. 15. Sea of Marmara: Quaiser, Zivanovic, et al., 2011. 16. Chagos Trench: Raghukumar, Raghukumar, et al., 2004. 17. Cape Cod, Massachusetts: Stoeck & Epstein, 2003. 18. Kagoshima Bay: Takishita, Miyake, et al., 2005. 19. Gulf of Mexico: Thaler, Van Dover, et al., 2012. *This figure was generated using GeoMapApp© and the actualized InterRidge Vents Database http://www.interridge.org/irvents/* (Beaulieu, 2010). For color detail, please see color plate section.

opportunistic pathogens of deep-sea animals (Van Dover, Ward, et al., 2007, Burgaud, Le Calvez, et al., 2009). This is emphasized by a recent report of fungal sequences in liquid from the pallial cavity of deep-sea hydrothermal bivalves (Savaudet, Gobet, et al., 2010). The yeasts and filamentous fungi retrieved (Burgaud, Le Calvez, et al., 2009) may also play a role in the decomposition of organic matter in such rich biomass environments.

Deep-Sea Sediments: A Tremendous Reservoir of Microbial Types including Fungi

Deep-sea sediments represent nearly two thirds of the Earth's surface and consequently the largest biome of the biosphere with a volume of 5.10^8 km^3 (*see* Fig. 15.3). However, the knowledge about this ecosystem and of the microbial-driven processes is still limited. Recent years have seen critical progress in exploration of the deep bottom mainly thanks to the International Ocean Drilling Program (IODP). The accessibility of deep-sea sediment samples provided crucial information about the microbial communities occurring and active at several hundred meters below the surface of the ocean floor (Roussel, Cambon-Bonavita, et al., 2008). Whitman, Coleman, et al. (1998) estimated that the marine subseafloor biosphere would host $3.5.10^{30}$ prokaryotic cells, but according to more recent estimates, the marine subseafloor biosphere would comprise 1/20th of all life on Earth or 5 to 15 percent of the Earth's microbial biomass (Kallmeyer, Pockalny, et al., 2009).

If the presence and activity of prokaryotes is becoming increasingly well documented, studies of eukaryotic diversity in deep-sea sediments remain sporadic. Pawlowski, Christen, et al. (2011) retrieved around 125,000 reads using 454-pyrosequencing on sediment samples. Among them, many phototrophs were detected, corresponding to dwelling organisms that sink to the bottom and can form up to 17 percent of the whole reads. Along with planktonic and metazoan reads, fungal sequences were retrieved but did not exceed 2 percent of the total assigned OTUs indicating their occurrence but at low abundance.

Takami (1999) isolated *Penicillium lagena* and *Rhodotorula mucilaginosa*, some ubiquist filamentous fungal and yeast strains, from Mariana Trench sediments at about 11,000-m depth. Several yeasts were isolated from deep marine sediments and described as novel taxa in the *Ascomycota* or *Basidiomycota* phyla (Nagahama, Hamamoto, et al., 1999; Nagahama, Hamamoto, et al., 2001; Nagahama , Hamamoto, et al., 2003; Nagahama, Hamamoto, et al., 2006; Nagahama, Abdel-Wahab, et al. 2008). The specific occurrence of fungi in deep-sea sediments is well documented for the Central Indian Basin (Damare, Raghukumar, et al., 2006; Das, Lyla, et al., 2009; Singh, Raghukumar, et al., 2010; Singh, Raghukumar, et al., 2011; Singh, Raghukumar, et al., 2012). Several filamentous fungi and yeasts were detected and were able to grow under elevated hydrostatic pressure. Ascomycetes were mainly represented by filamentous fungi and basidiomycetes by unicellular yeast forms (Singh, Raghukumar, et al., 2010). Jebaraj and Raghukumar (2009) isolated filamentous fungi and yeasts from marine sediments and indicated that several species of fungi were able to grow at close-to-zero dissolved oxygen levels and were actors in denitrification processes. This pattern was confirmed at St. Helena Bay where extracellular cellulases were synthesized by filamentous fungal isolates, thereby indicating their putative role in detrital decay processes (Mouton, Postma, et al., 2012). These fungi, mostly affiliated to *Aspergillus* and *Penicillium*, play an active role in denitrification, co-denitrification, and ammonification processes in the nitrogen cycle in marine sediments.

Fungal molecular signatures have been detected from shallow (350 m) to deep (3011 m) marine sediments particularly rich in methane hydrates in the south of China Sea (Lai, Cao, et al., 2007). Phylogenetic analyses using internal transcribed spacers as barcode marker revealed a fungal diversity composed of *Phoma*, *Lodderomyces*, *Malassezia*, *Cryptococcus*, *Cylindrocarpon*, *Hortaea*, *Pichia*, *Aspergillus*, and *Candida*. Recently, Edgcomb, Beaudoin, et al. (2011) revealed that fungal communities were dominant among microeukaryotes in marine subsurface sediments of the Peru margin. Analyses of DNA and cDNA sequences allowed description of genetic and functional diversity and revealed mainly some uncultured basidiomycetes and a few ascomycetes. Molecular approaches have revealed a large fraction of uncultured deep-branching fungi in deep-sea methane cold-seep sediments (Nagahama, Takahashi. et al., 2011) and deep-sea sediments (Nagano, Nagahama, et al., 2010). The discovery of a novel basal fungal group in deep-sea ecosystems suggests the presence of a reservoir of previously unknown fungal biodiversity.

Pooling culture-dependent and culture-independent data provides confirmation that fungi are present and metabolically active in marine sediments and could play a major role in biogeochemical cycles in the deep biosphere. The fungal diversity retrieved in deep-sea sediments and hydrothermal ecosystems may give clues regarding fungal evolution and diversification of early lineages.

Fungal Diversity in Deep-Sea Sediments and Hydrothermal Ecosystems Tells Us a New Evolutionary Story

As shown, few studies have investigated the fungal diversity in deep-sea sediments and hydrothermal ecosystems, but the sequences retrieved in these inventories raise many questions regarding their activity, ecological role, and even the evolutionary story of the basal fungal lineages retrieved in those two extreme environments. Most of those sequences are clearly divergent from described species. Using sediment samples, sequences of three different basal phyla were detected. Sequences of *Cryptomycota* were obtained from samples collected in the Mariana Trench (Nagano, Nagahama, et al., 2010). In the same way, organisms belonging to early diverging lineages appear to dominate the fungal communities retrieved in the methane cold seeps in the Sagami Bay (Nagahama, Takahashi, et al., 2011). Among them, five OTUs form a new phylogenetic clade close to *Blastocladiomycota* (Richards, Jones, et al., 2012) even if the maximal identity of this phylum was only of

91 percent. Moreover, sequences showing affinities with *Cryptomycota* were recovered from this kind of sediments. Finally, *Chytridiomycota* sequences were obtained from a methane cold seep located in the Gulf of Mexico (Thaler, Van Dover, et al., 2012). Regarding hydrothermal vents, only *Chytridiomycota* sequences have been harvested and these were found associated with *Bathymodiolus azoricus* mussels in hydrothermal vents at 860 - and 1,700-m depth (Le Calvez, Burgaud, et al., 2009). None of them present close phylogenetic relationships with described species. These results lead to the hypothesis of a diversification of fungi in deep-sea hydrothermal ecosystems. Hydrothermalism was likely mundane when fungi emerged during Precambrian (Robert & Chaussidon, 2006). Given the molecular clock estimates, the emergence and diversification of fungi in marine environments before land colonization is a reasonable hypothesis (Le Calvez, Burgaud, et al., 2009). And the flagellum loss (Liu, Hodson, et al., 2006) or losses (James, Kauff, et al., 2006) could be regarded as a possibility of a better dispersion and resistance of spores (Le Calvez, Burgaud, et al., 2009). Alternatively, the fungal emergence in freshwater ecosystems has been suggested (Richards, Jones, et al., 2012). The newly detected clades are initiating a new era in the understanding of fungal evolution and diversification of the early diverging lineages. Using high-throughput sequencing of genetic markers on several aquatic and marine ecosystems will certainly provide insights and settle this unsolved question.

Fungal Metagenomics: Predictions of Functions and Biotic Interactions

To date little is known about fungal diversity in hydrothermal ecosystems and the functions exerted are enigmatic. It is now possible to predict metabolic pathways from genes and thereby deduce lifestyles by a metagenomic approach (for review see Vandenkoornhuyse, Dufresne, et al., 2010), the metagenome being defined as the sum of the genomes of all organisms living in a given environmental sample. By applying this approach to a chosen sample the ecological functions of one fungal organism was aimed to be predicted in its habitat belonging to an unknown deep branching *Chytridiomycota* lineage (phylotype 1 in Le Calvez, Bugaud, et al., 2009). The sample processed corresponded to a biofilm on the outside of a *Bathymodiolus azoricus* shell. The working hypothesis was that fungi outside the mussel were heterotrophic.

Three 454-pyrosequencing runs were performed. The sequencing effort was checked to make sure it was sufficient to properly analyze the metabolic properties of the sample by applying genome recruitment tests (i.e., coverage tests from reference microorganisms in the metagenome). The metagenome consisted of a large majority of bacterial sequences.

Prediction of Fungal Functions in Hydrothermal Ecosystems from the Metagenome

Some 254 contigs were obtained (average length of 3400-base pairs). Approximately three fourths of those data contained metabolic information that allows producing hypotheses about fungal lifestyles and their ecological role in this ecosystem.

Amino Acid Metabolism Protein-coding genes involved in the synthesis and degradation of most amino acids were harvested: cysteine, lysine, histidine, glutamate, tyrosine, glycine, serine, threonine, leucine, valine, isoleucine, alanine, aspartine, and phenylalanine. Thus, it can be suggested that a pathogenic lifestyle could be rejected.

Carbohydrates and Energy Metabolism Protein-encoding genes involved in the pentose phosphate pathways, citrate cycle, and glycolysis were predicted from the annotation. Enzymes were found for the aerobic carboxylation of glucose, as well as enzymes involved in the anaerobic decarboxylation of pyruvate, a process characteristic of yeast-like metabolism. Thus, the studied fungus within the metagenome might be able to ensure both yeast-like and filamentous-like metabolism. Because enzymes involved in oxidative phosphorylation were predicted, the hypothesis of aerobic metabolism is reinforced. Signatures of genes involved in glycogen, formate, and pyruvate catabolism were also predicted. Surprisingly, no heterotrophic gene signatures were found, such as genes encoding for chitinases or glucosidases (typically found in fungal organisms). This might stem from the limited number of fungal contigs detected in the data set: the main carbon source of these organisms remains to be elucidated.

Along with a fungal antibiotic biosynthesis prediction, bacterial sequences involved in penicillin degradation and bacterial penicillin-binding proteins were found, suggesting biotic interactions between fungi and bacteria. It was suggested that the fungus produces and emits antibiotics into the environment (allelopathy mechanism) to improve its ability to compete in colonizing the ecological niche or habitat.

Global Analyses

Predictions of Autotrophic C Fixation Carbon assimilation in deep hydrothermal ecosystems relies on chemolithoautotrophy (Pimenov, Lein, et al., 2000). Many reads in the metagenome were assigned to ribulose biphosphate carboxylase (RuBisCo) that can fix CO_2, with water to form two molecules of phosphoglycerate from the substrate. Most autotrophic microorganisms, including hydrothermal-living organisms, use the Calvin Benson Cycle to assimilate CO_2 (Childress & Fisher, 1992). All the known enzymes in this cycle were found.

Methane Assimilation At the hydrothermal site studied, Lucky Strike, methane is emitted by the smokers within fluids (Pimenov, Lein, et al., 2000), and 115.2 μM has been detected (in the fluid) at this particular site (Desbruyères, Biscoito, et al., 2001). Methanotrophy is therefore supposed to be common.

As expected, the analyses of the metagenome data set revealed enzymes involved in methanotrophy. Enzymes transforming methane to methanol, methanol to formaldehyde, formaldehyde to formate, and formate to CO_2 were found.

As few reads were assigned to *Archaea,* it is hypothesized that, in this particular habitat, bacteria would mainly mediate methanotrophy. From our expert analyses, the two types of methanotrophic mechanisms (type I and type II) were hypothesized. However, only four reads were assigned to methane monooxygenase, the key enzyme driving the transformation of methane to methanol. There is thus discordance between the taxonomic analyses, which reveal a strong presence of methanogenic bacteria, and the functional predictions. A possible explanation could be the structure of this enzyme composed of three subunits, the genes of which could display considerable variation. Thus, there is a distortion between data to analyze and available sequences databases.

Concluding Thoughts and Future Directions

In this chapter the fact that fungi are living in the oceans and form diverse communities has been discussed. This feature clearly conflicts with the general dogma that fungi are exclusively terrestrial organisms. Recent data obtained in marine mycology introduce the possibility of a different story concerning fungal evolution. Loss of flagellum in higher fungi has been considered for decades as a signature of terrestrialization. An alternative hypothesis has been suggested recently by Le Calvez, Burgaud et al. (2009: 6416): "the loss of motile gametes in fungi was compensated for by the resistance and long-range dispersal of spores … and … this evolutionary innovation in eukaryotes should have led to colonization and longterm persistence in many new environments, including land. …" The better understanding of fungal kingdom evolution as a result of culture-independent molecular techniques has also led to the description of the *Cryptomycota* (Jones, Forn, et al., 2011a), most of which are found in marine environments. This newly described phylogenetic cluster led to the hypothesis of a new paradigm of fungal evolution (Jones, Richards, et al., 2011b) and suggested an earlier fungal emergence.

If it is now clear that fungi are living in oceans, the roles in ecological function remain poorly addressed. From comparisons with terrestrial species, it would be expected fungi to play a key role in nutrients cycling in oceans. Original results from a metagenomic analysis of an environmental sample containing a single *Chytridiomycota* phylotype are presented herein. From fungal gene predictions and expert annotations of contigs, the aim was to understand how this fungus was living in a hydrothermal ecosystem. Unexpectedly, the absence of a genomic signature of heterotrophy was noted From the genes the possibility of antibiotics production was predicted and bacterial counterparts in the metagenome able to bind or degrade these antibiotics were found, thus producing close biotic interactions. From this, it is also clear that oceanic fungi must be regarded from a biotechnological point of view. It is possible to culture and isolate a fraction of these marine fungi as demonstrated by Burgaud, Le Calvez, et al. (2009).

Future fungal research should place a special focus on marine fungi depending on different perspectives. One primary issue would be to better understand fungal diversity and evolution. This would permit to redefine the evolutionary history of the Opisthokonts, especially how and when animals and fungi have diversified. A second issue would be to understand the fungal ecological functions and biotic interactions occurring in marine ecosystems. A key question here would be to address whether there is a parallelism in between the known ecological functions exerted in terrestrial and marine ecosystems. If fungi emerged and diversified in marine environments before colonizing land, as suggested by Le Calvez, Burgaud, et al. (2009), symbioses from mutualistic to pathogenic biotic interactions have developed. It is also possible that marine specific lifestyle(s) exist(s) for fungi. These aspects have been poorly documented to date but genomics, metagenomics, transcriptomics, and metatranscriptomics will be important strategies. A third issue includes all possible applied aspects related to the use of these marine fungi including enzymes, organic carbon transformation and energy, and drugs. This will likely relate to the ability to culture such fungi or to analyze the nucleic acids data displayed.

More widely, marine mycology should be regarded as an emerging field of research that will bloom and will be drained by ideas and knowledge from land mycology, theoretical ecology, and evolutionary paradigms.

References

Abe F & Horikoshi K. 1995. Hydrostatic pressure promotes the acidification of vacuoles in *Saccharomyces cerevisiae*. *FEMS Microbiol Lett*. 130: 307–312.

Abe F & Horikoshi K. 1998. Analysis of intracellular pH in the yeast *Saccharomyces cerevisiae* under elevated hydrostatic pressure: a study in baro- (piezo-) physiology. *Extremophiles*. 2: 223–228.

Abe F & Horikoshi K. 2001. The biotechnological potential of piezophiles. *Trends Biotechnol*. 19: 102–108.

Abe F. 2004. Piezophysiology of yeast: Occurrence and significance. *Cell Mol Biol*. 50: 437–445.

Abe F. 2007. Exploration of the effects of high hydrostatic pressure on microbial growth, physiology and survival: Perspectives from piezophysiology. *Biosci Biotechnol Biochem.* 71: 2347–2357.

Alain K & Querellou J. 2009. Cultivating the uncultured: Limits, advances and future challenges. *Extremophiles.* 13: 583–594.

Alexander E, Stock A, et al. 2009. Microbial eukaryotes in the hypersaline anoxic L'Atalante deep-sea basin. *Environ Microbiol.* 11: 360–381.

Alker AP, Smith GW, et al. 2001. Characterization of *Aspergillus sydowii* (Thom & Church), a fungal pathogen of Caribbean sea fan corals. *Hydrobiologia.* 460: 105–111.

Amann RI, Ludwig W, et al. 1995. Phylogenetic identification and in situ detection of individual microbial cells without cultivation. *Microbiol Rev.* 59: 143–169.

Barghoorn ES & Linder DH. 1944. Marine fungi: Their taxonomy and biology. *Farlowia.* 1: 395–467.

Bartlett DH. 2002. Pressure effects on in vivo microbial processes. *Biochimica et Biophysica Acta.* 1595: 367–381.

Bass D, Howe A, et al. 2007. Yeast forms dominate fungal diversity in the deep oceans. *Proc Biol Sci.* 274: 3069–3077.

Beaulieu SE. 2010. InterRidge global database of active submarine hydrothermal vent fields, Version 2.0 (July 2011). Accessed May 16, 2013, at http://www.interridge.org/irvents.

Bernhardt G, Jaenicke R, et al. 1988. High pressure enhances the growth rate of the thermophilic archaebacterium *Methanococcus thermolithotrophicus* without extending its temperature range. *Appl Environ Microbiol.* 54: 1258–1261.

Bhakuni DS & Rawat DS. 2005. Bioactive metabolites of marine algae, fungi and bacteria. In: *Bioactive Marine Natural Products*, 1–25. New Delhi: Anamaya Publishers and Springer.

Biddle JF, House CH, et al. 2005. Microbial stratification in deeply buried marine sediment reflects change in sulfate/methane profiles. *Geobiology.* 3: 287–295.

Bills GF & Polishook JD. 1994. Abundance and diversity of microfungi in leaf litter of a lowland rain forest in Costa Rica. *Mycologia.* 86: 187–198.

Binder BJ & Liu YC. 1998. Growth rate regulation of rRNA content of a marine *Synechococcus* (Cyanobacteria) strain. *Appl Environ Microbiol.* 64: 3346–3351.

Bugni TS & Ireland CM. 2004. Marine-derived fungi: a chemically and biologically diverse group of microorganisms. *Nat Prod Rep.* 21: 143–163.

Burgaud G, Le Calvez T, et al. 2009. Diversity of culturable marine filamentous fungi from deep-sea hydrothermal vents. *Environ Microbiol.* 11: 1588–1600.

Burgaud G, Arzur D, et al. 2010. Marine culturable yeasts in deep-sea hydrothermal vents: Species richness and association with fauna. *FEMS Microbiol Ecol.* 73: 121–133.

Burgaud G, Arzur D, et al. 2011. *Candida oceani* sp. *nov.*, a novel yeast isolated from a Mid-Atlantic Ridge hydrothermal vent (22300 meters). *Antonie van Leeuwenhoek.* 100: 75–82.

Capon RJ, Ratnayake R, et al. 2005. Aspergillazines A–E: Novel heterocyclic dipeptides from an Australian strain of *Aspergillus unilateralis*. *Organ Biomol Chem.* 3: 123–129.

Chen Z, Song Y, et al. 2012. Cyclic heptapeptides, cordyheptapeptides C–E, from the marine-derived fungus *Acremonium persicinum* SCSIO 115 and their cytotoxic activities. *J Nat Prod.* 75: 1215–1219.

Childress JJ, Fisher CR, et al. 1991. Sulfide and carbon dioxide uptake by the hydrothermal vent clam *Calyptogena magnifica* and its chemoautotrophic symbionts. *Physiol Zoo.* 64: 1444–1470.

Childress J & Fisher C. 1992. Biology of hydrothermal vent animals: physiology, biochemistry, and autotrophic symbioses. *Oceanography and Marine Biology.* 30: 337–441.

Connelly DP, Copley JT. et al. 2012. Hydrothermal vent fields and chemosynthetic biota on the world's deepest seafloor spreading centre. *Nature Commun.* 3: 620.

Daims H, Lücker S, et al. 2006. Daime, a novel image analysis program for microbial ecology and biofilm research. *Environ Microbiol.* 8: 200–213.

Damare S, Raghukumar C, et al. 2006. Fungi in deep-sea sediments of the Central Indian Basin. *Deep Sea Research Part I: Oceanographic Research Papers*. 53: 14–27.

Damare S & Raghukumar C. 2007. Fungi and macroaggregation in deep-sea sediments. *Microb Ecol*. 56: 168–177.

Damare S, Singh P, et al. 2012. Biotechnology of marine fungi. *Prog Mol Subcell Biol*. 53: 277–296.

Damare S, Nagarajan M, et al. 2008. Spore germination of fungi belonging to *Aspergillus* species under deep-sea conditions. *Deep-Sea Research I*. 55: 670–678.

Das S, Lyla PS, et al. 2009. Filamentous fungal population and species diversity from the continental slope of Bay of Bengal, India. *Acta Oecologica*. 35: 269–279.

Dawson SC & Pace NR. 2002. Novel kingdom-level eukaryotic diversity in anoxic environments. *Proc Natl Acad Sci USA*. 99: 8324–8329.

Desbruyères D, Almeida A, et al. 2000. A review of the distribution of hydrothermal vent communities along the northern Mid-Atlantic Ridge: Dispersal vs. environmental controls. *Hydrobiologia*. 440: 201–216.

Desbruyères D, Biscoito M, et al. 2001. Variations in deep-sea hydrothermal vent communities on the Mid-Atlantic Ridge near the Azores plateau. *Deep-Sea Research Part I*. 48: 1325–1346.

Dupont J, Magnin S, et al. 2009. Molecular and ultrastructural characterization of two ascomycetes found on sunken wood off Vanuatu Islands in the deep Pacific Ocean. *Mycol Res*. 1131: 1351–1364.

Edgcomb VP, Kysela DT, et al. 2002. Benthic eukaryotic diversity in the Guaymas Basin hydrothermal vent environment. *Proc Natl Acad Sci USA*. 99: 7658–7662.

Edgcomb V, Orsi W, et al. 2009. Protistan community patterns within the brine and halocline of deep hypersaline anoxic basins in the eastern Mediterranean Sea. *Extremophiles*. 13: 151–167.

Edgcomb VP, Beaudoin D, et al. 2011. Marine subsurface eukaryotes: The fungal majority. *Environ Microbiol*. 13: 172–183.

Eloe EA, Fadrosh DW, et al. 2011. Going deeper: Metagenome of a hadopelagic microbial community. *PLoS One*. 6: e20388.

Fernandes PMB. 2004. How does yeast respond to pressure? *Braz J Med Biol Res*. 38: 1239–1245.

Fernandes PMB, Domitrovic T, et al. 2004. Genomic expression pattern in *Saccharomyces cerevisiae* cells in response to high hydrostatic pressure. *FEBS Lett*. 556: 153–160.

Fisher CR. 1995. Toward an appreciation of hydrothermal-vent animals: Their environment, physiological ecology, and tissue stable isotope values. In: *Seafloor Hydrothermal Systems: Physical, Chemical, Biological, and Geological Interactions* (eds. SE Humphris, RA Zierenberg, et al.), 297–316. Washington, DC: American Geophysical Union.

Follonier S & Zinn SP. 2012. Pressure to kill or pressure to boost: A review on the various effects and applications of hydrostatic pressure in bacterial biotechnology. *Appl Microbiol Biotechnol*. 93: 1805–1815.

Gadanho M & Sampaio JP. 2005. Occurrence and diversity of yeasts in the mid-atlantic ridge hydrothermal fields near the Azores Archipelago. *Microbial Ecol*. 50: 408–417.

Gao Z, Li B, et al. 2008. Molecular detection of fungal communities in the Hawaiian marine sponges Suberites zeteki and Mycale armata. *Appl Environ Microbiol*. 74: 6091–6101.

Gao Z, Johnson ZI, et al. 2010. Molecular characterization of the spatial diversity and novel lineages of mycoplankton in Hawaiian coastal waters. *ISME J*. 4: 111–120.

Gao SS, Li XM, et al. 2011. Conidiogenones H and I, two new diterpenes of Cyclopiane class from a marine-derived endophytic fungus *Penicillium chrysogenum* QEN-24S. *Chem Biodivers*. 8: 1748–1753.

Geiser DM, Taylor J, et al. 1998. Cause of sea fan death in the West Indies. *Nature*. 394: 137–138.

Gleason FH, Küpper FC, et al. 2011. Zoosporic true fungi in marine ecosystems : A review. *Marine and Freshwater Research*. 62: 383–393.

Golubic S, Radtke G, et al. 2005. Endolithic fungi in marine ecosystems. *Trends Microbiol*. 13: 229–235.

Gugliandolo C & Maugeri TL. 1993. Chemolithotrophic, sulfur-oxidizing bacteria from a marine, shallow hydrothermal vent of Vulcano (Italy). *Geomicrobiol J*. 11: 109–120.

Gutiérrez MH, Pantoja S, et al. 2011. The role of fungi in processing marine organic matter in the upwelling ecosystem off Chile. *Marine Biology*. 158: 205–219.

Hyde KD, Jones EBG, et al. 1998. The role of fungi in marine ecosystems. *Biodivers Conserv*. 7: 1147–1161.

Ishii A, Sato T, et al. 2004. Effects of high hydrostatic pressure on bacterial cytoskeleton FtsZ polymers in vivo and in vitro. *Microbiol*. 150: 1965–1972.

Jahnke RA & Jackson GA. 1992. The spatial distribution of sea floor oxygen consumption in the Atlantic and Pacific Oceans. In: *Deep-Sea Food Chains and the Global Carbon Cycle* (eds. GT Rowe & V Pariente), 295–308. Berlin: Springer.

James TY, Kauff F, et al. 2006. Reconstructing the early evolution of Fungi using a six-gene phylogeny. *Nature*. 443: 818–822.

Jannasch HW. 1989. Chemosynthetically sustained ecosystems in the deep sea. In: *Autotrophic Bacteria* (eds. HG Schlegel & B Bowien), 147–166. Berlin: Springer.

Jannasch HW & Taylor CD. 1984. Deep-sea microbiology. *Annu Rev Microbiol*. 38: 487–514.

Jebaraj CS & Raghukumar C. 2009, Anaerobic denitrification in fungi from the coastal marine sediments off Goa, India. *Mycol Res*. 113: 100–109.

Jebaraj CS, Raghukumar C, et al. 2010. Fungal diversity in oxygen-depleted regions of the Arabian Sea revealed by targeted environmental sequencing combined with cultivation. *FEMS Microbiol Ecol*. 71: 399–412.

Jones EBG, Sakayaroj J, et al. 2009. Classification of marine Ascomycota, anamorphic taxta and Basidiomycota. *Fung Divers*. 35: 1–187.

Jones EBG. 2011. Fifty years of marine mycology. *Fungal Divers*. 50: 73–112.

Jones MDM, Forn I, et al. 2011a. Discovery of novel intermediate forms redefines the fungal tree of life. *Nature*. 474: 200–203.

Jones MDM, Richards TA, et al. 2011b. Validation and justification of the phylum name Cryptomycota phyl. nov. *IMA Fungus*. 2: 173–175.

Johnson T & Sparrow F. 1961. *Fungi in Oceans and Estuaries*. New York: J Cramer.

Kallmeyer J, Smith DC, et al. 2008. New cell extraction procedure applied to deep subsurface sediments. *Limnology and Oceanography: Methods*. 6: 236–245.

Kallmeyer J, Pockalny R, et al. 2009. Quantifying global subseafloor microbial abundance: Method and implications. New York: Elsevier. In: *Abstracts of the 19th Annual V.M. Golschmidt Conference* 73(13S), A615, georefid:2010-036459.

Karl DM. 1995. Ecology of free-living, hydrothermal vent microbial communities. In: *The Microbiology of Deep-Sea Hydrothermal Vents* (ed. DM Karl), 35–125. Boca Raton, FL: CRC Press.

Kato C, Sato T, et al. 1995. Isolation and properties of barophilic and barotolerant bacteria from deep-sea mud samples. *Biodivers Conserv*. 4: 1–9.

Kato C, Li L, et al. 1998. Extremely barophilic bacteria isolated from the Mariana Trench, Challenger Deep, at a depth of 11,000 meters. *Appl Environ Microbiol*. 64: 1510–1513.

Khoa LV & Hatai K. 2005. First case of *Fusarium oxysporum* infection in cultured kuruma prawn *Penaeus japonicus* in Japan. *Fish Pathol*. 40: 195–196.

Kobori H, Sato M, et al. 1995. Ultrastructural effects of pressure stress to the nucleus in *Saccharomyces cerevisiae*: A study by immunoelectron microscopy using frozen thin sections. *FEMS Microbiol Lett*. 132: 253–258.

Kohlmeyer J. 1969. Deterioration of wood by marine fungi in the deep sea. *Materials Performance and the Deep Sea*. 445: 20–29.

Kohlmeyer J. 1977. New genera and species of higher fungi from the deep sea. *Rev Mycol*. 41: 189–206.

Kohlmeyer J & Kohlmeyer E. 1979. *Marine Mycology: The Higher Fungi*. London: Academic Press.

Kohlmeyer J & Volkmann-Kohlmeyer B. 2003. Fungi from coral reefs: A commentary. *Mycol Res*. 107: 386–387.

Lai X, Cao L, et al. 2007. Fungal communities from methane hydrate-bearing deep-sea marine sediments in South China Sea. *ISME J*. 1: 756–762.

Lauro FM, Chastain RA, et al. 2007. The unique 16S rRNA genes of Piezophiles reflect both phylogeny and adaptation. *Appl Environ Microbiol*. 73: 838–845.

Lauro FM & Bartlett DH. 2008. Prokaryotic lifestyles in deep sea habitats. *Extremophiles*. 12: 15–25.

Le Calvez T, Burgaud G, et al. 2009. Fungal diversity in deep-sea hydrothermal ecosystems. *Appl Environ Microbiol*. 75: 6415–6421.

Liu YJ, Hodson MC, et al. 2006. Loss of the flagellum happened only once in the fungal lineage: Phylogenetic structure of Kingdom Fungi inferred from RNA polymerase II subunit genes. *BMC Evol Biol*. 6: 74–87.

López-García P, Rodríguez-Valera F, et al. 2001. Unexpected diversity of small eukaryotes in deep-sea Antarctic plankton. *Nature*. 409: 603–607.

López-García P, Philip H, et al. 2003. Autochthonous eukaryotic diversity in hydrothermal sediment and experimental microcolonizers at the Mid-Atlantic Ridge. *Proc Natl Acad Sci USA*. 100: 697–702.

López-García P, Vereshchaka A, et al. 2007. Eukaryotic diversity associated with carbonates and fluid-seawater interface in Lost City hydrothermal field. *Environ Microbiol*. 9: 546–554.

Lorenz R & Molitoris H. 1997. Cultivation of fungi under simulated deep sea conditions. *Mycol Res*. 101: 1355–1365.

Macgregor RB. 2002. The interactions of nucleic acids at elevated hydrostatic pressure. *Biochimica et Biophysica Acta—Protein Structure and Molecular Enzymology*. 1595: 266–276.

Massana R & Pedrós-Alió C. 2008. Unveiling new microbial eukaryotes in the surface ocean. *Curr Opin Microbiol*. 11: 213–218.

Meganathan R & Marquis RE. 1973. Loss of bacterial motility under pressure. *Nature*. 246: 525–527.

Monchy S, Grattepanche JD, et al. 2012. Microplanktonic community structure in a coastal system relative to a *Phaeocystis* bloom inferred from morphological and tag pyrosequencing methods. *PLoS One*. 7: e39924.

Mouton M, Postma F, et al. 2012. Diversity and characterization of culturable fungi from marine sediment collected from St. Helena Bay, South Africa. *Microbial Ecol*. 64: 311–319.

Nagahama T, Hamamoto M, et al. 1999. Kluyveromyces nonfermentans sp. nov., a new yeast species isolated from the deep sea. *Int J Syst Evol Microbiol*. 49: 1899–1905.

Nagahama T, Hamamoto M, et al. 2001. Distribution and identification of red yeasts in deep-sea environments around the northwest Pacific Ocean. *Biomed Life Sci*. 80: 101–110.

Nagahama T, Hamamoto M, et al. 2003. *Cryptococcus surugaensis* sp. nov., a novel yeast species from sediment collected on the deep-sea floor of Suruga Bay. *Int J Syst Evol Microbiol*. 53: 2095–2098.

Nagahama T, Hamamoto M, et al. 2006. Rhodotorula pacifica sp. nov., a novel yeast species from sediment collected on the deep-sea floor of the north-west Pacific Ocean. *Int J Syst Evol Microbiol*. 56: 295–299.

Nagahama T, Abdel-Wahab MA, et al. 2008. *Dipodascus tetrasporeus sp. nov.*, an ascosporogenous yeast isolated from deep-sea sediments in the Japan Trench. *Int J Syst Evol Microbiol*. 58: 1040–1046.

Nagahama T, Takahashi E, et al. 2011. Molecular evidence that deep-branching fungi are major fungal components in deep-sea methane coldseep sediments. *Environ Microbiol*. 13: 2359–2370.

Nagahama T & Nagano Y. 2012. Cultured and uncultured fungal diversity in deep-sea environments. *Prog Mol Subcell Biol*. 53: 173–187.

Nagano Y, Nagahama T, et al. 2010. Fungal diversity in deep-sea sediments—the presence of novel fungal groups. *Fungal Ecol*. 3: 316–325.

Nagano Y & Nagahama T. 2012. Fungal diversity in deep-sea extreme environments. *Fungal Ecol*. 5: 463–471.

Nambiar GR, Raveendran K, et al. 2008. A glimpse of lignicolous marine fungi occuring in coastal water bodies of Tamil Nadu (India). *Comptes Rendus Biologies*. 331: 475–480.

Oger PM & Jebbar M. 2010. The many ways of coping with pressure. *Res Microbiol*. 161: 799–809.

Palhano FL, Gomes HL, et al. 2004. Pressure response in the yeast *Saccharomyces cerevisiae*: from cellular to molecular approaches. *Cell Mol Biol*. 50: 447–457.
Park SJ, Park BJ, et al. 2008. Microeukaryotic diversity in marine environments, an analysis of surface layer sediments from the East Sea. *J Microbiol*. 46: 244–249.
Pawlowski J, Christen R, et al. 2011. Eukaryotic richness in the abyss: Insights from pyrotag sequencing. *PLoS One*. 6: e18169.
Picard A, Daniel I, et al. 2007. In situ monitoring by quantitative Raman spectroscopy of alcoholic fermentation by *Saccharomyces cerevisiae* under high pressure. *Extremophiles*. 11: 445–452.
Pimenov NV, Lein AY, et al. 2000. Carbon dioxide assimilation and methane oxidation in various zones of the rainbow hydrothermal field. *Microbiology*. 69: 689–697.
Poulsen LK, Ballard G, et al. 1993. Use of rRNA fluorescence in situ hybridization for measuring the activity of single cells in young and established biofilms. *Appl Environ Microbiol*. 59: 1354–1360.
Quaiser A, Zivanovic Y, et al. 2011. Comparative metagenomics of bathypelagic plankton and bottom sediment from the Sea of Marmara. *ISME J*. 5: 285–304.
Raghukumar C & Raghukumar S. 1991. Fungal invasion of massive corals. *Marine Ecol*. 12: 251–260.
Raghukumar C, Raghukumar S, et al. 1992. Endolithic fungi from deep-sea calcareous substrata: isolation and laboratory studies. In: *Oceanography of the Indian Ocean* (ed. BN Deasai), 3–9. New Delhi: Oxford JBH Publishers.
Raghukumar C & Raghukumar S. 1998. Barotolerance of fungi isolated from deep-sea sediments of the Indian Ocean. *Aquatic Microbial Ecology*. 15: 153–163.
Raghukumar S & Raghukumar C. 1999. Marine fungi: A critique. *Aquatic Microbiology Newsletter*. 38: 26–27.
Raghukumar C, Raghukumar S, et al. 2004. Buried in time: Culturable fungi in a deep-sea sediment core from the Chagos Trench, Indian Ocean. *Deep-Sea Research I*. 51: 1759–1768.
Ravindran J, Raghukumar C, et al. 2001. Fungi in *Porites lutea*: Association with healthy and diseased corals. *Dis Aquat Organ*. 47: 219–228.
Richards TA & Bass D. 2005. Molecular screening of free-living microbial eukaryotes: Diversity and distribution using a meta-analysis. *Curr Opin Microbiol*. 8: 240–252.
Richards TA, Jones MDM, et al. 2012. Marine fungi: Their ecology and molecular diversity. *Annu Rev Mar Sci*. 4: 495–522.
Robert F & Chaussidon M. 2006. A paleotemperature curve for the Precambrian oceans based on silicon isotopes in cherts. *Nature*. 443: 969–972.
Roussel EG, Cambon-Bonavita MA, et al. 2008. Extending the sub-sea-floor biosphere. *Science*. 320: 1046.
Sakayaroj J, Pang KL, et al. 2011. Multi-gene phylogeny of the *Halosphaeriaceae*: Its ordinal status, relationships between general and morphological character evolution. *Fungal Divers*. 46: 87–109.
Sauvadet AL, Gobet A, et al. 2010. Comparative analysis between protist communities from the deep-sea pelagic ecosystem and specific deep hydrothermal habitats. *Environ Microbiol*. 12: 2946–2964.
Shearer CA, Descals E, et al. 2007. Fungal biodiversity in aquatic habitats. *Biodivers Conserv*. 16: 49–67.
Singer MA & Lindquist S. 1998. Thermotolerance in *Saccharomyces cerevisiae*: The Yin and Yang of trehalose. *Trends Biotechnol*. 16: 460–468.
Singh P, Raghukumar C, et al. 2010. Phylogenetic diversity of culturable fungi from the deep-sea sediments of the Central Indian Basin and their growth characteristics. *Fungal Divers*. 40: 89–102.
Singh P, Raghukumar C, et al. 2011. Fungal community analysis in the deep-sea sediments of the Central Indian Basin by culture-independent approach. *Microbial Ecol*. 61: 507–517.
Singh P, Raghukumar C, et al. 2012. Fungal diversity in deep-sea sediments revealed by culture-dependent and culture-independent approaches. *Fungal Ecol*. doi:10.1016/j.funeco.2012.01.001.
Stingl U, Tripp HJ, et al. 2007. Improvements of high-throughput culturing yielded novel SAR11 strains and other abundant marine bacteria from the Oregon coast and the Bermuda Atlantic Time Series study site. *ISME J*. 1: 361–371.

Stoeck T & Epstein S. 2003. Novel eukaryotic lineages inferred from small-subunit rRNA analyses of oxygen-depleted marine environments. *Appl Environ Microbiol*. 69: 2657–2663.

Stoeck T, Hayward B, et al. 2006. A multiple PCR-primer approach to access the microeukaryotic diversity in environmental samples. *Protist*. 157: 31–43.

Stoeck T, Bass D, et al. 2010. Multiple marker parallel tag environmental DNA sequencing reveals a highly complex eukaryotic community in marine anoxic water. *Mol Ecol*. 19: 21–31.

Takami H, Inoue A, et al. 1997. Microbial flora in the deepest sea mud of the Mariana Trench. *FEMS Microbiol Lett*. 152: 279–285.

Takami H. 1999. Isolation and characterization of microorganisms from deep-sea mud. In: *Extremophiles in deep-sea environments* (eds. K Horikoshi & K Tsujii), 3–26. Tokyo: Springer.

Takishita K, Miyake H, et al. 2005. Genetic diversity of microbial eukaryotes in anoxic sediment around fumaroles on a submarine caldera floor based on the small-subunit rDNA phylogeny. *Extremophiles*. 9: 185–196.

Takishita K, Tsuchiya M, et al. 2006. Molecular evidence demonstrating the basidiomycetous fungus Cryptococcus curvatus is the dominant microbial eukaryote in sediment at the Kuroshima Knoll methane seep. *Extremophiles*. 10: 165–169.

Takishita K, Yubuki N, et al. 2007. Diversity of microbial eukaryotes in sediment at a deep-sea methane cold seep: Surveys of ribosomal DNA libraries from raw sediment samples and two enrichment cultures. *Extremophiles*. 11: 563–576.

Thaler AD, Van Dover CL, et al. 2012. Ascomycete phylotypes recovered from a Gulf of Mexico methane seep are identical to an uncultured deep-sea fungal clade from the Pacific. *Fung Ecol*. 5: 270–273.

Usami Y, Ichikawa H, et al. 2008. Synthetic efforts for stereo structure determination of cytotoxic marine natural product pericosines as metabolites of *Periconia* sp. from sea hare. *Int J Mol Sci*. 9: 401–421.

Van Dover CL & Fry B. 1994. Microorganisms as food resources at deep-sea hydrothermal vents. *Limnology and Oceanography*. 39: 51–57.

Van Dover CL, Ward ME, et al. 2007. A fungal epizootic in mussels at a deep-sea hydrothermal vent. *Marine Ecol*. 28: 54–52.

Vandenkoornhuyse P, Dufresne A, et al. 2010. Integration of molecular functions at the ecosystemic level: Breakthroughs and future goals of environmental genomics and post-genomics. *Ecol Lett*. 13: 776–791.

Wang, F., Wang, J., et al. 2008. Environmental adaptation: genomic analysis of the piezotolerant and psychrotolerant deep-sea iron reducing bacterium *Shewanella piezotolerans* WP3. *PLoS One*. 3: 1–12.

Whitman WB, Coleman DC, et al. 1998. Prokaryotes: The unseen majority. *Proc Natl Acad Sci USA*. 95: 6578–6583.

Winter R & Jeworrek C. 2009. Effect of pressure on membranes. *Soft Matter*. 5: 3157–3173.

Xu Y, Nogi Y, et al. 2003. *Moritella profunda* sp. nov. and *Moritella abyssi* sp. nov., two psychropiezophilic organisms isolated from deep Atlantic sediments. *Int J Syst Evol Microbiol*. 53: 533–538.

Yayanos AA & Pollard EC. 1969. A study of the effects of hydrostatic pressure on macromolecular synthesis in *Escherichia coli*. *Biophys J*. 9: 1464–1482.

Zeng X, Birrien JL, et al. 2009. *Pyrococcus* CH1, an obligate piezophilic hyperthermophile: extending the upper pressure- temperature limites for life. *ISME J*. 3: 873–876.

Zengler K, Walcher M, et al. 2005. High-throughput cultivation of microorganisms using microcapsules. *Methods Enzymol*. 397: 124–130.

Zhang GF, Han WB, et al. 2012. Neuraminidase inhibitory polyketides from the marine-derived fungus *Phoma herbarum*. *Planta medica*. 78: 76–78.

Zobell CE & Cobet AB. 1964. Filament formation by *Escherichia coli* at increased hydrostatic pressures. *J Bacteriol*. 87: 710–719.

Zuccaro A, Schulz B, et al. 2003. Molecular detection of ascomycetes associated with *Fucus serratus*. *Mycol Res*. 107: 1451–1466.

Zuccaro A, Schoch CL, et al. 2008. Detection and identification of fungi intimately associated with the brown seaweed *Fucus serratus*. *Appl Environ Microbiol*. 74: 931–941.

16 The Biodiversity, Ecology, and Biogeography of Ascomycetous Yeasts

Marc-André Lachance

Department of Biology, University of Western Ontario, London, Ontario, Canada

Yeasts as Fungi

Yeasts are defined as fungi with a usually unicellular growth habit. They constitute small minorities in two fungal phyla, the Ascomycetes and the Basidiomycetes. Although yeasts and other fungi are phylogenetically intertwined, the approaches traditionally used to study yeasts have often resembled those of bacteriology more than those of mycology. This is a reflection of the divergent ecological adaptations of the two groups. Filamentous fungi penetrate solid substrates and frequently erect complicated devices aimed at dispersing spores. Yeasts prefer liquid or surface environments. Their thallus combines both growth and dispersal functions. Notwithstanding these important differences, it is hoped that many concepts that are relevant to the ecology and biogeography of ascomycetous yeasts will also find application for other fungi and contribute to setting the stage for fungal ecological genomics as a whole. At the least, thinking of other fungi as "non-yeasts" may help in the conceptualization of the filamentous forms.

Yeast Biodiversity

Counting Species

Biodiversity, in its simplest expression, is the number of existing species. The most recent edition of *The Yeasts, a Taxonomic Study* (Kurtzman, Fell, et al., 2011) listed the descriptions of more than 700 ascomycetous yeast species and reported on the existence of a tenth as many additional descriptions published too late to be incorporated in the monograph. This exceeds the total number of described species, including Basidiomycetes, reported in the previous edition of the treatise (Kurtzman & Fell, 1998), indicating that the number

The Ecological Genomics of Fungi, First Edition. Edited by Francis Martin.

of known yeast species has more or less doubled in a short interval. This remarkable increase is not due to a kind of 21st-century Holocene explosion but is the direct outcome of the advent of DNA sequencing and its application to species delineation.

Although the notion of Ascomycete is based on the formation of intracellular meiotic spores (ascospores) during the sexual life cycle, well more than half of all described ascomycetous yeasts are known strictly through their asexual state, causing them to have been assigned to a repository genus such as *Candida*. Asexual species cannot be delineated on the basis of reproductive isolation (Biological Species Concept). Furthermore a large proportion of sexually reproducing yeasts exhibit self-fertility as a result of homothallism or of rapid diploidization by conjugation of germinating ascospores. For this reason, until 1970, yeast species were delineated primarily on the basis of trivial differences in their responses to growth tests such as sugar assimilation or the utilization of nitrate as sole nitrogen source. In 1970, Bicknell and Douglas demonstrated that heteroduplex DNA formation could be used as an objective criterion for the detection of evolutionarily distinct yeast populations and applied this principle to show, *inter alia*, that several recognized species of *Saccharomyces* were in fact synonyms of *Saccharomyces cerevisiae*. DNA/DNA reassociation continued to have a considerable influence on species delineation during the following two decades but always suffered from being onerous. The next leap ahead was a direct result of the early genomics, which began its considerable impact on yeast systematics with the publication of barcode sequences for all known ascomycetous yeast species (Kurtzman & Robnett, 1998), almost coincident with the release of the first complete eukaryotic genome sequence, that of *S. cerevisiae* (Goffeau, Barrell, et al., 1996). The D1/D2 domains of the large subunit ribosomal RNA gene usually had just the right rate of divergence to make them ideal as a barcode for yeast species identification (Kurtzman & Robnett 1998). Furthermore, the empirical observation that polymorphic species rarely vary by more than three substitutions served as a new tool for recognizing new species. The ensuing avalanche of new species, still underway, is owed to the ease with which barcode sequences can be determined and queried against the constantly updated database of sequences. Whether or not the strict and exclusive application of the three-substitution criterion is in all cases justified is debatable but that is a matter for another forum (Lachance, Wijayanayaka, et al., 2011). An accurate species count of course hinges on the ability to circumscribe species accurately. Although the notion of species remains in flux, a broadly acceptable species concept may be at hand for all organisms based on the criterion of differential fitness (Hausdorf, 2011). Genomics will no doubt play a pivotal role in identifying groups of organisms that have entered distinct evolutionary paths.

The question of how many yeast species exist gets posed now and again (Blackwell, 2006, 2011) but tentative answers are rarely proposed. Using simple extrapolation of a saturation model, I have suggested that the total number ranges from 1,500 to 15,000 (Lachance, 2006), although the lower estimate has already been exceeded by the number of species descriptions now published. The higher bound, if correct, suggests that we currently have discovered about 10 percent of extant yeast species and that some 7,000 ascomycetous yeast species remain to be found. The total number of fungal species of any kind has been estimated to be in the order of 5 million (Blackwell, 2011).

The Yeast Tree

A deep understanding of biodiversity necessitates an accurate representation of the evolutionary history of species. Although the analytical methods required to assemble a reliable tree of all ascomycetous yeasts already exist, considerably more labor will be needed before the goal can be achieved. Rokas, Williams, et al. (2003) showed that a completely reliable phylogeny of *Saccharomyces* species can only be achieved with datasets consisting of 20 or more independent, orthologous genes. Currently available analyses that cover a broad taxonomic spectrum of the fungi (Hibbett, Binder, et al., 2007) or even all ascomycetous yeasts (Kurtzman, 2011) must therefore be regarded as provisional because they rely on incomplete data sets that may include up to a half-dozen protein-encoding genes, three adjacent (and therefore interdependent) nuclear ribosomal RNA genes, and various small subsets of the mitochondrial genome. Despite these limitations, it is now fairly well established that the phylum Ascomycota can be divided into the three subphyla Taphrinomycotina, Saccharomycotina, and Pezizomycotina (Fig. 16.1). Ascomycetous yeasts constitute nearly all of the Saccharomycotina, a large proportion of the Taphrinomycotina (*Neolecta* species form asci on fruit bodies), and none of the Pezizomycotina. These results effectively do away with the hypothesis that ascomycetous yeasts are reduced, convergent forms of an eclectic array of more complex fungi unlike basidiomycetous yeasts, which can be assigned to several orders dispersed in three fungal subphyla (Boekhout, Fonseca, et al., 2011). This may have to do with the fact that basidiomycetous meiotic spores are formed externally and often have the ability to proliferate autonomously before differentiating into one of a highly diverse collection of morphologies.

The enormous impact of phylogenetics has done little to solidify the simplistic concept of ascomycetous yeasts as ascogenous fungi with a distinctly unicellular phase because some species that are now regarded as yeasts are exclusively hyphal in growth habit (e.g., all *Ascoidea* species, some *Eremothecium* species) and half the ascomycetous yeast species do not form asci. These exceptions are not particularly problematic provided that users approach yeast taxonomy with some flexibility. More importantly, a reliable assignment of genera to meaningful families within the

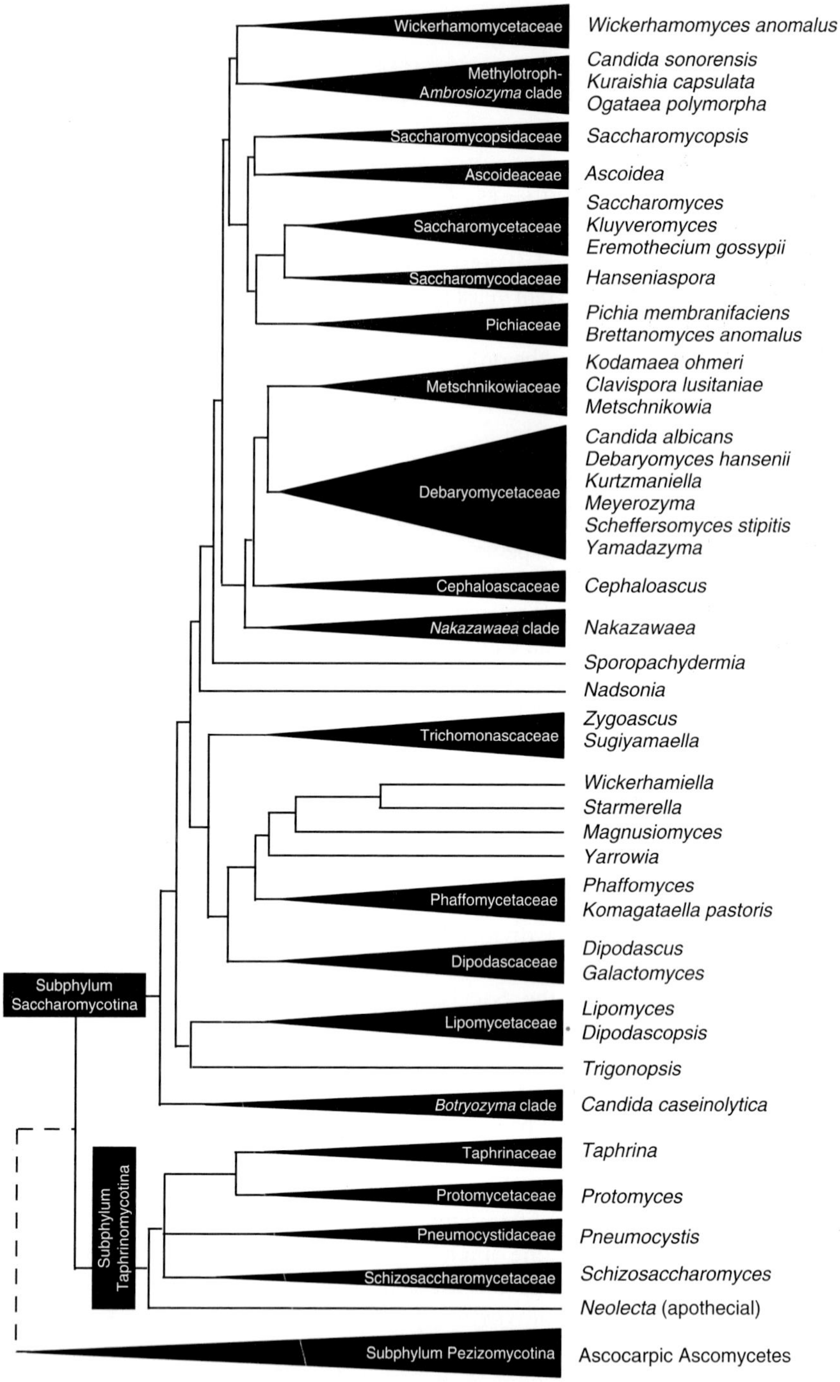

Figure 16.1 A phylogeny of ascomycetous yeasts. Modified from Kurtzman (2011). The genera or species listed were selected as typical representatives of each family or clade, as examples of better known taxa, or those discussed in the text. Yeasts are found in the subphyla Saccharomycotina and Taphrinomycotina. Species of *Neolecta* are not considered yeasts because they produce asci on the surface of club-shaped fruit bodies (ascocarps). The branch (*dashed*) connecting the Pezizomycotina is putative. Internal, but not terminal branches, are scaled.

Saccharomycotina and even in some cases the reliable assignment of species to genera are far from being a reality, and there is a pressing need for a thorough phylogenomic study. The vertiginous progress made in sequencing technology gives us confidence that phylogenies based on a sufficiently large sample of the genome are soon to be available (Haridas, Breuill, et al., 2011).

Determinants of Yeast Biodiversity

The nature of the forces that determine the composition of yeast communities remains to be elucidated. Communities are groups of species that share a particular habitat (Starmer & Lachance, 2011). What species are present in a community depends on two different kinds of factors, namely the biogeographic history of the community and the niche characteristics of its occupants. A community is defined in essence by its autochthonous members, which constitute a guild or group of organisms that fulfill certain roles in an ecosystem. Guild membership is not completely independent of taxonomic identity. It is affected by the phenotype of organisms, which for yeasts include morphology, nutrient utilization, stress resistance, release of bioactive metabolites, or other traits that may of course be correlated with taxonomy. Taken together, these attributes specify the fundamental niche of a yeast species, in other words the range of habitats that are likely to provide suitable growth conditions. Communities also contain allochthonous species, which are accidental occurrences and not guild members. Occupancy of a habitat by a particular species may enhance or diminish the ability of another species with a similar fundamental niche to cohabit. Such interactions define the realized niche of a species.

Bell (2001) and Hubbell (2001) offered a dramatically different view of community membership in which the species in a community are analogous to neutral alleles in a gene pool. When niche characteristics are disregarded, species composition depends on the rate of entry of each species, either by speciation (the equivalent of mutation) or by immigration (analogous to gene flow), and the rate of extinction (equivalent to allele loss by random genetic drift). The rates are affected by community size (as is drift) and the end result also depends greatly on the species composition of the metacommunity that serves as the source of immigrant species. The neutral model assumes that taxa are often interchangeable but does not ignore the fact that community members must first pass the test of adaptation, just as the Neutral Theory of Molecular Evolution does not deny the importance of natural selection. The relationship between community and metacommunity involves historical and geographic factors, which together constitute the purview of historical biogeography or phylogeography (Fontaneto, 2011).

This domain of research seeks to understand the diversification of species through time and space, combining the knowledge of phylogenetics and physical geography. Major themes of biogeography include the effect of continental drift and other large-scale geological events on allopatric species formation (vicariance biogeography) and the effect of the movement of organisms over large distances on their gradual differentiation (dispersal biogeography). New developments in genomics will be expected to inform not only the phylogeographic aspect of yeast diversity but also the genetic basis for yeast adaptation.

Little is known of the mechanisms by which individual yeast species are sometimes intimately associated with certain habitats. Attempts to understand these adaptations are the ultimate objective of yeast ecology.

Yeast Ecology

A Long-Neglected Field

The pervasive assumption that microbes are ubiquitous, recently reinvigorated by Fenchel and Finlay (2004), has had a considerable impact on the way practitioners of yeast systematics have exercised their trade. This is well illustrated by the fact that the fifth edition of *The Yeasts, a Taxonomic Study* (Kurtzman, Fell, et al., 2011) is the first in the series to include, in the description of each species, a section titled "Ecology." Perusal of that section will reveal that certain ascomycetous yeasts exhibit remarkable adaptations. Regrettably, a large proportion of species descriptions are based on a single isolate collected from a poorly defined source, such that little or nothing of interest can be said about the ecology of those species.

The ability to discuss mechanisms of adaptation of yeast species to their environments is also limited by the relatively small number of phenotypic characteristics available for yeast descriptions. These include cell size and shape; mode of cell division; dimorphic growth in the form of pseudohyphae (chains of buds) or true hyphae; the sexual cycle, if present; traits such as pellicle or floc formation in liquid media; presence or absence of growth on 36 or more carbon sources, one or more nitrogen sources, in the presence of inhibitors such as cycloheximide or concentrated salt or sugar, at various temperatures, or in the absence of vitamins or amino acids. Composition of wall polysaccharides and the length of the isoprenoid chain of the ubiquinone have also received some attention. As morphological and physiological attributes have become less and less important for the purpose of classification and identification, their retention as descriptors implies that they are of relevance in defining the fundamental niche of yeast species.

Ecological Bias: An Inordinate Fondness for Beetles

Assuming that it is true that only approximately 10 percent of ascomycetous yeast species are known, it is reasonable to wonder where the remaining 7,000 or so are to be found. Current knowledge is biased by geographic accessibility and the personal preferences of the most active biodiversity researchers. Insects are the most frequently reported source of ascomycetous yeasts, and among insects, beetles are by far the richest source of species, followed by drosophilid flies and bees (Kurtzman, Fell, et al., 2011). It is not a simple task, however, to determine whether these frequencies reflect true natural distributions or the bias of individual researchers. Natural substrates and localities that have received the most attention include decaying wood (e.g., Péter in Hungary; Grinbergs, Ramírez, González in Chile), insect frass or sap fluxes (Lachance, Phaff, Starmer, Wickerham in the New World; van der Walt in South Africa), cacti (Ganter, Phaff, Rosa, Starmer worldwide), beetles (Batra, Blackwell, Phaff, Suh in the New World; van der Walt in South Africa), floricolous insects including beetles and bees (Lachance, Rosa in the New World; Herrera, Herzberg in Europe), seawater (Fell, Hagler, Sampaio, van Uden worldwide), and soil (Bab'eva in Russia; Capriotti in Europe; van der Walt in South Africa). Geographic emphasis is now shifting from Europe, the New World, South Africa, and North Asia toward South and South-East Asia (Bai, Lee, Limtong, Nakase). The list given here does not address the usefulness of the collections in informing yeast biogeography or ecology. Regrettably, a considerable fraction of published data that had the potential to be useful in drawing biodiversity inferences are in the form of species descriptions supported by single isolates obtained from poorly defined sources, including unidentified plants or insects. Sampling strategies are not always formulated. More often than not, the approach is to screen large numbers of isolates, retain one exemplar of each new phylotype, and discard the rest. One can only hope that this state of affairs is about to change, although the matter is still the object of debate (Kurtzman, 2010; Lachance, 2011b).

The Fundamental Niche and the Realized Niche

In 1951 Wickerham proposed a number of chemically defined media (e.g., yeast nitrogen base, yeast carbon base) and a list of carbon and nitrogen sources that had the potential to be informative in yeast taxonomy. These were intended to improve yeast systematics by generating unique growth profiles for each yeast species. A number of additional compounds and growth tests gradually found their way into common usage. The resulting growth profiles had the potential to help us define the fundamental niche of yeast species. Some compounds included by various authors occur naturally in plants that serve as yeast habitats, and it is logical to assume that they might play a role in defining the fundamental niche of yeast species. For example, many yeasts are able to hydrolyse the β-glucosidic bonds of arbutin, salicin, amygdalin, aesculin, or β-methyl-D-glucoside, with the release of hydroquinone, benzylic alcohol, phenylacetonitrile, dihydroxycoumarin, or methanol, respectively, each of which has the potential to affect the ecological fitness of the yeasts themselves or that of other members of the community. However, there have been few attempts to draw parallels between these features and specific adaptations. Lachance, Metcalf, et al. (1982) did compare communities of yeasts found in exudates of various tree species in different North American regions. Although geography greatly affected the taxon composition of the communities, the taxonomic position of the trees themselves had a greater bearing on the average growth profile of the communities. In particular poplars, whose sap is known to abound in β-glucosides, harbored communities with unusually high numbers of yeasts capable of using cellobiose and salicin as carbon sources.

Model Organisms So far, clinical and food sources have been omitted, despite the enormous amount of attention these economically important matters receive. However, one cannot ignore the overwhelming interest that a small number of yeast species continue to attract in all fields of biology. A Google-Scholar search demonstrates that approximately 40 percent of publications having to do with yeasts are in fact about *Saccharomyces* species. A comparable proportion deals with *Candida* species and nearly half of these deal with *Candida albicans*. The vigorous research community that uses *S. cerevisiae* as the most important model eukaryote in genetics, biochemistry, molecular biology, and now genomics is laying claims to that species or its close relatives as model systems in evolution and ecology (Landry, Townsend, et al., 2006; Replansky, Koufopanou, et al., 2008; Dunham & Louis, 2011). In an attempt to correlate the ecology of *Saccharomyces* species with their phenotypic properties, Warringer, Zörgö, et al. (2011) found that variation in both genotype and phenotype within species had much more to do with the evolutionary history of populations than to habitat quality. The authors also commented on the importance of considering a fair sample of the species in formulating biologically meaningful generalizations, but stopped short of suggesting that the ecology of any yeast species is unlikely ever to be understood without taking into account yeast communities as a whole, to say nothing of other microorganisms that share the same habitats. Unfortunately, the evolutionary history of *S. cerevisiae* is largely the result of relatively recent human activity and has less to do with long-term selective adaptation. Warringer, Zörgö, et al. (2011) also expressed surprise at the considerable amount of variation observed in the species and incidentally observed that the battle horse of molecular biology, strain S288C, was the most unrepresentative of all the strains surveyed. In the same study, the close relative

Saccharomyces paradoxus, generally thought to be a "wild" species, was shown also to be structured on the basis of historical factors, although potentially adaptive phenotypic differences were also noted. *S. paradoxus* performed better on two traits, one of which was tolerance to oxalic acid, which may be a niche-defining factor for the species, but a mechanistic correspondence with habitat differences was not suggested. Sampaio and Gonçalves (2008) demonstrated that the presence of free sugar in the bark of various oak species affects the population densities of *Saccharomyces* species and that temperature responses of the various yeast species may underlie their differential distributions. However much more remains to be understood.

Interactions that affect yeasts in their natural habitats constitute their realized niche. Case in point, it is appropriate to single out one study where the environment of a yeast can profoundly influence its ecological fitness. Reuter, Bell, et al. (2007) observed that *S. cerevisiae* is carried in the guts of *Drosophila* primarily as separate ascospores, which is thought to favor sexual recombination under conditions of dispersal. This is in contrast to local populations of yeasts residing on their plant substrates, where sexual interaction is thought to take place primarily between spores of the same ascus. The authors performed a series of elegant experiments demonstrating that this is indeed the case and that the interaction of flies with the yeasts causes a major increase in heterozygosity. The role and importance of sexual reproduction in yeasts of the genus *Saccharomyces* is of particular interest given the propensity of mating to occur between sister spores. Ruderfer, Pratt, et al. (2006) estimated the rate of sexual outcrossing in *SS. cerevisiae* and *paradoxus* to be in the order of one in 50,000. Tsai, Bensasson, et al. (2008) examined the same question in *S. paradoxus* but dissected the question into finer components, concluding that sexual reproduction takes place approximately once for every thousand rounds of asexual reproduction but involved mostly sister-spore matings.

Unlike what is observed in the genus *Saccharomyces*, a large number of ascomycetous yeasts occur in nature as haploid, heterothallic mating types. It is fair to predict that in some of these species at least, sexual recombination will be shown to play a much more immediate role with respect to the realized niche. The considerable diversity of morphologies observed in other genera suggests that ascospores may also play roles that go beyond mere allele assortment, in particular as regards spatial or temporal dispersal.

Habitat Specificity Like any other organisms, different yeasts can be positioned along a multidimensional ecological gradient from generalist to specialist, terrestrial to aquatic, parasitic to saprobic. Generally speaking, habitats such as the phylloplane, the soil and seawater tend to abound in basidiomycetous yeasts, which prosper in conditions where nutrients are found at low concentrations (oligotrophy). Ascomycetous yeasts are better described as copiotrophs, preferring less dilute conditions, particularly with regard to carbon sources. The two subphyla also differ in characteristics that underlie different adaptations and modes of dispersal. Basidiomycetous species often are strictly oxidative, can be encapsulated, frequently accumulate carotenoid pigments, and may disperse actively through the formation of ballistospores. Ascomycetous yeasts often are facultatively fermentative, may use some form of filamentous growth for local dispersal, and release odorous compounds that facilitate long-distance dispersal by insects. These generalizations have multitudes of exceptions and should be viewed in that context. As will become evident in the following discussion, the fundamental niche of individual species or even groups of species is not easily defined. Arboricolous species can serve as a case in point.

Yeasts found in association with trees are often methylotrophic, a trait that is widespread among members of the aptly named methylotroph clade (*see* Fig. 16.1). Various species in this group include more than 31 *Ogataea* and 18 *Candida* species, most of which have been recovered at low frequencies from a variety of materials including sap fluxes, leaf surfaces, insect frass, wood products including tanning liquors, as well as some materials that have nothing to do with trees (Kurtzman, Fell, et al., 2011). Among these species figures *Candida sonorensis*, which is known from hundreds of isolates collected from necrotic cactus tissue worldwide. In addition to the assimilation of methanol, the ability to ferment pentose sugars is present in many tree-associated yeasts and is thought to contribute to defining their niche. Species of the smaller *Scheffersomyces* subclade (9 species), also found with regularity in decaying wood or beetle frass, do not assimilate methanol but are avid pentose fermenters. The digestive tract of wood- or fungus-associated beetles is replete with other yeasts such as those of the related *Yamadazyma* subclade (30 species), the *Wickerhamomyces* clade (30 species), the *Nakazawaea* clade (11 species), the *Meyerozyma* subclade (7 species), the *Kuraishia* subclade (6 species), and others for which a potentially adaptive physiological profile has not been identified.

Nutrition Flowers of various plants often serve as breeding grounds for a variety of insects that include drosophilids, bees, and nitidulid beetles (Lachance, Starmer, et al., 2001). These harbor extensive yeast communities that include members of the *Starmerella*, *Wickerhamiella*, *Kodamaea*, and *Metschnikowia* subclades. The first group, which is strongly associated with bees of various sorts (Rosa, Lachance, et al., 2003), consists of 20 species that share a narrow spectrum of nutrient utilization combined with a certain degree of osmotolerance. The second, with 18 species, exhibits strong endemism. Powerful extracellular lipolytic activity is often observed. The 15 species in the *Kodamaea* subclade often exhibit a propensity for filamentous growth. Particularly intriguing is the ability of *Kodamaea ohmeri* to release volatiles that attract their nitidulid beetle vector, *Aethina tumida*, and also cause an alarm response in bees (Torto, Boucias, et al., 2007). The yeast is considered an important contributor to the infestation of hives by the beetle and it is likely that analogous adaptations will be identified in other *Kodamaea* species that interact with other insects. The genus *Metschnikowia* and close relatives (47 species) contains several species that exhibit strong associations with nectar and nectarivorous insects. Of particular interest, *Metschnikowia reukaufii* dominates the digestive tract of bumblebees and the nectars of plants that they visit. This partnership has received considerable attention as of late (Herrera, Pozo, et al., 2012 and references therein). A growing subclade of *Metschnikowia* species exibits a strong affinity for small floricolous nitidulid beetles with which they appear to have co-speciated (Lachance, 2011a). Another *Metschnikowia* subclade contains species that are frequently isolated from fruit and yet another comprises species that appear to be parasitic to various aquatic invertebrates. Truly remarkable is the fact that most *Metschnikowia* species, regardless of habitat, share conserved morphologies and nutrient assimilation profiles. Moreover, the profiles are strikingly similar to those of species in the *Kodamaea* subclade as well as the *Kurtzmaniella* subclade, a moderately related assemblage of 13 species found in a variety of habitats, including nitidulid beetles of cactus flowers. Assuming that differences in the metabolic abilities of various yeast species do contribute to their habitat specificities, innovative ways of examining them will be needed. Genomics will no doubt play an important role in that quest.

Competitive Exclusion Many yeast species are capable of excreting substances that inhibit the growth of other yeasts. Killer factors or mycocins are thought to serve as agents of competitive exclusion in some yeast communities (Golubev, 2006). Pulcherrimin, an iron-binding pigment released by some species of *Metschnikowia* and *Kluyveromyces*, has strong inhibitory effects on yeasts and other fungi that share yeast habitats (Sipiczki, 2006). Members of the *Saccharomycopsis* clade (13 species) share the ability to penetrate cells of yeasts and other fungi (necrotrophic mycoparasitism), causing lysis and death (Lachance, Pupovac-Velikonja, et al., 2000). Intriguingly, members of the clade also share a defective sulphate uptake that seems unrelated to their mycoparasitic activity. These yeasts have been isolated from a variety of materials, including fruit, tree exudates, and beetle tunnels. Enzymes and toxic effectors produced by other microorganisms or invertebrates and plants that share or serve as yeast habitats certain contribute also to defining their realized niche. For example nectar appears to act as a filter that shapes the yeast community associated with nectarivorous insects (Herrera, Canto, et al., 2010).

Yeast Biogeography

The slogan (axiom) "Everything is everywhere" is simply false; "Everything is endemic" is a more meaningful starting point, a more meaningful axiom, if you wish (Williams, 2011).

Biogeography addresses the interplay between geological history and species formation and in a broader sense, the factors that affect the distribution of organisms on earth (Smith, 2007; Lomolino, Riddle, et al., 2010). The major themes of interest to biogeography include the respective roles of dispersal and vicariance (allopatry) in speciation, glaciations, continental drift, and the major ecological regions of the earth. The literature on this topic as it applies to yeasts is sparse, and in fact the existence of microbial biogeography has been questioned (Fenchel & Finlay, 2004), although this state of affairs is beginning to change with respect to microorganisms in general (Martiny, Bohannan, et al. 2006; Fontaneto, 2011) as well as yeasts in particular (Rosa & Péter, 2006; Ganter, 2011). As is often the case, considerable attention has been given to yeast species that are of interest because of their role in human health, food, or those that served as model systems in cell biology. As noted previously, Warringer, Zörgö, et al. (2011) recognized historical biogeography as a primary determinant of the genetic (and phenotypic) structure of *SS. cerevisiae* and *paradoxus*. They further identified, in both species, evidence of reduced fitness in the offspring of crosses involving some pairs of allopatric isolates, effectively contradicting the ubiquitist prediction that microorganisms are incapable of vicariant speciation. Biogeography implies that spatial scale is an important factor affecting the birth, life, and death of species. Koufopanou, Hughes, et al. (2006) studied the effect of scale on the genetic structure of populations of *S. paradoxus*, using a multilocus analysis, an approach that is greatly facilitated by the vast amount of genomic information available for *Saccharomyces* species. Genetic similarity was found to decrease with physical distance at all scales (isolation by distance), from centimeters to thousands of kilometers, leading to the conclusion that dispersal is largely local and that *S. paradoxus* in the long term is perfectly amenable to allopatric speciation. Goddard, Anfang, et al. (2010) used nine microsatellite markers to characterize 172 isolates of *S. cerevisiae* collected from numerous substrates in New Zealand. They concluded that a native population of the species exists, but that some lineages were introduced through French oak barrels. The same research group (Zhang, Skelton, et al., 2010) examined *SS. cerevisiae* and *paradoxus* isolates obtained from oak trees in New Zealand and concluded that the former species probably originates mostly from local vineyards, whereas the latter bears much similarity to European genotypes. They provided evidence that *S. paradoxus* was probably introduced to New Zealand through acorns brought by immigrants from Great Britain. These observations do raise the need for prudence, if *Saccharomyces* is to be regarded as an ecological model system because patterns that may be thought to be the result of processes that take place over evolutionarily time, measured in tens of million years, could easily be confounded by shorter-term, anthropogenic activities. For example, Ezeronye and Legras (2009) examined 23 strains of *S. cerevisiae* recovered from Nigerian palm wine using a variety of genetic markers and found that these and other West African isolates to be distinct from other populations including those of other African countries. Many studies focused on *S. cerevisiae* have unraveled unexpected amounts of genomic variation (e.g., Carreto, Eiriz, et al., 2008). In such cases, it is reasonable to wonder whether the variation is the result of long-term biogeographic factors or, instead, recent human activity.

Studies of *C. albicans* and related species have addressed the question of dispersal in this human pathogen. Fundyga, Lott, et al. (2002) provided strong evidence for isolation by distance when comparing microsatellite distribution in 13 cities worldwide. In particular, they observed that New World populations, represented by isolates from one Mesoamerican, one South American, and two North American localities were the most divergent, consistent with the primary east-west movement of humans in recent history. Forche, Schönian, et al. (1999) investigated the genetic structure of two atypical populations from Angola and Madagascar and found them to be similar to one another, but recognizably divergent from "typical" members of the species. They found no evidence that the divergence should be interpreted as a sign of incipient speciation. This state of affairs presents obvious parallels with human biogeography.

Biogeographic patterns have also been studied in yeasts that are not known to be human associated. The phylogeny of certain beetle-associated *Metschnikowia* species (11 described) is strongly suggestive of vicariant divergence between New World and Hawaiian species as well as dispersal-associated divergence among Hawaiian endemics (Lachance, Ewing, et al., 2005). More detailed studies of polymorphism in four of these species have shown that considerable allelic divergence can arise over short spatial distances in some cases (Lachance, Lawrie, et al., 2008), whereas in other cases extensive gene flow can be maintained over large distances (Wardlaw, Berkers, et al. 2009).

Concluding Remarks

Yeasts, as predominantly unicellular fungi, differ profoundly from other fungi in terms of ecological adaptation. Within the yeasts, species with ascomycetous affinities are known to interact with other organisms, in particular the plant-insect interface, but much of the nature of these associations remains to be elucidated. The application of genomics to these studies is bound to improve the knowledge of the evolutionary history of yeast species, including ecological and geographic components of the speciation process itself. One immediate goal is the generation of a comprehensive, accurate phylogeny. Another is the creation of tools that will enable us to identify differential adaptation and the basis for habitat specificity when it occurs.

Acknowledgements

I thank the Natural Sciences and Engineering Research Council of Canada for 33 years of continuous funding.

References

Bell G. 2001. Neutral macroecology. *Science*. 293(5539): 2413–2418.

Bicknell JN & Douglas HC. 1970. Nucleic acid homologies among species of *Saccharomyces*. *J Bacteriol*. 101(2): 505–512.

Blackwell M. 2006. How many yeasts? *Microbiol Today*. 2006 (November): 160–165.

Blackwell M. 2011. The fungi: 1, 2, 3 . . . 5.1 million species? *Am J Bot*. 98(3): 426–438.

Dunham MJ & Louis EJ. 2011. Meeting Point. *EMBO Rep*. 12(1): 8–10.

Boekhout T, Fonseca Á, et al. 2011. Discussion of teleomorphic and anamorphic basidiomycetous yeasts. In: *The Yeasts: A Taxonomic Study*, 5th ed. (ed. CP Kurtzman, JW Fell, et al.), 1339–1372. Amsterdam: Elsevier.

Carreto L, Eiriz MF, et al. 2008. Comparative genomics of wild type yeast strains unveils important genome diversity. *BMC Genomics*. 9: 524.

Ezeronye OU & Legras JL. 2009. Genetic analysis of *Saccharomyces cerevisiae* strains isolated from palm wine in eastern Nigeria. Comparison with other African strains. *J Appl Microbiol*. 106(5): 1569–1578.

Fenchel T & Finlay BJ. 2004. The ubiquity of small species: patterns of local and global diversity. *BioScience*. 54(8): 777.

Fontaneto D, ed. 2011. *Biogeography of Microscopic Organisms. Is Everything Everywhere?* Cambridge: Cambridge University Press.

Forche A, Schönian G, et al. 1999. Genetic structure of typical and atypical populations of *Candida albicans* from Africa. *Fungal Genet Biol*. 28(2): 107–125.

Fundyga RE, Lott TJ, et al. 2002. Population structure of *Candida albicans*, a member of the human flora, as determined by microsatellite loci. *Infection, Genetics and Evolution: J Mol Epidemiol Evol Genet Infect Dis*. 2(1): 57–68.

Ganter PF. 2011. Everything is not everywhere: The distribution of cactophilic yeast. In: *Biogeography of Microscopic Organisms. Is everything everywhere?* (ed. D Fontaneto), 130–174. Cambridge: Cambridge University Press.

Goddard MR, Anfang N, et al. 2010. A distinct population of *Saccharomyces cerevisiae* in New Zealand: Evidence for local dispersal by insects and human-aided global dispersal in oak barrels. *Environ Microbiol*. 12(1): 63–73.

Goffeau A, Barrell BG, et al. 1996. Life with 6000 genes. *Science*. 274(5287): 546–567.

Golubev WI. 2006. Antagonistic interactions among yeasts. In *Biodiversity and Ecophysiology of Yeasts* (eds. CA Rosa & G Péter), 197–219. Berlin: Springer.

Haridas SC, Breuill J et al. 2011. A biologist's guide to de novo genome assembly using next-generation sequence data: A test with fungal genomes. *J Microbiol Methods*. 86(3): 368–375.

Hausdorf B. 2011. Progress toward a general species concept. *Evolution*. 65(4): 923–931.

Herrera CM, Canto A, et al. 2010. Inhospitable sweetness: Nectar filtering of pollinator-borne inocula leads to impoverished, phylogenetically clustered yeast communities. *Proc Bio Sci*. 277(1682): 747–754.

Herrera CM, Pozo MI, et al. 2012. Jack of all nectars, master of most: DNA methylation and the epigenetic basis of niche width in a flower-living yeast. *Mol Ecol*. 21(11): 2602–2616.

Hibbett DS, Binder M, et al. 2007. A higher-level phylogenetic classification of the Fungi. *Mycol Res*. 111(Pt 5): 509–547.

Hubbell SP. 2001. *The Unified Neutral Theory of Biodiversity and Biogeography*. Princeton: Princeton University Press.

Koufopanou V, Hughes J, et al. 2006. The spatial scale of genetic differentiation in a model organism: The wild yeast *Saccharomyces paradoxus*. *Philos Trans R Soc Lond B Bio Sci*. 361(1475): 1941–1946.

Kurtzman CP. 2010. Description of new yeast species—is one strain enough? *Bull BISMiS* 1(1): 17–24.

Kurtzman CP. 2011. Discussion of teleomorphic and anamorphic ascomycetous yeasts and yeast-like taxa. In *The Yeasts: A Taxonomic Study*, 5th ed. (ed. CP Kurtzman, JW Fell, et al.), 293–307. Amsterdam: Elsevier.

Kurtzman CP & Fell JW, eds. 1998. *The Yeasts: A Taxonomy Study*, 4th ed. Amsterdam: Elsevier.

Kurtzman CP & Robnett CJ. 1998. Identification and phylogeny of ascomycetous yeasts from analysis of nuclear large subunit (26S) ribosomal DNA partial sequences. *Antonie Van Leeuwenhoek*. 73(4): 331–371.

Kurtzman CP, Fell JW, et al., eds. 2011. *The Yeasts: A Taxonomic Study*, 5th ed. Amsterdam: Elsevier.

Lachance MA. 2006. Yeast biodiversity: How many and how much? In: *Biodiversity and Ecophysiology of Yeasts* (ed. CA Rosa & G Péter), 1–9. Berlin: Springer.

Lachance MA. 2011a. *Metschnikowia* Kamienski (1899). In: *The Yeasts, a Taxonomic Study*, 5th ed. (ed. CP Kurtzman, JW Fell, et al.), 575–620. Amsterdam: Elsevier.

Lachance MA. 2011b. Microbial species descriptions: The importance of multiple strains. IUMS 2011—The Unlimited World of Microbes, Sapporo, Japan, September 6–10, 2011.

Lachance MA, Ewing CP, et al. 2005. *Metschnikowia hamakuensis* sp. nov., *Metschnikowia kamakouana* sp. nov. and *Metschnikowia mauinuiana* sp. nov., three endemic yeasts from Hawaiian nitidulid beetles. *Int J Syst Evol Microbiol*. 55(3): 1369–1377.

Lachance MA, Lawrie D, et al. 2008. Biogeography and population structure of the Neotropical endemic yeast species Metschnikowia lochheadii. *Antonie Van Leeuwenhoek*. 94: 403–414. doi: 10.1007/s10482-008-9258-7.

Lachance MA, Metcalf BJ, et al. 1982. Yeasts from exudates of *Quercus*, *Ulmus*, *Populus* and *Pseudotsuga*: New isolations and elucidation of some factors affecting ecological specificity. *Microbial Ecol*. 8(2): 191–198.

Lachance MA, Pupovac-Velikonja A, et al. 2000. Nutrition and phylogeny of predacious yeasts. *Can J Microbiol*. 46(6): 495–505.

Lachance MA, Starmer WT, et al. 2001. Biogeography of the yeasts of ephemeral flowers and their insects. *FEMS Yeast Res*. 1(1): 1–8.

Lachance MA, Wijayanayaka TM, et al. 2011. Ribosomal DNA sequence polymorphism and the delineation of two ascosporic yeast species, *Metschnikowia agaves* and *Starmerella bombicola*. *FEMS Yeast Res*. 11(4): 324–333.

Landry CR, Townsend JP, et al. 2006. Ecological and evolutionary genomics of *Saccharomyces cerevisiae*. *Mol Ecol*. 15(3): 575–591.

Lomolino MV, Riddle BR, et al. 2010. *Biogeography*, 4th ed. Sunderland MA: Sinauer.

Martiny JBH, Bohannan BJM, et al. 2006. Microbial biogeography: Putting microorganisms on the map. *Nat Rev Microbiol*. 4(2): 102–112.

Replansky T, Koufopanou V, et al. 2008. *Saccharomyces sensu stricto* as a model system for evolution and ecology. *Trends Ecol Evol*. 23(9): 494–501.

Reuter M, Bell G, et al. 2007. Increased outbreeding in yeast in response to dispersal by an insect vector. *Curr Biol*. 17(3): R81–R83.

Rokas A, Williams BL, et al. 2003. Genome-scale approaches to resolving incongruence in molecular phylogenies. *Nature*. 425(6960): 798–804.

Rosa CA, Lachance MA, et al. 2003. Yeast communities associated with stingless bees. *FEMS Yeast Res*. 4(3): 271–275.

Rosa CA & G Péter, eds. 2006. *Biodiversity and Ecophysiology of Yeasts*. Berlin: Springer.

Ruderfer DM, Pratt SC, et al. 2006. Population genomic analysis of outcrossing and recombination in yeast. *Nat Genet.* 38(9): 1077–1081.

Sampaio JP & Gonçalves P. 2008. Natural populations of *Saccharomyces kudriavzevii* in Portugal are associated with oak bark and are sympatric with *S. cerevisiae* and *S. paradoxus*. *Appl Environ Microbiol.* 74(7): 2144–2152.

Sipiczki M. 2006. Metschnikowia strains isolated from botrytized grapes antagonize fungal and bacterial growth by iron depletion. *Appl Environ Microbiol.* 72(10): 6716–6724.

Smith CI. 2007. Historical biogeography: The new synthesis. *Curr Biol.* 17(15): R598–R600.

Starmer WT & Lachance MA. 2011. Yeast ecology. In: *The Yeasts: A Taxonomic Study*, 5th ed. (eds. CP Kurtzman, JW Fell, et al.), 65–83. Amsterdam: Elsevier.

Torto B, Boucias DG, et al. 2007. Multitrophic interaction facilitates parasite-host relationship between an invasive beetle and the honey bee. *Proc Natl Acad Sci USA.* 104(20): 8374–8378.

Tsai IJ, Bensasson D, et al. 2008. Population genomics of the wild yeast *Saccharomyces paradoxus*: Quantifying the life cycle. *Proc Natl Acad Sci USA.* 105(12): 4957–4962.

Wardlaw AM, Berkers TE, et al. 2009. Population structure of two beetle-associated yeasts: Comparison of a New World asexual and an endemic Nearctic sexual species in the *Metschnikowia* clade. *Antonie Van Leeuwenhoek.* 96(1)1–15.

Warringer J, Zörgö E, et al. 2011. Trait variation in yeast is defined by population history. *PLoS Genet.* 7(6): e1002111.

Wickerham LJ. 1951. Taxonomy of yeasts. *Technical Bulletin No. 1029.* Washington, DC: United States Department of Agriculture.

Williams DM. 2011. Biogeography of microscopic organisms. In: *Biogeography of Microscopic Organisms. Is Everything Everywhere?* (ed. D Fontaneto), 11–31. Cambridge: Cambridge University Press.

Zhang H, Skelton A, et al. 2010. *Saccharomyces paradoxus* and *Saccharomyces cerevisiae* reside on oak trees in New Zealand: Evidence for migration from Europe and interspecies hybrids. *FEMS Yeast Res.* 10(7): 941–947.

Index

The Ecological Genomics of Fungi, First Edition. Edited by Francis Martin.

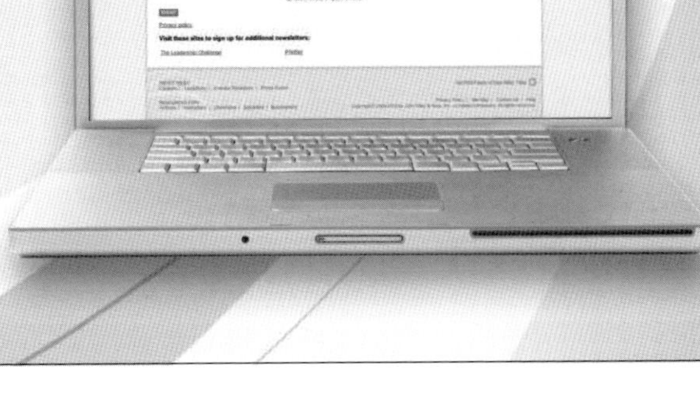

WILEY
Sign Up for An E-mail List - Edit Your Account - Unsubscribe

mierdochy

$ 205.75